VINDICATION

ALSO BY LYNDALL GORDON

T. S. Eliot: An Imperfect Life

Virginia Woolf: A Writer's Life

Shared Lives (A Memoir)

Charlotte Brontë: A Passionate Life

A Private Life of Henry James:
Two Women and His Art

VINDICATION

A Life of
Mary Wollstonecraft

Lyndall Gordon

HarperCollins*Publishers*

Endpaper Art: An illustration by William Blake for the second edition of Mary Wollstonecraft's *Original Stories from Real Life* (1791), depicting a character from the book, Mrs Mason, as a model for girls of the next generation.

HarperCollins books may be purchased for educational, business, or sales promotional use. For information, please write: Special Markets Department, HarperCollins Publishers Inc., 10 East 53rd Street, New York, NY 10022.

Published in Great Britain in 2005 under the title Mary Wollstonecraft by Little, Brown.

FIRST EDITION

Printed on acid-free paper

Library of Congress Cataloging-in-Publication Data
Gordon, Lyndall.
Vindication : a life of Mary Wollstonecraft / Lyndall Gordon.—1st ed.
p. cm.
"Published in Great Britain in 2005 under the title: Mary Wollstonecraft by Little, Brown"—T.p. verso.
Includes bibliographical references (p.) and index.
ISBN 0-06-019802-8
1. Wollstonecraft, Mary, 1759–1797. 2. Authors, English—18th century—Biography. 3. Feminists—Great Britain—Biography. I. Title.
PR5841.W8Z717 2005
828'.609—dc22
[B] 2005040237

05 06 07 08 09 RRD 10 9 8 7 6 5 4 3 2 1

I am . . . going to be the first of a new genus – I tremble at the attempt . . .

MARY WOLLSTONECRAFT TO HER SISTER EVERINA, 1787

For Siamon

CONTENTS

LIST OF ILLUSTRATIONS

PLATE SECTION ONE

Mary Wollstonecraft: The Courtauld Institute of Art, London
Gillray cartoon: courtesy of the Warden and Scholars of New College Oxford/Bridgeman Art Library
Margaret King: Houghton Library, Harvard University
George Ogle: The Courtauld Institute of Art, London
Joseph Johnson: By W. Sharpe after Moses Haughton. National Portrait Gallery
Henry Fuseli: National Portrait Gallery
A Vindication of the Rights of Woman: Mary Evans/The Woman's Library
Joel Barlow: Houghton Library, Harvard University
Ruth Barlow: Connecticut Historical Society Museum, Hartford, CT
Tom Paine: National Portrait Gallery
Helen Maria Williams: Miriam and Ira D. Wallach Division of Art, Prints and Photographs, New York Public Library
The Chinese Baths in Paris: established by Lenoir (w/c on paper), French School, (19th century) / Musée de la Ville de Paris, Musée Carnavalet, Paris, France, Lauros/Giraudon/ Bridgeman Art Library
George Caleb Bingham, *Daniel Boone Escorting Settlers through the Cumberland Gap*, 1851–52. Oil on canvas, 36½ × 50¼″. Mildred Lane Kemper Art Museum, Washington University in St. Louis. Gift of Nathaniel Phillips, 1890

PLATE SECTION TWO

William Godwin: Private collection
Swedish log: Riksarkivet, Stockholm
Risor: courtesy Knut Henning Thygesen, Risør, Norway
Elbe at Altona, 1790: courtesy of Staatsarchiv, Hamburg
Elizabeth Inchbald: National Portrait Gallery
Amelia Opie: National Portrait Gallery
Mary Wollstonecraft: National Portrait Gallery
Mary Shelley: Private collection
Fables Illustration: title page from Aesop, *Fables, Ancient and Modern*, 1805
Margaret Mount Cashell: by Edmé Quenedey, Paris, c.1801. Courtesy of Pforzheimer Library, NYPL
The Lover's Seat: Shelley (1792–1822) and Mary Godwin in Old St. Pancras Churchyard, 1877 (oil on canvas), Frith, William Powell (1819–1909) / Private Collection/Bridgeman Art Library
Claire Clairmont: Mary Evans Picture Library
Byron: Mary Evans Picture Library
Elizabeth Barrett Browning: Mary Evans Picture Library
Charlotte Brontë: National Portrait Gallery
Emily Dickinson: The Amherst College Library
George Eliot: National Portrait Gallery
Henry James: Mary Evans Picture Library
Virginia Woolf: Mary Evans Picture Library

The Wollstonecraft/Godwin/Shelley/Clairmont Family

Edward John Wollstonecraft
b. 1737?
1756
=
Elizabeth Dickson of Ballyshannon
d. 1782

Edward (Ned)
b. 1758? = 1778
Elizabeth Munday

issue emigrated
to Australia

Henry
b. 1761

Elizabeth
(BESS)
b. 1763
1782
=
Meredith
Bishop

Elizabeth
1783–4

EVERINA
1765–1843

James
1768–1806

CHARLES
1770–1817
emigrated to US
=
divorced
(1) Sarah Garrison
of New Orleans

Jane Nelson

= (2) Mary,
daughter of unnamed
New England clergyman

Gilbert IMLAY
of NJ and Kentucky
1754?–1828?
1793
=
MARY
WOLLSTONECRAFT (1)
1759–97
1797
=
William
GODWIN
1756–1836
1801
=
(2) Mary Jane Vial
('Mrs Clairmont')
1768–1841
==?

William
1803–32

Françoise (FANNY) IMLAY
('Fanny Godwin',
adopted by Godwin)
1794–1816

Harriet Westbrook (1)
1795–1816
=
Percy
Bysshe
Shelley
1792–1822
=
(2) Mary Wollstonecraft
Godwin
(MARY SHELLEY)
1797–1851

Charles
Gaulis
1795–1850

Clara Mary Jane
(CLAIRE) CLAIRMONT
1798–1879
1816
==
Lord Byron
1788–1824

Ianthe
1813–76

Charles
1814–26

unnamed
daughter
1815

William
(Willmouse)
1816–19

Clara
1817–18

Elena
Adelaide
1818–20
(adopted)

Percy
Florence
1819–89
=
Jane
Gibson
1821–99

Allegra
1817–22

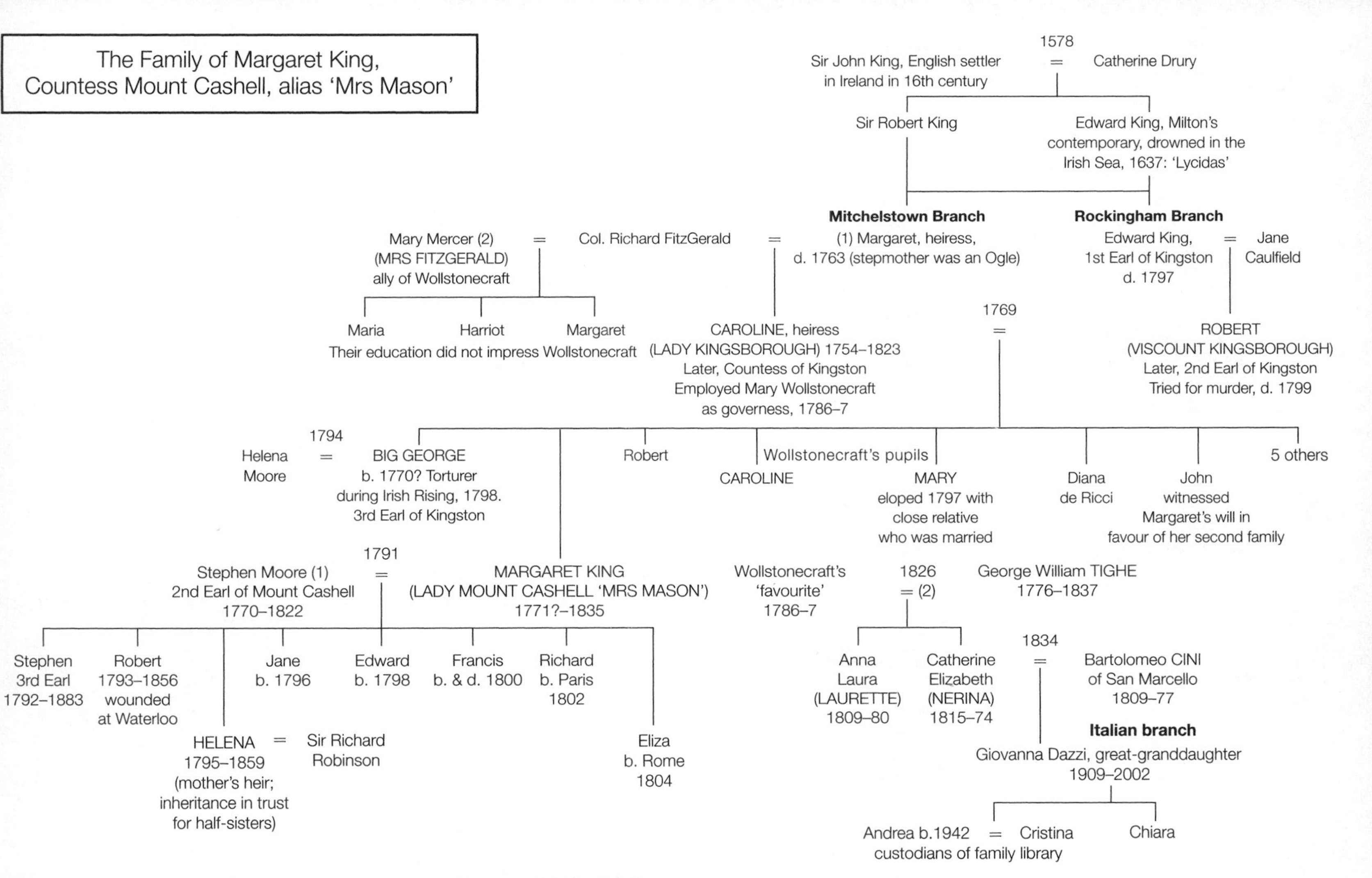
The Family of Margaret King, Countess Mount Cashell, alias 'Mrs Mason'
1578
Sir John King, English settler in Ireland in 16th century
=
Catherine Drury
Sir Robert King
Edward King, Milton's contemporary, drowned in the Irish Sea, 1637: 'Lycidas'
Mitchelstown Branch
Rockingham Branch
Mary Mercer (2) (MRS FITZGERALD) ally of Wollstonecraft
=
Col. Richard FitzGerald
=
(1) Margaret, heiress, d. 1763 (stepmother was an Ogle)
Edward King, 1st Earl of Kingston d. 1797
=
Jane Caulfield
Maria
Harriot
Margaret
Their education did not impress Wollstonecraft
CAROLINE, heiress (LADY KINGSBOROUGH) 1754–1823 Later, Countess of Kingston Employed Mary Wollstonecraft as governess, 1786–7
1769
=
ROBERT (VISCOUNT KINGSBOROUGH) Later, 2nd Earl of Kingston Tried for murder, d. 1799
1794
Helena Moore
=
BIG GEORGE b. 1770? Torturer during Irish Rising, 1798. 3rd Earl of Kingston
Robert
Wollstonecraft's pupils
CAROLINE
MARY eloped 1797 with close relative who was married
Diana de Ricci
John witnessed Margaret's will in favour of her second family
5 others
1791
Stephen Moore (1) 2nd Earl of Mount Cashell 1770–1822
=
MARGARET KING (LADY MOUNT CASHELL 'MRS MASON') 1771?–1835
Wollstonecraft's 'favourite' 1786–7
1826
= (2)
George William TIGHE 1776–1837
Stephen 3rd Earl 1792–1883
Robert 1793–1856 wounded at Waterloo
Jane b. 1796
Edward b. 1798
Francis b. & d. 1800
Richard b. Paris 1802
HELENA 1795–1859 (mother's heir; inheritance in trust for half-sisters)
=
Sir Richard Robinson
Eliza b. Rome 1804
Anna Laura (LAURETTE) 1809–80
Catherine Elizabeth (NERINA) 1815–74
1834
=
Bartolomeo CINI of San Marcello 1809–77
Italian branch
Giovanna Dazzi, great-granddaughter 1909–2002
Andrea b.1942
=
Cristina
custodians of family library
Chiara

1

VIOLENCE AT HOME

In December 1792 an Englishwoman of thirty-three crossed the Channel to revolutionary France. She was travelling alone on her way to Paris at a time when Englishmen like Wordsworth were speeding in the opposite direction – back to the safety of their country, in fear of the oncoming Terror. When, at length, Mary Wollstonecraft arrived at a friend's *hôtel*, she found it deserted, one folding door opening after another, till she reached her room at the far end. There she sat by her candle, knowing no one and unable to speak the language. The silence, in contrast to London, was eerie. As she looked up from the letter she was writing, eyes glared through a glass door. Looming through the darkness, bloody hands showed themselves and shook at her. She longed for the sound of a footstep; she missed her cat. 'I want to see something alive,' her pen scratched, 'death in so many frightful shapes has taken hold of my fancy.'*

That day she had seen the King carried past her window at nine in the morning, on the way to his trial. She records the stillness and emptiness of the streets, the closed shutters, the drums of the National Guard, her own assent to the 'majesty of the people', and the sight of 'Louis sitting with more dignity than I expected from his character, in a hackney coach going

*There are no numbered notes. For the reader who wants to know more, the source notes, starting on p. 456, contain additional material on contexts and issues.

to meet his death'. Violence had always roused her. As a child she had witnessed scenes of violence at home; she had heard 'the lash resound on the slaves' naked sides'; and now even Louis XVI called out her tears.

This will be the story of an independent and compassionate woman who devised a blueprint for human change, held to it through the Terror and private trials, and passed it on to her daughters and future generations. 'I am . . . going to be the first of a new genus,' Mary Wollstonecraft told her sister Everina, 'the peculiar bent of my nature pushes me on.'

She combined a dreamy voluptuousness with quick words, fixing brown eyes on her listener. The eyes didn't quite match, as though the right eye lingered in thought while the left drew one into intimacy with that thought. I want to dispel the myth of wildness: her voice was rational, deploring a fashion for 'romantic sentiments' instead of 'just opinions'. She wished 'to see women neither heroines nor brutes, but reasonable creatures'. Her earliest portrait presents a leader, austere in black, with powdered hair. Later portraits show the writer, her locks bound by a scarf, turning from her book to ruminate; and a sensible wife, auburn hair bundled out of sight, in the new, simple look of white muslin caught up under a rounded breast. She was pregnant at the time, but was always a large woman with a warm physical presence, unlike the bluestocking, the narrow female scholar of the eighteenth century.

Her husband, the philosopher and social reformer William Godwin, called her the 'firmest champion' and 'the greatest ornament her sex ever had to boast'. She was famous, then notorious. For most, her freedom to shape her life as she saw fit had to fade. Our society still repeats stories of doom, as though genius in a woman exacts a terrible end; as though it must be unnatural. Here, we test a different story, stripping the interchangeable masks of womanhood – queen of hearts, whore, waif – to seek out the novelty of what a 'new genus' implies: a new kind of creature who found her voice in a brief moment of historical optimism when, as Wordsworth put it, 'Europe was rejoiced,/France standing at the top of golden hours,/And human nature seeming born again.' Everything in

Mary's unsheltered life prepared her for the impact of the first heady phase of the French Revolution when all traditional forms of existence seemed ripe for change. At that moment, she stood ready to turn revolution towards a future for 'human creatures, who, in common with men, are placed on this earth to unfold their faculties'.

This pioneer of women's rights is even more a pioneer of character: in the secret mirror of her mind, the first of her kind. How did she shed, one by one, the stale plots that leach the 'real life' out of us? A 'new genus' needs a new plot of existence. Mary Wollstonecraft is, in this sense, rewriting her life for lives to come. Though she speaks of 'improvement' in the acceptable terms of her day, it's a grand design and, as such, vulnerable to those with the power to plunge her back into familiar scenes of wasted lives – wasted like her mother, prime victim of violence at home, the person for whom Mary the child felt her earliest, most instinctive and desperate pity. Virginia Woolf pictures a dauntless biographic creator: 'Every day she made theories by which life should be lived; and every day she came smack against the rock of other people's prejudices. Every day too – for she was no pedant, no cold-blooded theorist – something was born in her that thrust aside her theories and modelled them afresh.' She hails the French Revolution; then hates its bloodshed. She shuns marriage; then marries. We are tempted to criticise her inconsistency – and then remember that 'a foolish consistency is the hobgoblin of little minds'. To see Mary as shifting and rash would be to scale her down. Dimly, through the glare of celebrity and slander, it's possible to make out the shape of a new genus reading, testing, growing, but still uncategorised.

Each age retells this story; there have been invaluable portraits, from Godwin's 'champion' at the end of the eighteenth century to Mrs Fawcett's heroine for the suffragist Cause, and from Claire Tomalin's outstanding image of the wounded lover to Janet Todd's moody drama queen as seen through the exasperated eyes of her sisters. All present faces we can't forget. Yet there's also a face few see: that unnamed thing she feels herself to be. This biography will bring out the full genius of her evolving character as she projects from her generation to the next, unfolding with astonishing fertility from one kind of life to another. Each phase of her life is a new experiment – 'an experiment from the start', Woolf insists. There is an unprecedented authenticity in her voice and actions that cannot conform to standard scenarios.

Mary Wollstonecraft's unguardedness has made her an easy target. Godwin's *Memoirs* (set down with admiration for her spirit and pity for her sufferings) exposed her to attacks in the late 1790s, sustained through much of the following century, and renewed in our time. Horace Walpole, the gothic novelist, called her a 'hyena in petticoats'; John Adams, the second President of the United States, called her 'this mad woman', 'foolish', 'licentious'. It was said that the improper private life of the author of the *Rights of Woman* must discredit the book itself. In the opening year of our present century the *Times Literary Supplement* judged her 'little short of monstrous'. The time has come to probe the source of the slurs – promiscuity, folly, self-defeat – as we open up what's most enduring in this life: nothing less than a proposal to draw on women's skills in order to realise the full promise of our species – a more comprehensive purpose than feminist campaigns for the vote, opportunities and equal pay.

Part of the appeal of Mary Wollstonecraft is that she's fallible. She is proud and self-preoccupied, and does not suffer in silence. Does egotism detract from greatness, and is it more fallible in a woman? Is she too prone to collapse when she fears to lose the character she's bringing into being? What I hope to bring out is how her egotism and despair coexist with a pattern of extraordinary resilience. Ahead is always a new phase of experiment: a single young woman setting up on her own in London, resolved to earn her living by her pen; a journey far north to Norway to confront a captain accused of stealing a cargo of silver; a surprising marriage to a confirmed and cold bachelor. This will not be a story of defeat. She's struck down, it's true, by the counter-revolutionary temper of the 1790s (with the onset of the Terror and the Napoleonic Wars), but her honesty and eloquence, sustained by four women in the next generation, continue to re-emerge.

Though the late nineteenth century brought some revival of public interest in Wollstonecraft, the price paid was suppression of what were regarded as improprieties in her life. Another revival came a hundred years later with a new stage of the women's movement and a spate of biographies that exposed what was seen as sexual recklessness. Her attachments to men remained an embarrassment to late-twentieth-century feminists. Some discovered signs of prudery, and others saw in her domesticity a betrayal of her case for independence. The aim of her critics was not necessarily to kill her cause, but to appropriate it in limited terms. Scholarly fashion has locked her to the conduct books for girls and ephemeral pamphlets tossed out by the scribblers of the day. The effect has been to obscure what it is in her books that transcends her time.

Many of her issues presage the present: women's need to unfold their faculties as this knocks against the rock-face of their conflicting need for sexual commitment; the problems of communication between the sexes; long-term partnership in place of marriage; economic independence; the freedom to express desires without derision or loss of dignity; and, not least, the problems and triumphs of the single parent in the context of Wollstonecraft's belief that a child should not be left to the care of strangers. Her views on pregnancy, childbirth, breast-feeding and continuous parental closeness, unusual in her day, are strikingly modern. She speaks differently to us in this century, less on women's rights and more on both sexes striving to integrate private needs with family responsibilities. What sort of person is a desirable partner? What sort of arrangement should people devise for living together in a permanent partnership? Wollstonecraft is as interesting for her mistakes – her near-collapse into the familiar roles of unrequited lover, discarded mistress and unmarried mother – as for the imaginative solution she eventually finds.

In the course of an eventful life, on the scene of the most far-reaching revolution in history, Mary Wollstonecraft tried out a variety of roles. There was the constant danger that she would lose her way. In the 1770s, 1780s and 1790s she could have acted out a set of familiar scenarios: the uneducated schoolteacher; the humble governess; the scribbling hack; the fallen woman following a predictable course towards suicide; the practical traveller; the pregnant wife – yet each time she reinvents the role. How does she find the strength to transform stale plots of existence against overwhelming odds?

*

Her cause went back to her improvident and violent father. Mary was born on 27 April 1759 in a tall brick house in London. Master silk weavers, clustering near the Spitalfields market, had developed their skills from French Huguenots expelled a century earlier by Louis XIV. Wollstonecrafts had lived in London from the seventeenth century, though the bulk of the family was based in Lancashire. Mary's home was in Primrose Street, most of it now long flattened to make way for Liverpool Street Station. What was a residential area in the eighteenth century is now a scene of glassy office blocks and gliding cars. All that remains is a remnant of Primrose Street just north of the station; Spital Square with its market; a modest house that was the birthplace of John Wesley's mother Susanna in 1669; a more elegant Georgian house on a dingy parking lot, now headquarters of a society for conservation; and the local Church of St Botolph Without Bishopsgate where Keats was christened in 1795. Back in the 1750s, Wollstonecrafts were members of the church, and on 20 May 1759 Mary was christened there in the established Anglican faith. She was the eldest daughter of Edward John Wollstonecraft who fancied himself a gentleman, but was no gentleman at home. His unfortunate wife was Elizabeth Dickson who came of a family in the wine trade and connected with the landed gentry in Ballyshannon, County Donegal, in Ireland. In 1756 she had married Mr Wollstonecraft, at that time nearing the end of his apprenticeship to his father, a wealthy Spitalfields weaver who specialised in silk handkerchiefs.

Mary was not the favourite of either parent. Mrs Wollstonecraft favoured her eldest son, Edward Bland (called Ned). Mary's first autobiographical novel, *Mary*, presents a heroine who craves an object to love and whose mother disappoints her: 'the apparent partiality she shewed to her brother gave her exquisite pain – produced a kind of habitual melancholy'. A second autobiographical novel, *Maria, or the Wrongs of Woman*, labels the eldest brother of the narrator 'the deputy tyrant of the house', the result of his mother's doting. Mrs Wollstonecraft was 'harsh' with her eldest daughter, and determined to exact obedience. Lord Kames, a Scottish judge whose views were popular, said that 'women, destined by nature to be obedient, ought to be disciplined early to bear wrongs without murmuring'. It was not, then, unusual for Mary's early training to silence her voice.

She was made to sit in silence for three to four hours at a time, when others were in the room. Though, as a child, she did question the point of such an exercise, as well as submission to contradictory orders, Mary accepted her mother's reproofs as clues to what might win her love. She was avid for instruction and, given a father she could not respect, all the more attentive to her mother. For Mary to own up to her faults made her feel, she said, restored to her better self.

Mrs Wollstonecraft was softer with her younger daughters, the handsome Elizabeth (known as Eliza or Bess) and the robust Everina who lived to the age of eighty-five. There were three other brothers: James who entered the navy; Charles who emigrated to America; and Henry Woodstock who was apprenticed to an apothecary-surgeon. We can assume that Mrs Wollstonecraft practised wholesome methods since, unusually for that time, none of her children died, and the infant care Mary Wollstonecraft would advance in the 1790s may well have derived from her home. Yet, though the seven children had a good start, later on their morale would falter or fail in various ways. The two eldest were more fortunate: Mary with her irrepressible intelligence, and Ned as heir to a third of his grandfather's fortune.

Grandfather Wollstonecraft, the Spitalfields weaver, left £10,000 – a small fortune – when he died in 1765. The family, on that side, came from the enterprising middle class at a time when English manufacturers were about to acquire industrial power with Watt's invention of the steam engine, in 1765, followed by the appearance of a spinning machine in 1768 and the earliest experiments in electricity by Joseph Priestley, Benjamin Franklin, Humphry Davy, and, eventually, Michael Faraday. It was Mr Wollstonecraft's misfortune to marry and start a family just before the Industrial Revolution. Instead of pursuing the profitable enterprise of his father, it entered his head to rise in the social scale.

Though the British class system had minute gradations, the big divide came between worker and gentleman. The category of 'gentleman' covered a wide range. Elizabeth Bennet, heroine of *Pride and Prejudice*, contemplating her chances with the hero Mr Darcy, proud master of beautiful grounds at Pemberley (and therefore far above her embarrassingly vulgar mother and younger sisters), can defend herself as 'a gentleman's daughter'. What makes her father, Mr Bennet, a gentleman is not only that he has inherited a modest property: he does not work, and has the leisure to spend his days reading in his library.

When Mr Wollstonecraft came into his property he began to cultivate an air of leisure. It was not unknown for members of one class to cross into another, a residue of what was thought of as fine old English 'liberty' triumphing in the Glorious Revolution of 1688 when the people drove out the high-handed James II. This 'liberty' was not liberal – an oligarchy of nobles continued to rule the country well into the nineteenth century, and neither Dissenters, Jews nor Catholics, nor indeed women, could hold public office or take degrees – but what 1688 did achieve was the rule of law over the divine right of kings, and less rigid demarcations of class than in other European states. A gentleman could engage in making money, while a tradesman could buy land.

It was this social flexibility that encouraged Mr Wollstonecraft to rise out of the manufacturing class by acquiring a gentleman's blend of land and leisure. But leisure – extended with drink – does not fit the demands of farming. Mary's earliest memories, from the age of four, were of 'an old mansion with a court before it' near Epping Forest in Essex, north-east of London. Possibly, this was New Farm, close to the main Epping road on a contemporary map. There she preferred outdoor games with Ned, aged six, and Henry, aged two, to girls' games with dolls and baby Bess (born the summer before the family left London in the latter half of 1763).

In 1764, Mr Wollstonecraft moved to another farm close by, near a bend in a road called 'the Whale Bone', about three miles beyond Epping village, on the way to Chelmsford further along the route to the north-east. The local stopping-point was the Sun and Whalebone public house – Mr Wollstonecraft would have drunk there. Everina was born the following year, in 1765, and it was at this time that Mr Wollstonecraft, as chief legatee of his father, inherited three big houses, containing thirty apartments, back in Primrose Street. It is likely that income from the rents now paid for Mr Wollstonecraft's purchase of land eight miles from London, near

Barking, an Essex town with a thriving market and wharf on Barking Creek which flowed into the wide Thames at Gallion's Reach. Between the farm – it may have been Lodge Farm on the local map – and the river was uninhabited ground known as Barking Level, and south of the river lay the Marshes. Later, Mary recalled her 'reveries' as a child when she looked up from this flat land at the expanse of the sky and her sight 'pierced the fleecy clouds that softened the azure brightness'. Her sense of the Creator was born from her sense of creation. She had little religious teaching; it was the unspoilt natural world that woke her spirit.

From the time the family was settled in this part of Essex in the autumn of 1765, their 'convenient' house, their land, Mr Wollstonecraft's ease with his money and his willingness to deal with the crazed and paupers, opened the doors of their nearest neighbour. Mr Joseph Gascoyne had also moved from London trade into country gentility, and his brother Bamber Gascoyne was a Member of Parliament for several boroughs. Mr Wollstonecraft's increasing taste for the leisure of the landed classes was to bring him, in time, a reputation for idleness. The farm at Barking lasted three years: an eighteenth-century pastoral, outwardly intact but germinating disruption.

With no background or training in agriculture, Mr Wollstonecraft failed repeatedly, and sank lower with each move. He found consolation in the bottle and in the vulnerability of his wife, who took the brunt of her husband's blows. There were times when Mary, as a child, threw herself in front of her mother or thought to protect her by sleeping whole nights on the landing outside her parents' door. In her novel *Mary*, when 'Mary's' father threatens her mother, the daughter tries to distract him; when she is sent out of the room she watches at the door until the rage is over, 'for unless it was, she could not rest'. This fictional father, like her own, was 'so very easily irritated' when he was drunk 'that Mary was continually in dread lest he should frighten her mother to death'. Her compassion for her mother became 'the governing propensity of her heart through life'. This was complicated by a disturbing reflection of her father when she looked into her own nature: 'She was violent in her temper . . .'

Mr Wollstonecraft could be jolly and extravagantly fond of his children and pets; but his children sat tight at such times, frozen in their knowledge that their father's exuberance could burst into violence. Mr Wollstonecraft was a common kind of bully, the failure who picks on the vulnerable. Mary, not being weak, was not his prime victim. 'His passions were seldom directed at me,' she remembered when she was away from the family at the age of twenty. But his 'ungovernable temper' had been a 'source of misery'; her pity for his victims, her futile attempts to intervene, hovering as witness to abuse, would leave her depressed. This was worse than beatings. When her father beat her, Mary thought how stupid he was and, even as a child, showed her contempt. But this superiority, though it exercised her intelligence at an early age, writhed in the shadow of a mother trapped by marriage. In the supposedly civilised eighteenth century, women's legal status was, in fact, worse than it had been before the Norman Conquest. Then, the law had permitted married women to own land in their own right. The widow of Ealdorman Brihtnoth, the hero of the Anglo-Saxon poem on the Battle of Maldon, disposed of thirty-six estates, mostly inherited from her family. Such a wife had undisputed control of her 'morning-gift', her husband's present to her the day after the consummation of the marriage. If she wished 'to depart with her children', she could claim half the goods of the household, and defend her right in court.

Contrast this level of independence with *The Lawes Resolutions* of 1632, which allowed a man to beat 'an outlaw, a traitor, a Pagan, his villein, or his wife because by the Law Common these persons can have no action'. In 1724 Defoe's heroine Roxana, pursuing a lucrative career in prostitution, declares that 'the marriage contract was, in short, nothing but giving up of liberty, estate, authority, and everything to the man, and the woman was indeed a mere woman ever after – that is to say, a slave'. Sir William Blackstone's *Commentaries on the Laws of England* (1765–9) explains that 'the very being or legal existence of the woman is suspended during the marriage'; she is turned into a *feme-covert*, in plain words a 'covered' or 'hidden woman', obliterated in her legal protector. In the mid-1770s the young novelist Fanny Burney thought of marriage with dread: 'how short a time does it take to put an eternal end to a Woman's liberty!' she exclaimed, watching a wedding party emerge from a church.

A new marriage law, the Hardwicke Act of 1753, designed to clarify the legality of marriage, had the effect of tightening a wife's bonds. She had no right to her own property or earnings, nor to her children, no grounds for divorce, and no recourse to physical protection in the home. In effect, a woman, when she married, lost the basic right of habeas corpus; since she became the property of her husband, the law allowed him to do with her whatever he wished. Another century had to pass before an Assaults Act in 1853 could convict violent husbands, followed in 1857 by a new matrimonial court recognising women's right to release from abusive marriages.

Mr Wollstonecraft treated his dogs, as he did his wife, to the same unpredictable switches of mood. Once, hearing a dog's howls of pain, Mary's abhorrence became, she said, an agony. 'Despot' resonates like a repeated chord in the opening pages of Godwin's memoir of her childhood. In the second edition he modified that word, almost certainly at the wish of other Wollstonecrafts who were alive at the time. But most of Godwin's facts came from Mary herself in the last year or two of her life, so it's reasonable to assume that 'despot . . . despot . . . despot' had been her mature judgement of her father.

He never learnt from his mistakes: the less of the gentleman he became, the more he clung to that dream. In October 1768 when Mary was nine, the family, increased by a third son, James, travelled to the North of England: to a farm at Walkington, about three miles from Beverley, a trim town in Yorkshire near the sea. Beverley appeared to the Wollstonecrafts still in the light of social possibilities: 'a very handsome town, surrounded by genteel families, and with a brilliant assembly'. The terraced houses in the centre of town were filled with middle-class professionals and merchants. Here, after three years on the farm, the family moved with a new and last child, Charles. When he was born Mary was eleven, old enough to help with a baby. She would continue to love and help Charles. During the years in Beverley her eldest brother, destined for the law, was sent to a grammar school while Mary went to local day schools until she was fifteen and a half. She wrote later: 'I cannot recollect without indignation the jokes and hoiden tricks, which knots of young women indulge themselves in, when in my youth accident threw me, an awkward rustic, in their way.'

It was not only that she felt a rustic in town. A victim of domestic violence, especially a child, is isolated, an isolation enforced by the bully in order to preserve secrecy and control. His victims have to keep up appearances, so that the semblance of social life feels inauthentic. Set apart and awkward as Mary felt, there was a certain dignity and a longing to improve herself when, soon after she turned fourteen, she invited the friendship of a serious girl of fifteen called Jane Arden. The two girls took walks together on Westwood Common – Mary's 'darling Westwood' – where she felt at ease with the woods and windmills. Jane's movements were quick and active; her commands came forth as polite requests. She was the leader of a set who addressed one another with self-conscious civility.

Through Jane, Mary passed messages to other girls, suggesting they too might correspond with her. She laboured over a letter, tossing off quotations like the best-educated girl in the world, and then, with a child's frankness, ends abruptly: 'I wish you may not be as tired with reading as I am with writing. –' She spouted quotations whenever she got a chance, eager to prove herself literate to a friend of superior education: 'you know, my dear, I have not the advantage of a Master as you have,' she wrote. Jane's father John Arden, then in his mid-fifties, had been disinherited by his Catholic family for turning Protestant. As a man of education and wide intellectual interests, including astronomy and geography, he exerted himself as an itinerant lecturer demonstrating electricity, gravitation, magnetism, optics and the expansion of metals. Arden's civility showed up the furies of Mary's own father. She said, 'I shall always think myself under an obligation for his politeness to me.' When he invited Mary to join his lessons, Jane was first with the answers, but Mary led the way with questions. Arden had educated his daughter himself, and Mary was suitably impressed with Jane's understanding – but not too impressed to give way.

'Pray tell the worthy Philosopher, the next time he is so obliging as to give me a lesson on the globes [planets], I hope I shall convince him I am quicker than his daughter at finding out a puzzle, tho' I can't equal her at solving a problem.'

At fifteen, this girl already has a voice of her own. Her phrasing is 'spontaneous' (as she claimed), following an ideal of language based on the run of the speaking voice. A century earlier, Dryden had created a language that was clear and apparently artless, while John Locke had dismissed the affectation of unintelligible words as 'the covers of ignorance, and hindrance of true knowledge'. The Enlightenment, which they promoted, offered more direct modes of communication than the learned flourishes of an expensive education or the languid drawl of the pampered; it's a polished and playful manner with the offhand informality of a modern voice. Mary took to the 'downright' Yorkshire idiom: happiness was to feel 'so lightsome', sure 'it will not go badly with me' – phrases that she used long after she lived there.

In the course of her contact with the Ardens, her reading shifted from trite moralists to literature: Dryden's *Conquest of Granada* (1671–2) and Goldsmith's *Letters from a Citizen of the World* (1760–1) which she shared with Jane. Such books were above the ephemeral publications usually assigned to women: comedy, conduct books, platitudinous devotional rhymes and sentimental novels. It was through the Ardens that Mary discovered real books as the property of the professional middle class. She was a member of a stratum of the middle class that apprenticed its sons in the lesser professions (surgeon, not physician; attorney, not barrister), for the Wollstonecrafts fell just below the propertied class with access to the higher professions. As a keen-minded girl, alert to the ineffectual landed ambitions of her father, Mary had to find other ways to improve herself. One answer was reading, a conspicuous literacy in her early teens and sustained throughout her life; another answer was friends. Where other girls thought of hunting for husbands, Mary was determined to find a perfect friend.

Jane Arden could not live up to what Mary had in mind, and Mary often felt rejected. 'I spent part of the night in tears; (I would not meanly make a merit of it). I cannot bear a slight from those I love,' she blurted to Jane. 'There is some part of your letter so cutting, I cannot comment upon it.' She pressed forward with her feelings: 'I am a little singular in my thoughts of love and friendship; I must have the first place or none,' she explained. 'I own your behaviour is more according to the opinion of the world.'

Of course, a girl like this, so demanding, so disconcertingly open, had to be kept in her place. Jane indicated that Mary could not be 'first' as she had fondly expected.

It did not occur to her to hide these hurts when she found herself excluded. 'I should have gone to the play, but none of you seemed to want my company.' As ranks closed against her, she asked herself where she had erred. 'I have read some where that vulgar minds will never own they are in the wrong,' she told Jane, 'I am determined to be above such a prejudice . . . and hope my ingenuously owning myself partly in fault to a girl of your good nature will cancel the offence.' Mary was acting out what she saw herself to be – honest as well as steadfast – and inviting the same in return: 'I have a heart that scorns disguise, and a countenance which will not dissemble.'

Mr Arden soothed the situation by sending Mary an essay on friendship, which she copied out at once. The essay pictures the possibilities of two people who would be 'guardian angels to each other' and enjoy the benefit of a lifelong attachment that 'corrects our foibles and errors, refines the pleasures of sense and improves the faculties of mind'. To repeat these words to Jane was to restore hope of an ideal tie, a world apart from her degrading home.

'The good folks of Beverley (like those of most Country towns) were very ready to find out their Neighbours' faults,' Mary reminded Jane a few years later. 'Many people did not scruple to prognosticate the ruin of the whole family, and the way he [her father] went on, justified them.' So Mr Wollstonecraft's faults did not go unnoticed by townsmen and schoolmates, and during her teens Mary experienced the shame of her family's slide from respectability. Since there was no further point in secrecy, she acknowledged the problem to Jane: 'It is almost needless to tell you that my father's violent temper and extravagant turn of mind, was the principal cause of my unhappiness and that of the rest of the family.'

It's not clear why he decided to leave Yorkshire. Was this because of local gossip, failure, or the reason given by Godwin: that sheer restlessness in Mr Wollstonecraft tempted him to commercial speculation? In any case, he moved south with his family. We don't know whether he took or left behind a son of fourteen, Henry, apprenticed in January 1775 to Marmaduke Hewitt who had been mayor of Beverley. At this point Henry vanishes from record, and a sustained family silence suggests that something went wrong – something unmentionable, like madness or crime. It was out of character for the Wollstonecrafts never to mention him (accustomed as they were to exchange family news and troubles). He slides into one reply from Mary to Everina in the mid-1780s – some news of Henry had the effect of 'hurrying' Mary's heart – but the sentence shuts the door on whatever it was. In the last years of her life she did confide in Godwin, who kept the secret, avoiding Henry's name in his memoir.

When Mr Wollstonecraft reached London he took a house in Queen's Row, Hoxton, a village to the north of the city. Janet Todd speculates that if Henry Wollstonecraft had become mentally disturbed, he could have been placed in one of Hoxton's three major lunatic asylums, which could possibly have been the reason the family settled there for a year and a half. It was during this spell in Hoxton, when Mary was sixteen to seventeen, that she formed a strange friendship.

Next door to the Wollstonecrafts in Queen's Row lived a clergyman called Mr Clare who had a taste for poetry. It was said that he looked rather like the frail, disabled poet Alexander Pope, who had died in 1744 and was therefore, thirty years later, still within living memory. Mr Clare seldom went out, and boasted a pair of shoes which had served him for fourteen years. Little is known of Mary's connection with this recluse, but it seems that he and his wife took to her as surrogate parents. She stayed with them for days, sometimes weeks, and said, 'I should have lived very happily with them if it had not been for my domestic troubles, and some other painful circumstances, that I wished to bury in oblivion.' It was impossible to turn her back on her mother's abuse, but she did benefit in another way. This 'amiable Couple', as she called them, 'took some pains to cultivate my understanding (which had been too much neglected) [;] they not only recommended proper books to me, but made me read to them'. The Revd Mr Clare became a kind of private tutor, and it may have been now that Mary, warmed by his affection and benevolence, learnt a vital lesson for her future: how to teach. At some point along the way she came to understand that thoughts ripen best in a climate of individual care that she later called 'tenderness'.

One day, Mrs Clare took Mary to Newington Butts, a village just south of London. They came to the door of a house that was small but carefully furnished, neat and fresh. A young woman of eighteen, slender, elegant, was dishing out food to the younger members of the household – the youngest, a boy called George, was fifteen. Mary had never seen such delicacy as the way the young woman took charge of her sisters and brothers. Godwin tells us that the 'impression Mary received from this spectacle was indelible; and, before the interview was concluded, she had taken, in her heart, the vows of eternal friendship'.

This was Frances Blood, known as Fanny, who combined domestic responsibility with a remarkable gift for drawing. Like Mrs Wollstonecraft, the Bloods were Irish: they came from Cragonboy, County Clare. There were Bloods who owned land, some given by Charles II to an ancestor, Colonel Thomas Blood, who had served his king as a spy. The London Bloods were well bred and hospitable, but dreadfully poor. Matthew Blood, Fanny's father, like Mary's, had been an idle drinker who had squandered the sums he'd gained through no effort of his own – in his case, the substantial dowry of his wife, Caroline Roe. He had fled his creditors, first to Limerick, later to Dublin and London.

The present mainstay of these parents and their seven children was Fanny's professional work as an artist: her meticulous drawings of wild flowers. They were published by William Curtis in his *Flora Londinensis*. Curtis was a demonstrator in botany to the Company of Apothecaries, and had founded a garden where he was cultivating five hundred different species to observe each stage of their growth. He employed a number of artists, some of whom, like Fanny, did not sign their work, but the style was uniform: thin black outlines and scientific detail of every part of the flower and fruit, with washes of colour as delicate as nature's own. The artists drew from 'living specimens most expressive of the general habitat or appearance of the plant as it grows wild', foxgloves in Charlton-wood, broom on Hampstead Heath, and violets found in watery ditches 'on the right hand side of the Field Way leading from Kent-street Road to Peckham' (the last not far from Fanny's home). The point of the exercise was to establish each species of indigenous plant in the environs of London. In the first volume, published in 1777, two hundred and sixteen plants were named using the recent classification of botanical species by Linnaeus. Their uses in medicine, agriculture and rural economy were listed beside a full-page illustration. A second volume, on the same lines, was published in 1798.

Fanny's gifts extended to other arts: she played and sang, read literature, and wrote with the grace and application she brought to all she did. Since Fanny and Mary lived on opposite sides of London, they wrote to each other when they could not meet at the Clares. These letters have not survived, but later Mary told Godwin that those she received were better worded and more correct than her own. In Beverley and under the guidance of Mr Clare, Mary had devoured books with a thirst for knowledge, but until she met Fanny, she had not thought of writing as an art. This struck her now with the excitement of possibility – a passion to excel. Fanny, who was two years older, agreed to become her instructor, and so began a friendship based on learning. Mary's Linnaean language of 'genus' and her metaphor for herself as an 'opening flower' may have come from her contact with Fanny's botanical vocabulary. What Mary called Fanny's 'masculine understanding and sound judgment', set off by 'every feminine virtue', bound her friend to her 'by every tie of gratitude and inclination'. To live with Fanny and hear her 'improving' conversation seemed 'the most rational wish'.

Minimal schooling in Yorkshire can't account for the level of Mary's reading and the force of her style. Who taught her to write with such directness and conviction? Was it John Arden who gave her a lesson or two? Was it the Revd Mr Clare in Hoxton who took an interest in her mind? Was it possibly someone obscure like the Master of the Merchant Taylor's School, the Revd Mr Bishop, whom she and Fanny used to meet at the Clares? Was it the verbal finesse of Fanny Blood? No one can say, but Mary came to believe that 'a genius will educate itself'.

What education came her way was nothing like the classical training of boys in public schools. The flourish of classical allusion was one of the ways gentlemen signalled their control of power to one another as members of a club closed to women and other lesser orders of society. The counter to this was not the battery of handbooks on girls' education that continued to stress obedience as the prime quality to be cultivated. Coming into being in the late eighteenth century was an odd new creature whose urgent intelligence transforms miseducation into education. Mary's curtailed schooling had an advantage: it never occurred to her not to think for herself. Another century had to pass before universities opened their doors to women, but already Mary Wollstonecraft was shaping the quality the best teachers of the future would look for: a searching, not obedient, intelligence. She presents a rare case of an intelligence that did not receive the impress of given moulds. Godwin found her untouched by 'the prejudices of system and bigotry' and with a spirit that 'defended her from artificial rules of judgement'. Here were shoots of a new form of life: a girl who conceived her freedom to grow the fruit she was made to bear – and what else is genius in the making?

2

'SCHOOL OF ADVERSITY'

When Mr Wollstonecraft lost money, his solution was to move, not work. In the spring of 1776 he hauled his family west across England to Wales, where he farmed – or attempted to farm – in Laugharne on the Pembrokeshire coast. Often, he jaunted off to London on the pretext of business. Mary, mindful of Fanny's efforts to support her family, proposed leaving home to take up a post. Mrs Wollstonecraft wept – a working daughter was a comedown – and begged her not to go. Some financial disaster happened at this time, too shaming to reveal, though Mary did refer, tight-lipped, to her father's 'misconduct' followed by 'a keen blast of adversity'. She and her sisters were duty-bound to advance him whatever they had. Since the sisters would try to recover this money, their father must have asked for a loan, not a gift, but Mary had no hope that she would ever get it back. 'I have therefore nothing to expect,' she said in 1779 at the age of twenty.

Money was to remain a problem for the rest of Mary's life, as for everyone else in the family except for Ned, who combined his advantage as heir with professional prospects as an articled clerk in a London law firm on Tower Hill. Ned now took over the Primrose Street houses, and passed on the rents. In *The Wrongs of Woman* the eldest brother of the narrator, articled to an attorney, assumes a 'right of directing the whole family, not excepting my father. He seemed to take a peculiar pleasure in tormenting and humbling me; and if I ever ventured to complain of this treatment to either my father or mother, I was rudely rebuffed for presuming to judge of the conduct of my eldest brother.' This was 'forwardness' in a girl.

Ned's status embodied, for Mary, the unfairness of the patriarchal system. Her grandfather had made an heir of his eldest grandson, excluding his granddaughters. Her father, having squandered his own funds, called on those daughters to surrender what provision had been made for their futures – possibly part of their mother's marriage settlement or a bequest from their Dickson grandfather. Girls without dowries would be unlikely to marry, and would have to work for a living in a society which barred middle-class women from all kinds of career except teacher, governess or paid companion. Mary Wollstonecraft would try all three.

Soon after the Wollstonecrafts returned to London, Mary obtained a post in Bath as companion to a widow who was known for her temper. A succession of companions had left. Mrs Dawson belonged to the class Mr Wollstonecraft had failed to join: the landed gentry. She was born Sarah Regis, the daughter of the late Balthazar Regis, Chaplain to the King and Canon of Windsor. For the gentry and nobility, Bath was a centre of fashion and the second capital. Mary thought herself as a country girl attached to nature, more ill at ease than charmed as the coach drove through the long course of streets from the Old Bridge, amidst the dash of other carriages, the heavy rumble of carts and drays, the bawling of newsmen, muffin-men and milkmen. Though she admired the buildings of Georgian Bath as 'the most regular and elegant I have ever seen', the surrounding landscape could not compare with the 'romantic' vistas of Wales.

Mrs Dawson received Mary in the house of her grown-up son William, in the prime position of Milsom Street. Mary's childhood training in self-effacement and deference was now put to the test. If a companion 'cannot condescend to mean flattery,' she soon discovered, 'she has not a chance of being a favorite; and should any of the visitors take notice of her, and she for a moment forget her subordinate state, she is sure to be reminded of it . . . She must wear a cheerful face, or be dismissed.' A companion was paid no more than £10 to £20 a year, and was looked on as little better than an upper servant. Servants resented the drawing-room place of the paid companion, and treated her as a spy. To add to this, Mary missed Fanny, and brooded over her mother, wishing that the recent financial 'storm' would persuade her father 'to see his error, and act more prudently in future, and then my mother may enjoy some comfort'. Vain hope. By this stage Mrs Wollstonecraft had lost interest in her own fate, a state of apathy her daughter took to be indifference.

Throughout her time with Mrs Dawson, Mary longed to 'haste away', but managed to hold out and even to head off her outbursts – less alarming than Mr Wollstonecraft's violence. Now and then, Mrs Dawson owned that no one had managed her better. After three months Mary was able to acknowledge her employer's 'very good understanding', and since she was a woman who had 'seen a great deal of the World', Mary thought 'to improve myself by her conversation, and . . . endeavor to render a circumstance (that at first was disagreeable,) useful to me'. From the age of eighteen to twenty she had the sense to mop up knowledge in whatever form it came her way: 'A mind accustomed to observe can never be quite idle, and will catch improvement on all occasions.' This will to pluck a shoot of improvement from the thorn of 'adversity' was often less blatant than her glooms and groans, yet always there.

The main trial proved less Mrs Dawson's temper than Mary's own depression during her eighteen months in Bath. It was no place for a middle-class girl with a grandfather who had been in trade, a lost portion, and no friends to see her into society. During the first months, Mrs Dawson took her only twice to the Assembly Rooms, where she sat amongst 'Strangers'. 'A young mind looks round for love and friendship,' she thought, 'but love and friendship fly from poverty . . . The mind must then accommodate itself to its new state, or dare to be unhappy.' To 'dare' was a drama never far from Mary's sights, while the passivity of 'accommodation' acted as a blight on her active nature. As necessity forced her along the narrow track of a dependant's life, unhappiness took hold. Her vivacity seemed to have 'gone forever'; her present course was leading her to 'a kind of early old age'.

For such pain, Bath's warm springs – thought to have healing powers – could offer no cure. The town, climbing gracefully up a hill, is built around the Roman bath where invalids, in brown costumes, hung suspended in the steam like wilting mushrooms, as the ill passed their infections unknowingly to one another and to floating hypochondriacs. Nearby was the Pump Room where ladies took their 'turns' about its limited space, parading their finery at the right hour of the morning. There, Mary first came into contact with women of the upper classes. 'In the fine Lady how few traits do we observe of those affections which dignify human nature!' she exclaimed. How disheartening to see before her eyes what the eighteenth century whispered in private when the gallantries of mixed company were done, and gentlemen like Lord Chesterfield picked up their pens to advise their sons ('Women are but children of larger growth . . . A man of sense only trifles with them, plays with them, humours and flatters them'). Contemplating such specimens of her sex, and longing in vain for a sign of 'moral beauty', Mary felt her soul 'sicken'. It was cold comfort to pride herself on the triumph of reason: 'I am persuaded misfortunes are of the greatest service, as they set things in the light they ought to be view'd in.' She would repeat this principle, out-staring her weaker self: 'In the school of adversity we learn knowledge and control our inconsistent hearts.' But her weaker self was not so easily subdued.

It can only have been in Bath that she first met a handsome flirt, thirteen years older than herself. Joshua Waterhouse was the son of a yeoman farmer in Derbyshire, who had crossed class divides when he entered as a sizar* at St Catherine's College, Cambridge, in 1770. He had been ordained as a priest and elected a Fellow of his college in 1774. It was not an age when Fellows took on many pupils or absorbed themselves in research, but idleness attained new heights in Waterhouse, who did the rounds of the watering-places in the company of a titled friend. Well-spoken and stylish, master of the smothered sigh and downcast look, he appealed to women. He amassed piles of love-letters – enough, he liked to boast, to cook a

*An undergraduate who received an allowance from the college, enabling him to study. A sizar used to perform certain duties now performed by college servants.

wedding feast – though his fellowship forbade marriage. Mary's letters to Waterhouse have vanished in a sackful of others, but she did once describe her efforts to reform him: 'I knew a woman very early in life warmly attached to an agreeable man, yet she saw his faults; his principles were unfixed . . . She exerted her influence to improve him, but in vain did she for years try to do it.' In the process, Mary learnt something about herself that was hard to accept: it was driven home that however much she prided herself on good sense, her biological nature was, and would always be, as much prey to passion as that of lighter women. In fact, she acknowledged ruefully, the chaste woman who took a man seriously was 'most apt to have violent and constant passions, and to be preyed on by them'. She was honest enough to record what reason deplored: the 'extreme pain' of unexpressed desire. Next to guilt, she thought, the greatest misery was to love a person whom her reason could not respect.

Played on and lonely, Mary held fast to old ties. It came to her ears that the Ardens were living in Bath, and she hurried to see them in St James's Street. She found Jane's father and sisters at home, not Jane herself who had been employed since 1775 – before she was seventeen – as governess to the six daughters of Sir Mordaunt Martin of Burnham in Norfolk. To be a governess had not been Jane's wish, yet she had found it in her to accommodate. However much she missed her Yorkshire home and sisters, she found, as did her father, sufficient happiness in the exercise of her intellect and the importance she gave to education. So began a teaching career that was to last sixty years. Jane Arden was a born teacher who endeared herself to her pupils in an affectionate and cultured family. Through them she met Captain Nelson – later, Admiral – who gave her a list of paintings she should know, including a Rubens of Mary bathing Christ's feet with her tears, which the family took Jane to see at Houghton, the seat of the Earl of Orford. This sort of attention meant much to Jane with her longing, like Mary's, for knowledge. Mary wrote to her in the spring of 1779 'by way of a prelude to a correspondence'. She looked to Jane for a counter to sentimental dreams: 'I should be glad to hear that you had met with a sensible worthy man, tho' they are hard to be found – .'

It was to Jane that Mary confided the 'blast of adversity' resulting from her father's misconduct. Confessions of gloom ('Pain and disappointment have constantly attended me since I left Beverley') alternate with unconvincing efforts to act out the role of compliant paragon. It's often assumed that Mary Wollstonecraft was a born depressive, though at this time her cause for gloom is patent in her father's losses and the inferior positions for single women.

Her depression was temporarily relieved by a visit to Southampton that summer. Sea bathing, recommended by a doctor, refreshed her, as did Southampton hospitality: 'I received so much civility that I left it with regret.' Depression hadn't 'frozen' her feelings for others, she was relieved to find; attachments sprang up with all her old warmth. No complaint of Mrs Dawson ever crossed her lips; yet away from her, Mary revived. The problem lay in the post itself: the companion who treads the round of another's life.

In April 1780, Mary moved with Mrs Dawson to her family home on St Leonard's Hill in Windsor. If anything, she found Windsor more frivolous than Bath: '– nothing but dress and amusements are going forward; – I am the only spectator'. Women sported enormous 'heads' as much as two feet high. In order to cover the wire and cushions of the structure, hair from the top of the head was pulled upwards, teased or 'frizzled', then dressed with plumes for added height, while hair from the back and sides was arranged in tiers of curls like horizontal sausages curving round the neck. The whole structure was then smothered with white powder. As Mary put it, 'truth is not expected to govern the inhabitant of so artificial a form'.

It took prolonged efforts to look as theatrically artificial as the Quality aimed to be. The main ingredient of face powder was a lethal white lead (which did for two beauties of the age, the Gunning sisters). Mary was aware that 'white' was 'certainly very prejudicial to the health, and can never be made to resemble nature', since it took away the glow of modesty, affection, or indeed any expression. This chalky mask was set off by spots of rouge, and black 'patches', sometimes at the corner of the mouth, meant to be alluring. Mary disagreed with contemporary opinion that bathing was unhealthy: people disguised their smell with pomatum (a fragrant ointment) which she found 'often disgusting'. The many layers of women's clothes – the stays, the wide hoop (balancing on either side of the hips so that the feet had to process in a stately way to avoid seesaw sways), the skirts under satin panniers – all fastened at the back, and so required the help of a lady's maid. Without a maid, Mary dressed with a plainness that made her appear 'a poor creature'. She added, in the defensive tone she often adopted when she found herself in a weak position, 'to dress violently neither suits my inclination, nor my power'.

Although privately she defied fashion as a badge of slavery, certain concessions to society were unavoidable: she did powder her hair, and did wear stays (marked with an 'MW'). It was indecent for any woman to appear without encasing her flesh in a whalebone cage that lifted the breasts and held the frame upright from shoulder to thigh.* Stays limited a woman's movements: when she read, she could not recline but must hold the book upright; when she curtsied, she sank from the knees with rigid back. The head, rising from the cage on the pliant column of the neck, could turn from side to side, the lowered lids or widened eyes transmitting the coded signals of their class.

'I am particularly sick of genteel life, as it is called,' Mary told Jane, ' – the unmeaning civilities that I see every day practiced don't agree with my temper; – I long for a little sincerity, and look forward with pleasure to the time when I shall lay aside all restraint.'

Then, too, her continued stagnation in a 'state of dependance' weighed on her, and she began to think anything would be better. Later, Mary Wollstonecraft would analyse the effect of dependence on women's natures; it was now that she herself experienced a creeping indolence. In this state, she could almost wish to delude herself with foolish hopes that might enliven slow hours, but home had taught her that marriage would soon prove a disappointment.

Her chief consolation was St George's Chapel. Although

---

*Ballet, developed in the eighteenth century, retains that lift of the diaphragm and tightened hips, setting off the carriage of head and arms.

Wollstonecraft has often been taken for an atheist, she remained all her life in the established faith. 'I go constantly to the Cathedral,' she reported from Windsor, 'I am very fond of the Service.' Later, when the cathedral was cleaned, she felt it lost something of its sombre grandeur. She would enter with 'the measured pace of thought'. For her, principles could not be wholly imposed; they had to be affirmed from within by an expansive soul reaching out to the great questions: 'Life, what art thou? Where goes this breath? This *I*, so much alive? In what element will it mix, giving or receiving fresh energy?' Her faith looked to a benevolent deity suggested by the sublimities of nature, a deity of forgiveness, not hell.

In the spring and summer of 1780 the King and royal family were in residence at Windsor Castle. Mary granted that George III was a family man, liking his children about him, but when he 'killed three horses the other day riding in a hurry to pay a visit', he too lost her respect.

'I cannot bear an unfeeling mortal,' Mary observed of the King. 'I think it murder to put an end to any living thing unless it be necessary for food, or hurtful to us. – If it has pleased the beneficent creator of all to call them into being, we ought to let them enjoy the common blessings of nature, and I declare no thing gives me so much pleasure as to contribute to the happiness of the most insignificant creature.'

At the other end of the scale of significance was 'the principal beau' of Windsor, none other than the youthful Prince of Wales (the future Regent and, later, George IV) who wore makeup and drenched himself in scent. Born in 1762, he was Mary's contemporary, one for whom she had no time, nor for local girls who hung on his smiles – 'forward things' said older women, pecking away at the reputations of those the Prince deigned to notice. These dramas served to keep 'envy & vanity alive'. Mary was decidedly not one of the Windsor women who dreamt of impossible romance. She planned a future with Fanny Blood. 'This connexion must give colour to my future days,' she told herself. She knew her resolve to put Fanny before all others would appear 'a little extraordinary', but was prepared to defend it as a reasonable alternative to marriage as well as 'the bent of my inclination.'

That spring she visited Fanny in Town. In the post coach she enjoyed the 'entertaining and rational' conversation of a physician and his well-travelled son.

At the same time she was anxious over Fanny's weak health and thankful to find her somewhat better. They 'passed a comfortable week together, which knew no other alloy than what arose from the thoughts of parting so soon'. She clung to the prospect of another and longer reunion: 'to that period I look as to the most important one of my life'.

Her spirits returned whenever she was freed from being the paid companion. Once, when Mrs Dawson was away and she had 'the whole house to range in', her mood lifted as she supped on bread and grapes. She mused on Fanny, drank Jane's health 'in pure water', and relished her solitude. Bent in the poor light of her candle she sits sideways at a chest of drawers, making pale characters on her page with ink so watered that she fears her writing is too faint to read – and so she fades from sight into the late shadows of a summer night.

That summer of 1780, the Wollstonecrafts moved to Enfield. It was then a rural place ten miles to the north of the outlying north London villages of Hoxton, Hackney and Newington Green, an area of scattered country houses. It was typical of Mr Wollstonecraft to incur the expense of a new house while still committed to his rent in the south London village of Walworth. Mary asked why he was paying two rents at once – she 'cannot divine the reason' – in the tone of one who knows that sense will never prevail. The disruptive elements in the family are plain: the father a spendthrift; the mother dispirited and cold; the eldest son assuming the role of family 'despot'; and now the second sister, Bess, taking Mary's place as the daughter at home, became jealous of a sister who was out in the world. She accused Mary of 'condescension' and forgetfulness about her family, offered with ironic compliments.

'You don't do me justice in supposing I seldom think of you,' Mary protested, 'the happiness of my family is nearer to my heart than you can imagine – perhaps too near for my own health or peace – for my anxiety preys on me.' She had not heard from her mother, and imagined a further withdrawal into harshness. 'Some time or the other, in this world or a better she may be convinced of my regard . . .' This was nearer the bone than she imagined, for Mrs Wollstonecraft turned out to have dropsy – fluid retention in the legs, a sign of heart failure. In late summer or autumn, Mary went home as nurse. At first her efforts were received with gratitude; then, they were taken for granted as her mother deteriorated slowly over two years.

'I was so fatigued with nursing her,' Mary complained to Jane at the end of this period, that she herself had become 'a stupid creature', hard to rouse, and barely in the land of the living. Peering in the mirror she detected, at twenty-three, 'the wrinkles of old age'. Mr Wollstonecraft's temper did nothing to ease the ordeal, especially when his wife's lingering led him to suppose her illness was fancied. Mrs Wollstonecraft's last words were 'a little patience and all will be over'. Those words would return to haunt Mary during her own trials, and she would repeat them yet again in *The Wrongs of Woman*:

> I shall not dwell on the death-bed scene . . . or on the emotion produced by the last grasp of my mother's cold hand; when blessing me, she added, 'A little patience, and all will be over!' Ah! . . . how often have those words rung mournfully in my ears – . . . My father was violently affected by her death, recollected instances of his unkindness, and wept like a child . . . My father's grief, and consequent tenderness to his children, quickly abated, the house grew still more gloomy or riotous. . . My home every day became more and more disagreeable to me.

Mrs Wollstonecraft died in April 1782. Soon after, Mr Wollstonecraft married the housekeeper, Lydia, whom Mary despised but who probably saved the family a lot of trouble by taking on a ruined man and contriving small economies. He retired to Laugharne in Wales where he remained red-faced, rash and needy – ever on the point of death, but indestructible.

At this point the family scattered. Everina went to stay with Ned and his 'agreeable' wife, Elizabeth Munday, at no. 1 St Katherine's Street behind the Tower of London. Bess became engaged to a shipwright called Meredith

Bishop from Bermondsey, across the river from the London Docks – not far from Ned. James, aged fourteen, went to sea, and only Charles, aged twelve, stayed with his father after a short spell at Ned's home.

Mary moved in with the Bloods. From 1782 until the late autumn of 1783, she lived in their house at 1 King's Row in the village of Walham Green, two or three miles to the west of Chelsea, near Putney Bridge on the Thames. It's not clear how she supported herself. She may have lent a hand in a little shop Fanny and Mrs Blood kept for a while; she certainly helped Mrs Blood with needlework, a common way for poor, respectable women to earn a living. It meant, though, punishing hours, strain on the eyes when a gown had to be hemmed by candlelight, and starvation pay. Though her line in chat could not engage Mary's mind, Mrs Blood was 'our' mother. Mary had the emotional benefit of her transfer to Fanny's family, whose acceptance soothed the scar left by her real mother's partiality for Ned. She also drank in talk of Ireland. Half-Irish herself, she began to speak of 'the dear County of Clare' as though she knew it.

'The women are all handsome, and the men agreeable; I honor their hospitality and doat on their freedom and ease, in short they are people after my own heart – I like their warmth of Temper, and if I was my own mistress I would spend my life with them.'

Mrs Blood damped down this burst of enthusiasm with motherly warnings: the men were 'dreadful flirts'; a visiting girl must beware not to leave her heart 'in one of the Bogs'.

The Bloods sometimes entertained an Irish cousin called Neptune, about nine years older than Mary. His attentions led her to expect a declaration, but nothing definite was said. Neptune Blood was a snob. The poverty of the London Bloods put them beneath him; and so, too, the warm, attractive but penniless young woman who was part of the household. 'Few men seriously think of marrying an inferior,' she saw. Such a woman can be deceived 'until she has anticipated happiness, which, contrasted with her dependant situation, appears delightful. The disappointment is severe; and the heart receives a wound which does not easily admit of compleat cure.' Mary did not hide her resentment. Later, when she cooled down, she judged that she had been 'as much to blame in expecting too much as he in doing too little – I looked for what was not to be found.' Her cure co-opted reason instead of the usual misery of silent brooding. The latter was Fanny's lot.

Fanny had long been in love with an Irishman called Hugh Skeys. His hesitation left her disappointed (in the old sense of rejected); and since he continued to court her in his on-off way, disappointment was renewed over some eight years. Her favourite song was 'In a Vacant Rainy Day You Shall Be Wholly Mine'. This 'canker-worm', Mary saw, was 'lodged in her heart, and preyed on her health'. All Fanny's earnings went to her family; she could offer a man nothing beyond her grace, artistic gifts and selfless character. It was considered foolish – even culpable, by a man's family – to take a bride without a dowry.* Could Skeys bring himself to ignore public opinion, another drawback lay in the fluctuations of Fanny's lungs: her troubled health was no good prospect for the sapping effects of yearly childbirth, the lot of almost all married women. Childbirth was full of dangers, and for ailing women like Fanny the chance of dying, very much higher. Fanny's longing for Skeys put friendship second. Mary, running to Fanny with eagerness, stopped short in the face of Fanny's pain. She hated to obtrude her affection, or receive a love warmed by the poor fuel of gratitude – she still longed to be first.

Bess married in October 1782; so did one of the Arden sisters. In confidence to Jane, Mary painted a bride's prospects. One month after the honeymoon 'the raptures have certainly subsided', together with 'the dear hurry' of wedding preparations and 'the rest of the delights of matrimony' – all are 'past and gone and have left no traces behind them, except disgust: – I hope I am mistaken, but this is the fate of most married pairs.' Her own resolve was unshaken: 'I will not marry, for I dont want to be tied to this nasty world.' As a single woman she looked forward to pursuing her

*In Roman law it was illegal: *nullum sine dote fiat conjugium* (let no marriage be made without a dowry). If the bride's family could not pay the groom, the marriage could not take place, and previous understandings were nullified. Olwyn Hufton, *The Prospect Before Her: A History of Women in Western Europe 1500–1800* (1997).

'own whims where they lead, without having a husband and half a hundred children at hand to teaze and controul a woman who wishes to be free'.

Bess Wollstonecraft's marriage turned out more disastrous than anything Mary predicted. It had been approved by her eldest brother as effectively head of the family. Bess had been married in Ned's parish church near the Tower. She had needed a home and protection, and her bridegroom's 'situation in life was truly eligible', as the family put it to one another. He and his father built lighters (flat-bottomed barges) to unload cargo ships. The Bishops were well supplied with money.

Bess's baby, called Mary, was born ten months after the wedding, on 10 August 1783 – a difficult birth followed by melancholia alternating with 'fits of phrensy'.* Late that year Meredith called in Mary to nurse her sister. At first, Mary tried practical remedies: she held Bess in her arms; she took her for a drive in a coach, and sent the baby across the river to Everina (with an instruction to 'Send the child home before it is dark'). Nothing Mary tried seemed to help. If anything, she found Bess's wandering mind more of a worry than her previous 'raving'. 'Her ideas are all disjointed,' Mary reported to Everina, 'and a number of wild whims float on her imagination and unconnected fall from her – something like strange dreams when judgement sleeps and fancy sports at a fine rate – .' This fits the report of the 'lovely maniac' in *The Wrongs of Woman*: 'a torrent of unconnected exclamations and questions burst from her'. When she's put in a madhouse, it's said that she had been married against her inclination to a rich man, 'and in consequence of his treatment, or something which hung on her mind, she had, during her first lying-in, lost her senses'.

Mary had to watch Bess every moment until she herself felt the infection of mental illness: 'Poor Eliza's situation almost turns my brain – I

*This would now be diagnosed as postpartum depression, but the label belies continued ignorance about this condition. Most cases would recover, but not all: for instance, in the 1840s Thackeray's wife fell into depression after the birth of her second child, and remained in an asylum for the rest of her life.

can't stay and see all this misery – and to leave her to bear it by herself without anyone to comfort her is still more distressing.'

This was the mounting crisis as autumn died into winter: was Bess falling into permanent madness? Was her sister to believe Bess when she said that her husband was the cause? 'She seems to think she has been very ill used.' Whatever the exact meaning of 'ill-use', Mary certainly believed that Bishop thought only of 'present gratification' – coded words for clumsy sex with no pleasure in pleasing his wife. No word exists in manmade language for marital sex that verges on rape, but it's possible that Bess, tenser and more refined than Mary, shared her sister's homegrown shudder at the prostitution of wives in marriage. Mary listened with her usual attentiveness to what her sister was saying: 'she declare[s] she had rather be a teacher than stay here'.

Mary listened also to her brother-in-law. At first, these sessions left her unsure what to do. She could pity Bishop his disturbed wife, but attempts to explore the crisis soon revealed an uncomprehending boor, one of those people at ease with money – he lent £20 to rescue the bankrupt Bloods 'very properly without any parade' – but blocked off when it came to intimacy. At length Mary lost hope of getting through to a made-up mind: 'it will ever press forward to what it wishes regardless of impediments and with a selfish eagerness believe what it desires practicable tho' the contrary is as clear as the noon day . . . My heart is almost broken with listening to B. while he *reasons* the case – I cannot insult him with advise – which he would never have wanted if he was capable of attending to it.' Instead of hearing his wife's needs, Bishop rationalised 'fixed conclusions from general rules'. Mary's witness of her father's violence and the deterioration of her mother into an unloving and unlovable victim alerted her to other forms of unseen abuse in the situation of the apparent patient. Had Bishop been violent, had his abuse been visible, we can assume that Mary would have acted sooner. Her hesitations suggest a situation in which abuse was invisible – sexual roughness or the kind of covert tyranny that twists a wife's character. These are so close to contemporary notions of wifely compliance that words, again, did not exist – and silence still reigns a century later when novels like *Daniel Deronda* and

The Portrait of a Lady open up the psychic torments of a tyrannical marriage. Protest, in such cases, could only take the forms of passivity, breakdown or, in the worst cases, what would appear as inexplicable derangement. When Mary came to nurse Bess she was sliding into derangement, and craving rescue. Bess managed to convey that her sole hope of sanity was to leave her husband.

Though Mr Bishop looked like a normal young man, Mary became increasing alarmed by an impregnable thickness. 'Only a miracle can alter the minds of some people,' she saw. 'To the end of the chapter will this misery last.' Her brother-in-law guarded his fortress against the damage in his vicinity. His friend John Skeys, a brother of Fanny's suitor, was convinced by Bishop's air of baffled innocence. Mary, who herself was 'confused' at times by Bishop's version, could not blame Skeys: 'For I that know and am fixed in my opinion cannot unwaveringly adhere to it.' She had no illusions as to whose version would be believed if Bess could neither get well nor stay with her husband. Socially, as well as legally, right was on his side.

As the frenzies gave way to more settled disturbance, Mary asked what she should do in urgent letters to Everina (still living in her brother's house and therefore best placed to plead their sister's case). 'In this case,' Mary wrote, 'something desperate must be determined on – do you think Edward [Ned] will receive her – do speak to him – or if you imagine I should have more influence on his mind I will contrive to see you . . . To be with Edward is not desirable but of the two evils she must chuse the least.'

That could have been a respectable solution: Bess could have appeared to be making a family visit, and would have been able to take her baby with her. For whatever reason, Ned failed to come up with an offer of temporary refuge. Any different refuge meant open defiance, and it would compound the defiance to take the baby who legally belonged to her father. Though Mary is usually pictured acting alone, in fact the three sisters faced the problem together. On the one hand, there was Bess deteriorating to the brink of madness, there was her sisters' fear of permanent damage, and daily evidence of her husband's intractability. On the other hand, there was the law of the land that bound wife to husband as property. The law allowed a husband to lock his wife at home for life.

Given the degree of mental illness, Mr Bishop might obtain a doctor's order to shut his wife in a lunatic asylum. The historian of marriage Lawrence Stone sets out this possibility: 'One of the most terrible fates that could be inflicted on a wife by a husband was to be confined . . . in a private madhouse . . . where she might linger for months, or even years. The mere threat of such confinement, which was frequently used by angry husbands in the eighteenth century, was enough to strike terror. In eighteenth-century England, this fear hung over every wife . . .'

At twenty-four, Mary was under this pressure 'to snatch Bess from extreme wretchedness'. The time had come to act, not talk. If Bishop got wind of it he would dismiss Mary and forbid her further contact with his wife. If they got away, he was likely to pursue them. The law allowed him to kidnap his wife against her will. Discreetly, Mary questioned a man called Wood, a friend of Bishop, about his responses: he was either a 'lion, or a spannial' was the answer. News of the lion was disquieting, but did not deter Mary and Everina from planning their sister's escape.

Late in December Mary sent a message to Everina asking her to be secret and warning Ned not to 'expostulate' on Bess's behalf in case Bishop scent the level resistance to him had reached. While Bishop took to bed with a passing fever, almost all his wife's clothes were smuggled out of the house with assistance from Everina and Ned's wife. Fanny was invited to stay for a few days in Bermondsey: as a guest going about her own affairs, she conveyed parcels to a brush-maker called Lear in the Strand; there, Everina would collect them. She found a place to store the bundles – not, presumably, Ned's house. Its location was not written down, for obvious reasons. Mary also had the forethought to have some 'shirts' made – washable undergarments – since fresh things would be beyond their means once Bess was on her own. As the plan advanced Mary noted that Bess grew better 'and of course more sad – '.

Their plan, set for mid-January 1784, was to switch coaches so as to throw Mr Bishop off their trail as they made for Hackney, a village to the north of London, beyond Hoxton – an area known to Mary from earlier days. In the eighteenth century Hackney was a centre for market gardening, with a nursery for ferns, camellias and roses. Coaches rolled along the narrow curve of Church Street and stopped at the Mermaid, pre-eminent amongst the hostelries, and scene of political and parish meetings or the annual feasts of London tradesmen. The Mermaid itself was too public for runaways. Mary had chosen a lodging-house opposite, looking out on rural quiet at the back. In the dimness of swaying coaches as the sisters crossed the river and rumbled north, Bess's knuckles were in her mouth. Mary feared she would go into 'one of her flights, for she bit her *wedding ring* to pieces'. She sighed over her baby, five months old, in a way that filled Mary with love and pity.

'The poor brat it had got a little hold on my affections,' Mary granted, 'some time or other I hope we shall get it.'

Leaving Bishop was less questionable than leaving the child behind. It's thought Mary never mentioned this episode to Godwin out of shame for her actions; more likely Godwin himself suppressed the facts in view of the illegality of the escape. Nor must we forget that when Godwin wrote his memoir, Elizabeth Bishop was still in her mid-thirties, needing as a teacher to preserve an impeccable reputation and therefore opposed to any memoir at all. As for the baby, we must remember, too, that according to law a wife who abandoned a marriage would have to relinquish her offspring. This tied most mothers to marriage, however disastrous. Another consideration was that when Mary and Bess left the Bishop house in January 1784, they had but £3 between them. Bess was not the only loving mother who convinced herself her child was better off with a wealthy father to provide for her.

Their hideaway was the house of a Mrs Dodd, who looked forbiddingly 'wild' when she opened her door.

'Heaven protect us,' Mary thought.

She introduced herself as Miss Johnson. Bess, less alert to danger, quietened down once the journey was over, while Mary's heart 'beat time' with every carriage that rolled by, and a knock at the door threw her 'into a fit'.

'I hope B[ishop] will not discover us,' she wrote immediately in a shaky hand to Everina, 'for I could sooner face a Lion.' Whenever the door opened she expected to feel his panting breath. 'Ask Ned how we are to behave if he should find us out[,] for Bess is determined not to return[–]can he force her – but I'll not suppose it – yet I can think of nothing else – She is sleepy and going to bed[;]my agitated mind will not permit me – Don't tell Charles or any creature – Oh! – let me entreat you to be careful – for Bess does not dread him now as much as I do – .'

Oddly, it was Bess who was composed. She wrote a 'proper' letter to her husband, sent via Everina, their intermediary in matters great and small from the problem of clean linen to the question of whether Mr Bishop could be induced to agree to a separation.

In the meantime, they stayed in hiding. The day after the escape they lay about drained – nursing assorted aches, Mary joked, like languid ladies. She joked, too, that in her fear she had *almost* wished for a husband to protect her. At night, she was hot with a fever caught from Bishop; Bess too, though increasingly rational, was suffering from headaches which Mary put down to lack of exercise. They did not dare go out, but took heart from small mercies: their room was comfortable, and Mrs Dodd turned civil, reassured by lodgers with genteel aches and languor who directed letters to their attorney at a respectable address in Town.

The sisters continued to keep their heads down as the news spread. Mr Bishop reacted with angry 'malice', according to Mary; had he been hurt or repentant, she could have felt for him, but as it was, he relieved her of compassion. Reports reached Mary that people blamed her as 'the *shameful incendiary* in this shocking affair of a woman's leaving her bed-fellow – they "thought the strong affection of a sister *might* apologize for my conduct, but that the scheme was by no means a good one" – In short 'tis contrary to all the rules of conduct . . . for the benefit of new married Ladies.' One Mrs Brook let it be known that 'with grief of heart' she gave up Mary's friendship.

They were not entirely bereft of sympathy. Good Mrs Clare came in the rain from Hoxton as soon as she heard, offered a loan and cautious advice – she was too responsible not to raise the question of reconciliation – and, returning home, sent the runaways a pie and some wine. Fanny entered into Bess's grief at leaving her child, and sent a note to John Skeys, through her brother George, begging Skeys to find out how the child did. Bishop was making the baby the centre of his outrage, and Mary had to concede this point. He sent Bess no word of the child, and his answer, through Skeys, was cool and unsatisfactory, with no perception of his own part in his wife's breakdown. The message was that 'poor', ill-done-by Bishop was puzzling his head as to how to effect a reconciliation, and hoped, if so, to make his wife happy.

Bess refused to return. Her voice sounds through Mary's reports with a force – frenzied, determined – of her own. Since Mary was blamed, then and since, for her 'monstrous intervention', it's important to note that it's Bess who willed this end to her marriage, not Mary, who at times was unsure what to believe or do. Mary's part was to carry through her sister's decision, glad to see a Bess no longer in Bishop's power. For the look of 'extreme wretchedness' hanging over her sister's face started to clear within days of leaving her husband. Her recovery vindicates this decision; so does the fact that Bess never regrets it – however vehemently she complains of everything else. The alternative would have been to compel her to bear with Bishop, a man so forceful, Mary remarked, that he would make even Ned 'flinch'.

The next question was what two destitute young women were to do for their support. Their concerned doctor, Saunders, had given the sisters ten guineas to tide them over their crisis but could not undertake to support them further. Mr Blood invited them to live in his house, and this, of course, they could not accept.

A more appealing suggestion came from Fanny: she, Bess and Mary could live together and earn their living from painting and needlework. Mary seized on this plan with dreams of helping Fanny with her commissions. Such amateur enthusiasm stopped Fanny short: Mary had no conception of the professionalism required. Fanny explained this to Everina with a firmness that shows her to have been different from the weak figure who has come down to us. Fanny had her feet on the ground (apart from indulging Mary's tendency to hypochondria). Her letter documents the economic plight of young middle-class women who had self-respect and consideration for one another, but no money or means of support:

Walham Green
Feb 7, 8th 1784

My dear Everina,

The situation of our two poor girls grows more and more desperate – My mind is tortured about them, because I cannot see any possible resource they have for a maintenance, now that Dr S[aunders] begins to waver from his friendly professions. The letter I last night received from Mary disturbed me so much that I never since closed my eyes, and my head is this morning almost distracted. – I find she wrote to her brother informing him that it was our intention to live all together, and earn our bread by painting and needle-work, which gives me great uneasiness, as I am convinced that he will be displeased at his sisters being connected[?] with me; and their forfeiting his favour at this time is of the utmost consequence. – I believe it was I that first proposed the plan – and in my eagerness to enjoy the society of two so dear to me, I did not give myself time to consider that it is utterly impracticable. The very utmost I could earn, one week with another, supposing I had uninterrupted health, is half a guinea a week, which would just pay for furnished lodgings for *three* persons to pig together. As for needle-work, it is utterly impossible they could earn more than half a guinea a week between them, supposing they had constant employment, which is of all things most uncertain. This I can assert from experience, for my mother used to sit at work, in summer, from *four* in the morning 'till she could not see at night, which with the assistance of one of her daughters did not bring her more than half a guinea a week, and often not quite that; and she was generally at least one third of the year without work, tho' her friends in that line were numerous. Mary's *sight* and health are so bad that I am sure she never could endure such drudgery; and you may recollect that she was almost *blinded* and sick to death after a job we did for Mrs Blensley when you were there. As for what assistance they could give me at the prints, we might be ruined before they could arrive at any proficiency in the art. – I own, with sincere sorrow, that I was greatly to blame for ever mentioning such a plan before I had maturely considered it; but as those *who know me* will give me credit for a good intention, I trust they will pardon my *folly*, and inconsideration. As I believe you will readily perceive that Mary and Eliza can never prudently embark in the above-mentioned plan, I will venture to mention to you one plan Mrs Clare lately proposed, which, if practicable, might provide a refuge from poverty, provided B[ishop] cannot be brought to allow his wife a separate maintenance. You will probably be shocked when I tell you this plan is no other than keeping a little shop of haberdashery and perfumery, in the neighbourhood of Hoxton, where they may be certain of meeting encouragement. Such a shop may be entirely furnished for fifty pounds, a sum which I should suppose might be raised for them, if it was mentioned to your brother . . .

I beseech you to let me hear from you as soon as possible – for I am impatient to know whether there is the least prospect of comfort for our dear girls. – Believe me to be, dear Everina

Yours sincerely

F. Blood

This was an indirect plea to Ned on his sisters' behalf. It failed to produce any help for a shop, though it's likely Ned did exert himself to bring about a legal separation for Bess. Divorce was out of the question: only four women had succeeded in the last two hundred years. It required an Act of Parliament costing seven to eight hundred pounds and proof of incest, sodomy, bigamy or the invalidity of the marriage. When Milton had argued for divorce on the grounds of incompatibility, he had seen it as a man's problem: there was no conception that a woman might suffer in the same way. Mr Bishop declined to maintain an absentee wife. So again the runaways faced the question – how were virtuous single women to survive?

3

NEW LIFE AT NEWINGTON

Rescue arrived at the neediest moment in the shape of a widow called Mrs Burgh – an aged and unknown fairy godmother who suddenly appears on the scene, bearing out the truth that half the good in the world is done by those who lie in unvisited tombs. How Mrs Burgh came upon the runaways is not recorded, possibly through Mrs Clare, but in the course of 1784 she turned two homeless young women, without capital or experience, into owners of a school. She brought this off despite the fact that Mary could offer none of the usual accomplishments of genteel education: no French, no skills in music and drawing, and no fancy needlework. 'I shall ever have the most grateful sense of this good old woman's kindness to me,' Mary said afterwards.

At first, she and Bess took lodgings in the north London village of Islington with 'expectations' of establishing a school. It was not to be, though encouraged by Mrs Burgh's nephew Mr Church – 'Friendly Church,' Mary called him – a '*humane*' businessman who lived in the village. They soon discovered the number of competing schools in the area.

Mrs Burgh then stepped in and persuaded Mary to start a school further out, two miles north of London where she herself lived at Newington Green, a community of merchants and Nonconformists. The sole record comes from Everina who recalled that 'there through her exertions they in the course of two or three weeks obtained near twenty scholars'. If each pupil paid the current rate of £1 a quarter, it meant they would get about £20 for thirteen weeks. Since there were two of them, this was still below the minimum for survival, the half-guinea a week that, in Fanny's experience, was the most a professional woman could expect. It's likely that Mrs Burgh helped them with more than advice.

A school in a healthy spot, beyond the range of the city's noxious air, soon attracted lodgers as well: Mrs Campbell who enrolled three children in the school, Mrs Morphy, and Mrs Disney who enrolled two sons. Though Mrs Disney came from a prominent Dissenting family, she 'daubed' her face, Mary noticed disapprovingly. As the school expanded, Mary rented a larger house; she was also able to provide a post (and home) for Everina, and take on an assistant, Miss Mason. Mary sometimes referred to her as 'poor Mason', as though some misfortune were common knowledge – in most such cases the parents had lost their fortune, so that instead of fulfilling her destiny as a marriagable 'lady' the daughter was compelled to work as teacher or governess. Given her own past, Mary was sympathetic to the drop in status for a girl who must work. The sturdy 'Mason' (as the Wollstonecrafts called her) had a 'clearness of judgement' not overburdened with sensibility. Her bluntness was wholesome, not wounding. The Wollstonecraft sisters talked of Mason long after she left the school in July 1785, but when Mary remembered her in after years it was not as an intellectual companion.

It had always been Mary's dearest plan to live with Fanny; she had invited Fanny to join them in Islington. This may have been when Fanny confided her longing to be released from her dependent family; and Mary, who promptly found and readied a new home, was then put out when Fanny had scruples about abandoning her parents. Mary recalled this episode with an edginess that might seem excessive if it weren't for the unnoted fact that, at about this time, the Bloods moved from the southern outskirts of London to Islington. Their motive can only have been to join Fanny, which meant that they joined Mary as a drain on her new life before it could secure itself. This would explain her fret. She loved the Bloods like a daughter, but for a young woman about to start a school it was not a moment to take on the perennially unemployed Mr Blood, his troublesome daughter Caroline, and son George who was prone to skip from one job to another in a train of mishaps. The moves of Fanny and her family in 1784 look like a chessboard chase. If this is so, then one reason for leaving Islington would have been to shake off the Bloods who stayed on there when Fanny moved, once more, to join Mary in Newington Green.

It was a place of substantial Georgian houses along the Roman road that led north from London. The houses were not in the usual ribbon formation; they encircled a well on a green where sheep grazed, fenced by a low rail with scattered elms at different stages of growth. A forest with ferns and flowers divided this village from its neighbour Stoke Newington, where Mary found a friend in James (later, Sir James) Sowerby, who, like Fanny, was a botanical artist. He specialised in fungi, and provided models for the British Museum. This bachelor was Fanny's exact contemporary and two years older than Mary. She mentions his visits to her in the summer of 1785, his financial help in 1786, and a long letter from him in 1787. Many other accomplished people had gravitated to the area, amongst them Daniel Defoe a century earlier, the Jewish writer Isaac D'Israeli, and the family of Anne Stent who married James Stephen, an early activist in the fight to abolish slavery. The misty chilliness of Newington Green, its gnarled trees and fragrant shrubbery were later recalled by Edgar Allan Poe who went to school there.

At the time the Wollstonecraft sisters settled in the village it was a bastion for Dissenters, who worshipped in a small church built in 1708 on the north side of the Green (the oldest Nonconformist church still in use in London). When Mary attended services, she thought it too plain; as an Anglican she preferred architectural grandeur. Since the Civil War in the seventeenth century there had been a proliferation of dissenting sects, drawing the bulk of their numbers from the poorer classes, but there were also Dissenters whose forebears had been ennobled by Cromwell, and others who'd grown rich during the Commonwealth. These gravitated to the area of Newington Green. So Mary Wollstonecraft encountered here the well-to-do edge of radical Protestantism, and its cutting-edge intellectually in the form of Dr Richard Price, a Welsh divine who preached political doctrines of liberty and equality. Dr Price's congregation at

Newington Green included Mrs Burgh, and he had been a close friend of her late husband, the Revd James Burgh.

Mrs Burgh's husband had been a Calvinist Scot, educated at St Andrews, who had opened an academy in Stoke Newington and then moved to the Green. There he had reigned as schoolmaster from 1750, his pupils attending the sermons of Dr Price. He had published his *Thoughts on Education*, followed by an array of educational and political tomes.

This schoolmaster might have appeared a formidable personage to follow, but Mary Wollstonecraft took a line which contradicted Mr Burgh's demeaning education for girls, summed up two years later in her *Thoughts on the Education of Daughters* (her title looks back to Burgh's, as Burgh's looks back to Locke's *Thoughts Concerning Education* (1693). Burgh had held that a girl should know just enough arithmetic to do household accounts, and just enough geography to converse with her husband and his friends. Boys were generally trained to block tenderness as a form of weakness. The only emotion Burgh had encouraged was patriotism – no different, in this, from most educators. The schoolmaster writing so busily had not seen that the education he had meant to extend and refine had been skewed to feed the very materialism he deplored, that of a predatory nation moulding an elite of fighters and colonisers.

Mary Wollstonecraft refused to shape her pupils to fit predetermined forms; she asked herself what girls learnt that left them lisping like infants and parading themselves in clothes whose 'unnatural protuberances' bore no relation to the shape of the female body. Ever since Bath and Windsor, she had deplored the triviality of female accomplishments: the tinkling on the harpsichord, and pride in landscapes touched up by a drawing master. It infuriated her to hear ladies bleating received opinions: 'I am sick of hearing of the sublimity of Milton, the elegance and harmony of Pope, and the original, untaught genius of Shakespear.' Such bleaters knew 'nothing of nature' and 'could not enter into the spirit of those authors'. Her cure was simple: 'I wish them to be taught to think.'

As a thinker herself, Wollstonecraft stressed the ungendered possibilities of the mind – the 'mind', she repeats, wondering how it might come into its own. The answer came from her own history of self-education: agency must be transferred from teacher to pupil. The teacher can't 'create' a child's mind, she said, though 'it may be cultivated and its real powers found out'. Basically, 'it must be left to itself'. She was speaking as a disciple of Rousseau, who had enraged Burgh in the 1760s when he proposed that a child should follow nature, unwarped by formal education till the age of twelve. Mary Wollstonecraft did not put this literally into practice – it would have made her school redundant – but did grasp the crux of Rousseau's theory when she urged pupils to look into 'the book of nature', and banned rote learning: 'I have known children who could repeat things in the order they learnt them, that were quite at a loss when put out of the beaten track.' Instead, she taught them to combine ideas, comparing things similar in some ways and different in others. Then too, where Burgh stuffed his language with Greek and Latin tags, Wollstonecraft cut through to the heart of matter, dismissing 'words of learned length and thund'ring sound' designed to cow the common reader: 'A florid style mostly passes with the ignorant for fine writing; many sentences are admired that have no meaning in them.' Milton had established a verbal league table with Anglo-Saxon monosyllables at the bottom and Latinate words at the top; the graver the subject, the more sonorous the language. Though Wollstonecraft did read Milton, her own practice favoured Enlightenment ideals of simplicity and clarity.

Her primary aim as a teacher was to elicit an authentic character in place of sameness. The same things, she thought, should not be taught to all: 'Each child requires a different mode of treatment.' Nor were pupils urged to display uniform manners. In place of affectation, she encouraged naturalness: 'Let the manners arise from the mind, and let there be no disguise for the genuine emotions of the heart.' Other schools had a fixed code of manners, not for the good of the pupil but to promote an image of the school. Mary, on the contrary, put the weight on what she called 'temper', extending the benefits of a home education by those who knew the child best. Boarding establishments were schools of 'vice' and 'tyranny'. There, vicious children were prone to 'infect' a number of others, while love and tenderness remained undeveloped in the absence of domestic affection.

Home, then, was central to Wollstonecraft's education; she did not compete with parents for control. Burgh had laid it down that boys should be removed to boarding schools to avoid the 'weakening' effect of maternal love; attachment to parents, he thought, should be a matter of 'principle', not instinct. Again, Wollstonecraft opposed this: she realised that the most important education of all begins with a baby's mouth on the mother's breast, responding to 'the warmest glow of tenderness'. This grants mothers the central role in education. Her insistence on breastfeeding went against fashionable practice in her youth when it had been customary amongst the upper classes to send infants away to be cared for – in many cases, neglected – in the country. Jane Austen's family was amongst those who followed this practice: Jane spent her first two years in a local cottage, the idea being that a child returned home when she was ready to be civilised. Mrs Austen must have had a superior arrangement, for all her infants survived. She was less well advised when, in 1783, she sent her daughter Jane, aged seven, to boarding-school. Girls in most schools of the time were poorly fed, callously treated, and in many cases succumbed to illness. It was only through the initiative of a fellow-pupil, Jane's cousin, who managed to send an alarm to the Austens, that a sick Jane was fetched away.

Mary Wollstonecraft ran her school along entirely different and what were then innovative lines: she had a maternal attentiveness to the physical as well as mental needs of a child; she was committed to wholesome food; and her methods were flexible. Godwin tells us that she 'carefully watched symptoms as they rose, and the success of her experiments; and governed herself accordingly'. She was confident in her theories of education without pressing them too hard. She did believe in moral discipline, but not in the first place as a set of rules to be enforced, rather as a child's imitation of tender parents whose principles take root in its earliest apprehensions. So, unlike other schools, Wollstonecraft's did not disconnect the mind from domestic affections.

Was Mrs Burgh aware of Mary's deviance from her husband's regime? If so, did she mind? It's inconceivable that she would have backed Mary had she not been impressed with her ideas. Mr Burgh has his place in *The*

New Dictionary of National Biography; his dour voice drones on in his tomes in the manner of those too well informed to be aware of the person who listens – the occupational hazard of a schoolmaster. Marital sex, Burgh believed, should be curtailed. It is our duty, of course, to 'support the species', but abstinence at other times is to be desired. Women are vain creatures who should not obtrude their prattle on educated men. Beauty is nothing more than a 'mass of flesh, blood, humours, and filth, covered over with a well-coloured skin'. Men's admiration always contains a 'filthy passion'. A wife must obey her husband because of the 'superior dignity of the male-sex, to which nature has given greater strength of mind and body, and therefore fitted them for authority'. These were his words in 1756, three years into his childless marriage to Hannah Harding, who appears in *The New Dictionary of National Biography* and in *The Dictionary of British Radicals* only in her capacity as Mr Burgh's wife, yet her help to Mary Wollstonecraft in the last four years of Mrs Burgh's life, from 1784 to 1788, may be now more significant than any other fact about the Burghs.

It's unlikely that the Revd Mr Burgh would have approved as his successor an untrained young woman of twenty-five, pursuing intuitions instead of tried methods, opinionated and disinclined to curb her eloquence. And yet Mrs Burgh more than approved Mary Wollstonecraft; she came to treat her, Mary felt, 'as if I had been her daughter'. All this suggests that Mary Wollstonecraft spoke to some part of Hannah Burgh that the schoolmaster had silenced. We might speculate further that Mrs Burgh's feeling for Mary as 'daughter' found an answer in Mary's maternal deprivation. At seventeen she had taken to Mrs Clare; she had loved 'our mother', Mrs Blood; and now a third 'mother', richer and well connected, became her benefactress.

The school in Newington Green put Mary in a position to provide for her sisters and house Fanny – an asset to the school with her graceful manners and expertise in botany. This was the independence Mary had hoped for: Everina rescued from dependence on Ned; Bess freed from a husband who threatened her sanity; Fanny released from the exhausting demands of her family. In planning for others Mary exercised the emotional and practical responsibilities of an eldest sister, stretching those skills beyond the circumscribed role of the daughter at home. For the following eight years, she accustomed her sisters and the Bloods to her exertions on their behalf. 'I love most people best when they are in adversity,' she remarked to George Blood, ' – for pity is one of my prevailing passions . . .' Benevolence was the top virtue in eighteenth-century England; in Mary it shed the tone of patron, and took on the warmth of affection.

As it happened, Jane Arden became a teacher so much in the same style that both may well have looked back to the encouraging schoolroom of John Arden, with its blend of benevolence and enquiry. In 1784, the same year that Mary's school opened in Newington Green, Jane opened her own boarding-school in her home town of Beverley, and went on to publish grammars as well as a travel book filled with botanical observation and reflections on the lives of purposeful women. Jane said: 'When I think that happiness . . . depends in a great degree on education, I most deeply feel the importance of the duties which I have to fulfil.'

Mary too knew teaching as a passion, and even better, as a relationship. 'With children she was the mirror of patience,' Godwin testifies. 'Perhaps, in all her extensive experience upon the subject of education, she never betrayed one symptom of irascibility . . . In all her intercourse with children, it was kindness and sympathy alone that prompted her conduct . . . I have heard her say, that she never was concerned in the education of one child, who was not personally attached to her, and earnestly concerned not to incur her displeasure.'

It was different with her sisters. Sometimes they exasperated her, when Bess moped with her nose in the air or Everina seemed too light and casual to make an effort. Though Mary could make equals and superiors feel small, she never took advantage of those in subordinate positions as pupils or servants. She passed on to pupils the quality that prompted her from the age of fourteen: to have the courage to say what you know. 'Indeed,' she said, 'it is of the utmost consequence to make a child artless, or to speak with more propriety, not to teach them to be otherwise.' If she was ignorant in certain areas, she knew what *not* to teach. Her pupils were *not* taught to feign raptures they had not felt. They were *not* taught 'pompous diction'. They were *not* taught 'artificial' manners or 'exterior' accomplishments. They were *not* to read in order to quote, nor were they to choose books on the basis of celebrity. Wollstonecraft's radical programme was designed to free a child's tongue; children were invited to tell stories in their own words. Her initiatives began with education, keen to retrieve human endowments the schoolroom shuts off.

As she tried out these ideas, in her mid-twenties, Mary presided over a group of women who were supporting themselves entirely on their own. Lacking dowries, they were marginal to the dominant society, but as long as their school flourished they found a place in a larger marginal community of Nonconformists. Newington Green was no ordinary village. It was high-minded, politicised and literate: full of subscribers to published sermons, and supporters of America in its War of Independence. No letters survive from Mary's first year at Newington Green, yet since this was the period when she became politicised, we must enter the experimental hothouse of ideas in which she lived.

One person in Newington Green whom she would later recall with particular gratitude (together with Mrs Burgh) was the Revd Dr Price. Though she continued to attend Anglican services, she did also hear his political sermons. He was a thinker of many parts: a mathematician and economist, as well as political philosopher. He preached liberty as part of a programme of moral perfection, a religious utopianism stressing the divine image implanted in our nature. His humanitarian ideas were far-sighted: he dreamt of abolishing war and planned an international tribunal for settling disputes, but at the time Mary Wollstonecraft came under his influence his keenest thoughts were concentrated on the future of America. He even went so far as to declare that 'next to the introduction of Christianity among mankind, the American Revolution may prove the most important step in the progressive course of human improvement'. The political core of what Wollstonecraft put forward after her contact with Dr Price reflects his thinking in relation to the new-formed United States.

In August 1775, George III had declared the American colonists to be in a state of rebellion, and sent troops. When nine hundred British soldiers had fired on seventy Americans at Lexington, Thomas Rogers, a banker in Newington Green, put on mourning. Later that year Dr Price wrote a pamphlet in favour of the rebels, *Observations on the Nature of Civil Liberty*. It sold sixty thousand copies when it was published in February 1776 (reinforcing the impact of Thomas Paine's *Common Sense* which argued the case for a republic), and is said to have encouraged the American Declaration of Independence on July 4th. Britain, Price argued, could not win this war. His main point, though, was that Britain was in the wrong because political authority derives from the people, and is limited by natural rights and the common good. He held that there were no grounds for justifying imperialism.

Anonymous letters threatened Price with death. He couldn't have cared less for threats when it came to the cause of truth and liberty, but there was no egotism in his politics. His eyes had the keenness of intelligence, not the expressiveness of personality. He lived simply, and gave a fifth of his income to charity. A modest man, thin, in a plain black coat, with a shy bend to his back, he was never unkind or uncivil. So revered was Price by artisans and market women that when he trotted through London on his old horse he could hear orange-women calling, 'There goes Dr Price! Make way for Dr Price!' His objections to war and the corruption of the ruling class were shared by the artisan class in London (including the poet William Blake), by manufacturers and traders in the Midlands and North West, and also by the poor who provided the soldiers for the American war. The country lost the labour of at least a hundred and fifty thousand men for eight years. All these productive but disfranchised classes were struck by the American experiment in democracy, and crowds came to hear Dr Price.

In 1781 the British surrendered to Washington at Yorktown. The mother country had lost one hundred thousand soldiers and £139 million in its attempt to hold on to the thirteen American colonies. Britain was forced to recognise the United States by the Treaty of Versailles on 3 September 1783, and Congress ratified a final peace treaty in January 1784. That year, at the very time Mary Wollstonecraft and company arrived in Newington Green, Dr Price was penning another influential pamphlet, *Observations on the Importance of the American Revolution.* It was designed initially for American leaders – George Washington, Benjamin Franklin (an old friend of Price), Thomas Jefferson, John Adams and John Jay – who all entered into correspondence with Price over his astute recommendations for the States' future in peacetime. Franklin, the American Minister in Paris, gave the Comte de Mirabeau, a future leader of the French Revolution, an introduction to Price on 7 September 1784. Mirabeau had written an attack on the hereditary nobility for the radical London publisher Joseph Johnson. To fill out the volume, Mirabeau included Dr Price's *Observations*, together with his own 'Notes Détachées sur l'ouvrage de M. le Docteur Price'. Joseph Johnson then brought out an English translation of Mirabeau's commentary on Price.

The early and mid-1780s were a time of extraordinary optimism: the American victory, the newly freed nation, was greeted by supporters as a victory for all mankind. It resonated for classes blocked by existing institutions, who wished to shake off a hereditary ruling class and extend the rights of those not represented in Parliament. In April 1784 Figaro's speech, vaunting the ingenuity of a servant, shocked and thrilled audiences when Beaumarchais's play *Le Mariage de Figaro* opened in Paris. An English Dissenter and playwright, Thomas Holcroft, attended ten successive performances in order to memorise the play for the London stage. Political and theatrical radicalism was linked with a growing faith in the perfectibility of human nature proclaimed by Dr Price from the pulpit. Since political institutions shape our nature, the time had come, he said, 'when the Dissenters in England have more reason to look to America, than America had to look to them'. Americans were applying ideas in the unprecedented setting of New World republicanism, where ideas had the chance to be different things altogether. This phase when Mary Wollstonecraft was putting into practice her radical ideas in education was also the phase in which she bent an ear to this remarkable pastor.

It's perhaps not surprising that her germinating soil was at the margin of society. Mary had no contact with the metropolitan milieu of the 'Bluestocking Club', the fashionably learned women whose soirées were attended by Mrs Garrick, wife of the foremost actor of the age, Edmund Burke the parliamentary orator, Sir Joshua Reynolds the great portraitist, the conservative Fanny Burney soon to be waiting on the Queen, the equally conservative Hannah More (who would be one of Wollstonecraft's most determined critics) and Horace Walpole (another of her future critics, the politely sneering author of *The Castle of Otranto*, a novel creaking with lifeless gender types), all brought together by the voluble, faintly absurd Mrs Vesey, wife of a Member of Parliament. It was not these celebrities of the capital but the outlying milieu of Dr Price, his friend Mrs Burgh and other Dissenters who met every fortnight in one another's homes, who sowed in Mary Wollstonecraft the seed of 'rights' to life and liberty.

How was Mary transformed from a young woman in hiding into the political thinker she was by the time she left Newington Green in the autumn of 1786? The answer does not lie in her surviving letters, nor will it do to fill this gap with the insistent self-pity of Mary's lamentations over one difficulty or another. From February 1784 to September 1786 there remain only six letters to George Blood, and two to her sisters, a small fraction of what she would have written over the course of two and a half years. The surviving letters never speak of the educational and ideological interests she was developing. When Mary writes to George, she's advising a young scamp who, when he wasn't abandoning jobs for long spells of idleness, assisted in a haberdasher's in Cheapside. He was responsive to Mary, his eyes danced, and she was fond of him as a 'sister', but this was no thinker – and, not surprisingly, her bond with George does not reflect that side. He helped her in so far as he allowed her to confide her setbacks and troubles, and she must have confided also in Mrs Burgh's nephew Friendly Church, for he told her that she would 'never thrive in the world'. But Church was wrong. When Mary said, 'my harassed mind will wear out my body', she did not expect imminent death. Cries and sighs were the commonplaces of eighteenth-century sensibility, introduced in the 1740s by Richardson's hugely successful novels *Pamela* and *Clarissa*, whose heroine pits her integrity against those who control the world (exploiters, bullies, rakes). Clarissa is forever having her laces cut when she falls in a faint. She exhibits the virtue of weeping. It rebukes the heartless and vindicates her honesty. Mary's sighs signal in the same way the honest, unprotected woman at odds with the world, a position she shared with her sisters. Her letters to them too leave out the stimulus of the Dissenters' bid for rights in the face of political exclusion, and, more immediately, the impact of the American model as it came to her through Dr Price.

Richard Price was sixty-one when Mary met him. A portrait of this time by Benjamin West shows a thinker beside a bookcase. His lined forehead and hollow cheeks are framed by a full-bottomed wig. His expression conveys the calm and sweetness of spirit of those whose strength comes from within. Repeatedly, Mary would be drawn to the gentleness of confident men (as unlike her father as it was possible to be). This kind of gentleness is far from weak: in Price it shows an edge in his rather formidable dark eyebrows, the right raised interrogatively, which together with the keen eyes behind the spectacles convey a quiet authority.

He came from Llangeinor in Glamorganshire, the son of a Calvinist minister called Rice Price who was harsh and bigoted to the extent that his son rebelled. The son took up the rational faith of Unitarians who stressed the ethics of compassion, denying miracles and 'superstitions' in favour of Christ's humanity. Richard Price never became fully Unitarian in so far as he retained a sense of Christ's divinity, but a sectarian label is unimportant beside his embrace of Christ's non-violence. He rejected religions with histories of coercive violence, Islam and Catholicism. Protestantism he thought little better, while pagans with their lewd and cruel deities sanctified the worst human traits. It was better not to believe in a deity, he thought, than to project a punitive being.

Price moved to London in about 1740 to study for the ministry at Moorfields Academy, run by his uncle Samuel Price, with help from Isaac Watts the hymn-writer. It was a time when Dissenting academies were the real centres of education in England, while the established seats of learning, Oxford and Cambridge, dozed through the eighteenth century. In 1758 Price settled at Newington Green as preacher. After twelve years there he began to preach in the mornings to the much larger congregation of

Unitarians at the Gravel Pit Meeting House in Hackney. A wider public read his pamphlet against the American war. Congress invited Price to become a citizen in 1778, and Yale University conferred on him, together with Washington, the degree of Doctor of Law on 24 April 1781. Price declined the invitation to leave England, but declared that he looked 'to the United States as now the hope, and likely soon to become the refuge of mankind'.

In the early 1780s reform was in the air; fellow-feeling for revolutionaries was marked at all levels of society, except perhaps the lowest. Aristocrats in the circle round the leader of the Opposition, Charles James Fox, and William Petty Shelburne (created Marquess of Lansdowne in 1784) joined associations like the London Friends of the People, as did the gentry and manufacturers all over the country. Even politicians like Pitt and Burke, who were to lead a counter-revolution in the 1790s, as yet appeared potential reformers. Lord Shelburne, who was Price's patron, became Prime Minister from March 1782 until February 1783, and offered Price the post of Private Secretary. Price served as unofficial adviser on government finance, proposing a 'sinking fund' to cope with the national debt, while Shelburne brought the American war to an end.

One of the foremost American admirers of Dr Price was the future second President, John Adams. The principles and sentiments that Price expressed had been, said Adams, 'the whole scope of my life'. In 1785, aged fifty, Adams arrived in London as the first American Ambassador to the Court of St James. The Court and diplomatic community were taken aback when Adams and his wife Abigail chose to join the congregation of Dr Price at the Gravel Pit Meeting House. 'This is the 3d Sunday we have attended his meeting,' Abigail Adams wrote to her son John Quincy Adams on 26 June 1785, 'and I would willingly go much further to hear a Man so liberal so sensible so good as he is. He has a Charity which embrases all mankind.'

Price provided a refuge from the hostility the Adamses encountered that summer for having the 'impudence' to defend their country. Stared at or ignored, they had to withstand a repeated assumption that America would 'of course' return to the fold when it was weary of independence.

British statesmen refused to comprehend Adams when he declared there was 'a new order of things arisen in the world'. Journalists and American refugees took to 'snearing at [John Adams] for having taken Dr. Price as Father confessor', Abigail reports in August; and by September decides that she prefers the society of Dr Price to balls, card parties and 'titled Gamesters'. She wondered repeatedly at his diffidence, so different in manner from an American's air of independence. 'If I live to return to America, how much shall I regret the loss of good Dr. Price's Sermons,' Abigail said during her return voyage three years later. 'I revered the Character and Loved the Man. Tho far from being an orator, his words came from the Heart and reached the Heart.'

Revolutionary rhetoric, in the mouth of John Adams, began with the premise, 'all Men would be Tyrants if they could'. Abigail Adams suggested an application her husband had not considered. In March 1776, she had asked him to 'remember the Ladies' in the new code of laws yet to be formulated.

'If particular care and attention is not paid to the Laidies,' Abigail warned, 'we are determined to foment a Rebellion, and will not hold ourselves bound by any Laws in which we have no voice, or Representation.'

'As to your extraordinary Code of Laws, I cannot but laugh,' her husband replied. The danger of revolution, he explained, is its knock-on effect for disobedient children, Indians and Negroes. 'But your Letter was the first Intimation that another Tribe more numerous and powerfull than all the rest were grown discontented.'

Abigail Adams was to become an outspoken admirer of Mary Wollstonecraft. They were moving in the same religious milieu in the London of the mid-1780s, and possible evidence of their proximity is the fact that the Adamses spelt her name 'Woolstoncraft', as it's pronounced in England. It's conceivable that they sat in the same congregation when they heard Dr Price. His belief that American liberty derived from the Glorious Revolution of 1688 was the source of Mary Wollstonecraft's hopes of the American experiment. 'The Anglo-Americans having carried with them the principles of their ancestors, liberty appeared in the New World with renovated charms and sober matron graces,' she wrote when she came to contrast the American Revolution with the Terror unleashed by the French. At the outset of his Presidency, Adams underlined these words. 'I thank you Miss W.,' he remarks in the margin. 'May we long enjoy your esteem.'

Although Price looked on the American Revolution as 'the fairest experiment ever tried in human affairs', the real test would be to sustain it. In 1784–5 he warned its leaders that the collected wisdom of confederation, not war, should settle disputes. On 1 February 1785 he received a gracious acknowledgement from Jefferson, who in that year replaced Franklin as American Minister in Paris. 'I have read [your *Observations on the American Revolution*] with very great pleasure, as have done many others to whom I have communicated it. The spirit which it breathes is as affectionate as the observations themselves are wise and just.'

Price replied to Jefferson, slipping in another warning between the graces of this exchange:

> Newington Green M<u>ch</u> 21<u>st</u>, 1785
>
> Dear Sir,
>
> . . . Your favourable reception of the pamphlet which I desired Dr Franklin to present to you cannot but make me happy; and I am willing to infer from it that this effusion of my zeal will not be ill received in America. The eyes of the friends of liberty and humanity are now fixed on that country. The united states have an open field before them, and advantages for establishing a plan favourable to the improvement of the world which no people ever had in equal degree.
>
> . . . The character, however, of popular governments depending on the character of the people; if the people deviate from simplicity of manners into luxury, the love of shew, and extravagance[,] the governments must become corrupt and tyrannical . . .
>
> I am, Sir, your most obedient and humble servant,
>
> Rich*d* Price

It's not entirely possible to explain the success of America (in contrast with other revolutions, where thugs take over from intellectuals). Certainly, there was Washington's ability to unite states with divided interests. During those first vital decades, from the 1770s to almost the end of the century, he had the standing to control dissensions. When the war came to an end in 1783 his 'Circular to the States' was designed to channel the energies of rebellion from potential anarchy towards subordination to government. Dr Price had his ear and those of other American leaders at the start of their debate on the Constitution. Where the Declaration of Independence had been idealistic, even utopian, the Constitution was realistic about the power-grabbing element in human nature. It was brilliantly balanced, with curbs on power on every side. Yet it had two flaws.

The lesser was the failure to grant equality to women: I say lesser, because until Wollstonecraft publicised what would be for more than a century to come the contentious issue of women's rights, no power remotely considered such a thing – woman being what the Bible termed 'the weaker vessel'. The second and unforgivable flaw in the American Constitution was, of course, the perpetuation of the slave trade, what Wollstonecraft called 'the abominable traffick'. In her later writings on the rights of woman, she repeated the common link between the positions of women and slaves. For this reason the long-forgotten words of Dr Price against slavery in 1784–5 are of the utmost importance to the way Wollstonecraft began to think.

'The negro trade cannot be censured in language too severe,' Price warned American leaders in 1784. 'It is a traffic which . . . is shocking to humanity, cruel, wicked, and diabolical.' Until the States should abolish it, they would not deserve their liberty. In this one respect he believes he can recommend to the States the example of his own country from the time of Judge Mansfield's ruling in 1771. 'In Britain,' said Dr Price, 'a negro becomes a freeman the moment he sets his foot on British ground.'*

---

*In truth, Judge Mansfield had been a protector of property and no friend to slaves. He had intended his ruling in favour of freedom in the case of one man (who actively opposed his enslavement on British soil) to have no repercussions for slavery in general, but many slaves had taken the following judgement as a cue to leave their owners: 'no Master ever was allowed here to take a Slave by force to be sold abroad, because he had deserted from his Service, or for any other Reason whatever; we cannot say the Cause set forth by this Return [of a man to slavery] is allowed or approved of by the Laws of this Kingdom . . .'.

Jefferson sent a dodging reply on 7 August 1785 to this anti-slavery plea. He appears to urge Price to speak out further against Southern opposition to emancipating slaves, but in reality distances the issue when he suggests that Price address his comments to young men who might influence some 'future' debate on the question. What he doesn't say is that he will take it up himself. Nor does he question himself as slave-owner.

That same summer Price confronted John Jay, New York president of a newly formed Society for Promoting the Manumission of Slaves (whose members had declined a proposal that they begin by freeing their own slaves):

> Newington Green, nr. London
> July 9, 1785.
>
> Dear Sir,
>
> . . . It will appear that the [American] people who have struggled so bravely against being enslaved themselves are ready enough to enslave others: the event which has raised my hopes of seeing a better state of human affairs will prove only an introduction to a new scene of aristocratical tyranny and human debasement: and the friends of liberty and virtue in Europe will be sadly disappointed and mortified . . .
>
> I am
> Your most obedient and humble servant,
> Richard Price.

Nothing was more important than a plan of education in framing a new state, Price advised Americans in what was, by far, the longest part of his pamphlet. His educational ideas followed Wollstonecraft's, or hers his. To Price in 1784, the 'secret' of education had yet to be found, but he points to the 'turn' given to the mind by the child's earliest impressions, which Wollstonecraft in a sense answers when she points to a mother's breast. Price, too, attacked formal learning that teaches *what* to think instead of *how* to think, and 'perverts' minds with 'the jargon of the schools'. This 'puffs up' a child. Pride and dogmatism he called 'the worst enemies to improvement'. Wollstonecraft, sweeping away words of thund'ring sound, the commonplaces of received opinion and other obstacles to communication, takes this to a logical conclusion when she shifts the power from teacher to pupil. It's as though Price was invoking Wollstonecraft when he said, 'I am waiting for the great teacher.'

As long as she taught in Newington Green, Wollstonecraft found a mentor in a mind of this calibre: its eighteenth-century trust in continuing enlightenment, its compassion for victims and its commitment to liberty – all qualities she could share.

Not far from the Green was the village of Shacklewell where lived a young Anglican priest, aged twenty-two, who served three parishes in London. The Revd Mr John Hewlett was the son of a gentleman, owner of Chetnole in Dorset and Milborne Wick in Somerset. He befriended Mary, who was three years older, and to some extent confided in her – or she was intuitive enough to guess his troubles – for she pitied his marriage at twenty to bossy Elizabeth Hobson of Hackney: 'Poor tender friendly soul how he is yoked!' It was a waste of a good man. At the height of his association with Mary, Hewlett was admitted as a sizar at Magdalene College, Cambridge, in January 1786 at the age of twenty-four – it would take all of ten years to take his degree in divinity while he ran his own boarding-school in Shacklewell. In the next century he became known as a biblical scholar and Chaplain to the Prince Regent, but when Mary met him in 1784 he mixed with editors and writers of the capital. Hewlett took Mary to visit the Great Cham of English literature Samuel Johnson, then seventy-five, scarred by scrofula and ailing from a stroke the previous year. He received her kindly in Bolt Court, seeming to propel himself forward by a constant roll of his head and body, and jerking his majestic head as he made his pronouncements as one speaking from on high. His voice was loud and deliberate; his slovenly clothes and convulsive motions of hands, lips, feet and knees at odds with graceful words.

Dr Johnson loathed Mary's hero. As a pensioner of the government he opposed radicalism, in particular the politics of Dr Price. Once, in Oxford, when Price entered a room, Dr Johnson left it. 'All change is of itself an evil,' he said. Johnson approved the war against the States, decried equality, and looked with equanimity on a beggar, pronouncing that some must be miserable so others might be happy.

It says something for both Johnson and Mary Wollstonecraft that they did not clash. Dr Johnson is now recognised as more ambiguous than some of his pronouncements appear: a conservative political writer who in certain ways opened up a brave new world for more radical voices – brave, say, in his refusal to be gulled by the hot air of aristocratic amateurs. Then, too, Johnson could put by his occasional misogyny – he had long enjoyed cordial relations with several 'Blues'. The biographer of Fanny Burney, Claire Harmon, detects a romantic tenderness in his later encounters with women that suggests a suppression of his sexual feelings with a resigned awareness how repellent his scars and twitches might be. He treated Mary 'with particular kindness and attention', had a long conversation with her, and asked her to come 'often'. Despite overt differences, they actually had much in common.

Both remained staunch rationalists at a time when taste embraced the cult of sensibility. Though the prestige of the pre-Romantic Grey was at its height, Dr Johnson looked back to the rational incisiveness of Pope who 'thought himself the greatest genius that ever was' and had the felicity 'to rate himself at his real value'. Wollstonecraft, like Johnson, trusted the common reader; both shunned obscurities; both were devout Anglicans; and both had deep veins of melancholy. There was no social prohibition against melancholy; no need to hold back. Melancholy in the eighteenth century was admired, particularly in men, as a sign of sensitivity. Dr Johnson called melancholy 'a rust of the soul', and his cure was work. For Wollstonecraft and Dr Johnson (as for many writers), melancholy coexisted with ambitious effort: Johnson's with his *Dictionary* and *Lives of the Poets*; Mary's with her case for women's education. Both were compulsive educators who wished to generalise about human nature; their minds turned on nothing less than our species. Johnson thought of his essays as 'lessons'. He sanctioned novels as 'lectures of conduct and introductions into life', which perfectly describes the novels Mary was to write – not ordinary novels because, like Dr Johnson's *Rasselas*, fiction is a shell for philosophical polemic.

Unfortunately, Dr Johnson died in December 1784 before Mary could take up his invitation to visit again. For a long time she revered his memory, read his posthumous *Prayers and Meditations*, and, in her anthology for women readers in 1789, included one of his last poems expressing his preference for 'constant nature' over contrived charms.

To impress men like Johnson and Price was no mean feat for an obscure schoolmistress in her twenties with no languages or formal accomplishments who as yet had published nothing. It's plain that Mary Wollstonecraft's originality invited attention from those discerning enough to see it. Meanwhile, fussing lodgers and the gossips of the Green – the busy Smallweeds – were doing what they could to choke her enterprise. Yet, throughout her hard second year at Newington Green, she did retain the support of Mrs Burgh, Friendly Church, Miss Mason, James Sowerby, the Revd Mr Hewlett and Dr Price. And all through the troubles that lay ahead, the shape of an alternative was coming into being in the shadow of the American Revolution: a woman's claim to life and liberty.

At the time John and Abigail Adams joined the congregation of Dr Price in London, Mary Wollstonecraft too heard that gentle voice urging reform on the model of the Americans, who had 'established forms of government favourable in the highest degree to the rights of mankind'. The rights of womankind were no more than a logical step from what Dr Price preached.

4

A Community of Women

At first the school flourished. The staff held together as a community of women, Mary's dream during her depressed time as companion to Mrs Dawson and the grim years nursing her mother. In August 1784 came news that Bess's baby had died a few days short of her first birthday. We don't know how this affected her and her sisters, but Fanny's presence would have been a comfort. Her love for Mary extended in a playful way to 'the girls', as her sisters were called. And the energy Mary put into trying out radical views of education had, for parents, a reassuring complement in Fanny's tact. Her humour would have teased 'the girls' into compliance, as it would have caught the drip of complaints from the boarding mothers. But towards the end of that year came an ominous warning: Fanny spat blood.

Doctors advised her to leave the damp of England for a southern climate. Since she had no money, the only way to provide for a voyage and subsistence was to join her uncertain suitor Hugh Skeys who had remained a merchant in Lisbon for some years and was now willing to marry her.

When Mary discussed this possibility with Fanny, she was not thinking of her own wishes or the interests of the school; she was weighing the best course for her friend. In favour of marriage was Lisbon's sun, provision for Fanny, rest from work, and the hope that an end to her miseries over this man would prolong her life. Against this was a voyage through the stormy

Bay of Biscay; and the more worrying fact of Fanny's unfitness to bear children. Mary told Godwin, later, that she advised Fanny to choose Lisbon because she was certain that Fanny would die if she stayed in London, and any chance for improvement – if not recovery – should be tried.

So, Fanny sailed for Lisbon early in 1785, and married there on 24 February. A month later she was pregnant, which may account for 'an extreme depression of spirits' at the start of a letter that goes on with brave humour to the 'dear lasses' at Newington Green:

Lisbon
March 30th 1785

> . . . A letter, however stupid and uninteresting in itself, needs no apology when conveyed to you by an agreeable young man, such as I hope the bearer of this will prove to be. I assure you I find him a tolerable *flirt*, tho' I have been but twice in his company; and, if such an animal as *I* am could engage a little of his attention, to what a degree of vivacity must he be animated by *the assemblage of irresistable charms* he will meet at N[ewington] Green. – You are to know that his name is Brockbank – that he has spent a considerable time in Spain – and has a brother, a watch-maker in London, where I suppose he is going to settle. – – By next June, I hope to send you another flirt – Mr Jeffray, a phisician – but fear you will think him too grave, and too ugly – yet, if you are not carried away by prejudice in the first interview, he will afterwards, probably, steal into your favour, as he has done into mine. I have given a description of him to Mary; and she is, I hope, already prepared to love him. – He leaves Lisbon in about a week, where he has gained great reputation by some instances of uncommon judgement in his profession – yet, with all the merit in the world, I fear his diffidence and sincerity of temper will ever impede his getting forward in a world where impudence and hypocrisy seldom fail of success. – I shall greatly regret his departure – and the more so, as he spends as much time with me as he possibly can. – He is my only phisician – and by his advice I quitted the country a few days ago, and find myself already much recovered; the spitting of blood being quite stop'd, and my cough very

trifling. – I shall remain in town (at Mr Windhorst's) a month or two, as I find it agrees with me, even tho' I play the rake here, and have a crowd of visitors almost every evening – I think there is no end o' them and I shan't return the visits of half of them. – – Oh! It just occurred to me, (and 'tis well I recollected it before my paper was filled up) that Bess desired a description of Skeys – I have then only to tell her – from the experience of five weeks – that he is a good sort of creature, and has sense enough to let his cat of a wife follow her own inclinations in *almost* every thing – and is even delighted when he sees her in spirits enough to coquet with the men, who, to do them justice, are not backward in that way. – Skeys's picture was more like him than pictures in general are – but he is much fatter, and looks at least ten years older than it. – He has been a dreadful flirt among the damsels here, some of whom I could easily perceive were disappointed by his marriage – but I have – completely – metamorphosed him into a *plain* man – and I am sorry to add, that he is too much inclined to pay more attention to his wife than any other woman – but 'tis a fault that a little time, no doubt, will cure. – Well, girls, are you almost tired? – – I knew you would – and now that you too are convinced of my inability to entertain you, will not I suppose desire – – Yes, yes; I know you love me, and will be sincerely glad to receive an epistle, now and then, from your affectionate

Frances – (Heigh ho!) *Skeys*

Fanny sounds content with marriage, and it's hard to decide whether Mary was sceptical, jealous, or declaring a hidden truth when she thought Skeys unworthy of her friend. He seemed to Mary strangely passionless, a lover whose feeling bestirred itself in Fanny's presence, but so long as the woman he loved was far away, willing to bear years of separation. Mary, whose love for Fanny intensified in her absence, wondered at Skeys for his blindness to Fanny's declining health and long struggle to support her family. In Mary's analysis, he went about well wadded with pride, content to look fondly on his chosen wife without perceiving her fidelity as more than his due.

When Dr Jeffray reached England, Mary received him like an old friend.

'He is such a man as my fancy has painted and my heart longed to meet with – his humane and tender treatment of Fanny made me warm to him.' If Fanny had directed Jeffray to Newington Green as a potential husband for one of the Wollstonecrafts, nothing came of it. Sooner or later they all discovered the reluctance of even kind men like Dr Jeffray and Sowerby to link their fortunes to penniless women, however attractive – and the Wollstonecraft sisters were very attractive, Mary with her full breasts and curly auburn locks, and Everina with her young vivacity. Mary tried to stay her disappointment that the news the doctor brought from her friend proved little more than a letter of introduction. Until July she hoped Fanny's health would improve, but by August Fanny's letters made it plain that Lisbon was failing to heal her lungs.

'She is still very ill and low spirited, a poor solitary creature,' Mary informed George Blood. She was tugged between the school and a growing responsibility to nurse Fanny through an ordeal bound to be complicated by consumption. Early in September 1785, Mary resolved on the journey to Portugal.

'I have many difficulties to overcome,' she said, 'yet I am not intimidated tho' worried almost to death.' She faced down opposition from a busybody on the Green, Mrs Cockburn. This opposition was not disinterested. Mrs Cockburn and her husband took lodgers, which meant that Mary had to contend with a business rival who spied an opportunity in Mary's departure, and moistened her lips on the moral luxury of warning her in advance. Pretending concern for prospective lodgers, Mrs Cockburn let it be known that if Mary took off, she would warn three 'very advantageous' ones against rooming with the Wollstonecrafts.

'I'm not to be governed in this way.' Mary brushed Mrs Cockburn aside.

After all, her sisters were there. She would not be the first to leave a school in assistants' hands. Everina had recently 'grown indefatigable in her endeavors to improve herself' and assist in school and house. But at twenty-six, Mary was too inexperienced to realise the damage that Smallweeds can do.

Fortunately, Mrs Burgh backed Mary so keenly that she almost quarrelled with Mrs Cockburn. Mrs Burgh supplied a loan for Mary's passage (though Mary believed the money came from Dr Price). No offer came from Hugh Skeys. At no point did he enquire into her expenses, and she couldn't bring herself to mention them. Later, she commented that his behaviour had been 'uniform' – tight-fisted – throughout his wife's ordeal.

The voyage to Lisbon took thirteen days, relatively fast for a sailing ship, but autumn gales were high and rough waves broke through Mary's porthole. It was precarious to move while the vessel rolled from side to side. She fixed her attention on a dying fellow-passenger, holding him up as he coughed and gasped for breath – his consumptive paroxysms lasted for hours at a time. She didn't expect him to survive the journey. On a Monday in November they landed in Lisbon, the very day Fanny gave birth to a puny boy. Mary was just in time for the delivery. She at once found baby William a wet-nurse, for Fanny appeared too depleted to live. Writing in intervals of hope, broken by days of fear and despair, Mary relayed the crisis to her sisters:

> My dear Girls
> I am now beginning to awake out of a terrifying dream – for in that light does the transactions of these two or three last days appear – Before I say any more let me tell you that when I arrived here Fanny was in labour and that four hours after she was delivered of a boy – The child is alive and well and considering the *very very* low state Fanny was reduced to she is better than could be expected[.] I am now watching her and the child, my active spirits has not been much at rest ever since I left England. I could not write you on shipboard the sea was so rough – and we had such hard gales of wind the Capt. was afraid we should be dismasted – I cannot write to-night or collect my scattered thoughts – My mind is quite unsettled – Fanny is so worn out her recovery would be almost a resurrection – and my reason will scarce allow me to think 'tis possible – I *labour* to be resigned and by the time I am a little so some *faint* hope sets my thoughts again a float – and for a moment I look forward to days that will, alas! I fear, never come – I will try tomorrow to give you some little regular account of my journey – tho' I am almost afraid to look beyond the present

> moment – was not my arrival Providential? I can scarce be persuaded that I am here and that so many things have happened in so short a time – My head grows light thinking of it –
>
> Wednesday night. [She returned to this on] Friday morning
> Fanny has been so exceedingly ill since I wrote the above I intirely gave her up – and yet *I could not* write and tell you so, it seemed like signing her death warrant – yesterday afternoon some of the most alarming symptoms a little abated and she had a comfortable night – yet I rejoice with trembling lips – and am afraid to indulge hope, she is very low – Her stomach is so weak it will scarce bear to receive the lightest nourishment – in short if I was to tell you all her complaints you would not wonder at my fears . . .

Mary poured life into her friend and prayed for her. She ends her letter at Fanny's bedside saying that, were it not for prayer, 'I should have been mad before this – but I feel that I am supported by that Being – who alone can heal a wounded spirit . . .' It was now she came to know 'the lot of most of us to see death in all its terrors, when it attacks a friend; yet even then we must exert our friendship'.

Fanny died on 29 November. Mary closed Fanny's eyes and cut a lock of her hair to make a ring. She had lost her 'best earthly comfort'. Ten years later, she still remembered 'looks I have felt in every nerve, which I shall never more meet. The grave has closed over a dear friend, the friend of my youth; still she is present with me, and I hear her soft voice . . .'

The day after Fanny was buried – 'by stealth and in darkness' as a Protestant in a Catholic country – her brother George arrived to take up a post Skeys had arranged for him with the British consulate at St Obes, near Lisbon. At once, George abandoned his post and returned to Dublin. This was his way, careless of embarrassment to those who had backed him. Earlier that year he had abandoned the post with the haberdasher in London – on that occasion too, he had run off to Dublin where he had remained unemployed and dependent on friends of the family. The Clares, who had recommended him, were furious; as were his employers, Mr and Mrs Poole; as was his landlord, an attorney called Palmer, for murkier reasons.

Palmer had tried to pass off a client, a Mrs Jones, as a clergyman's widow whose son would have been entitled to a pension. This fraud, George had witnessed. He himself had been in a scrape with a girl called Mary Ann, and was dreading that this would go against him when another girl, Mr Palmer's servant, named him as the father of her child. (It was a common means of survival for poor women to get pregnant and name a father who, if caught, was obliged to marry the woman and support her child. This was encouraged by the parish, who otherwise had to bear the cost of a bastard.) The accusations of Palmer's girl had been loud enough to collect a mob that pursued George through the streets of London. When he succeeded in throwing them off, a group of thugs turned up at Mary's school, led by the fraudulent Mrs Jones who hoped to discredit a witness for the prosecution. In tow came the pregnant servant, wailing her injuries at a local alehouse. The scandal raced round the Green. George denied sex with Palmer's girl. Mary and his family believed him, fairly it seems in this instance. (When George put himself beyond the reach of the law, by leaving England, the girl shifted the blame to Palmer himself – by this time, in prison.) All the same, George was bad news on the Green, where it was felt Mary indulged him. Mary dismissed this as the tattle of fools. Loyal as ever to the Bloods, she trusted George's 'honest heart', and liked him the more for his willingness to let her improve him. Not once did she blame him for leaving her with the burden of supporting his parents. Even when Mary had been hastening to Fanny's side, she made sure that Mr and Mrs Blood would not be in want while she was away. But the Clares, who had linked their name to a scamp, did not forgive George. Mary resented this, and her attachment to the Clares cooled.

She remained in Lisbon for almost a month. The Portuguese seemed to her 'the most uncivilised nation in Europe' in their treatment of women and workers. She rode along the banks of the Tagus feasting her eyes on the river's magnificence; gazed at historical paintings in churches; and inspected the ruins of the earthquake of 1755 that had killed between ten and twenty thousand, and provoked a debate on God's beneficence in

Voltaire's 'Poème sur le désastre de Lisbonne' (1756). During this month she almost certainly took charge of the baby, who was alive when she left on about 20 December. After she sailed, William was handed over to Skeys's sister-in-law, and soon died. Skeys felt neglected by his brother, he complained in a letter sent after Mary – an implied contrast with her attentions. The extent of his gratitude for the expense Mary had incurred for a return passage to Portugal, for nursing his wife and child, and her consideration in staying on, was to offer her a gown. She would have been glad of a gown, but Skeys never sent it.

The return voyage, prolonged by winter storms, took a month. Mary recalled the heaving seas in her novel *Mary* written eighteen months later, where the heroine goes on deck to survey the 'contending elements' as the vessel rises on a wave and descends into 'a yawning gulph'. The squalls rattle the sails, which are taken down. Every so often, when the wind dies away, 'the wild undirected waves rushed on every side with a tremendous roar'. Where the other passengers are appalled, 'Mary' faces into the storm: as her soul mounts and sinks with the waves, 'she felt herself independent'.

En route the ship sighted a French vessel in difficulties. If the more extensive detail of *Mary* is to be trusted, the ship was dismasted and drifting, its rudder broken. The sailors were starving, and begged to be taken on board. The English captain, with his own rations running low, was inclined to sail past. Mary wouldn't have it. Assuming the manner George Blood called 'the princess', she threatened to report the captain if he did not rescue the Frenchmen – and so he did. *Mary* re-imagines the scene: 'They bore down to the wreck; they reached it, and hailed the trembling wretches: at the sound of the friendly greeting, loud cries of tumultuous joy were mixed with the roaring of the waves, and with ecstatic transport they . . . launched their boat, and committed themselves to the mercy of the sea. Stowed between two casks, and leaning on a sail, [Mary] watched the boat, and when a wave intercepted it from view – she ceased to breathe, or rather held her breath until it rose again.' When at last the boat arrives alongside, 'Mary' tends the rescued men, and joins them in thanking 'that gracious Being' who 'rode on the wings of the wind, and stilled the noise of the sea; and the madness of the people – He only could speak peace to her troubled spirit!' Then she sings – as well as she can recall – the Hallelujah Chorus from Handel's *Messiah*: 'The Lord Omnipotent reigned, and would reign for ever, and ever!'

At length, in January 1786, Mary arrived back in Newington Green. A shock awaited her: the school seemed about to collapse, and the atmosphere was 'very disagreeable'. Boarders had left or were leaving, the last being Mrs Disney. She and the Wollstonecrafts had quarrelled, and the two Disney sons were already installed as 'whole' boarders at the rival establishment of Mrs Cockburn. Without boarders, the house on the Green was too expensive, and, to make matters worse, Mrs Morphy, victim of some unexplained misfortune, had departed without paying her bill, as did the Disneys whose debt (roughly £70–£80), if paid, could have rescued the school. Mary had no hope that Mrs Disney would pay and, it appears, no means of retrieving the money – obviously, she didn't dare extend her loss with legal fees. On top of all this, Mary felt for the Bloods who must suffer with her. 'I am determined they shall share my last shilling,' she declared, but was not 'yet' able to pay their last quarter's rent. Skeys exasperated her with a 'very short unsatisfactory' letter apologising for not sending a promised sum for his father-in-law.

'It would have been particularly acceptable to them at this time,' she reported to George, 'but he is prudent and will not run any hazard to serve a friend – indeed delicacy made me conceal from him my dismal situation, but he must know I am embarrassed.' Three weeks later further letters from Skeys arrived, minus the awaited 'trifle'.

'I am certain a few pounds would not make any difference in his affairs,' Mary thought, 'yet why should I be surprised – did he not neglect Fanny – .'

To keep the school going, and probably also to help the Bloods, she borrowed heavily from Mrs Burgh; also from Friendly Church, from the recently married Sowerby, and from a number of neighbours on the Green including a musician called Mr Hinxman, a vague man who could ill afford it. Meanwhile, George, who might have supported his parents, was again living off friends in Dublin. He implored Mary to keep his most recent defection a secret from his parents, who believed him still employed in

Portugal. George was ready enough with bright suggestions. Why not open a school in Ireland? She explained that without financial backing it would be impossible to attract pupils; also, that she owed it to her creditors to find a safer means to pay them back.

George urged Mary to run off to Ireland, leaving her debts behind. 'Nothing should induce me to fly from England,' she said. 'My creditors have a right to do what they please with me, should I not be able to satisfy their demands.'

Her first concern was to exert herself to save the school before it dwindled 'to nothing'. In her present low spirits, exacerbated by fits of anxiety, this was particularly difficult: ''tis a labour to me to [do] any thing – my former employments are quite irksome to me'. She felt 'haunted' by the 'furies' of debt. 'Let me turn my eyes on which side I will, I can only anticipate misery – Are such prospects as these calculated to heal an almost broken heart – The loss of Fanny was sufficient of itself to have thrown a cloud over my brightest days.'

Even while Fanny was alive, her absence had been hard to bear. Bess and Everina could not compare as companions, and Mary's spirits had sagged when weeks had passed without a letter. 'How my social comforts have dropped away,' she echoed Dr Johnson's elegy for a friend. The worst of her grief was to acknowledge that friendship, which she still called the '*cordial* of life', had been imperfect: 'A person of tenderness must ever have particular attachments, and ever be disappointed,' she said, 'yet still they must be attached . . .' Now, depleted of dreams, Mary had to press on day by day in a failing school, dreading the hour she must face that failure and almost wishing it would hasten, so struggle and dread might end.

During the winter of 1786, there seemed no point to her continued life, and there were times she wished she had drowned at sea. 'I have lost all relish for pleasure – and life seems a burthen almost too heavy to be indured,' she confessed on 4 February. 'My head is stupid, and my heart sick and exhausted.' In this state, a month after her return, ills began to fasten on her body: a pain in her side 'and a whole train of nervous complaints'. Her eyes felt strained, her memory lapsing – signs, she thought, of decline. She hoped it would not be slow. This wish to die goes beyond grief: she suffered from depression, manifesting often as irritability, hypochondria and self-pity. These are typical after-effects of a danger that has called up vigilance, protectiveness, fear and anger – emotions Mary had felt as a child when she had tried, and failed, to protect her mother or dog from her father's fist. That 'agony' of helplessness, encoded in childhood, remained a recurrent threat to this active young woman. As witness to Fanny giving birth, she had again failed to save a victim she loved.

One of the Revd Mr Hewlett's sermons of this time seems to address her depression. The house of mourning, he urged, 'is replete with instruction'. After 'the calls of friendship led you to take a last farewell of those you loved' and 'you viewed the last struggles of nature, saw the shades of death gathering around' and felt your 'own weakness', do not forget, Hewlett counselled. Consider, meditate, *study* scenes of sorrow. The source of knowledge is experience. Not the teachings of others, not example, but the few events that befall us make a more lasting impression on the heart: 'Keep thy heart with all diligence, for out of it are the issues of life.' At a time when women were taught it was inconsistent with their nature to draw comprehensive conclusions from private experience, Hewlett encouraged her and others to shape, not blunt, traumatic memory. His idea of the uses of grief came from the Bible (Deuteronomy 4: 9): 'Only take heed of thyself, and keep thy soul diligently, lest thou forget the things which thine eyes have seen, and lest they depart from thy heart all the days of thy life.'

Another consolation was the music of Handel. His 'sublime harmony' raised Mary 'from the very depths of sorrow', she said. 'I have been lifted above this little scene of grief and care, and mused on Him, from whom all bounty flows.'

Although she did manage to retain eleven pupils, by April it was clear that however hard she tried, she could not make up her debt by trying to rebuild a large establishment. Pupils flocked to prosperous schools, which appeared to have no need for them. There was nothing for it. She must close her boarding concern, and continue in a smaller way with day pupils. But what would become of her dependants: moody, handsome Bess, unfit to fend for herself, yet tied for life to a cast-off husband; light-spirited

Everina, only twenty-one and susceptible to a 'Lothario'; and the pathetic Bloods who were once more penniless?

From the beginning of May, Mary was preparing herself to 'plunge again into some new scene of life'. In mid-June, still in Newington Green, she moved into cheap lodgings at a Mrs Blackburn's, and went on teaching the residue of eleven pupils. She sold the school's furniture and everything else she could spare, paid off the servant, and lived with 'rigid' economy – too stringent, she said, for her sisters, perhaps her way of letting them know that she could no longer support them. Everina went back to Ned on Tower Hill. Bess lingered on with Mary until mid-July when Mrs Burgh found her a lowly teaching post at Mrs Sampel's school in Market Harborough, Leicestershire. The town lay in a valley surrounded by hills white with sheep. In the outlying hamlets, peeping from behind bushes, cottages were largely made of mud; some of stone. Bess was surprised to find these clean and comfortable, with woodbine and roses twined over their windows. Rural walks were an escape from a strict Presbyterian school. Her fellow-teachers were obsessed with hell, and damned plays and all books except for the Bible – a far cry from the enlightened Dissent of Newington Green.

This was hard for Bess, who was less spiritual than Mary and with a keener taste for the '*delightful little elegances of life*'. She pictures the provincial narrowness in a letter to Everina: the lugubrious quoting from the Bible from morn to night, the four services on Sundays, her homesick awakenings, and lonely evenings in her attic room with Everina's portrait for company. 'Oh! How my heart pants to be free –' was her secret cry, while she schooled her face. She felt 'shut out from all society, or conversation whatever, I cannot make myself understood here; had I an inclination so to do, praying is their only amusement, not forgetting eating, and *Marr[y]ing*, and so on – the idea of parting from *a husband* one could never make them *comprehend* . . .' At the same time, she spells out the economic case for marriage: 'Oh! that you had a good Husband, to screen thee from those heart-breaking disagreeables . . .' Even now, Bess did not regret leaving Meredith Bishop. She's astonished when fellow-teachers can't comprehend that a bad husband is worse than none.

To Mary's relief, Bess held up. After more than two months she sent a reassuring and 'very affectionate' letter which, Mary replied, 'was a cordial to me, when my worn-out spirits required a very potent one – Indeed my dear girl I felt a glow of tenderness which I cannot describe – I could have clasped you to my breast as I did in days of yore, when I was your nurse – . . . I was pleased to find you endeavor to make the best of your situation, and try to improve yourself – You have not many comforts it is true – yet you *might* have been in a much more disagreeable predicament at present.'

The most intractable of Mary's problems had long been the Bloods and their feckless brood. Mr Blood was a mite sobered by the death of Fanny who, sick as she was, had taken his place as breadwinner. He did now exert himself to find work, and had the promise of a caretaker's job in the Church of Ireland. Yet, without money, how could he get there? Mr Blood was not backward in putting this plea to Mary, though she was stretched to breaking point, as he must have known. As always, she took a daughter's responsibility, and consulted the minister at Shacklewell how she might contrive to earn a little extra.

Mr Hewlett suggested that she write a book, and Mary came up with her *Thoughts on the Education of Daughters*. He took her proposal to Joseph Johnson, who was handling his own volume of *Sermons* and whose print shop in St Paul's Churchyard was near Hewlett's town parish of Foster Lane. Hewlett's face shone with 'sensibility and goodness' when he returned to Mary with the publisher's offer of ten guineas. Unhesitatingly, she turned over the advance to the Bloods to pay their fare to Dublin. No sooner had they departed than a letter came from Skeys lamenting his 'inability to assist them', and dwelling on his 'own embarrassments'.

Now, when the day's classes were done, night after night through the late spring and summer of 1786 Mary's pen travelled over a sheaf of paper, and as it did so, her sense of purpose returned. 'Whenever a child asks a question,' she writes, 'it should always have a reasonable answer given it. Its little passions should be engaged.' Her bias was rational: children's heads should not be filled with 'superstitious accounts of invisible beings, which breed strange prejudices and fears in their minds'. The force of her writing, its unposturing directness, is plain. As she moved from her portrait of a girl reader discovering a taste of her own, to her portrait of an unmarried teacher braving the rigours of self-reliance, she drew on her experience of the last two years. Her *Education* looks deceptively slight; it was the fruit of long thought tested in a school of her own.

Rising out of this book is the portrait of a professional woman. Though Wollstonecraft draws on her experience, it's not a self-portrait, more a possibility. Here is her first attempt at the alternative life plot that could bring into being an exemplar of her sex. The first thing she learns at her mother's breast is the 'warmest glow of tenderness'; and the next most important lessons go back to 'The Nursery', the title of the first chapter. Unlike Mary, the model girl is not subject to excessive restraint in the nursery; affection calls out her 'amiable propensities'. This loved child becomes a reader from her earliest years, searching out books that improve her whole being. She cultivates the intelligence to judge for herself, ignoring the craven chorus of those around her and dissociating her sensibility from fictional heroines 'so different from nature'. Later, *Northanger Abbey* would mock the gush and tremors of girls who imitate fictional heroines. Wollstonecraft precedes Jane Austen when she dissociates genuine sensibility from affectation: 'those who imitate it must make themselves very ridiculous'. Discipline is not imposed from above; it grows spontaneously from the secure ritual of the nursery, encouraging the responsiveness of the small child. Wollstonecraft in no way slights the domestic character of women – traditional achievements in nurture and emotional literacy are the heart of her model – but the outcome of training is realistic self-reliance, displacing the old model of passive dependence.

One of the hardest trials, Wollstonecraft warns in a chapter on 'Love', is the single woman's irrational susceptibility to unsuitable men. At twenty-seven, her advice is tough, born of her own disappointments with Joshua Waterhouse and Neptune Blood: 'The passion must be rooted out, or continual excuses that are made will hurt the mind.' Wollstonecraft's confiding manner is free of the loftiness of men's advice books, which took it upon themselves to school females in proper femininity, though ostensibly *Thoughts on the Education of Daughters* belongs to that genre. An editorial note to excerpts printed in the *Lady's Magazine* commends their 'many judicious observations'.

Daughters' education had not always been quite as constricted as in the era of advice books. Katherine Parr, the last of Henry VIII's wives, was the daughter of a learned mother, published two books, and dared to debate with the King on risky questions of theology – nearly costing herself a horrible death when Henry began to fume: 'A good hearing it is, when women become such clerks, and much to my comfort to come in mine old age to be taught by my wife.') The intellectual training of Elizabeth I made study fashionable for aristocratic women, especially in the circle around Mary Sidney, Countess of Pembroke (sister of the poet Sir Philip Sidney).

In the seventeenth century during the English Civil War 'bold impudent huswives' emerged, and 'preacheresses' who would 'prate' an hour or more, some sanctioned by Dissenting sects like the Quakers and Levellers. Charles I employed a learned woman, Basua Makin, to teach his children, but at the licentious court of Charles II women were treated as toys and their standing declined. In the 1660s Basua Makin said that a female scholar was looked on like a strange comet that bodes mischief. In *The Female Vertuosos* (1693) Sir Maurice Meanwell complains that 'now a-days Wives must Write forsooth, and pretend to Wit'. His sister Catchat protests that women's wit is innate, not pretended. ''Tis the partial, and foolish Opinion of Men, brother, and not our Fault hath made it ridiculous nowadays.' Catchat is in fact ridiculed as a single woman past her first youth who is out to catch a man. Sir Maurice: 'A woman's wit was always a Pimp to her Pleasures.'

During the eighteenth century advice or courtesy books, stressing obedience and manners, became increasingly popular with the rising middle classes. These books are wordy and leaden beside Wollstonecraft's, whose briskness is deliberate: a counter to the 'affected' style of others' advice, 'designed to hunt every spark of nature out of [girls'] composition'. Wollstonecraft said she would ban such books on the grounds of style alone. Her emphasis on domestic training does not rule out public life. She believes that nursery instincts like tenderness, if empowered by the right training to think and act, could one day redeem the world. The enormity of this claim widens the contrast between forbidding advice (the most influential advisers being Lord Halifax in 1688 and Dr James Fordyce in 1766) and the moral independence of the young Wollstonecraft.*

Women a little older than Wollstonecraft – those born in the early to mid-1740s – tended to concur in the subordination of their sex. Mrs Barbauld, a writer from Newington Green, who had run a school in Suffolk, laid it down that girls 'must often be content to know that a thing is so, without understanding the proof'. They 'cannot investigate; they may remember'. Wollstonecraft's ideas for girls' education burgeoned in a context where millions of girls were taught to memorise, not to think. She mocks one of Mrs Barbauld's poems, a cascade of clichés which likens women to 'DELICATE' flowers, free from toil, 'born for pleasure and delight ALONE'. Mrs Barbauld's concluding lesson is that 'Your BEST, your SWEETEST empire is – TO PLEASE.' Even Hannah More, a purveyor of popular pieties and leading member of the 'Blues', believed that the 'bold, independent, enterprising spirit' encouraged in boys should be suppressed in girls. Wollstonecraft recognised that it was through such misteaching that 'daughters' internalised their subjection. Education was therefore central to her message.

At the time she completed the book, a new phase loomed. The Bloods had invited her to live with them in Ireland, but she refused: 'I must be independant and earn my own subsistence, or be very uncomfortable.' Nor could she forget her debts: she had hoped to save, but by early July, after two or three weeks of her Spartan regime, saw she could not. To pay off her debts, the only course was to become a governess. Dr Price, with the help of Mrs Burgh, alerted a friend at Eton, the Revd John Prior, an assistant master at the school, where he lived with his wife. Recommended by the Priors, Mary had several offers, amongst them a post in Wales and another with a noble family, the Kings, in Ireland. This last offer seemed hand-

---

*Jane Austen too was less than submissive to Fordyce. His *Sermons to Young Women* is the book Mr Collins insists on reading aloud to the Bennet girls in *Pride and Prejudice*.

some: £40 a year. (It's been suggested that this was a poor sum, reflecting Mary's educational limitations. But we need only compare it with the £30 accepted half a century on in the late 1830s by Charlotte Brontë who, as a governess, could offer French and highly developed skills in art as well as teaching experience.) Mary took up the Irish offer. Her plan was to use half her salary against her debts. Paying off £20 a year meant that she would have to work at least four years, and this was not to be the end of it, for she had a further plan to support Bess. Years of dependence seemed to stretch ahead as she sighed over the prospect of social isolation between employers on the one hand and servants on the other: the usual position of the governess. Mary Wollstonecraft was coolly realistic. The most she hoped for was civility.

As summer waned she took lessons in French, and looked through her clothes in dismay – they were worn, unequal to a proper appearance amongst grandees. Miss Mason came for two days to help her make a coat ('I do not know what I should have done without her,' Mary said), while Mrs Cockburn, having won the contest for lodgers, could afford the conciliatory gift of a rather weird blue hat. Regretting the modish gown Skeys had promised, Mary was driven to ask George if he could send her fabric for an old pattern Mrs Blood could supply.

Beneath this wave of activity rolling her towards Ireland were undertows tugging her back to the deep sea of inertia. Some unkindness could make her turn, sick, from social ties. One night her friend seemed to beckon her towards death like Hamlet drawn to the ghost of the beloved dead: 'I dreamt the other night I saw my poor Fanny, and she told me I should soon follow her[.] I am sick of the world, "'tis an unweeded garden" – . . . I want a friend[.] I am now *alone* and my heart not expanded by the usual affection preys on itself. I can scarce find a name for the apathy that has seized me – I am sick of every thing under the sun – . . . all our pursuits are vain . . .' Part of this sickness was self-despair. Repeatedly she questioned her affections, 'which are too apt to run into extremes' – apt to carry her 'beyond the pitch which wisdom prescribes', then fall into 'apathy'. Two things she trusted to carry her through: she held by her understanding, and her faith never faltered: 'He has told us not only that we *may inherit* Aeternal life but that *we* shall be *changed* if we do not perversely reject the offered Grace.'

A lawsuit to do with a '*lapsed legacy*' remained unsettled in late August. Ned had brought a court case against a man called Roebuck, the senior partner in a firm of insurance brokers called Roebuck & Henckell of 49 Threadneedle Street, for an annuity or trust which the Wollstonecrafts claimed was theirs. It's likely this was the 'fortune' Mary spoke of losing in her late teens – what she and her sisters had loaned to their father during the financial 'storm' associated with the family's move to Wales in 1776–7. An outcome in favour of the Wollstonecrafts would wipe out the bulk of Mary's debts – she would need to remain a governess for no more than a year. Correspondence flew back and forth without much result; Ned was rude and unhelpful; the day school went on through August; the pupils teased their distracted teacher; and then, in the midst of her daily life, Fanny's image would blot it out.

Her debts pressed on her conscience: certain people could not wait for what they had lent her. One day, when a creditor was rude, her confidence as a teacher faded into the squirm of an unprotected female. It seems that Dr Price absorbed certain of these debts. 'He has been uncommonly friendly to me,' she told Bess. 'I have the greatest reason to be thankful – for my difficulties appeared insurmo[u]ntable.' Mrs Burgh, too, showed endearing solicitude, inviting Mary and Hewlett to dine, and offering a further loan to set Mary's mind at rest about some remaining creditors, including the vulnerable Hinxman. 'Mrs Burgh has been as anxious about me as if I had been her daughter – I have paid all my trifling debts and bought all the things I think absolutely necessary –,' she reported to Bess. 'You have no conception of Mrs Burgh's kindness.'

As she packed her bag, and took leave of the faithful Mason and benevolent Mrs Burgh, it was the end of her Newington years – underlined by the death of Mrs Price that same September. For Dr Price, too, was leaving the Green. 'I am at present literally speaking on the wing,' she wrote to Bess at the end of the month. Her farewell letter was 'the last . . . from this Island'.

*

The spring and summer of 1786 saw the end of the community of women, and the start of her writing life. This resurgence after loss, failure, mourning marks the inception of a new story. It's not her father's narrative of moving from place to place, nor the Bloods' narrative of running away, nor – still tempting to Mary – the eighteenth-century tragedy of female independence: Clarissa dying in protest against her abuse. For Mary would repeatedly find the strength to go on in the face of disaster. In doing so, she was never idle and ever mindful what she owed to others – a responsible alternative to persisting master-plots of greed and power. As a friend, Mary groans over loss; as educator and writer, loss turns to gain. Her book confirms her vocation as an educator, for though she never taught in school again, everything she wrote from this time was driven by the transforming passion of the born teacher.

It's often assumed that greatness is produced by circumstance. We say that domestic violence gave rise to rebellion; we say that adversity in Mary's late teens set her apart as a middle-class girl who must work to survive; yet neither, on its own, can explain greatness. In fact, these conditions usually produce victims like Mary's mother or frustrated strugglers like her sisters. It's impossible to explain genius, but certainly, in the case of Mary Wollstonecraft, it was not thrust upon her. She would describe herself later as 'a strange compound of weakness and resolution!' When it came to resolution, she showed extraordinary vigour; her groans were the obverse of daring, in part its cost in genuine suffering, in part a protective shield for missions too embryonic to expose. These don't appear in her letters, where she relieves her feelings during the hard times she certainly endured. In reading her letters, then, we must not allow the volume of the groans to muffle the rising voice of an educator: a voice designed to be heard beyond the range of one failing school on a village green in 1786.

Though her *Thoughts* were delivered in the acceptable guise of innumerable guides to girls' education, the message itself cut through the feminine model of weakness and passivity. Mary's new post now offered an opportunity to try her alternative model on three young members of the Irish aristocracy: the eldest daughters of Lord and Lady Kingsborough.

5

A Governess in Ireland

The first stage of Mary's journey to Ireland took her to Eton College in Windsor. Stepping down from the coach early in October 1786, she was welcomed by Mr Prior who had taught classics in the school since 1760. He was the Master of a red house opposite the west doorway of the chapel, and his family also owned a boarding-house on the south-east corner of Keate's Lane, where Mary may have stayed. Orders were to await the arrival of George and Robert King, aged sixteen and thirteen, the eldest sons of her employers, who would travel with her post-chaise via the Welsh port of Holyhead to Dublin, and from there to Mitchelstown Castle, County Cork, another hundred and thirty miles to the south.

At the time, it was not uncommon for boys to be away from school during term. A message came that the Kings were 'actually on the road' to Eton; they were expected hourly. But day after day passed. Possibly, they had instructions to escort the new governess, a proper attention to a young woman about to cross into unknown territory. Certainly, this explanation never occurred to Mary, accustomed as she was to move about alone. For a traveller to Portugal who had braved the Bay of Biscay, the Irish Sea presented no fears.

Waiting over two or three weeks for boys who never appeared, she became uneasy, then impatient and put out by the strange ways of Etonians. All appeared to 'move in the same round', permitting no boy to 'fly off to any other sphere'. Two of the boys in Mr Prior's house were younger sons of the King family: Edward who had been at the school for five years, together with his elder brothers, and Henry who had joined them in 1785. Laughter at private jokes exploded around her. 'Witlings abound,' she wrote to Everina on 9 October, 'and *puns* fly about like crackers, tho' you would scarcely guess they had any meaning in them, if you did not hear the noise they create – .' Wit and politeness appeared to have banished love, ' – and without it what is society?' In tears over a 'tender unaffected letter' from Everina, she proposed a visit from her sister, and then, when Mrs Prior could not accommodate Everina as well, Mary almost fainted and had to be nursed. Depressed to be once more on course as a subordinate, she found herself in the nursery of this ruling class, and excluded by its lingo. Eton alerted her to her place as foreigner in her own country.

Two weeks were enough to see how the school of that day reared a warped specimen with an undeveloped heart. Boys, removed from home to school, developed instead their physical strength, competitive aggression and caste solidarity, preparing them to wrest an empire from inferior races. Future rulers were schooled in Spartan conditions to endure hardships far from home. Mary Wollstonecraft is already a revolutionary, original, far-sighted, when she identifies the problem as domestic atrophy: the disempowering and exclusion of the mother. In theory, the House Master and Mrs Prior provided a substitute home; in practice, Mary saw, this didn't work: the boys were repressed with the Master, and rampant the moment they left his presence. After a silent dinner they would swallow a hasty glass of wine, 'and retire', she observed, 'to ridicule the person or manners of the very people they have just been cringing to'. The masters seemed indifferent to morals. Mary overheard them saying that 'they only undertook to teach Latin and Greek; and that they had fulfilled their duty, by sending some good scholars to college'.

Public schools were rough places in the eighteenth century. Boys of the upper classes were trained to numb sensitivity to bullies and defer to their seniors, biding their time until they could take their turn as top dogs. Part of the ethos was caste loyalty: not to 'sneak' – not to expose the defects of the system to outsiders. This was still the brutal period before Dr Arnold, headmaster of Rugby, introduced fair play ('cricket') and chastity as attributes of manliness in the 1820s – and chastity, through those long, shut-up nights in dormitories, was not to be expected. Boarding-schools, Mary said, were 'hotbeds' of furtive sex. They pushed boys into 'libertinism . . . hardening the heart as it weakens the understanding'. She deplored the 'vice' of senior boys, linked as it was to 'the system of tyranny and abject slavery' of younger boys.

Her voice is so forthright that it's easy to miss what she holds back: a hint of a sea change at the age of twenty-seven during these weeks of enforced idleness. She has 'so many new ideas of late', she tells Everina, she 'can scarcely arrange them'. She is plunged 'in a *sea* of thoughts'.

On the packet to Dublin, she fell in with a clergyman fresh from New College, Oxford. Henry Dyson Gabell, aged twenty-two, was bound for the household of John O'Neill of Shane's Castle near Antrim, where he was to serve as tutor. Two years later he would enter on duties as an Anglican priest in Wiltshire. The men who appealed most strongly to Mary in these early years were alert to the soul. They did not impose the falsities of gallantry. Like Dr Johnson, Dr Price and the Revd Mr Hewlett, young Gabell heard 'the *tone* of melancholy', a balm after her sojourn with the witlings. One issue she longed to discuss was why the deity – 'the Searcher of hearts' – should burden people (like herself) with a sensitivity that seemed to have no use in society, and only complicated the effort to ready the soul for heaven. As with Dr Johnson, it didn't put her off that Gabell's politics were opposed to hers. In the late 1790s he would use the pulpit to preach counter-revolution. In the meantime, he took the view that the masses should be kept down in 'fat contented ignorance'.

'The appetites will rule if the mind is *vacant*,' Mary argued.

Gabell countered that our reasoning is often fallacious and our knowledge conjectural.

Mary had to concede this. But it seemed to her also true that 'flights into an obscure region open the faculties of the soul'. Afterwards, when she had to mute her mind to some extent in the company of Right Honourables, she recalled this opportunity to flex her eloquence.

In Dublin she was welcomed by Mrs Blood, George Blood and a family friend called Betty Delane who had known Fanny. It was a relief to find old Blood at last in comfortable circumstances. After a few days, a civil and unexpectedly kind butler from the castle arrived to escort Mary on the long last lap of her journey. She would have enjoyed it, she said, had she not been in tears all the way at the prospect of the journey's end. Dr Price's talk of 'aristocratical tyranny and human debasement' had warned her what she would find.

She approached her destination along a road that ran parallel to the northern slopes of the Galtee Mountains – a boundary to the Kingsboroughs' lands. It was a landscape of peaks and drops, with torrents of water falling between birch and whitethorn trees on the mountainsides. After six or seven miles the road turned over a hill, opening a vista of their other side. Mary looked down on a long plain, bounded to the south by the furred, dark-green tops of the Knockmealdown Mountains. At the centre of this plain, fed by its rivers, was Mitchelstown.

They were approaching the spreading elegance of a stately home. It had a square Palladian centre with wings on either side, a massiveness stretched out rather than high – height was provided by the position of the house on a rise. It backed on to a gorge with a river below. In the fourteenth century and for the next three hundred years a castle had provided a lookout over the surrounding countryside. It had been destroyed (except one tower) and rebuilt in 1645 with an upstairs gallery seventy feet long, overlooking the Galtees. Lord Kingsborough had recently added a storey and moved what had once been the adjacent medieval village of Villa Michel, or Baile Mhisteala, to a safer distance, the present site of Mitchelstown. All that remained of the old village was a graveyard, its mossy, rain-washed stones there to this day.

Between what continued to be called the 'castle' and the town – between, that is, the Protestant Ascendancy and Catholic natives – was the buffer of a spreading Georgian square of two-storey houses, named Kingston College.* Protestant occupants had been brought in to fill this buffer zone nine years prior to Mary's arrival.

*College in the sense of a resident community. A dated stone says August 1780, but construction had begun well before.

As she drove through the new-built town, on through the new square, and reached the gates in the wall that surrounded the castle, she felt cut off in a setting contrary, she said, 'to every feeling of my soul'. It struck her as a prison: 'I entered the great gates with the same kind of feeling as I should have if I was going into the Bastile.' The prison in Paris had come to stand for an oppressive regime; its thick walls that made it impossible for an inmate to be heard meant live burial: walls rearing up as a frontier between being and nonexistence. Mary's apprehension of a Bastille appears excessive – the Wollstonecraft proneness to moaning, and moaning here in the midst of luxury – but apprehension did have some basis in North Cork, a settler part of Ireland with a long history of violence, where tension between rulers and ruled was reflected in the layout: the distanced town and the containment of the castle with its new-built barrier of a ten-to-twelve-foot wall.

In contrast with its austere façade, the interior of the castle was ornate, with elaborate plaster mouldings. The Rape of Proserpina decorated the ceiling of the hall. The butler would have conducted Mary upstairs to the drawing-room and gallery on the first floor. Waiting to greet her was Lady Kingsborough; her stepmother Mrs FitzGerald, with three grown-up daughters; and what appeared to Mary a '*host*' of Ascendancy ladies. All were examining her minutely.

As a 'solemn kind of stupidity' froze her 'very blood', the children, her charges, presented themselves. 'Wild Irish' was her first impression. The eldest, Margaret, was fourteen or fifteen, tall for her age, with a high pointed nose and eyes so pale as to appear almost colourless. Her brown hair fanned out on either side of her face, frizzled in fashionable disarray and ending in locks curled about her neck. Her gaze had the directness of a girl who nerves herself not to fear. Next came her prettier sister, Caroline, aged twelve; and last, Mary who was only six, with blue eyes and abundant dark-brown locks like her mother's. The three had planned to drive out this English governess, or at least amuse themselves by giving her a hard time. She would be shut up in the schoolroom with these aliens from whom she could expect little mercy for her meagre French and nonexistent accomplishments. It crossed her mind that the Priors had oversold her.

After these none too promising introductions, she had to contemplate the parade and diversions of high life. Politeness required her to hear this set 'decide with stupid gravity, some trivial points of ceremony, as a matter of the last importance', and join in jollities about the getting of husbands. It was an effort to hide her dislike. Save me from the 'volubles,' she sighed a few months later, longing for time to think and read.

She did not see the castle in Georgian terms of ceremonial, grace and order; she questioned its deference to custom and public opinion. Her faith went beyond the formal Anglicanism which, above all, defined membership of the Ascendancy; to her, faith meant something close to her idea of genius: reaching beyond the frontiers of consciousness towards divine intentions – the sort of talk she continued to exchange in letters to Henry Gabell. She was in her element with clergymen who could sift the higher claims of the moral life, or hear the sweet, sad music of humanity. Lady Kingsborough was charitable, more so than most, but had not the remotest notion of the interior drama – soulfulness crossed with Hamlet – in which her governess specialised. It upset the latter, in the way it embittered Hamlet – and she was always quoting Hamlet – to have to hide her true self. Without losing her professional effectiveness, Mary nursed 'that within which passes show'. There were, anyway, no exact words for her drama of imprisonment or invisibility. Here again is a glimpse of a new form of life, sprouting apart from the flower of the Ascendancy, in the concealed ground of Mary's gloom. No amount of consideration from her employers could make up for their failure to know who, in this sense, she was.

Her room at the back of the castle looked out across the gorge and dark hills to the Galtees in the distance. This prospect and its solitude would please her in the months to come, but that first night she felt the common isolation of the governess, set apart from her employers and from the servants dancing to a fiddle below.

The next day or soon after, Mary was called to an interview with Lady Kingsborough who lay in bed with a sore throat. The first sight that met her eyes was numbers of dogs reclining on cushions. Throughout the interview, their mistress caressed and watched them with the most assiduous care, lisping out 'the prettiest French expressions of ecstatic fondness in accents that had never been attuned by tenderness'. Her display of French prodded Mary's weak spot. So did her regret that Mary could not teach 'fancy work'. Lady Kingsborough had assured Mrs Prior who in turn had reassured Mary of Lady K's opinion that preceding governesses had neglected the children's minds, attending only to the ornamental part of education which 'ought ever to be a secondary consideration'. Where that message had soothed Mary, the present about-face was calculated to deflate assurance, and she was further disconcerted to find her ladyship's eloquence equal to her own. Lady Kingsborough's manner of condescending to give an opinion reminded her disagreeably of the Revd Hewlett's overbearing wife. Her ladyship was clever as well as pretty, and civil enough to wish every attention should be paid to the new governess. Together with Mrs FitzGerald, she set an example of consideration the castle followed. 'Every part of the family behave with civility – nay, even with kindness,' Mary had to admit when she had been there a little over a week.

Yet though she was surprised into gratitude, instinctively she distrusted Lady Kingsborough. The wife, mother and human creature seemed swallowed up by a rouged and 'factitious' femininity. Mary saw her as a product of 'an improper education' who could utter 'nonsense' in infantile accents 'to please the men who flocked round her'. Again, as in Bath and Windsor, Mary had to put a lock on thoughts and feelings if she was to retain her post. She practised forbearance, and banished the 'contending strain' from her voice. 'I am thought to have an angelic temper,' she comments wryly to Everina during her third week. Being angelic took its toll: her mind languished, her responses were blunted – replaced by indifference. The buzz of good-humoured attention to whatever she had to say, even the wonder she aroused, gave no pleasure – she was not flattered by the admiration of people whose judgement was 'of the grosser kind'. Two feelings remained acute: irritation with her own sex – the boisterous, unmeaning laughter and bickerings of 'silly females' – and persistent alienation: 'I am an exile – and in a new world.'

Cut off from friends and family affections, she picked up her pen – and found some release. Alone at night beside her fire, listening to the wind or lifting her eyes to the hills, she would dash down letters where she freed herself to say what she saw and heard. Her position was not unlike that of a spy – an unwilling but penetrating spy – who infiltrates the closed world of the enemy. During her first weeks at the castle she would have taken in the facts of its history and inhabitants, and assessed what she was up against at the start of a campaign to draw the daughters of the enemy into her camp.

A certain John King, born in Staffordshire, had made his way to Ireland during the reign of Elizabeth, and there married Catherine Drury, grand-niece of the Lord Deputy of Ireland, in 1578. In return for military services, Elizabeth knighted him and granted the lease of Boyle Abbey in the county of Roscommon. By the end of the sixteenth century Elizabethan entrepreneurs like John King made up less than two per cent of what was to them a strangely 'unreadable' population, a bardic culture which continued to possess its land in its poetic and legendary aspect. During the reign of James I, Sir John acquired so much land that for nearly three hundred years the Kings would remain amongst the wealthiest and most influential of the Protestant families. He settled in Dublin, and fathered six sons and three daughters. His fourth son, Edward, was Milton's friend at Christ's College, Cambridge – Milton wrote his great elegy *Lycidas* after Edward King was drowned in the Irish Sea in 1638 at the age of twenty-five. He was the brother of Margaret King's great-great-great-grandfather.

During the curtailed reign of James II (1685–8), the Catholic King granted power to the Catholics in Ireland. The Protestants became uneasy, and remained so even after Catholic forces were overthrown at the battle of the Boyne in 1690. That victory began the 'long' eighteenth century in Ireland. Its separate Parliament, entirely Protestant, took a harder line against the Catholic majority than did the English Parliament. There were laws excluding Catholics from the vote and intermarriage, as well as some very petty and unenforceable laws: a Catholic, for instance, might not own a horse worth more than £5.

Meanwhile, the Protestant establishment became increasingly rich in land, and none richer than the forebears of Mary Wollstonecraft's employers. In 1658 John King's grandson, another John King, married an orphaned child-heiress, Catharine Fenton. Another calculating marriage took place a century later: on 5 December 1769, when sixteen-year-old Robert King, heir of the 1st Earl of Kingston, married his fifteen-year-old cousin Caroline FitzGerald, the richest heiress in Ireland,* their combined estates made them the greatest landowners in the country. Caroline's Mitchelstown estate of 100,000 acres, extending from North Cork into the counties of Limerick, Tipperary and even Kerry in the west, brought in £42,000 in rents a year (worth approximately £2 million today). Those whose families had owned land for generations *felt* Irish, yet to the Catholic majority – amounting to seventy-five per cent of the population – they remained English, colonisers who copied English manners, furnishings and clothes, and sent their sons to English schools.

Robert King, Viscount Kingsborough, had only recently left Eton when he married. The month before, in November 1769, his father the Earl had engaged an Eton master, John Tickell, to continue his son's interrupted education. These actions bear on another curious fact: the family fudged the birth dates of the couple's first two children. Burke's *Peerage* gives 8 April 1771 as birth date of their eldest child, George. The year has to be wrong, because a letter from Robert to his father on 7 July of that year reveals that Caroline was about to give birth to their second child. This was Margaret, named after Caroline's dead mother. It was not uncommon for the nobility to have records altered. If George's birth date was shifted to one year later, Margaret's had to be shifted accordingly to 1772.

Caroline combined delicate features with two extravagant beauties: her large, slanted eyes, rather wide apart, with creamy lids, and a tremendous

*Caroline inherited her lands in 1761 at the age of seven, from her King grandfather through her mother Margaret, who died in 1763. Margaret had been the daughter and heiress of James, Lord Kingston of Mitchelstown Castle. Caroline's father was Colonel the Right Hon. Richard FitzGerald of Mount Offaly, Co. Kildare. She was mistress of her lands for her lifetime, after which they were to go to her eldest son.

head of hair that needed no fashionable props. In a portrait she has twined ribbons through two long curls that hang down from her unpowdered pompadour. She and Robert King were intended for each other: if, on the strength of this, Robert seduced Caroline and she was pregnant, a hasty marriage would have been the only solution to the birth of George some four months later. This would fit the fact that the bridegroom's schooling was cut short. It was not customary for boys to leave Eton at sixteen to be married. If these are signs of a cover-up, another fact would fit too, for George was born away from home in the quiet village of Chelsea, at a distance from the centre of London.

Whatever the truth, this marriage had a bad start. To the Earl, Caroline mattered only on three counts: she was moneyed, she was pretty, and she was there to 'breed'. Though she bore three children in four and a half years of marriage, when there came a lull, the Earl wondered at the cause of her 'not having children as fast as at first we had reason to expect'. The cause is not far to seek: Caroline and Robert did not get on.

As Mary Wollstonecraft would later grant, Caroline King was fundamentally well meaning in her unhappy situation. She had lost her mother at the age of nine; at seventeen she was already a mother of two, having been transplanted at fifteen into her husband's family who expected her to 'breed' whether she was loved or not. After Margaret's birth Caroline did not conceive for eighteen months, which could mean that she refused her husband so long as she had to live with his parents.

At length, Robert agreed to live in London. There, Caroline soon conceived again, and her husband reassured his father they'd stopped rowing since they left: 'You very well know how unhappy Caroline & I were when we were with you.' The Earl replied that he had been so very shocked at their misery 'that I have often wished the match had never taken place, but it was then too late & I kept my mind to myself. I am glad, at any rate, that mutual dislike has ceas'd, & I hope it may continue so.'

They returned to Ireland when Caroline came of age in 1775, and settled at Mitchelstown. Lord Kingsborough came back with a rage for building. The reconstruction of his wife's property was to occupy the rest of his life. When an English agriculturalist, Arthur Young, toured Ireland in 1776, he had seen at Mitchelstown 'the face of desolation'. The poor subsisted on potatoes and milk, as elsewhere, and the potato-driven soil had been exhausted and left to weeds all over the estate. Women weeded for no pay, and men would walk ten or twelve miles to a fair to sell a lamb for 3s 6d – a sure sign of poverty. Their cabins were made of mud and thatch with no chimney – Young's sketch shows smoke coming out of the entrance, the sole aperture, at the front. When Young became steward of the estate in 1777, he tried to persuade Kingsborough to do away with middlemen (the 'gombeen men'), who exploited the peasants.

Young was defeated by a leading middleman, Colonel James Thornhill, who was related to Lady Kingsborough. By now she had been ill matched for ten years, with numerous children, though she was only twenty-five. Arthur Young, aged thirty-eight, was also unhappily married (to the sister of Fanny Burney's stepmother from Lynn in Norfolk). He amused Caroline with chess in the evenings, and Mrs Thornhill insinuated to Kingsborough that love was in the air. Young was dismissed, but his work remained. The estate was divided into agricultural fields, each with a name, including a Ghost's Field, while sheep, cattle and deer were farmed virtually up to the castle door.

Kingsborough has been praised for his building and planting. Certainly, he did well to establish schools, a library, a market square and a plantation of four hundred thousand mulberry trees for the production of silk which provided opportunities for employment and brought, for a time, a measure of prosperity to the peasants. But the layout had also a darker purpose. Since his lordship's wall had cut through the old main street, he threw up a new one at a distance, later named 'George Street' after his godfather, the King. At one end is King's Square with its Georgian church; at the other stands another church which I assumed must be Catholic – but on closer inspection turned out to be Protestant. This means that whenever a tenant lifted his eyes in George Street he saw, planted in his path in either direction, the God of the landlord to whom he owed his family's subsistence. And looking north, he saw a road straight as logic sweeping through the new town, on through the centre of King's Square and up the rise to the 'castle' at the top, encircled by its mortar wall. In short, it's a power statement: a remade landscape spells out the Protestant Ascendancy. It speaks at once of arrogance, beauty, disciplined taste – and, lurking under a perfect order, a twitch of danger. Was it fear of an untamed populace going back to the days of old John King: elusive, slipping over their bogs, fluid in their arrangements, somehow uncontainable in the confusion of Elizabethan campaigns? Or was it the danger of insecure superiors who might forsake civility for the sword? In 1783 when Kingsborough stood for the Irish Parliament, his men fought numerous duels – no more than what was expected.

Kingsborough was also commanding colonel of the Protestant-led Volunteers, the Mitchelstown Independent Light Dragoons, formed in 1780 in fear of an American or French invasion. He paraded and drilled in a scarlet uniform with silver epaulettes. The Volunteer movement helped the Irish Parliament achieve legislation without British interference in 1782. In the mid-1780s Mary Wollstonecraft was seeing the Ascendancy at the apogee of its power before it began to be eroded by the impact of the French Revolution. Irish Georgians had what R. F. Foster has called a more 'gamy' flavour than their English counterparts, more ferocious, more given to a 'savagery of mind', amplified by the insecurity of their political position and 'balked' nationality. Foster quotes Clonmell, a Lord Chief Justice, who thought 'a civilized state of war is the safest and most agreeable that any gentleman, especially in *station*, can suppose himself in'. Mary was not to know that Lord Kingsborough's men had horse-whipped nine supporters of his opponent in the run-up to his parliamentary election. The violence resurfaced in 1797 when, as we shall see, Kingsborough shot a member of his family, and was intensified by the atrocities of the Viscount's heir, 'Big George' (the elder of the boys who had failed to meet Mary Wollstonecraft at Eton), during the Irish Rising in 1798. The militia George took over from his father, the Mitchelstown Light Dragoons, was part of the notorious North Corks whose conduct has been described as 'stupidity and cowardice in the face of danger; ferocity and outrage in the wake of victory'. George's portrait has been removed from the Provost's Lodge at Eton, and relegated to the art storeroom – not to be lost. It is, after all, a painting by Romney.

*

Though Mary Wollstonecraft did not witness acts of brutality, she was struck by a curious absence of sensibility amongst the inhabitants of the castle. Compassion, central to her teaching, was not part of their education. Girls, she observed, were harnessed to '*cart*loads' of history – this puzzled her rather more than it need puzzle us, for history until recently has been the story of dominant groups. To aggrandise families and augment power was the subliminal message of what girls of this class were put to read; it conditioned them to accept their fate. In their late teens they must assent to a marriage in which love had no part. Mrs FitzGerald's three daughters were, as their brother laughed to Mary, 'just off to market'. She was astonished to find that Maria, Harriot and Margaret FitzGerald read no novels. This lack seemed to Mary to have left them devoid of the interior life of sensibility that led to genuine 'refinement', not seen in manners alone. Nor had cartloads of history diminished their silliness.

The FitzGerald girls were also schooled in modern languages. They appear in Mary Wollstonecraft's first novel as girls whose 'minds had received very little cultivation. They were taught French, Italian, and Spanish; English was their vulgar tongue. And what did they learn? Hamlet will tell you – words – words. But let me not forget that they squalled Italian songs in the true gusto. Without having any seeds sewn in their understanding, or the affections of the heart set to work, they were brought out of the nursery, or the place they were secluded in, to prevent their faces being common: like blazing stars, to captivate Lords.' There's a similar critique of the miseducation of girls in Jane Austen's errant Maria Bertram, who 'had learned languages with facility and been taught to set a very high value upon her knowledge of history and chronology'. She makes a disastrous marriage (with parental assent) to a wealthy fool.

It was not customary to send daughters away to school if a family could afford a governess at home. Since a girl's purpose in life was to marry to advantage and 'breed' as many male heirs as possible, she was given an education in manners and the sort of accomplishments – music and modern languages – that would show well in the marriage market. Wollstonecraft was, as we know, unfit for such training as well as being ideologically opposed to it, and this added uneasiness to dislike when Lady Kingsborough made her descents on the schoolroom.

As a governess with advanced views on the necessity for women's education, Wollstonecraft recognised at once that the eldest daughter, Margaret, had 'a wonderful capacity but she has such a multiplicity of employments it has not room to expand itself – and in all probability will be lost in a heap of rubbish miss-called accomplishments. I grieved at being obliged to continue so wrong a system.' Lady Kingsborough ruled the girl 'with a rod of iron', while, in Mary's opinion, 'tenderness would lead her any where'. Margaret later confirmed the 'baneful effects' of 'unkindness & tyranny', and her relief to be treated with respect. Mary planned to expand Margaret's knowledge by taking her to visit 'the poor cabbins'. But she was careful, at first, to placate her mother.

At Eton, Mary had heard gossip to the effect that his lordship had been extravagant in a way that hurt his wife. He had granted an annuity of £50 to one of Mary's predecessors, Miss Crosby, dismissed when she was said to be attracting the attentions of her master. Fifty pounds for doing nothing was £10 more than Mary was due to receive for a year's work. The amount itself is suspicious, as though Kingsborough had to compensate Miss Crosby for an injury to her reputation that would make it impossible to find another post. Mary Wollstonecraft, in turn, could not be unaware of the roving eye of Lord Kingsborough, ready for 'a little *fun* not refined'. It didn't take her long to see the vacuity of a society marriage: a couple who met only in bed, he a little drunk, she with her head full of compliments from admirers at the card table. In this way the Kings had produced seven sons and five daughters. Their last child, James, was born the year Mary came to the castle, probably before she arrived, since the birth is not mentioned in her letters.

'I am treated like a gentlewoman,' she wrote, 'but I cannot easily forget my inferior station – and this something betwixt and between is rather awkward – it pushes me forward to notice.' Where the previous governess had been treated as a servant, Mary received extraordinary notice from the family.

The castle was ever more pleased with what it saw. The new governess was dignified and devout; she kept to herself because she craved privacy, but the effect was to stimulate curiosity. She was gratified to find herself treated as a gentlewoman, but mystified, even put out, by numerous daily visits from Ascendancy ladies. In fact, she was soon complaining to her sister of too much company. This protest tells us of her unintentional charisma. The fact that she continued to attract conservatives – Dr Johnson, Henry Gabell, and now Irish grandees – tells us also that she expressed her views with disarming civility. These soon overturned the regimentation Lady Kingsborough had imposed on her daughters. But Mary's biggest gain was the confidence of her charges.

Within five days, they gave up their plan to drive out the governess. A prohibition against novels had led to surreptitious reading of shallow ones. Mary's policy was to free the girls to read what they liked, and govern them solely through their affections, with the result that they felt uneasy about reading what their governess despised. Her ladyship had to concede the benefit, and surrendered control to Mary.

It was decided that Margaret, who had exhibited certain 'grave faults' – too shaming to be named, possibly incurable Mary thought – was to be given special attention. 'She is to be always with me,' Mary reports to Bess on 5 November, while Margaret breaks in on the letter with happy offerings.

'. . . I have just promised to send her love to my sister – so pray receive it.' Mary speaks with amused patience, bent on winning her pupil's affection. 'My sweet little girl is now playing and singing to me – she has a good ear and some taste and feeling – I have been interrupted several times since I began this last side.'

What Mary Wollstonecraft taught Margaret over the next few months would lead her later to abandon class and country for her own experimental course that would allow her to develop a medical practice, decades ahead of the first professional women doctors. Margaret's was to be an original life – a sequel to come – and its inception in the ideas and methods of her governess is fortunately recorded in Wollstonecraft's *Original Stories from Real Life* (1788). It's another education book for girls, using fiction as a form of sermon. At night in her room, when the day's duties were done, Wollstonecraft was reading letters on education by a French governess, Mme de Genlis; also her tales of castle life, *Les Veillées du château*, which taught 'that luxury dazzles none but fools, and does not produce one real delight; for nothing is more troublesome than magnificence'.

Real Life concurs. One of Wollstonecraft's stated aims was to counteract the effect of parents preoccupied with frivolities. The King girls had been brought up largely by servants, offset by their mother's severity that required formal obedience to a code of public conduct. This code taught the grace of decorum and civility, yet because it was imposed by a distant parent it did not reach daughters through their minds and feelings. This was Jane Austen's point when she dramatises the distant and ultimately ineffective severity of Sir Thomas Bertram whose elder daughter runs off with a rogue. Both *Mansfield Park* and, earlier, *Real Life*, show the futility of schooling daughters in manners without any real encounter with morality. To teach the King girls virtue, their governess found that she had to 'explain the nature of vice. Cruel necessity!' She wished to give them the means to recognise the 'profligate Lord', and to distinguish dignity from the 'state' of the fine lady. Margaret was quick to apply this to her parents and their milieu.

Mrs Mason, the guardian and governess in *Real Life*, teaches that 'friendship and devotion', not money, not power, is 'what principally exalts man'. Emotional literacy has to be the fount of responsible public action; 'compassion' must replace the vulgarities of 'prejudice'. Margaret always saw this as her book; her education. It was an education in seeing – seeing, for instance, that the lower classes and all lower forms of life were no less alive than aristocrats. Neither at home nor at school had Margaret's father and eldest brother learnt the most important lesson of social existence: not to injure others. A woman or child doesn't have to witness horse-whippings and the tortures perpetrated (in years to come) by Margaret's brother George, in order to sense callousness. In middle age Margaret set down a record of her youth where she recalls that 'the society of my father's house was not calculated to improve my good qualities or correct my faults; and almost the only person of superior merit with whom I had been intimate in my early days was an enthusiastic female who was my governess from fourteen to fifteen* years old, for whom I felt an unbounded admiration . . .'

There are two girls in *Real Life*, undergoing an equivalent to the secret curriculum in the schoolroom of Mitchelstown Castle. The elder, the Margaret figure, is called 'Mary'. She is Margaret's age, plain, slothful, and indifferent to cleanliness. Her younger sister Caroline – like the real Caroline, aged twelve – is vain of her much-praised beauty, which makes her affected. Mrs Mason regrets their insensitivity. They should have learnt otherwise in the nursery. In order to reach these hardened older girls, she devises a new system, addressing the senses as 'inlets to the heart'.

Sixty years before Henry Thoreau lost his post as schoolmaster for holding lessons in the Concord woods, Mrs Mason teaches the girls during exploratory walks each morning. Outdoors, they see 'real life' for themselves, moving from ants whose lives and social structures can be observed, to birds shot by callous boys, to humans low in the social scale, who are not to be stared at and derided if they are disabled, or ignored if poor and unfortunate. These 'stories' are discoveries of other forms of life. Some of the characters tell their own stories or they are told as flashbacks – opening up the lives of dying wives, starving children, brutalised dogs and human fodder for wars. All are disregarded by military commanders, landlords, rulers, or owners. This is an attack on power itself, and the elder of the pupils takes it on. 'I wish to be a woman, said Mary, and to be like Mrs Mason . . .'

So Mary Wollstonecraft took the King girls beyond what their mother permitted: condescension to objects of charity. She thought it might be too late to wake compassion in girls habituated to non-seeing. There is no knowing the impact, if any, on the two younger girls, Caroline and Mary, whom Miss Wollstonecraft judged 'middling', but in the course of 1786–7 Margaret's conscience, politics, and ultimately her future were permanently transformed.

*According to her real rather than fudged birth date, Margaret would actually have been a year older, fifteen to sixteen.

Wollstonecraft alerted her pupils to falsehood in tones of voice, and in movements of hand or head. These were no less than 'lies'. 'I never, to please for a moment, pay unmeaning compliments, or permit any words to drop from my tongue that my heart does not dictate,' she assured them.

In this way the schoolroom questioned the drawing-room, in particular their mother's train of strong expressions with no meaning attached, as well as Lady K's preoccupation with ceremonial and outward shows. Miss Wollstonecraft remarked that all people have a sense of the sublime, but for many ignorant people the sublime is the grandeur of aristocrats. It may be all that 'narrow souls' can imagine.

She 'lived' at night when her charges were asleep. Her daytime duties were extended by elaborate preparations for bed. She had to pin up three girls' locks so that curls would cling to their necks in the morning, and bathe three young faces in milk of roses (at 10s 6d a bottle). Then, she would sit up till midnight or two in the morning, writing letters or reading. It was primarily through books and pamphlets that advanced thinkers of her generation came to withdraw their assent to the social system, a precondition for revolution. At this time she read Rousseau, who gave new meaning to the word 'nature'. Followers of Rousseau dreamt of a return to 'nature' where they might cultivate an authenticity impossible in society. This coexisted with talk of 'rights', but Mary Wollstonecraft was seeking something more than political rights. There had been many women before her who had argued for rights, going back to Christine de Pisan in medieval France, followed amongst others by Mary Astell in the 1690s and Lady Mary Wortley Montagu earlier in the eighteenth century. Where 'rights' could transform the external conditions of people's lives, 'nature' offered a chance to shape existence from within.

She began to plan her novel *Mary*, which would 'develop a character' different from Richardson's Clarissa, or Rousseau's Sophie whose perfections 'wander from nature'. Her aim was to demonstrate 'the mind of a woman, who has thinking powers': such a creature 'may be allowed to exist' as a fictional *possibility*. As she taught Margaret, Caroline and little Mary, that shadow of possibility moved inside her, feeling its way on to the platform of action in the character of 'Mary' – a survivor of setbacks.

*

One person in the castle detected something of this hidden purpose. Soon after her arrival, Mary noticed a handsome man in his early forties with a pale, rather melancholy face. George Ogle was neatly dressed without being too fine. His high forehead was accentuated by a neat wig, immaculately plain with a tight roll over each ear, and tied back in a queue. There was no lace at his wrists, and he was too sensible to spend time tying an elaborate cravat. He wore his well-cut coat and narrow shoes casually as though they had grown on him. His attention to the new governess marked her out as a person of intelligence. When the relations of landlords and tenants were discussed, he was willing to castigate the landlords as 'great extortioners'. He spoke with firmness and enthusiasm, emphasising his superlatives, with the air of an orator. Two fingers of his right hand poised on his hip, he would bend forward slightly on his right foot to make a case for Irish independence – not in a 'national' sense, but upholding the right of the Irish Parliament to make its own laws independent of England. At this time he was seen as a reformer, though later he was to fall out of favour with English reformers. An upholder of the established Church and opposed to Catholic emancipation, including ownership of property, as were most 'patriots' of the Ascendancy, he nonetheless declared that he hated no man for his faith. As he spoke with lifted chin, he surveyed the room with large, alert eyes; his eloquence set him apart from those Mary called 'the volubles'.

Enquiry revealed that he was none other than the author of 'On the Banks of the Banna', a popular song of the day, beginning 'Shepherds, I have lost my love.' (It was famous enough to be parodied in the caption of a 1790s cartoon about the change in women's fashions: 'Shepherds, I have lost my waist.') Mary relayed his identity to Everina in some excitement. Unexpectedly, she had discovered a man of sensibility, even 'genius', in the Kings' circle.

George Ogle of Bellevue, County Wexford, had been born in 1742 into a wealthy family. His uncle Samuel Ogle had been colonial governor of Maryland. His father had been a man of letters whose works had included imitations of Horace and a continuation of the squire's tale from the fourth book of Spenser's *Faerie Queene*. His son joined the Whigs, and in 1768 became Member of Parliament for Wexford (on the coast to the east of Mitchelstown). The 1770s saw him doing the rounds of fashion in Bath (never crossing the obscure path of Mary Wollstonecraft). In 1779 he attacked Fox and the Opposition in England for not resisting with greater force the Prime Minister Lord North's coercive policy in Ireland. In Dublin he was thought to 'shed a lustre on every society in which he moved', combining the attractions of a scholar with elegance. His statue in St Patrick's Cathedral, Dublin, memorialises him as 'a perfect model of that exalted refinement which in the best days of our country characterized the Irish gentleman'. In 1783 he was admitted to the Irish Privy Council, and in 1784 became Registrar of Deeds (a post whose purpose was to keep land out of Catholic hands) at a salary of £1300 a year. Mary's £40 a year is a measure of the distance between them, laid down by current ideologies of class, position and gender. She was ready to demonstrate how little that distance need matter to true minds.

'In solitude were my sentiments formed,' is the sort of thing Mary would confide, 'they are indelible, and nothing can efface them but death – No, death itself cannot efface them, or my soul must be created afresh . . .'

Ogle's understanding encouraged her to say more, so she took the risk of exposing her sense of purpose.

'The same turn of mind which leads me to adore the Author of all Perfection – which leads me to conclude that he only can fill my soul, forces me to admire the shadows of his attributes here below; and my imagination gives still bolder strokes to them. I know I am in some degree under the influence of a delusion but does not this strong delusion prove that I myself "am *of subtiler essence than the trodden clod*": these flights of imagination point to futurity; I cannot banish them.'

In *Mary*, Ogle appears in passing as a middle-aged man of 'polished manners, and dazzling wit' whom 'Mary' meets when she is sunk in grief after her dearest friend dies in Lisbon. She is cheered by his attempts to draw her out; he, struck by her location of genius in the soul's power to speak truly. *Mary* offers an idealised self-portrait seen through his eyes: 'She glanced from earth to heaven, and caught the light of truth. Her expressive countenance shewed what passed in her mind, and her tongue was ever the faithful interpreter of her heart; duplicity never threw a shade over her words or actions.' He is forced to question his lightweight opinion of women.

Ogle was married to a benign woman, Elizabeth Moore, who greeted Mary with a smile when they met at the castle. They were accompanied by her sister Miss Moore. She, too, appeared friendly, with sense as well as beauty. Mary learnt that this sister-in-law had inspired Ogle's more recent song, 'Molly Asthore'.

There must have been a distant connection between his branch of the Ogle family and Caroline King's stepgrandmother, Dame Isabell Ogle. Mary noticed that Lady Kingsborough took a possessive view of George Ogle, and that he pleased the women in the castle with expected gallantries. In his company Mary felt her faculties unfold, while others played cards. Their talk went unmarked, for there was nothing of gallantry in Ogle's address to her, and nothing alluring in the clothes and manners of the governess – her intent face, unstudied gestures and seriousness had no part in a lady's repertoire. In Mary's cool view, a 'fine lady' like Caroline King had no idea how to interest a thinking man.

Ogle's admiration did Mary no harm with the King family: on the contrary, the Irish regard for what Mary called 'genius' – the ferment of the mind, in which the governess was seen to hold her own – warmed the castle to a degree she had not thought possible. 'The whole family make a point of paying me the greatest attention,' she wrote to George Blood on 4 December, ' – and some part of it treat me with a degree of tenderness which I have seldom met with from strangers.' And yet, she continued to feel alien: 'a strange being'. Her strangeness was not so much a difference of class or nationality; rather, she sensed some quickening in herself without having words to explain it. Seen by her own light as a mutation of sorts, a creature in the making, it's inevitable that, lacking the companionship of a like creature, she should 'vainly pant after happiness', lose hope, and expect to end the experiment.

Fixed still on Fanny's death, she begged Fanny's brother George to supply, if he could, a similar support. 'About this time last year I closed my poor Fanny's eyes – I have been reviewing my past life – and the ghost of my former joys, and vanished hopes, haunted me continually – pity me – and excuse my silence – do not reproach me – for at this time I require the most friendly treatment.' Her breaths are palpable: 'I want the tender soothings of friendship – I want –' she cried to Everina, 'you must read my heart . . . – it is not to be described . . .' It's this inexpressible 'want' between the words that others, locked in custom, could not be expected to hear.

'I have no just cause for complaint,' she had to acknowledge. 'Everything which humanity dictates is thought of for me – .'

Help could only come through the warmth of friendship, and friendship was what she could not expect from the castle – had it seen any sign of her sufferings. It never did. Mary performed her daily tasks with outward verve, while privately she took the measure of her performance, encouraged a little by her appeal for the children: 'I am some times so low spirited, I think anything *like* pleasure will never revisit me – I go to the nursery – *something like* maternal fondness fills my bosom – the children cluster about me – one little boy who is conscious that he is a favorite, calls himself my son – At the sight of their mother they tremble and run to me for protection – this renders them dear to me – and I discover the kind of happiness I was formed to enjoy.'

Thoughts of Fanny led her to visit Fanny's uncle, aunt and cousins in Tipperary, not far off to the north, across the Galtees, a country town that reminded Mary of Beverley. As a child Fanny had lived there during Mr Blood's first financial setbacks. Fanny's aunt had a look of Mrs Blood, but her husband, Archdeacon Baillie, was coldly correct. Their silent daughter was highly cultivated like Fanny, but in too strained a way. After a few days of disappointing cordialities, Mary returned to the castle.

There, she found herself in a position of trust. Lady Kingsborough's stepmother Mrs FitzGerald wished to consult Mary about her only son, lazy, violent Gerald, who at fifteen was close in age to Margaret. So intractable was Gerald FitzGerald that school was out of the question. He was hanging about the castle during this term. His father had left him a fortune of his own, and his mother was ready to offer £100 a year to a tutor, preferably a married clergyman, who would house the boy and improve his temper. At about this time, Mary received 'a very civil note' from her publisher with a gift of his new volume of Cowper's poems. She now wrote to him to ask, with a view to Gerald, if Mrs Barbauld and her clergyman husband were taking pupils, as planned, in Hampstead, and if the Revd Mr Hewlett was succeeding with his school in Shacklewell. Lady Kingsborough requested Hewlett's new *Introduction to Spelling and Reading* for the children; and Mary added requests of her own for Hewlett's *Sermons*, for Charlotte Smith's *Elegiac Sonnets* and for her own 'little volume' (*Thoughts on the Education of Daughters*) due out early in the New Year. Nothing further is known of Gerald, except that, later, he ran through three wives and distinguished himself as begetter of ten sons.

Trust in Mary's methods was heightened by her skills in the sickroom. Soon after she arrived, an epidemic had swept the castle: all the children fell ill. Their mother's visits, Mary noted, were awkwardly formal: the girls lay silent under their covers while her ladyship babbled to her pets. Then, towards the end of December, Margaret's temperature rose frighteningly. 'My poor little favourite has had a violent fever – and can scarcely bear to have me a moment out of her sight,' Mary wrote to Bess. It must have been a disease of the lungs – pleurisy or pneumonia – for Mary feared Lady K's strenuous commands would drive her daughter 'into a consumption'. Mary was able to soothe the mind of her patient. She had an instinctive understanding of the interdependence of physical and mental wellbeing. Her *Education of Daughters* advocates the study of 'physic', for soothing without knowledge, she believed, could do more harm than good.

There came a moment when Margaret's life was 'dispaired of', and the result of this crisis, when she began to recover, was to 'produce an intimacy' between this governess and her employers which 'years of toil might not have brought about'. Within two months, the enemy who had entered the castle was now, she could not deny, 'a GREAT favourite in this family'.

6

THE TRIALS OF HIGH LIFE

The court case Ned Wollstonecraft had brought against Roebuck went in the Wollstonecrafts' favour. At first Mary rejoiced: her share would enable her to pay off her debts after another half-year. Then, in the course of the early months of 1787, Ned's silence bore it home that he would retain all gains for himself, apart from a small annuity for Everina – not enough to keep her, but an incentive to leave his home. Mary and Bess got nothing whatever. Nor was any explanation forthcoming. Mrs Burgh urged Mary to contact her brother, but she could not bring herself to plead.

'It is so severe a disappointment,' Mary said, 'I endeavor not to think of it.' What she could not face was the prospect of going on indefinitely in her present post. Yet, as always, blows to an unwanted way of life were not wholly unwelcome. They prompted shoots of a different kind to push out in other directions. One is just visible at the close of Mary Wollstonecraft's letter to Joseph Johnson in December 1786: an apparent afterthought about her 'state of dependance'. Its luxury, she tells her publisher, could never outweigh her loss of liberty, and without liberty, she would die.

By the end of January, Mary had been almost immured within the castle walls for three months. Then came orders to join the family in Town for the season: she was to go on ahead with the children. Her thoughts at once turned to the prospect of reunion with her kind. 'Is Neptune in

Dublin?' she asked George Blood. She had not forgotten this Irishman who had made up to her in London.

While she readied herself for the move, Mary warmed to a French novel *Caroline de Lichtfield* that offered an alternative to traditional marriage. It starts out as a Beauty and the Beast story of a teenage heiress whose father marries her to a disfigured favourite of the Prussian King. On the day of her wedding, she writes her new husband a letter, asking to be allowed to return to her surrogate mother until she feels mature enough to enter on the duties of marriage. A sensitive reply from her 'monster' husband, Waldstein, frees her to do so. She learns that he is a man of generous character, and as time goes by he becomes less hideous: his disfigurements heal; he walks more upright. Love stirs when Caroline and Waldstein start living in the same house, though not as man and wife. Slowly, they come to know each other as Caroline shares her artistic skills, while her husband introduces her to studies 'which are too generally neglected in the education of women'. He reads to her, watching as her face 'assumed the passion or imbibed the wisdom of the writer'. This is a man to draw out a woman's character.

For all her independence, Mary Wollstonecraft at twenty-seven had led a life of total chastity. No man, apart from Neptune Blood, had sought her out, partly because she had no dowry and partly because she took a bleak view of marriage. She had long ruled out the terminations of the wedding-bell plot, but a form of love where mutual education begins with marriage took her back to it.

In the first week of February, she and the children settled into Lord Kingsborough's redbrick townhouse at 15 Merrion Square on the south side of the River Liffey. Mary had 'comfortable' apartments and a 'fine' schoolroom, and was invited to use the main drawing-room – the one with the harpsichord and long windows on the first floor. 'Here is no medium!' she had to concede. In addition, she was assigned a parlour for receiving her '*Male* visitors'. But no male came. The Ogles called to welcome her, as did the Bloods' friend Betty Delane whose lively conversation, Mary said, 'diverted – nay, charmed my little Margaret'; but the private parlour stood empty. Neptune Blood was known to be in Dublin. She had sent him word (through George) of her arrival, but the days went by and Neptune made no move. Mrs Blood said she never saw him – despite the hospitality he had enjoyed in London.

It came to Mary's ears that Neptune had enquired after her, yet still he did not call. The weeks passed. Then, one evening early in May, after three months in Town, Mary spied him at the Rotunda, a charitable concert hall at the corner of Sackville and Great Britain Streets. Neptune looked startled to see her in the party of 'a Lady of *quality*' – not the neglected governess, and still less the penniless lodger with the Bloods who'd helped out as seamstress three years before. Here was a handsome young woman who had submitted to the attentions of hairdresser and milliner, and was mingling with fashion in pleasure grounds similar to London's Vauxhall. As Neptune approached with new-found alacrity, Mary stepped back in sudden revulsion: this was the snob who, after all, had left her in London; who in the past three months had not found the time to visit. Face to face, she turned her head and looked past him, flashing a look of 'ineffable contempt' from the corner of her eye. The snub sufficed. When they saw each other again in the Green Room at the playhouse, Neptune did not attempt to speak.

The social strata of Dublin, with a population of only 180,000, struck Mary as simpler than those of a metropolis: 'there are only two ranks of people'. A governess, with her ambiguous status, was bound to shift up or down. Mary's standing with the King family – signalled by their invitation to join them in their drawing-room – drew her, now and then, into high society. Years later, when she was travelling in Scandinavia and could look back on life in Paris as well as in London and Lisbon, she recalled Dublin as the most hospitable city she had known. The strength of the Ascendancy lay partly in its willingness to co-opt brains in the form of lawyers and other professionals, but it did look down on trade. Mary – and her sister Bess too – shared this prejudice. As children they had learnt this one lesson from Mr Wollstonecraft: to scrape off the taint of trade. Mary's letters from Ireland never speak of her mother's kin in the wine trade. There were kin in Cork, not too far from Mitchelstown Castle, but she did not contact them. Her visible connections were with the Church of Ireland – the impeccable Archdeacon Baillie – in Tipperary. In Dublin, she was prickly about her status, and resented her dependence on Lady K whose grand manner reminded her how precarious was her social position. A signal from Lady K could banish her to the underclass.

Mary consoled herself with the company of Betty Delane, whose quickness matched her own. Betty lived with her tubercular eldest sister Susanna and rather snaky brother-in-law, the English painter Robert Home, at 48 Little Britain Street, a few doors from George Blood at 96 (the home of his employer, Brabazon Noble). When the two women talked of Fanny, it was a comfort to share their tears. Mary was pleased, too, by a gift from George Blood, a newly published set of Shakespeare's plays edited by Dr Johnson.

Books alone spoke to her secret self: 'I commune with my own spirit – and am detached from the world – I have plenty of books,' she told Everina. 'I am reading some philosophical lectures, and philosophical sermons – for my own *private* improvement. I lately met with Blair's lectures on genius taste &c &c – and found them an intellectual feast.' Hugh Blair was Professor of Rhetoric at Edinburgh University, and in 1785 had published his *Lectures in Rhetoric* in three volumes. It was to be another hundred years before universities in England opened their doors to women, and several more decades before some could take degrees; for Wollstonecraft, avid for higher education in the 1780s, Blair provided a first-rate substitute. During long winter nights in her room she put herself through his rigorous course in literature, genre and style. 'Genius' is the word echoing in her thoughts in 1787 as she gazed from Blair into the mirror of Rousseau. '*L'exercice des plus sublimes vertus élève et nourrit le génie,*' she noted, and 'a genius will educate itself'. Blair's distinction between genius and taste confirmed her own distinction between the power she sensed in herself and Fanny's refinement. Genius, Blair assured her, was the higher power, a flair that might be improved by art and study, but not attainable by art and study alone. Enlightenment rules of correctness, regularity and accuracy were not enough. 'The rays must converge to a point, in order to glow intensely,' Blair urged.The result will be a simple, rapid, 'torrential' language, close to oratory.

*

The grandparents of the King girls, the Earl and Countess of Kingston, still lived on the other side of the Liffey in the north-east quarter of Dublin at 3 (now 15) Henrietta Street, the home of Robert and Caroline King during the fraught early years of their marriage. Here their daughter Margaret was born and stayed for spells till she was eighteen months. It was a double-fronted house, the first on the left as the family coach turned into a short street, sixteen houses in all, rising up a hill towards the Primate's garden (the site for Kings Inn, now the Law Library). The Kings' house had been built in about 1740 at the height of the street's Georgian grandeur. Earls and other grandees of the Protestant Ascendancy lived there behind tall, black 'pig-iron' railings. Across the road was John Ponsonby, one-time Speaker of the Irish House of Commons, and still influential in political society. The dark doors and the windows painted a creamy Portland stone colour faced one another across the unpaved street. In a reception room on the ground floor of no. 3, Shakespeare and Milton (in papier-mâché mouldings) looked down on the heir and his wife, the grandchildren and their governess, from either side of the ceiling, while the four seasons held their own in each corner. From the top of the house was a spreading view of Dublin: St Patrick's Cathedral, Dublin Castle, the law courts, and the hills beyond.

In the wider society of Dublin, Mary's manners appeared 'gentle, easy, and elegant, her conversation intelligent and amusing, without . . . apparent consciousness of powers above the level of her sex'. At one large gathering a lady singled her out, and entered into a long conversation. Afterwards the lady, curious about this eloquent young woman, enquired about her – and found, to her mortification, that she was only 'Miss King's governess'. That this aristocrat did not detect Mary's status tells us she could pass as part of Lady Kingsborough's circle. It tells us too that Mary's English accent, with a dash of Yorkshire, was close to the received registers of metropolitan or regional pronunciation. Mary later related this story as a joke; at the time, though, she fumed inwardly at women who looked down on one who worked. Ideas of equality took root in this semi-silent rebellion against her subordination, the sap of her superiority rising in a ferment of suppressed rage or depression.

It sometimes happened that Lady Kingsborough ordered Mary to appear in the drawing-room, and Mary would try to excuse herself. She could not bear to 'stalk in to be stared at', and Lady K's '*proud* condescension' added to her discomfort.

'I begged to be excused in a civil way – but she would not allow me to absent myself – I had too, another reason, the expence of hair-dressing, and millinery would have exceeded the sum I chuse to spend in those things. I was determined – .'

Lady K, equally determined, offered Mary a poplin gown and petticoat, which she refused. This made Lady K 'very angry'.

Calm was restored when Mrs FitzGerald came to Mary's rescue. Lady K had to ask Mary's pardon, and consent to her staying, if she chose, in her room.

To some extent she contrived to keep herself warm with small triumphs and self-congratulation. When the company of lords and viscounts was unavoidable, she queried what happiness was for one endowed like herself with 'a greater refinement of mind' and 'a keener edge to the sensibility nature gave me – so that I do not relish the pleasures most people pursue – nor am I disturbed by their trifling cares'. Her proximity to a cliff-edge of contempt made it hard to keep underground. She felt a rising in the throat; she suppressed her 'starts'; and then one Sunday in February, when she was in church with the family, her 'starts' became visible. Something in the service 'hurried' her emotions. They broke through her clamp on public conduct. Lady Kingsborough led her out of the family pew, and directed the coachman to Mary's friend in Little Britain Street. Once in the house, Mary fell into an uncontrollable 'fit of trembling as terrified Betty Delane – and it continued in a lesser degree all day'.

Lady Kingsborough resolved to consult the family physician when he came to examine the lingering effects of Margaret's illness. Mary, who knew her own problem was mental not physical in origin, resisted; she feared too the cost of a consultation with a fashionable doctor. Her ladyship, alert enough to the wants of peasants but unfamiliar with Mary's middle-class cares, would have taken this resistance to be a creditable show of stiff upper lip. In the event, her ladyship had her way, and the physician diagnosed 'a constant nervous fever'. It was necessary, of course, for Mary to restore her clamp in company. There was always a danger that truths would burst out at some unguarded moment, if only through Margaret whose outbursts against her mother were raising the temperature at home. Margaret, by now, had gone over to the radical camp, and was venting her young steam. What could rise only halfway in Mary's throat, Margaret could utter with increasing fearlessness.

Letters alone freed Mary's voice. She corresponded with her sisters, Mrs Burgh, Mrs Prior, James Sowerby, and 'poor Mason' whose situation was 'truly deplorable'. She asks repeatedly for news of Newington Green, Dr Price and 'poor dear Hewlett'. Her misery in the surviving letters is insistent, yet her manner remains dignified: not the assumed dignity of rank, but that of a person who respects the dignity of every creature. She used humour ('I am worried to death by dogs') to deflect fury at the skewed model of a mother nursing her Irish wolfhounds.

She was still confounded to find how sensibility plus 'genius' produced 'misery'. To hold to 'genius' before it had yet been of 'use' often required more courage than she could find. This she confided to the Revd Henry Gabell (having sent him a copy of her book). Misery of this kind can appear self-indulgence, though it's more like the self-doubt that can kill imagination when it sends out roots in the parching soil of incomprehension. Wollstonecraft prayed to 'the Searcher of hearts' to balance her '*peculiar*' wretchedness with something else.

Another confidant was again her publisher. She had met Joseph Johnson before she left for Ireland, when she had delivered the manuscript of her *Education* to his print shop in St Paul's Churchyard. Johnson, whose initial manner towards her had been a little stiff, now proved ready to hear her plaints, though still cautious. He urged her to find her cure in present tasks. Her reply stressed her powerlessness. This letter is calmer than her first, but no less bent on an alternative existence:

> Dublin, April 14 [1787]
>
> Dear Sir,
>
> I am still an invalid – and begin to believe that I ought never to expect to enjoy health. My mind preys on my body – and, when I endeavour

> to be useful, I grow too much interested for my own peace. Confined almost entirely to the society of children, I am anxiously solicitous for their future welfare, and mortified beyond measure, when counteracted in my endeavours to improve them. – I feel all a mother's fears for the swarm of little ones which surround me, and observe disorders, without having power to apply the proper remedies. How can I be reconciled to life, when it is always a painful warfare, and when I am deprived of all the pleasures I relish? – I allude to rational conversation, and domestic affections. Here, alone, a poor solitary individual in a strange land, tied to one spot, and subject to the caprice of another, can I be contented? I am desirous to convince you that I have *some* cause for sorrow – and am not without reason detached from life. I shall hope to hear that you are well, and am yours sincerely
>
> Mary Wollstonecraft

Lady Kingsborough seemed 'more haughty' in Dublin. Her daughters feared her. One 'little girl' – probably the six-year-old Mary – sobbed herself 'sick' when she had to accompany her mother on a week's visit.

On Lady K's birthday, Mary was tugged into 'the mighty important business of preparing wreaths of roses for a birth day dress'. On such occasions 'the whole house from the kitchen maid to the GOVERNESS are obliged to assist, and the children forced to neglect their employments'. The event of the season was the vice-regal ball at Dublin Castle in March. Charles Manners, the Duke of Rutland, had been appointed Lord Lieutenant of Ireland in 1784 at the age of thirty. Pitt, the British Prime Minister, had been initially in favour of parliamentary reform for Ireland, but Rutland backed the Ascendancy, and charmed it further with the magnificence of his entertainments at Dublin Castle. Mary was cast in a servile Cinderella shade. Assisting her ladyship to adorn herself, she had to hear 'fulsome' civilities, a language of 'untruths' stretched out in a train of strong expressions 'without ideas annexed to them'. 'The conversation of this female can't amuse me,' Mary confided to Everina. 'I try to entertain her with the result that I have more of her company.'

Some have marvelled at Lady Kingsborough's graciousness and Mary's ingratitude. Mary did not take kindly to condescension – she was not to be appropriated as another pet in her ladyship's possession. This antagonism blinded her to a vein of benevolence in Lady Kingsborough: what Mary experienced as demands could have been a conscientious effort on the part of her employer to carry out whatever the physician had suggested to cure the 'nervous fever' of the governess. The uncommon interest the Kings, FitzGeralds and Ogles all showed in Mary Wollstonecraft suggests that her character roused their curiosity. This was no ordinary governess, and Lady Kingsborough was no ordinary employer in having the sense to make the most of Mary's merits. After their first interview in her ladyship's bedroom at the castle, there's no sign Mary was held to account for her lack of genteel accomplishments. Where a lesser employer might have carped, this one simply hired an array of extras – masters in singing, dance and modern languages – as soon as her daughters arrived in Dublin. Then, too, Lady Kingsborough would not have taken Mary about had she not been proud of her. Mary's advanced educational ideas served as a foil to the vanity that Mary thought she detected in Lady Kingsborough. Privately, she thanked heaven that she 'was not so unfortunate as to be born a Lady of quality', as she went along to theatres and concerts.

What she did enjoy was Handel: in the spring of 1787 there were Handel concerts at St Werburgh's Church where the Lord Lieutenant had his private pew beside the organ. Swift, the great Dean of St Patrick's Cathedral and author of *Gulliver's Travels*, had been baptised in this church in 1667. Its interior dates from 1759 when it was rebuilt after a fire, with black and white paving extending down the centre aisle and across the front of the chancel. The main Anglican church, St Patrick's, was built outside the city walls; St Werburgh's was built inside, in the old part of Dublin, amongst the tenements south of the Liffey. This did not deter fashion from attending concerts, enhanced by the exceptional acoustics. There, on 2 and 3 May, they heard a group of amateur musicians commemorate the recitals Handel had given in Dublin in 1741–2. Caroline Stuart Dawson, talented granddaughter of Lady Mary Wortley Montagu, sang. Mary admired her voice – and pitied her marriage to Lord Portarlington: a '*nonentity*'.

When it came to theatre, Mary found Dublin inferior to London, with touring English actors in undistinguished plays. She disliked contrived plots and false emotion, preferring the 'almost imperceptible progress of the passions, which Shakespeare has so finely delineated'. In London she had ridiculed Calista, who retires to a cave until 'her Tears have wash'd her stains away'. Mary had heard this 'unmoved', while thrilling 'beyond measure' when Lear says, 'I think this lady/ To be my child Cordelia', and Cordelia answers, 'And so I am, I am.' Cordelia's tormented fidelity may be found in Mary herself, as well as the purity of statement in one who is 'true' in a distorted world.

When a masquerade was announced, Mary at first wished to go; then changed her mind when Lady Kingsborough took this up as a scheme of her own and produced tickets for Mary and Betty. When Mary refused, saying she could not afford a costume, Lady K lent her a black domino: a loose cloak with a half-mask for the upper part of the face. She planned to take Mary and Betty to the houses of several 'people of fashion'. They proved 'a much admired group', Mary reported afterwards to Everina. Betty went as a forsaken shepherdess, Lady Kingsborough wore a domino with a smart cockade, and George Ogle's sister-in-law Miss Moore was in the guise of a female savage recently arrived in high society from one of the newly discovered islands in the North Pacific. As it was taken for granted that this stranger could not speak the language, the black domino was appointed her interpreter. This gave Mary 'an ample field for satire', as she put it: ' – this night the lights[,] the novelty of the scene, and every thing together contributed to make me *more* than half mad – I gave full scope to a satirical vein – .' Words flew, unstifled, from her lips as, masked, she took on the Ascendancy – and the Ascendancy, masked, allowed her to do so, even warmed to it as a kind of theatre. It was impossible not to be aware that she was something of a star in the citadel of the enemy.

That spring, the balance of power shifted somewhat in Mary's favour. Lady K continued to be diverted by her conversation, and became, Mary noticed, 'afraid of me'. Rivalry came into play when George Ogle lent Mary a batch of his 'pretty stanzas'. She thought they had 'really great merit'. On the evening of 23 March she supped with Mrs Ogle and her sister till midnight, and two mornings after, called on them again. That evening, 25 March, the Earl of Kingston came to dine with his son, and expressed a wish to see Miss Wollstonecraft. Lady Kingsborough sent up repeated invitations, which Miss Wollstonecraft declined. She was in low spirits, unfit for the drawing-room. Eventually, Lady Kingsborough came up herself, bringing Mrs Ogle and Miss Moore to add their pleas. Mary had finally to make herself ready.

When the gentlemen joined the ladies after dinner, Ogle at once crossed the room to sit with Mary, and paid her some 'fanciful compliments'. His latest offering was a definition of genius – their favourite subject:

> Genius! 'tis th'etherial Beam, –
> 'Tis sweet Willy Shakespear's dream, –
> 'Tis the muse upon the wing, –
> 'Tis wild Fancy's magic ring, –
> 'Tis the Phrenzy of the mind, –
> 'Tis the eye that ne'er is blind, –
> 'Tis the Prophet's holy fire,
> 'Tis music of the lyre
> 'Tis th'enthusiast's frantic bliss –
> 'Tis anything – alas – but this.

Mary could be sad no longer. Kingsborough came near and caught her eye – a look that made her blush, as it brought back her own night-time reflections on her single state: 'Is it not a sad pity that so sweet a flower should waste its sweetness on the *Dublin* air, or that the Grave should receive its *untouched* charms . . . an Old Maid – "'Tis true a pity and 'tis pity 'tis true" – Alas!!!!!!!!' Kingsborough's look told her that she was young, unattached; and though not pretty in the dressy style of his wife, she stood out with her healthy, unrouged cheeks and hastily piled locks, amongst rows of dressed and powdered heads. And was she not holding forth to a man widely regarded as the epitome of all that was civilised in the Ascendancy, who encouraged her vivacity of mind and displayed his pleasure in her company for all to see?

At this moment Lady Kingsborough signalled Mary to go. Mary ignored the signal, refusing to be forced to appear and then dismissed in so peremptory a way. The reason was not far to seek: Lady K had chosen Ogle for her 'flirt'.

Ogle's renewed attentions to Mary were no longer the uncomplicated cordialities of the castle. Some move he made cast her into a 'painful quietness which arises from reason clouded by disgust'. She saw a 'sensualist' – not the 'purest' man of his time, as his epitaph declares. Ogle had 'serious faults', Mary confided to Everina. Disturbed by this shift yet convinced of his goodness, her eyes followed flights of virtue 'on the brink of vice', observing a comet-like figure shooting off-course into confusion.

One day in May 1787, when Mary was again with the Ogles, her 'starts' broke through her clamp. This was about the time that she had to give up any hope that her brother would part with the portion due to her from the lawsuit. Could Mary's starts have to do with this disappointment? Or her disillusion with Ogle, or needs of her own? To the alarmed Ogles, she ascribed her starts to her mortification on discovering weaknesses in herself like those of people she despised. If she confessed this to the Ogles, it's likely she confided other problems: her brother's unscrupulous greed, the prospect of continued servitude, and her craving a more purposeful life. She hinted to Everina of 'schemes which are only in embryo'. Until she was free of debt, she could take no '*active* step'.

Meanwhile, as Mary went about with Lady Kingsborough, she stored impressions for her novel, shunned her employers when she could and shut the door of her room in order to read. Solitary reading for women was seen as antisocial and selfish, a sign of self-indulgence bordering on the moral dangers of discontent and excess. Letters reveal that Wollstonecraft did suffer as an intellectual (reading Rousseau's *Émile*), surrounded as she was with the emptiness of aristocratic occupations (five hours preparing for balls) and idle conversation ('Lady K's animal passion fills up the hours which are not spent dressing'). Though treated far better than Charlotte Brontë was to be as governess to middle-class manufacturers, they felt a similar contempt and frustration. Both were sensitive to slights and alert to nuances of manners. The difference was that where Brontë longed to be a lady, Wollstonecraft, impervious to class, managed to intimidate Lady Kingsborough, who had the wit to recognise superior endowment in a social inferior.

Wollstonecraft had already formulated her ideas about education before she began reading *Émile* in March 1787. Its proposal to let a child loose on mountain crags had been published to a furore of resistance in 1762. (The book was burnt by public executioners in France and Switzerland, and Rousseau went into exile, partly in England.) In the seventeenth century Locke had defined a child as a blank slate on which education inscribed the formula for the man. Rousseau granted the child the right to discover its own nature. Education should come from within and move outwards towards the social contract. But this did not apply to girls, and Rousseau did not extend Émile's liberties to Sophie, his ideal mate. When he comes to Sophie, Rousseau begins to sound like an old pimp: his Sophie is a coquette formed to please men. Wollstonecraft termed his style libertine, training women to be the devious playthings Rousseau believes them to be. In contrast, Wollstonecraft's authenticity is manifest in a style disdaining the shifty language of insinuation. Her crucial advance over Rousseau is to plant reason in growing girls. It's not an arid exercise; it germinates in a soil fertilised with affection. Where the theoretic Rousseau (who abandoned his own children to institutions) believed it futile to reason with a child, Wollstonecraft taught through reason. A rational girl in the process of discovering her own nature is going to call into question the established model of femininity. This was the basis of the ideological conflict between Mary Wollstonecraft and Lady Kingsborough, trained in the old way and in Mary's eyes a woman whose brains had been irretrievably trivialised. It became Mary's mission to rescue Margaret King from this fate.

'A fine girl,' Mary pronounced with pride. 'I govern her completely.'

This was not entirely true. She could not control Margaret's outbursts against her mother. What Mary was actually saying was that Margaret never turned her temper against her governess, for Margaret was now a disciple. Lurking in the walled enclosure of Anglo-Irish aristocracy, a young volcano was firing up, ready to erupt.

As summer approached, Mary encouraged Lady Kingsborough to go abroad. A consolation for the loss of her inheritance, and for continuing a governess, would be to travel: to contemplate new scenes as a release from the narrowness of the female life. Plans for a Grand Tour with Mary and the children began to go forward. Their first stop, on crossing the Irish Sea, was a spa in Bristol.

At Bristol Hot Wells, from June to August, the gulf between Lady Kingsborough and Mary deepened. After taking unusual pains to draw Mary into the pastimes of high society, Lady K finally lost patience with a governess who cast a cold eye on these efforts. She took to introducing Mary in a manner that put her in her place. Her tone warned society not to be deceived: this dignified and articulate young woman of twenty-eight, who was travelling with the family, should not be mistaken for a friend; this was a dependant who worked for a living. Mary was stung repeatedly, yet she was indeed no friend to the Ascendancy – with the exceptions of Mrs FitzGerald, the Ogles and, above all, her favourite pupil whom she called fondly 'my little Margaret'.

It's at this moment that we catch Mary Wollstonecraft in the act of greatness. She was not a born genius; she became one. Here is someone with ordinary abilities transgressing the limits of ordinariness. Throughout her period as governess, and moving behind the constraints of that position, are hints of enterprise: the sea of ideas at Eton which she can't set in order; her studies at night; and the 'schemes' that she cannot reveal to Everina. In Bristol, the purpose sounds again when she reminds Everina of her identity as an Author, and hints of some 'writing'. To be great, neither innate ability, nor ideas, nor ready words, nor shafts of criticism were enough: there had to be the character to press on. She had the will to rise again when prospects appeared to fade and life felt untenable. She can confess a death wish ('I long to go to sleep – with my friend [Fanny] in the house appointed for all living') at the very time that a new form of life bursts its chrysalis: 'to make any great advance in morality genius is necessary,' she writes, ' – a peculiar kind of genius which is not to be described, and cannot be conceived by those who do not possess it'.

The advancing genius; the fading nerve: we might say the contradiction is the character, or we might see a woman strung out between extremes.

Her letters do vent the extremes, but letters can be misleading, especially those like Wollstonecraft's which appear so confessional. They have suggested a pattern of collapse and failure, but to read them collectively in the context of her actions indicates the reverse: a pattern of renewed purpose. The letters do state once more how '*flat, stale* and *unprofitable*' the things of this world appeared, and her impatience to leave its '*unweeded* garden'. Yet Hamlet's intellectual melancholy, alienation and almost suicidal inability to take action, join with Wollstonecraft's sense of purpose. We see energy interfused with melancholy, and in some way, the one depended on the other. Her sister Bess acted out the melancholy without the counterforce. This is the difference: on Mary's lips, Hamlet's words laid claim to heroic possibilities, even as they proclaimed her powerlessness as a temporary, even necessary, phase. To see this merely as self-dramatisation for its own sake is to lose sight of latent powers that Mary Wollstonecraft had to bring to bear in ways she had not yet determined.

There was a daily balance to her Hamlet role in the steady occupations of teaching and reading. In June, she was studying a very sober book, *The Principles of Moral and Political Philosophy*. Its author William Paley asks how 'adventitious' rights (the rule of one over another in a manmade system) might be distinguished from 'natural rights' (what people who found themselves on a desert island would be entitled to claim, the right to life and liberty, to air, light and water). Though a manmade system may be capricious and absurd, it would be a 'sin' to oppose it, Paley argues, because God wills civil society to exist for the happiness of mankind. Paley takes his reader to the border of revolution, then leads him away, and likewise with issues of gender: 'Nature may have made and left the sexes of the human species nearly equal in their faculties, and perfectly so in their rights'; but, he goes on, to guard against competition with men, which women's equality would produce, Christianity rightly enjoined obedience on the wife in a marriage. Mary's unqualified respect for Paley in June 1787 should not surprise us. She was still faithful to the Anglican Church, and this fits her loyal Cordelia aspect. Yet her prime trait of compassion, the visibility for her of suffering individuals, drew her at the same time to a Dissenter like Dr Price, who brought religion and revolution together.

Christianity preached resignation to the earthly lot, and in 1786–7 Wollstonecraft was repeating this, albeit with difficulty. Since resignation reinforced the status quo, some radicals like Godwin lost their faith during the revolutionary course of the 1780s. The radicalism of Mary Wollstonecraft differed from theirs in that her politics did not lose sight of the soul. In remaining a Christian, she was tugged between the claims of human rights and those of an otherworldly faith that emphasised the virtue of suffering.

Suffering seems to be the only course open to the heroine of Mary Wollstonecraft's first novel, which she was writing in Bristol that June. She draws on herself in a way so close to actual event as to leave us uncertain what is fiction. Title and subtitle, *Mary, A Fiction*, appear a contradiction. An 'Advertisement' owns at the outset that 'the soul of the author is exhibited and animates the hidden springs'. Using the author's own name and those of her parents ('Edward, who married Eliza'), *Mary* starts with parents and childhood, and continues in the measure of biographic record. 'Mary's' mother is untender and sickly. She favours her son and ignores her daughter, who learns to seal off her real life as a thing apart. Wollstonecraft blends her own mother with Lady Kingsborough as an heiress who 'carefully attended to the *shews* of things'. 'Mary's' correspondence with a friend called Ann trains her in taste and correctness, but Ann is indifferent to 'Mary's' attachment because she pines for a lost love. In Lisbon, 'Mary' cares for this friend who is dying of consumption in a sanatorium, and can't articulate her fears because it would be like pronouncing a sentence of death.

The plot of *Mary* is designed to expose the futile plots of women's lives, drawing on lives the author could validate and distilling their pathos to make this point: Fanny's decline as a result of the dowry system, Mrs Blood's squandered patrimony, and the weak position of Bess in a damaging marriage. The novel makes a case against marriage, from which the only escape is death: the closing words tell us that 'Mary' was 'hastening to that world *where there is neither marrying*, nor giving in marriage'. In 1748 the protracted death of Richardson's Clarissa, refusing to paper over her rape with the legality of marriage, had vindicated the independence of a virtuous woman for a generation of readers across Europe and America. A callous father marries 'Mary' to a man who repels her. The one 'Mary' could have loved, Henry – a man who values her intelligence – is a dying, dim figure. No established form of life can answer 'Mary's' need for learning, philosophy and a meaningful existence, expressed in soliloquies that mark the growth of her 'original' mind.

She is, then, a lone phenomenon as she moves – restless – with no institutional habitation. When she communes with the wind 'which struggled to free itself' from whatever 'impeded its course', she 'rejoiced in existence, and darted into futurity'. Hidden in her consciousness are phantom forms of purpose, frail shoots of a buried life – too shaded, too yellow, to burgeon in the light of day, but a possible answer to Rousseau's artificial Sophie.

The only real character in this novel is 'Mary' herself, with the interest of a self-portrait: a young woman who is 'tender and persuasive' in conversation, yet whose subdued energy shows in her quick movements and a flash of contempt from her eyes such as 'few could stand'. 'Mary's' prime trait is her 'uncommon humanity'. Her 'knowledge of physic' leads her to 'prescribe' for an apparently dying woman, and her practical sense tells her that physic is not enough: cleanliness and wholesome food are essential for recovery. More than half a century before Florence Nightingale cleaned up army wards at Scutari, 'Mary' makes the connection between dirt and disease. She understands, too, the links between physical and mental illness: the efficacy of 'the healing balm of sympathy' as 'the medicine of life'.

This young woman does want a sexual tie. There are transient likings for men, 'but they did not amount to love'. Her preference, like that of the author, is for philosophic 'men past the meridian of life'. 'The society of men of genius delighted her, and improved her faculties. With beings of this class she did not often meet; it is a rare genus.' Her own has a rarity to match. The Advertisement for *Mary* spells out its intention 'to develop a character different from those generally portrayed'.

It's a character whose philosophic questionings blend Hamlet with the sermons of Dr Price – 'Le Sage' as Wollstonecraft called him. She herself '*lived*', she says, in Hamlet's 'witching time'. 'I think and think, and these reveries do not tend to fit me for enjoying the *common* pleasures of this world,' she had scribbled in March. Failing biographic plots crumple around 'Mary's' soliloquies, as she ruminates on life's purpose, on eternity, immateriality and happiness, in the long shadow of her friend's death: 'Still does my panting soul push forward, and live in futurity, in the deep shades o'er which darkness hangs. – I try to pierce the gloom, and find a resting-place, where my thirst of knowledge will be gratified, and my ardent affections find an object to fix them.'

Fiction was a strategic choice: 'Without arguing physically about *possibilities* – in a fiction, such a being may be allowed to exist.' It's a being apart from the model of fine ladies, trusting to 'the operations of its own faculties'. And it's drawn from real life: 'the original source'. The real Mary differed from 'Mary' in her resilience. Unlike 'Mary', she was not married, not tied for life, on course for early death. Yellow shoots, transplanted in a more favourable place, could turn green.

'I have been lost in stupidity,' she told herself at Bristol Hot Wells. Impatience brought her again to a psychic verge where she contrived to sit tight, listening to the chat of 'some people of quality'. A letter from Betty Delane reminded her what true quality was: 'a truly elegant friendly epistle'. She missed Betty. 'Lords are not the sort of beings who afford me amusement,' she wrote, ' – nor in the nature of things can they.' Her thoughts turned to Ogle, who had certainly amused her with his verse and cleverness; she had been sorry to see him 'sink into sensuality'. Lonelier in Bristol, she felt less inclined to conceal her superiority – never far from the surface. Her refusal to play the obligatory game of self-deprecation would have jarred an employer. Dismissal, it seemed to her, was in the air, and it could be sudden. Again, rescue came at a critical moment.

In June 1787 Mary had an offer of money from a mystery donor. To spare her feelings, it was offered as a loan but was really 'a present'. Godwin hints that he's keeping the lid on a secret when he tells us he has 'reason to know' that Mary's debts were repaid. For the first time in her life, she felt 'rich', and this meant she could end her servitude as governess. Mary told Godwin that she left when the Kings ruled out a tour of the Continent. Godwin is the closest we get to Mary's voice outside her writings – her intimate, eager tones filtered through his calmly factual narration. He implies that Mary might have stayed on with the King family and complied more readily had they offered her the opportunity of travel; failing this, she wished to leave on the strength of the donation, backed by her novel and plans for further writing.

Who freed Mary? She agreed not to divulge the name of the donor. Speculation has it that it was Lord Kingsborough or Joseph Johnson, but given Mary's indifference to Lord K as a hunter of '*fun*', and the unlikeliness of any mystery about the known generosity of her publisher, another candidate seems to gaze past our heads. He has the 'genius' to discern Mary's promise; he stands with two fingers on his hip in the stance of the perfect gentleman at the height of the Ascendancy; and seats himself beside her when the gentlemen join the ladies after the Earl's dinner in Merrion Square – real scenes predating fictional scenes in *Pride and Prejudice*: a hero of high society distinguishing the only woman minus social credentials in the drawing-room; the annoyance of the hostess that a gentleman can prefer this nobody to her own high-born and fashionable self; followed by his nameless largesse when our heroine runs into trouble. Who then but that admiring – and wealthy – Irish statesman with a need to redeem himself in Mary's eyes? Ogle was, she reflected, 'the only R[igh]t Honourable I was ever pleased with'.

The pretext for her dismissal in August was Margaret's quarrels with her mother, and the girl's unconcealed lament when her governess was away. Margaret's allegiance remained staunch, and after Mary departed, they corresponded in secret.

7

VINDICATION

When Mary was dismissed, she did not go to her sisters. She did not inform Ned Wollstonecraft nor raise her voice to plead for what he owed her. This was not a moment to waste her strength in futile struggle, and experience had taught her to hide her windfall from her brother's legal jaws. She did not return to Newington Green, nor contact friends there including her benefactress, Mrs Burgh. At this point, they appeared in the light of her debts, people who had a right to expect her to stay in a safe post – if not the Kings, then like employers. She could not, as yet, take the risk of returning large sums until she had secured some new income. What she hoped to do had to be tested in secret to see if it would work; the decisiveness of her moves tells of long planning.

An inn in Bristol saw a lone young woman board a coach that trundled slowly away over country ruts along the route to London. Once there, she made her way through the narrow, dark lanes of the City, past the London Coffee House, towards St Paul's Cathedral on Ludgate Hill. On the north side of the churchyard, to the left of the sweep of steps leading to the lofty entrance of St Paul's, she found the door of no. 72, a three-storey building with an alley on one side and a yard on the other. Here, Mary entered the print shop of Joseph Johnson, alongside some forty other publishers who clustered in Georgian houses round the great baroque pile of the church: this place, and adjoining it, Leadenhall Street, Paternoster Row and Ave Maria Lane, had been the centre of the book trade since the seventeenth century, and remained so until the Blitz wrecked almost the whole of the area to the north of St Paul's. An eighteenth-century drawing shows a bustle of business and shoppers, not far from the Fleet Prison, Newgate Prison with its public hangings, and the skewed old justice of the Old Bailey which looked away from extortion, seduction and other gentlemanly misdemeanours while it savaged the minor crimes of subordinates – the poor and women – as though this were God's order.

Mary had been here a year before. This time, she brought Johnson news of *Mary*, as well as the prospect of her children's stories. He received her with astonishing '*tenderness* and humanity', and offered her refuge in his rooms above his shop for as long as she needed to stay. The second and third storeys with walls at odd angles were used for storage as well as living-quarters. This was a bachelor's establishment – none too comfortable, none of the luxuries of Merrion Square or Henrietta Street – but to Mary it brought mental comfort. For it was during the three weeks she stayed here in the late summer of 1787 that her publisher heard her hopes, and assured her that if she worked hard, she could support herself. She was to be a writer and translator, and later a contributor and editorial assistant for Johnson's forthcoming *Analytical Review*. He would prove the best friend she ever had.

Joseph Johnson (Mary called him 'Little Johnson' to distinguish him from Dr Johnson) was the son of a Baptist farmer at Everton, near Liverpool. As the younger of two sons, asthmatic, with a delicate frame and bookish tastes, farming was not for him. He came to London at the age of sixteen in 1754, only months before Dr Johnson defined a patron as 'one who looks with unconcern on a Man struggling for Life in the water and when he has reached ground encumbers him with help'. These famous words were addressed to Lord Chesterfield. 'The notice you have been pleased to take of my labours, had it been early, had been kind; but it has been delayed until I am indifferent and cannot enjoy it; till I am solitary and cannot impart it; till I am known and do not want it.' This letter signalled the demise of patrons like Pope's 'Bufo' – 'fed with soft dedication all day long'. The bookseller's bond with the author became paramount from the later eighteenth century. Toadying to patrons is spoofed by Laurence Sterne in his 1760s novel *Tristram Shandy*, where the narrator's thanks come, belatedly, halfway through the text – outdated flourishes, forced, absurdly over-the-top. Joseph Johnson served his apprenticeship at the very onset of this shift, and by the end of his life in 1809 was known in London as 'the father of the book trade'.

He learnt the trade in an age when the five hundred titles published a year in the 1750s rose to seven hundred and fifty by the 1790s. During the Restoration, reading had been largely confined to a sophisticated circle; during the late Stuart years it had extended to merchants and rich citizens of London; and now, during the Hanoverian decades of the eighteenth century, the reading public was drawing in the aspiring middle class and ladies' maids, avid for advice books on conduct, manners, letter-writing and mental improvement. Dr Johnson observed in 1758 that 'the knowledge of the common people of England is greater than that of any other vulgar'. In the early 1760s the young Joseph had a bookshop in Fish Street Hill (opposite the Monument marking the spot where the Fire of London had broken out a century before). Then he had a partnership in Paternoster Row and helped to publish Joseph Priestley's experiments with electricity, as well as his *Essay on the First Principles of Government* (1768), arguing the state's obligation to provide for the good of the greatest number of citizens. It impressed Jefferson, as future writer of the American Declaration of Independence. In 1770 a fire destroyed Johnson's stock, which was not insured. Friends set him up on his own at 72 St Paul's Churchyard, where he remained for the rest of his life.

In 1777 he brought out an anonymous volume on the *Laws Respecting Women, as they Regard Their Natural Rights*, and in 1780, at the height of the American war, the political essays of his friend Franklin. Many booksellers specialised, like Egerton's Military Library and Taylor's Architectural Library. Joseph Johnson went in for politics and theology (he became the official distributor of Unitarian works), but didn't limit himself when he saw a winner, and his list grew to include medical and surgical books, more science (Humphry Davy and Thomas Malthus), educational books and a lot of literature. He was the first to publish the poems of William Cowper in the early 1780s, and not only did he hearten the poet when initial sales were poor, but volunteered an extra £1000 when recognition came. He was one of the cooperative group who commissioned Dr Johnson's *Lives of the Poets*. In the year he befriended Wollstonecraft, he brought out the first English edition of William Beckford's oriental tale, *Vathek* (1787), which became something of a private cult for same-sex lovers. Later, he was to publish early Wordsworth and Coleridge. He also provided Blake with much-needed work as an illustrator in the early years of his career; though Blake's mysticism divided him from the religious rationalists in Johnson's circle, he shared their vision of a new earth. At a time when large book-shops carried the newspapers, and rivalled coffee-houses as social centres, Johnson's buzzed with an extraordinary range of innovation.

He himself belonged to the radical Club for Constitutional Information, in an age when the House of Commons represented only fifteen thousand voters, a half of one per cent of the adult male population of Britain. What made his politics appealing to Mary Wollstonecraft was what had appealed to her in Dr Price: compassion, sharpened by Dissenters' exclusion from power. When Bess Wollstonecraft eventually met Johnson, she noticed his simplicity of heart and the ray of '*tenderness*' in his dark eyes. He sympathised with slaves, Jews, women, chimney-sweeps, maltreated animals, press-gang victims and those who fell foul of the game laws. When the pamphleteer of the American Revolution, Thomas Paine, was tried (*in absentia*) for high treason, Johnson testified in his defence; and when Paine was arrested for debt, Johnson rescued him. He always maintained the practicability of compassionate conduct.

This publisher succeeded in his business without the sacrifice of quality. He had the gift of discerning greatness – lasting greatness – while still in the making. Happy Mary Wollstonecraft! He proved more a father than her own, as she was quick to acknowledge. 'I never had a father or a brother. You are both to me.' At the age of twenty-nine, she had a new 'plan of life': to live entirely by her pen as few women had done without a supporting career in theatre or the drudgery of low-grade publications. Her first effort was an edge-of-existence fable called 'The Cave of Fancy'. A sage (more aged than her own 'Le Sage', Dr Price) commands the entrance to a subterranean cave of spirits. The spirits come forth to educate an orphan girl, Sagesta, whom the sage has adopted. It doesn't seem to have set her back that this was too high-flown and unfinished.

'I am determined!' she told Johnson. 'Your sex generally laugh at female determinations; but let me tell you, I never resolved to do any thing of consequence, that I did not adhere resolutely to it, till I had accomplished my purpose, improbable as it might have appeared to a more timid mind.'

Independence was Mary's watchword; she felt chained by obligation. So she said to Johnson, recalling Lady Kingsborough's offers of high society with a graciousness that could have kept her in place. The very fact she could say this to Johnson tells us that she felt *no* obligation of this kind to her publisher (which suggests again that her windfall came from someone else). Johnson was less an employer than a mentor, and as she talked to him, Mary adjusted her aims. The pressing question was: should debts, sisters or work take precedence?

Mary did not separate professional from private life. She still wished to be a mother to her sisters, and in September set out to visit them in order to assess their needs. Her first stop was Henley in Oxfordshire where Everina was teaching at Miss Rowden's school. Of the two sisters, Everina had a first claim, as Mary saw it: she had nursed Bess and rescued her from her damaging marriage in 1783–4; it was now Everina's turn. Mary saw her favourite sister wasting her youth and high spirits in an inferior school.

Filled with schemes and possibilities, Mary was buoyant as she tramped along the Thames, listening to the falling leaves or the sound of a watermill, while her mind strayed 'from this *tiny* world to new systems'. When she returned to the school at mealtimes, she tried out her 'real' stories on the pupils. 'They think me *vastly* agreeable,' she told Johnson, and proceeded to write *Original Stories from Real Life* in the course of that autumn.

On her return to London, Johnson saw her onto the coach bound for Leicester. When its black curtains were closed, she found herself in a group of businessmen who bored her with the tricks of their trade. Mary had dismissed all memory of trade for the life of the mind; she was not just averse to business as a way of life, but prophetic, as was Dr Price, on the power of commerce to infect the body politic. At Market Harborough, Bess still languished as a teacher locked into narrow piety. Mary had to break it to her that no help was at hand. There were 'painful emotions', for Mary could not even promise to pay for Bess to join her in the holidays. She tried to hearten this unhappy sister with a hope of finding a situation nearer her, and a year later, did manage to place her in a cosmopolitan school in Putney, first in a paying position as parlour-boarder, then as teacher.

While Mary was away, Johnson found her a small terrace house at 49 George Street (now Dolben Street, with a few remaining Georgian houses) on the Surrey side of the Thames, an area newly accessible to the City with the construction of the first Blackfriars Bridge. Mary moved in at Michaelmas with little more than a bed and table. There she sat in her spectacles – finding them dim when an organ under her window played '*for tenderness formed*' and '*welladay my poor heart*', trying to curb her 'fancy' and 'live to be useful – benevolence must fill every void in my heart'. Affectionate letters from Margaret King filled her 'with all a mother's fondness', and sometimes vexed her with 'childish complaints' – forgetting her own propensity to complaint. She missed the girl's 'innocent caresses' and wondered if Margaret might one day cheer her childless old age. As with Fanny, she told herself it was reasonable for those who shunned marriage 'to love a female'.

In fact, solitude at this stage gave her the privacy she needed in order to write; and to free her further, Johnson engaged a servant, a member of his family from the country. He also introduced Mary to another children's writer, Mrs Sarah Trimmer, whom she visited while preparing her own stories. In November, she handed Johnson the manuscript of *Mary*. His care for her continued to amaze, and she expressed her gratitude with characteristic directness.

'Without your humane and *delicate* assistance, how many obstacles should I not have had to encounter – too often should I have been out of patience with my fellow-creatures, whom I wish to love! – Allow me to love you, my dear sir, and call friend a being I respect.'

'You can *scarcely* conceive how warmly and delicately he has interested himself in my fate,' she repeated to Everina, 'whenever I am tired of solitude I go to Mr Johnson's, and there I me[e]t the kind of company *I* find most pleasure in.' Most afternoons, she walked from her house along the thoroughfare of Great Surrey Street, and across Blackfriar's Bridge to Johnson's print shop. On the way, she passed Albion Mill, the first mill to use rotary power from steam – an advance on water- and windmills of the period – built in 1786 on the Surrey side of the bridge. Here, as Mary came up to the bridge, she would have had a panoramic view of London. To the west lay Westminster Abbey and the squares of Mayfair; downstream was the Billingsgate fish-market, the spires of City churches, the distant Tower; and across the way, her own landmark: the dome of St Paul's floating above crooked lanes and coffee-houses.

From August to November 1787, Mary kept her secret, even from her sisters. So long as the Kings remained in England people assumed she was still with them, merely taking time off to see her family. Mary had promised to pay the debt to Mrs Burgh in instalments of half her annual salary. She meant to continue this scheme, for she sent Mrs Burgh £20 at the time she was setting up house and found it hard to hand that sum over. She feared to reveal her new life until it was an accomplished fact, but her main reason, it appears from her actions, was a resolve to divert what she could spare from her windfall to Everina, and repay her debts in the gradual way she had undertaken. Only in November, when the Kings returned to Ireland without her, did her situation emerge – and even then she urged her sisters to say nothing to Ned or anyone else. Everina was made to promise that if she came to stay over Christmas she would comply with Mary's 'whim', and not mention her 'place of abode or mode of life'. Mary was 'vehement' in her wish not to appear as an author.

'You can conceive how disagreeable pity and advice would be at this juncture – I have too, other cogent reasons,' she confided to Everina when she broke the news. 'I am then going to be the first of a new genus – I tremble at the attempt . . . My undertaking would subject me to ridicule – and an *inundation* of *friendly* advice, to which I cannot listen – I must be independant . . . This project has *long* floated in my mind. You know I was not born to tread in the beaten track – the peculiar bent of my nature pushes me on.'

When Mary Wollstonecraft made this declaration on 7 November 1787, the word 'nature' implies a thing for which there was no word: a generic mutation with a need to 'seed', grow, and live out a new narrative of its own. She was taking up a position as an unchaperoned but respectable young woman living on her own and often the sole woman in a group of intellectuals and radicals. London grew from 500,000 to 800,000 in the course of the eighteenth century, a city large enough for opportunity, but small enough to retain a sense of community. Johnson's circle, including two of Mary's friends from Newington Green, Dr Price and John Hewlett, met for dinner on Sundays or Tuesdays in his crooked upstairs room.

'I often visit his hospitable mansion,' Mary said, 'where I meet some sensible men, at any rate my worthy friend – who bears with my infirmities.' In fact, she was in better health, she said, 'than I have enjoyed for some years'.

This doesn't mean that she ceased to mourn for Fanny, but grief no longer undermined her, or only now and then, as when she heard of fortune smiling on Skeys. The thought of him sickened her and brought back 'past tumultuous scenes of woe – thinking of that dear friend – whom I shall love while memory holds its seat'.* She went to hear a sermon Hewlett wrote especially for her, 'on the recognition of our friends in a future state; the subject was affecting, and rendered more so by his tremulous voice'.

Once again, she pitied Hewlett his 'domestic vexations', and now saw Sowerby too go under, with her infallible Austen eye for the tendency of able men to mate with inferiors: 'Mrs S[owerby] is sunk into a mere nurse, her little stock of beauty not vivified by a soul, is flown – flown with the childish vavicity [*sic*] which animated her youthful face, and inspired her animal gambols – the sporting of lambs and kittens is pleasing when it does not occur too of[ten]. She is entirely the mother – I mean a fo[oli]sh one – and the child of course is not properly managed.' Domestic life in Newington Green was gradually receding from the City perspective of a professional woman in a man's world. Mrs Burgh, duly informed of this venture in November 1787, was not invited to comment. She died in 1788.

**Hamlet*, I, v, 96-7: '...while memory holds a seat/ In this distracted globe'.

From November 1787 to December 1792 Wollstonecraft flourished with Johnson's support. This period in her life was relatively free of the resistance to women who pushed beyond the bounds of their sex – ridicule of the kind that Pope had unleashed on Anne Finch when her poems were published in 1713 and again on Lady Mary Wortley Montagu in *The Dunciad* in 1728. There was a common assumption that a woman who published her writings violated her modesty. Anonymity was a cover, and Mary Wollstonecraft used it before her books were established; and further cover could be provided by menfolk, crucial to Jane Austen as she ventured into the arena of publication. Mary Wollstonecraft was fortified in a similar and in some ways more effective manner by closeness to her publisher and the clubbability of his circle. It was a form of protection that looks forward to Virginia Woolf writing from the citadel of the Bloomsbury set and fortified by a press of her own in the first half of the twentieth century.

Johnson's milieu included the German-Swiss artist of nightmares, Henry Fuseli; Cowper, whose poetry pitied those who are beyond effectual aid; the mathematician John Bonnycastle, long-headed like a horse dipping into a nosebag; and a doctor from St Thomas's Hospital and author of *Elements of the Practice of Physic* called George Fordyce, who had made two lasting observations – that the body's temperature is constant whatever the external conditions, and that digestion is a chemical not mechanical process. Then there was Thomas Paine, whose return to England from America in 1787 coincided with Mary's arrival in the group. Paine, like Fuseli, had an international reputation. He was an Englishman of Quaker family who had settled in Philadelphia in 1774 as editor of the *Philadelphia Magazine*, which proposed international arbitration, and more rational laws for marriage and divorce. Paine was the only revolutionary leader in America to promote women's rights in a sustained way. *An Occasional Letter on the Female Sex* (1775) broaches this issue: 'Man with regard to [woman], in all climates, and in all ages has been either an insensible husband or an oppressor.' Five weeks after Paine attacked slavery, the first American anti-slavery society was founded in Philadelphia. After the British slaughter of Americans at Lexington, Paine had been the first to preach independence and republicanism, and when fortunes were low during the Revolutionary War, his words had heartened the new nation. He had written by night, after long marches, by the light of a campfire: 'These are times that try men's souls . . . Tyranny, like Hell, is not easily conquered . . .' During the Revolutionary War, he had served on Washington's staff and acted as Foreign Secretary to the insurgent Congress.

One other celebrity in Johnson's circle was the large, expansive naturalist Erasmus Darwin (grandfather of Charles Darwin). His *Loves of the Plants* (1789) saw propagation as part of a natural, not divine, plan. 'Nature' opposed 'culture' as well as the Church's emphasis on the soul above the body. Dissenters did not demean the body, and Wollstonecraft concurred; she had opinions on nutrition, cleanliness, our susceptibility to 'tenderness' in sustaining mental health, malpractice in childbirth, the advantage of breast-feeding, and the increasing awareness of an infant's need for intimate, hands-on parental care.

What did Wollstonecraft do when dinner was over? Johnson's dinners were at five in the afternoon. Did she linger after dark? Did any of the company see her home in a coach? It's unlikely: Mary did not invite gallantry. From the autumn of 1787 till that of 1791 she lived frugally, with no furniture to call her own. She could not afford new clothes, nor have her hair dressed in the puffed and powdered disarray of the day. Fuseli called her undressed hair 'lank'; she was 'a philosophical sloven'. This is the image of the unfeminine intellectual, dear to misogynist tradition. (Nearly a century later Henry Adams would warn his intellectual wife that a woman who reads Greek must dress well.) None of Wollstonecraft's portraits bears out Fuseli's slur; her hair is soft and curly, and she appears meticulously clean and neat (in accord with her belief in cleanliness as 'an excellent preservative of health' in an age when bathing was a rare event, fleas infested even the rich, and vermin nested in the built-up structures of their hair). One of the two or three women in Johnson's circle, the novelist Mary Hays, confirms the 'charm' of her looks and manners: 'her form full; her hair and eyes brown; her features pleasing; her countenance changing and impressive; her voice soft, and, though without great compass, capable of modulation'. Since Mary could not have looked less a 'philosophical sloven', the function of the slur was to undermine a woman who enters the male citadel. Rare women who succeeded in past centuries usually had fathers who empowered them. Elizabeth I is the most obvious example; and she had the additional advantage of the foremost Cambridge tutors of the day. Thomas More created a school in his house for his daughters, whose brilliance made a stir in the sixteenth century. Wollstonecraft presents the rare spectacle of a woman undertaking this on her own through her commitment to self-education and her extraordinary self-possession.

Mary Wollstonecraft may best be seen as she saw herself, a 'solitary walker', unafraid of footpads in the shadows of St Paul's or lurking under the girders of Blackfriar's Bridge. Oil-lamps appeared every two hundred yards or so, but between lay a considerable stretch of darkness – settling on the City from late afternoon in winter. She would have heard, behind her, the faint rattle of a coach turning a corner, the clop of horses, and the night watch cry the hour. So this independent young woman wrapped in her cloak slips into darkness, taking her way to the next lamp, her step on the bridge fading in the direction of George Street.

Within three months, she rescued Everina. At Christmas, Everina left Miss Rowden's school, and in February 1788 Mary sent her to Paris. Fuseli helped to find lodgings with a Mlle Henry in the rue de Tournon, Faubourg Saint-Germain. Initially, the stay was to be for six months, but in the end Mary supported Everina abroad for two years, trusting a thorough training in Parisian French would secure her sister a superior post.

Johnson later testified that over the next few years Mary spent at least £200 on her family. In May 1788 she announced to George Blood that she would gain £200 that year – this, only six months into her new 'plan of life'. It can't be proved, but £200 may have been close to the 'present' Mary had received in June the previous year. Her early books were not amongst the big sellers of the day: only *Real Life* went into a second edition, and that after three years. If her first advance is a fair indication of what she might expect to earn, it's simply not possible that she could have earned much more than £100 for the two books she published and the three she began, together with a few reviews, in the course of 1788. Her prediction of £200 (set against the £40 she had earned as a governess) was more by way of vindicating what would have seemed to others a rash venture. If the sum of £200 were already in hand, its existence would have had to be secret (except from Johnson), or people with claims on Mary – creditors and the improvident members of her family – would have snapped it up at once. All the same, she did manage to support each member of her family as their needs arose. What she didn't do was destroy herself by trying to fund them all at once. Mary did not invite her sisters to live with her, since this would have meant taking on two dependants whose presence interfered with work, as she learnt that Christmas. Johnson would have set boundaries for her survival as a writer, and when she exceeded them out of pity for one or other member of her family, he did advance small sums – as, later, when she had to fit Everina out for a post in Ireland.

As the months and years passed, Mary did manage to support herself by writing, and all her publications succeeded. *Original Stories from Real Life* was published (anonymously) in 1788. Here, we recall, a governess called Mrs Mason introduces two girls to the realities of poverty, hunger and child mortality. Their best education lies not in accomplishments with a view to the marriage market – in practice a property market – but in responsible fellow-feeling for the obscure, the rude, the weak and misused. Blake illustrated the second edition in 1791. His woodcuts break with earlier eighteenth-century styles, the rococo ornamentation associated with aristocratic frivolity in Watteau or the property-portraiture of houses, estates, horses, dogs, guns, wives and heirs. Blake's illustrations for *Real Life* have the purity of his poem 'Visions of the Daughters of Albion', a protest against the existing order that 'inclos'd [a woman's] infinite brain into a narrow circle'. In perfect harmony with Mary Wollstonecraft, he shifts inward and lights up states of being: pity, grief, devotion. His illustrations are wordless sermons. Mrs Mason appears in white, a stand-in for God; her arms are stretched out to form a cross, but without the usual iconography of pain: this model is not the martyr inducing voyeuristic emotions of horror. She has the purer, altruistic appeal of a teacher who wants to be of 'use'. Her image is tender; her arms reach out and bless her pupils – 'Mary' and

'Caroline' on either side – who fold their arms across their breasts like converts in prayer.

Johnson's tactic was to publish first the irreproachable stories (praised for 'solid piety and virtue'), before he brought out (also, anonymously) the radical *Mary* soon after, in the spring of 1788. Meanwhile, Mary herself, more a survivor than her fictional namesake, turned to yet another genre: translation.

'I really want a German grammar, as I intend to attempt to learn that language,' she wrote to Johnson in the autumn of 1787. '. . . I ought to store my mind with knowledge – the seed-time is passing away. I see the necessity of labouring now – and of that necessity I do not complain; on the contrary, I am thankful that I have more than common incentives to pursue knowledge, and draw my pleasures from the employments that are within my reach.'

In practice, she found it harder than she had expected to mop up enough Italian to cope with a translation of a difficult hand which Johnson offered first. 'I cannot bear to do any thing I cannot do well,' she told him, ' – and I should lose time in the vain attempt.' By March 1788 she was 'deeply immersed' in the study of French with a view to translating Jacques Necker's *De l'importance des opinions religieuses.* Once she made a start, Johnson trained her to prefer cohesion to literal translation.

'My dear sir,' she wrote, 'I send you a chapter which I am pleased with, now I see it in one point of view – and as I have made free with the author, I hope you will not have often to say – what does this mean?'

Soon, she said, initial difficulties began 'imperc[ep]tibly [to] melt away as I encounter them – and I daily earn more money with less trouble'. She thought of translation as 'study', and to be paid to study as she breathed the fragrant gale of spring, made her 'excellently well'. Johnson published *On the Importance of Religious Opinions* at the end of 1788. The Advertisement owns to 'some Liberties . . . taken by the Translator, which seemed necessary to preserve the Spirit of the Original'.

By then she had secured a good contract to translate a 1782 collection of German tales for children, *Moralisches Elementarbuch* by Christian Gotthilf Salzmann. Wollstonecraft felt an affinity for this author. As in *Real Life*, morality is taught not as precept but through experience, as when a child with toothache is unsure whether he can accept the offer of a cure from a Jew. His parents, who are absent, have taught him that Jews are not to be trusted. The child in the end allows the Jew to fill the tooth, admitting his fear and prejudice. The Jew gives a rational answer: not all Jews are good, he says, but 'in every religion there are good people'. Wollstonecraft adapted the text to suit English children, and given the current dispossession and extermination of native Americans as the frontier pushed westwards, she thought it necessary to introduce an episode 'to lead children to consider the Indians as their brothers'.

When she refers to 'learning' German in 1789 she was at work on this translation, while keeping her sights on a larger aim: an encounter with a culture where, she heard, 'the people have still that simplicity of manners, I dote fondly on'. In the winter of 1790–1 Johnson published this work as *Elements of Morality* in three volumes, followed by a second edition with fifty plates, some by Blake. This translation took longer – not surprisingly, since German was new to her, and it's therefore all the more remarkable that Salzmann was so pleased that, later, he translated her *Rights of Woman* and, still later, Godwin's memoir of Mary Wollstonecraft.

During these healthy, hard-working years, she undertook abridgements: Maria Geertruida de Cambon's *Young Grandison*, a series of letters characterising the model child, translated from the Dutch, which Wollstonecraft 'almost rewrote', and Johann-Caspar Lavater's unconvincing treatise on the pseudo-science of physiognomy, the analysis of character from facial features. At some point Wollstonecraft, not one to hide her doubts, lost this commission to a rival translator aided by Lavater's friend Fuseli. Yet another project, begun in mid-1788, was to compile selections of prose and verse for *The Female Reader*, including a few extracts from her own *Education* and *Real Life*, Cowper's abolitionist poem 'On Slavery', and Mrs Barbauld speaking as one woman to another: 'Negro woman, who sittest pining in captivity, and weepest over thy sick child: though no one seeth thee, God seeth thee; though no one pitieth thee, God pitieth thee: raise thy voice, folorn and abandoned one . . .'

There were as yet no schoolbooks for girls apart from conduct books.

Most girls of the time were protected, even deflected, from serious books. Johnson thought Wollstonecraft introduced 'some original pieces' – unusual enough for him to take the precaution of publishing this anthology under the name of a man, an elocution teacher called Mr Cresswick. Elocution – the classroom method of recitation – was not a practice Wollstonecraft would have encouraged, given her emphasis on inward growth rather than performance. Her aim was innovative: the first anthology for and about women, and in part, too, *by* women, with a view to a high-flying education. The editorial approach to women's writing is participatory, not exclusive, and offers a longer-term model than women-only collections (useful as they have been for purposes of retrieval and reconstruction of gender since the 1970s). Wollstonecraft's educative aim is more ambitious. Towards the end of the first century when numbers of women took up the pen, and just before the emergence of Jane Austen (fourteen at the time, reading at home, and the perfect recipient for this book), Wollstonecraft assumed that women (Mme de Genlis, Mrs Trimmer, Mrs Barbauld, Mrs Chapone, Lady Pennington, Miss Carter, Charlotte Smith and herself*) could start to hold their own beside the Bible, Shakespeare, Richardson, Dr Johnson and Cowper. Those heights are everywhere present, a stimulant, not a threat, like an extract from Edward Young about 'Dying Friends' as mentors of inward revolution:

> And shall they languish, shall they die in vain?
> Ungrateful, shall we grieve their hov'ring shades,
> Which wait the revolution in our hearts?
> Shall we disdain their silent, soft address . . .

Johnson not only provided work and published everything Wollstonecraft offered him, he was able to pluck out her thorns, as he

---

*Surnames only are used for male writers, while females are identified by the prefix 'Miss' or 'Mrs' or 'Lady——'. Wollstonecraft resists this identification for herself: she gives her surname alone in the list of contents, and 'M. Wollstonecraft' elsewhere for extracts from her *Education*. Her extracts from *Real Life* remain anonymous. She does not draw on *Mary*, and I've wondered if Johnson ruled it out as too subversive for schoolgirls.

proudly recalled: 'During her stay in George Street she spent many of her afternoons & most of her evenings with me. She was incapable of disguise. Whatever was the state of her mind it appeared when she entered, & the tone of conversation might easily be guessed; when harassed, which was very often ye case, she was relieved by unbosoming herself & generally returned home calm, frequently in spirits. [Fuseli] was frequently with us.'

She looked on Johnson as 'the only person I am *intimate* with', to whom she owed an apology for her moods: 'I have often been very petulant. – I have been thinking of those instances of ill-humour and quickness, and they appeared like crimes.' Five years on, she still blushed to think 'how often I . . . teazed you with childish complaints, and the reverses of a disordered imagination'.

Johnson had a Unitarian friend, Thomas Christie, a doctor from Montrose in Scotland, who gave up medicine for literature in 1787 after travelling on horseback around the country for six months meeting writers like Erasmus Darwin in Derby, Anna Seward in Lichfield and Priestley in Birmingham. At the time Mary joined the circle, Christie persuaded Johnson to bring out the *Analytical Review*. It began in May 1788, promoting religious toleration and the extension of the vote. For the first issue, Wollstonecraft chose *A Sermon Written by the late Samuel Johnson, LL.D., for the Funeral of His Wife* from a selection of 'trash', taking a stand against her publisher who would have opposed so obdurate a pensioner of Pitt's government. For Wollstonecraft, the soul took precedence over politics.

'I seemed (suddenly) to *find* my *soul* again – It had been for some time I cannot tell where,' she explained. 'I felt some pleasure in paying a just tribute of respect to the memory of a man – who, in spite of his faults, I have an affection for – I say *have*, for I believe he is somewhere – *where* my soul has been gadding perhaps . . .'

Wollstonecraft was the first woman to take up short-notice professional reviewing as a substantial part of her income. (Once, anonymously, she reviewed her own translation of Necker – favourably, it need hardly be said.) Though in the main she followed the magazine's policy of objective summary together with extracts, she gradually introduced evaluation, and scorned the *Analytical*'s tame rival the *Critical* for its *un*critical oozings over established fame.

However provocative her content, the fact is that up to the end of the 1780s Wollstonecraft kept within the bounds of what were thought of as lesser genres, open to women (conduct books, children's books, fiction, translation). But in 1790, backed still by her publisher, she entered the male preserve of politics.

The occasion was the outbreak of the French Revolution. With the storming of the Bastille on 14 July 1789, Cowper believed that he spoke for the nation at large in his 'Address to the Bastille': 'Ye Dungeons, and ye Cages of Despair! – / There's not an English heart that would not leap, / To hear that ye were fall'n at last . . .' It seemed the start of a new age of rights for all. Jefferson's law of religious toleration passed for the state of Virginia in 1786 had encouraged Dissenters to campaign for a repeal of laws excluding them from public office (the Corporation Act of 1661 and the Test Act of 1673). The first reading of the bill in the British Parliament won a good deal of support, though not the majority needed to carry it; then, support had dropped with the second and third readings. The vogue for reform that had touched the ruling elite in England in the early 1780s had given way to renewed conservatism, reinforced by the impact of what had been lost with the colonies in America and also by fear of advancing notions of human rights in America's ally, France. Though Dr Price reminded Parliament of Dissenters' loyalty during the Jacobite uprising in 1745, this now seemed distant, and recent memory blamed Dissenters for disloyalty during the late war in America. British rulers were unnerved by events in France: the overthrow of the *ancien régime* (the nobility and higher clergy) by the mass of the people, represented by middle-class thinkers in a reforming National Assembly.

The most unnerving event was the women's march on Versailles on 5–6 October 1789: a starving mob from the markets of Paris who advanced on the King and Queen, demanding bread. There were reports of rabble invading the apartments of a hated Queen cut off in inaccessible opulence from 'real life'; of Louis XVI, Marie Antoinette and their children forced to move to Paris; of the danger to the royal family (averted by the King's promise to side with the Revolution). All this alarmed the British landed classes in their newly enclosed estates. There had been sixty-four Enclosure Acts between 1740 and 1749, and 472 between 1770 and 1779, with the effect of driving the poor off what had been for centuries common land, and bringing an immense increase in riches to owners of estates, who controlled or were themselves the Members of Parliament who initiated and carried through these legalised appropriations. New boundaries – ditches, hedges and walls (including the new wall around Mitchelstown Castle) – were a visual reminder of the power of the landlord to exclude outsiders from territory over which he now exercised sole rights. Apart from changing the landscape for ever, the result was to wall off the landed from the labouring classes. So, the excluded were feared, and a counter-revolution settled on Britain, postponing urgent electoral reform for another forty years.

It was in this divisive political context that Richard Price preached his sermon, *On the Love of Our Country*, at the Meeting House in the Old Jewry in London on 4 November 1789. Six editions appeared in three weeks. Officially, the sermon, invited by the Revolution Society, was a centenary tribute to the Glorious Revolution of 1688 in which Dissenters had played a vital part. Dr Price's motives were twofold: given the failure of Dissenters' pleas for rights in Parliament (defeated by 122 votes to 102 in the spring of that year), he was reminding the powers that be of the peaceful liberties the country owed to them. At the same time he was celebrating the American and French Revolutions, and the right of the people to choose and 'cashier' their rulers. Price went on to prophesy further revolutions to complete the blood-free dissent of 1688: 'Behold kingdoms . . . starting from sleep, breaking their fetters, and claiming justice from their oppressors! Behold, the light . . . after setting America free, reflected to France and there kindled to a blaze that lays despotism in ashes . . .'

An opponent of Price, Edmund Burke, criticised the language of inspiration co-opted for the political sphere, and talked sarcastically of the spread of political preachers as 'an addition of nondescripts to the ample collection of known classes, genera and species, which at present beautify the *hortus siccus* [the dry garden] of dissent'. Burke also claimed Price was sanctioning the mob rule of the women's march on Versailles. Price insisted that he had referred, rather, to the fall of the Bastille. On its first anniversary he had toasted a United States of the World.

Burke, at sixty, had long been the greatest orator in the Commons, never quite accepted by the grandees of his Whig party who looked on him as an Irish adventurer. He had risen out of the professional class into the landed gentry, and always represented the property interests of the gentry together with its blend of classical humanism and chivalry. His chivalric focus on Marie Antoinette ruffled Wollstonecraft and Paine, who remarked famously that in mourning the plumage, Burke forgot the dying bird. Burke, who had loathed Price's late patron Lord Shelburne, condemned Dr Price with rhetoric that would have an impact on the course of history. His *Reflections on the Revolution in France*, published on 1 November 1790, slates the international, apocalyptic narrative of Dr Price in favour of the English historical narrative which is seen to be an unbroken contract between generations across time. The so-called 'Glorious Revolution', Burke argues, was merely a glitch in the continuum of hereditary monarchy. Burke would never see the future possibility of what Wollstonecraft calls a 'new genus', whether it be a person or a nation, because, for Burke, identity is cumulative; it comes from the past. 'Upon that . . . stock of inheritance we have taken care not to inoculate any cyon alien to the nature of the original plant.' Above all, Burke warns of mob rule and the breakdown of civilisation. Nineteen thousand bought his treatise in 1790–1; thirty thousand by 1793. The fear disseminated by this scare encouraged a counter-revolutionary coalition, and its combined might, in turn, fomented wartime terror in France and the climate of suspicion that led to the fall of the moderate Girondin party. Power passed to the extremist Jacobin party, led by Marat, Danton and Robespierre who had no compunction at shedding blood – and so, the guillotine went to work.

When the Terror struck, Burke seemed vindicated. Yet during the earlier phase of the Revolution, from 1789 until a massacre in September 1792, its outcome looked promising to the circles of Dr Price and Joseph Johnson, as to most Britons. The streets of Paris were tranquil, according to Thomas Christie who, armed with introductions from Dr Price, spent the first six months of 1790 in Paris analysing the new order as it was promulgated by the National Assembly. His *Letters on the Revolution of France*, published by Johnson in 1791, used facts to counter Burke, but nothing could budge the alarm once it took hold. Burke split the Whigs when he crossed the floor of the Commons to join Pitt's side, and the Whig leader, Fox, wept to see it. Scores of writers set out to answer Burke, the most famous being Tom Paine in 1791–2, who links monarchy with war-making, conflicts that 'unmake men' – undo their natural benevolence: 'Man will not be brought up with the savage idea of considering his species his enemy.' But ahead of Paine, in fact the first protest, within four weeks, came from Mary Wollstonecraft at the end of November 1790.

While writing *A Vindication of the Rights of Men*, Wollstonecraft stopped in the middle. She went to Joseph Johnson, who had been printing her work page by page as she wrote it, and told him she could not do it. Johnson's response was perfectly judged: he reassured her that he would be willing to ditch the printed sheets. Eased and piqued, Wollstonecraft resolved to press on.

The pulpit voice of Dr Price had taught her that truths 'coming warm from the heart . . . find the direct road to it'. Her first target was the untruth of what we now call 'spin': social wrongs wrapped in Burke's 'flowers of rhetoric': 'Words are heaped on words, till the understanding is confused.' Wollstonecraft identifies the dodges of rhetoric as the most dangerous enemy of human rights. Her answer is the dissenting voice of truth, above the puerile tit-for-tat of party debates that 'narrow the understanding and contract the heart'. This is no hasty pamphlet. Wollstonecraft is taking on the whole edifice of power. Her adversarial heat appears to come off the pulse, but it's under control, pruning the millenarian dreams of Dr Price in favour of facts, logic, exposure.

A Vindication of the Rights of Men exposes a politician as a vain, verbose, self-interested climber. In the form of a letter, it confronts Burke with his secret motives. 'You have raised yourself by the exertion of abilities, and thrown the automatons of rank into the back ground,' she concedes. 'But, unfortunately, you have lately lost a great part of your popularity: members [of Parliament] were tired of listening to declamation or had not sufficient taste to be amused when you ingeniously wandered from the question . . . You were the Cicero of one side of the house for years; and then to sink into oblivion . . . was enough to . . . make you produce the impassioned *Reflections*.' Another of Burke's motives was to redeem his loss of face over his treatment of George III during the King's bout of madness in 1788–9. The Prince of Wales, manoeuvring to be king, had gained Burke's support with the bribe of the post of Paymaster-General. When Pitt introduced a bill to limit the Prince's powers, Burke had hurriedly collected statistics from mental institutions suggesting that, at the age of fifty-five, the King was unlikely to recover. Burke tried to convince the Commons that the hand of God was 'hurling' George III from the throne. When the King recovered, Burke's self-interest did not go unnoticed in the press, and Wollstonecraft contrasts his tears for the French King with 'unfeeling disrespect and indecent haste' in trying to oust his own. 'You were so eager to taste the sweets of power, that you could not wait till time had determined whether a dreadful delirium would settle into a confirmed madness; but, prying into the secrets of Omnipotence, you thundered out that God had *hurled him from his throne* . . . And who was the monster whom Heaven had thus awfully deposed, and smitten with such an angry blow? Surely as harmless a character as Lewis XVIth . . .'

Wollstonecraft's aim was not to weigh Burke's conduct, she tells him – 'it is only some of your pernicious opinions that I wish to hunt out of their lurking holes; and to show you to yourself stripped of the gorgeous drapery [of rhetoric]'.

This is not recognisably a woman's voice, nor is it gentlemanly. Yet it's not a vulgar voice. In fact, it can't be placed according to the traditional registers of class or gender. It takes its eminence rather from the dignity of Reason holding up a mirror to irrational sentiment. Burke has no tears, she notes, for men taken by violence to fight wars, nor for people who can hang for stealing £5; his tears are reserved for 'the downfall of queens'. She cites his demand that the poor 'must respect that property of which they cannot partake', and look for justice in the afterlife. 'This is contemptible hard-hearted sophistry, in the specious form of . . . submission to the will of Heaven,' Wollstonecraft retorts. 'It is, Sir, *possible* to render the poor happier in this world, without depriving them of the consolation you gratuitously grant them in the next.'

Burke opens the way to an attack on laws that put property before morality. Should ecclesiastical revenues, extorted in times long past, continue in the hands of the clergy 'merely to preserve the sacred majesty of Property inviolate . . .'? To politicians, 'an abolition of the infernal slave trade would not only be unsound policy, but a flagrant infringement of the laws (which are allowed to have been infamous) that induced the planters to purchase their estates'. Slavery on these estates 'outrages every suggestion of reason and religion' and is a 'stigma on our nature'.

The tide of her eloquence rises as she comes to the current issue of the slave trade. 'Is it not consonant with justice, with the common principles of humanity, not to mention Christianity, to abolish this abominable traffic.' While Burke mourns the pageant of the French aristocracy, 'the lash resounds on the slaves' naked sides . . . Such misery demands more than tears.' Nowhere is her eloquence more affecting than at this point where, dramatically, words appear to fail, and the reader stops short at two lines of silent dashes.

A Vindication of the Rights of Men was published anonymously at the low price of 1s 6d, half the cost of Burke's pamphlet. All the best journals of the day discussed it. The author, aged thirty-one, sent a copy to the sixty-year-old historian Catharine Macaulay, who said how it pleased her that the *Rights of Men* 'should have been written by a woman and thus to see my opinion of the powers and talents of the Sex in your person so early verified'. Wollstonecraft also sent a copy to Dr Price in Hackney, and though sick and near to death, he replied on 17 December that he had 'not been surprized to find that a composition which he has heard ascribed to some of our ablest writers, appears to come from Miss Wolstonecraft. He is particularly happy in having such an advocate; and he requests her acceptance of his gratitude for the kind and handsome manner in which she has mentioned him.'

A newcomer, William Godwin, appeared at Johnson's table on 13 November 1791. He looked pale, with thin, pale hair, a long, straight nose with an enquiringly tilted tip, and a pursed mouth – its slightly jutting lower lip belied the paleness of his presence. Talk turned to Voltaire and religion. Wollstonecraft would not have agreed with Godwin's atheism, and a quarrel developed that spoilt the dinner. She was at this time writing a larger, more daring book than her *Vindication of the Rights of Men* called *A Vindication of the Rights of Woman.* Godwin had little interest in women's emergence, and had come, that evening, to meet Paine who was due to leave for France.

'I had little curiosity to see Mrs Wollstonecraft',* Godwin said, 'and a very great curiosity to see Thomas Paine.' Paine turned out to be rather quiet; Mary, not. 'I . . . heard her, very frequently when I wished to hear Paine,' Godwin repeated.

He had glanced at the first *Vindication*, 'displeased as literary men are apt to be, with a few offenses against grammar and other minute points of composition' – the common response of a thinker who sights a woman of no importance advancing on his territory. (Virginia Woolf spoofed such rebuffs when she pictures the dismay of an educated man to find the housemaid browsing in his library and Cook composing a Mass in B Minor.) Godwin was three years older than Mary, a bachelor aged thirty-five, who earned his living by the skill of his pen. He was hardworking, prolific, and soon to formulate ideas of political justice that were to make him the leading radical thinker for his own and the next generation. The Romantics didn't have to be revolutionaries (as Byron, for instance, was not) to respect the stand Godwin took against the injustice of power. His arguments were rational to the point of coldness – not a man to cultivate disciples, though he had them. Godwin prided himself that though he disagreed with 'Mrs Wollstonecraft', he did grant her independence of mind. It annoyed him to find this act of justice was not reciprocated. 'Mrs Wollstonecraft', he heard, made it known that she disliked him. He would have condoned disagreement, but dislike was irrational. When they met two or three times in the course of 1792 they agreed no better.

The second *Vindication* (published in January 1792 with the author's

---

*'Mrs' was an honorary title, assumed when a woman passed marrying age or, as in this case, had made up her mind not to marry.

name) asks legislators to turn their attention to women, demanding 'JUSTICE for one-half of the human race'. The French constitution of 1791 did away with aristocratic privilege in the name of the Rights of Man, but denied women equal citizenship. Following a *Declaration of the Rights of Woman* by French feminist Olympe de Gouges in 1791, Wollstonecraft's *Vindication of the Rights of Woman* deplores the relegation of her sex to a state of 'ignorance and slavish dependence', encouraged to see themselves as silly creatures trapped in sensibilities.

An alternative in the late eighteenth century, the woman as preacher, was visible in society – not in Wollstonecraft's immediate milieu, but taken for granted by the Dissenters with whom she mixed. A memorable instance was Elizabeth Evans, George Eliot's aunt whom she would immortalise as the preacher on the green in *Adam Bede*. Her eloquence is of a different kind from doctrinal preaching, in that it is attentive to the lives of her listeners, and at the same time she displays none of the affectations of femininity: 'there was no blush, no tremulousness, . . . no casting up or down of the eyelids, no compression of the lips, no attitude of the arms that said, "But you must think of me as a saint." She held no book in her ungloved hands, but let them hang down lightly crossed before her, as she stood and turned her grey eyes on the people. There was no keenness in the eyes; they seemed rather to be shedding love than making observations; they had the liquid look which tells that the mind is full of what it has to give out, rather than impressed with external objects.' Such women, whose faith granted them a speaking role, did not formulate this as a cultural expression; it remained for Mary Wollstonecraft to shape public words to conscious effect.

As many before her, she scorned the wiles women had to adopt for the marriage market, and gallantry as the manipulations of male ascendancy. True civility, she believed, can only exist between equals. This sweeps aside a distinction between decent and rakish gallantry championed by the mid-century philosopher David Hume and other members of the Scottish Enlightenment who did, in fact, go some way towards Wollstonecraft's position. They did look to women as bearers of social sympathies, and concede women's apparent weakness as a consequence of external pressures like education. To Wollstonecraft, though, gallantry in any form is condescension; it puts women in a false position – reality and rhetoric diverge – and she calls on 'reasonable men' to eschew it. In February 1792 she told Everina of a proposal: 'my book &c &c has afforded me an opportunity of settling *very* advantageous in the matrimonial line, with a new acquaintance; but entre nous – a handsome house and a proper man did not tempt me'.

The writer Joan Smith has pointed to the similar enclosures of land and women's bodies in the eighteenth century. Lord Hardwicke's Marriage Act was designed to ensure the transmission of property to the rightful heir in the next generation. Women's bodies were therefore policed by laws preventing adultery. As Dr Johnson put it: 'The chastity of women is of all importance, as all property depends on it.' From 1753 no marriage was legal except the formal, indissoluble one that controlled the passage of property. The Act outlawed all alternative forms of union: cohabitation by mutual agreement, like the partnerships of today, or clandestine marriage without parental consent, an escape route for a girl whose father or guardian was forcing on her a suitor of his choice. The new marriage laws were really property laws: a wife, her money and her children being the property a man gained in perpetuity – divorce was virtually impossible. A legal decision of 1782 entitled a husband to beat his wife with a stick no thicker than a thumb, and an earlier judgement by Sir Matthew Hale established that rape in marriage was no crime.

Wollstonecraft had witnessed the dangers of these laws in her father's licensed violence to her mother, and in her sister's indissoluble marriage to Meredith Bishop. 'From the respect paid to property flow, as from a poisoned fountain,' she wrote in the *Vindication* of 1792, 'most of the evils and vices which render this world such a dreary scene to the contemplative mind.' The injustice of English law for women was to be the subject of a sequel. In the meantime, she questioned the association of morality with women's subordinate behaviour and sexual reputation, as a misconception of morality itself – the focus on women's sins and virtues deflecting attention from a wider morality of tolerance and compassion ignored by those in power.

Women's bodies were not only turned into sites for public morality, they were also pathologised. Wollstonecraft opposed a system which defined women as weak, and women themselves for foolish complicity. They tended to pore over such nonsense as *The Ladies Dispensatory; or, Every Woman her own Physician* (1740, republished many times): 'The delicate Texture of a Woman's constitution . . . subjects her to an infinite Number of Maladies, to which Man is an utter Stranger,' women were told. 'That lax and pliant Habit, capable of being dilated and contracted on every Occasion, must necessarily want that Degree of Heat and Firmness which is the Characteristick of Man.'

Wollstonecraft's alternative model of womanhood resembles the rational, moral Mrs Mason in *Real Life*. Much that is sober in this model was designed to counter the coquette trained to live through her sexuality, and we need no longer debate what seemed to modern women prudish advice that they should surrender expectations of sexual pleasure in favour of friendship in the course of marriage. Friendship was – and remains in certain cultures – an obvious way a wife can rescue her self-respect from sex-based subordination reinforced by legal, religious and educational disabilities. This is really not about prudery nor about Wollstonecraft herself; it's about coping with a lifelong prospect of inferiority.

At the heart of her argument is the revolutionary idea she had first put out as a practising teacher in her *Education*. Here, again, she insists on an education in domestic affections as opposed to governance based in contests of power. Domestic affections cut across distinctions of gender, offering a basis for a common morality. It is easy to pass over this domestic ideology as a plea for old-fashioned femininity, but nowhere does she dare more the judgement of her reader. By domesticating an aggressive order she wants to change the whole world. The resistance then and now comes from a fear of feminisation as effeminacy, including fear of the 'queer', rather than what Wollstonecraft actually advocates: the political empowerment of gentleness, nurture, compromise and listening – all traits which the civilised of both sexes already share.

Two other questions remain at issue: sex education and the unresolved nature of gender. The transparency prompted by the Enlightenment led

Wollstonecraft to propose sex education instead of filling children's heads with 'ridiculous falsities'. Although she is of her time when she warns against homosexuality and masturbation, she is beyond her time in her 1790 preface to *Elements of Morality*, where she asks parents 'to speak to children of the organs of generation as freely as we speak of the other parts of the body, and explain to them the noble use, which they were designed for'. There is no precedent for Wollstonecraft's proposal, nothing beyond Locke's general recommendation that children's curiosity should be answered. At the time, the human sex organs were revealed in the only explicit book, *Aristotle's Complete Master-Piece*, a seventeenth-century compilation that claimed to be intended mainly for practitioners of midwifery, and as such 'for public benefit'. It was not intended for 'some lacivious and lewd persons' who might ridicule 'the secrets of nature'. In fact, the book was read widely, and reprinted about three times a decade. What was permissible for midwifery was not permissible for advice literature, the popular eighteenth-century genre from which almost all Wollstonecraft's writings, including the second *Vindication*, stem. For Wollstonecraft to fuse sex education with the proper-lady tradition of the advice book was shocking to many of her generation, as to several generations that followed. Even now, sex education in schools can be minimal and often embarrassed, but back in 1790 dignified words came easily to Wollstonecraft.

Children, she said, should know about the 'germ of their posterity, which the Creator has implanted in them for wise purposes'. Two years later, in the second *Vindication*, she builds on what children already know: 'Children very early see cats with their kittens, birds with their young ones, &c. Why then are they not to be told that their mothers carry and nourish them in the same way? . . . Truth may always be told to children, if it be told gravely; but it is the immodesty of affected modesty that does all the mischief; and this smoke heats the imagination by vainly endeavouring to obscure certain objects.' This truth should accompany instruction in anatomy and medicine, 'not only to enable [women] to take proper care of their own health, but to make them rational nurses of their infants, parents, and husbands; for the bills of mortality are swelled by the blunders of self-willed old women, who give nostrums of their own without knowing anything of the human frame'.

These recommendations coincided with Erasmus Darwin's demonstrations of the sexual characteristics of plants. His *Loves of the Plants* was a sensation when it came out in 1789, so much so that Johnson offered him the extraordinary sum of £1000 (equivalent to today's six-figure advance) for a sequel. In 1791 the two parts were collected as *The Botanic Garden* with a frontispiece of Flora adorning herself with the help of the elements, designed by Fuseli. Fire – implying the heat of passion – holds up a mirror to the goddess of nature. This book was designed to overthrow unscientific myths about the source of life. Wollstonecraft was familiar with the Linnaean classification that lay behind Darwin's popular verses – she would have seen Fanny's botanical drawings with their delicate watercolour shadings of pistils and stamens, outlined in firm, accurate black lines. In Darwin's eroticised science, the dramas of fertilisation are upheld by hard fact in copious footnotes. *The Loves of the Plants* gave new impetus to the traditionally female subject of botany, leading girls towards scientific knowledge through the amorous play of flowers, personified as wantons and virgins:

> With secret sighs the Virgin Lily droops,
> And jealous Cowslips hang their tawny cups.
> How the young Rose in beauty's damask pride
> Drinks the warm blushes of his bashful bride.

Nature proves at odds with the propertied control of female bodies:

> Each wanton beauty, trick'd in all her grace,
> Shakes the bright dew-drops from her blushing face;
> In gay undress displays her rival charms,
> And calls her wandering lovers to her arms.

Not surprisingly, Darwin's natural world driven by sex was, to many, an improper study for women for whom scientific study was, anyway, thought unsuitable.

Wollstonecraft reports on the response of an unnamed writer to a 'lady who asked the question whether women may be instructed in the modern system of botany without losing their female delicacy?' The unnamed writer declares that if the lady had 'proposed the question to me, I should certainly have answered – they cannot'. To Wollstonecraft, this presents 'a gross idea of modesty'. She adds caustically, 'Thus is the fair book of knowledge to be shut with an everlasting seal!'

A backlash against botanising girls inflamed by the loves of plants continued in Richard Polwhele's *The Unsex'd Females* (1798):

> With bliss botanic as their bosoms heave,
> Still pluck forbidden fruit, with mother Eve,
> For puberty in sighing florets pant,
> Or point the prostitution of a plant;
> Dissect its organ of unhallow'd lust,
> And fondly gaze the titillating dust . . .

Conservatives like Polwhele were infuriated with Mary Wollstonecraft who, they thought, promoted immodesty. In fact she urged 'natural' modesty as opposed to its affectation.

Where most saw human nature as unchangeable, revolutionaries argued that political institutions made men and women what they are; change institutions, discard the mind-forged manacles, and you change human nature itself. Wollstonecraft dedicated the *Rights of Woman* to the French politician Talleyrand, who had just then drawn up a report on the need for national education for girls as well as boys. Talleyrand justified the usual bias when he claimed girls' inferiority as 'the will of nature'. Wollstonecraft refutes this: femininity is not the creation of nature, she says; it is the enfeebled consequence of miseducation. She deplores the sickly delicacy that stifles 'the natural emotions of the heart'; the slippery language of sensibility ('pretty superlatives dropping glibly from the tongue'); and false refinement cultivated in 'a premature unnatural manner'. Wollstonecraft is knocking against advice books like Dr Gregory's *Legacy to His Daughters*, where it's thought indelicate for women to have what Wollstonecraft calls 'the common appetites of human nature'; even a wish to marry was indelicate, and should be concealed. Wollstonecraft blamed women's susceptibility to rakes on the equally pernicious influence of the sillier romantic fictions. Women, she argued, live by a stock of worn-out narratives, unable to reshape their lives. She laments there was not as yet 'a road open by which they can pursue more extensive plans of usefulness and independence . . . Women might certainly study the art of healing and be physicians as well as nurses. And midwifery, decency seems to allot to them . . . They might also study politics, and settle their benevolence on the broadest basis.'

On the face of it, Wollstonecraft forecasts the way we live now; but at a deeper level of implication she is raising the longer-term issue of women's 'nature': that word underpins her enterprise as much or more than 'rights'. Her private letters and political writings are attuned to states of mind; all assume that right action comes from within – from an educated capacity to judge in a way that breaks the constricting mindsets of her sex. She herself demonstrates 'natural strength', eloquence not meekness, and, later, passion in place of the 'unnatural coldness' women were taught to cultivate. She questions Rousseau's belief that women exist to please men and that works of genius are beyond their capacity. Her preferred model is Catharine Macaulay, whose *Letters on Education* (1790) resisted 'the absurd notion, that the education of females should be of the opposite kind to that of males. How many nervous diseases have been contracted? How much feebleness of constitution has been acquired, by forming a false idea of female excellence . . .' Girls were trained to suppress their natural energy, and play the babe: 'to lisp with their tongues, to totter in their walk, and to counterfeit more weakness and sickness than they really have, in order to attract the notice of the male'. Coeducational day schools, open to all classes, would help to eliminate what is false in our differentiation of boys and girls.

Above all, the *Rights of Woman* proposes insight and sympathy as an alternative basis for political action. 'Brutal force has hitherto governed the world,' Wollstonecraft observes. 'Man accustomed to bow down to power in his savage state, can seldom divest himself of this barbarous prejudice . . .'

Military heroics are no longer wanted: 'It would puzzle a keen casuist to prove the reasonableness of the greater number of wars that have dubbed heroes . . . I sincerely wish to see the bayonet converted into the pruning-hook.' Future hopes may lie with 'moral agents' who have not accustomed themselves to brutal force: 'It is time to effect a revolution in female manners – time to restore them their lost dignity – and make them, as a part of the human species, labour by reforming themselves to reform the world.'

The second *Vindication* sold in the region of three thousand copies. Not a large number compared with the two hundred thousand copies of Paine's *Rights of Man* sold by 1793, yet Wollstonecraft's name became known throughout the land. It was the first demand for women's transformation to enter the mainstream of British and American politics. The *Rights of Woman* was debated by groups of women in public meetings in the British provinces. In Glasgow, Mrs Anne Grant wrote to a Miss Ourry: 'I have seen Mary Wollstonecraft's book which is [so] run after here, that there is no keeping it long enough to read leisurely.'

Jane Austen's early novella *Catherine* is dated August 1792, eight months after the *Rights of Woman* appeared. Wollstonecraft's ridicule of women's inflated gush and plaything education reappears in Austen's ridicule of Camilla: 'those years which ought to have been spent in the attainment of useful knowledge and mental improvement, had been all bestowed in learning drawing, Italian, and music . . . and she now united to these accomplishments, an understanding unimproved by reading'. This is Wollstonecraft's *un*charmed voice, and it sounds again in Catherine's protest against marriage as a form of prostitution for the orphaned Miss Wynne, who is sent off by relations to marry a stranger in India. Silly Camilla sees a romantic adventure, but Austen, at seventeen, already knows better. The prospective husband could turn out a tyrant or a fool or both, and to marry though 'infinitely against [Miss Wynne's] inclinations had been necessitated to embrace the only possibility that was offered to her, of a Maintenance'. A generation earlier, this had been the fate of Jane's orphaned and unhappy aunt, Philadelphia Austen.

Wollstonecraft's voice sounds once more when Elizabeth Bennet will not grovel to imperious Lady Catherine de Burgh in *Pride and Prejudice*. '"Upon my word", said her Ladyship, "you give your opinion very decidedly for so young a person."' Elizabeth's 'abominable sort of conceited independence' makes her the butt of the Bingley sisters, who play out the idle, empty-headed model of femininity. 'Her manners were pronounced to be very bad indeed, a mixture of pride and impertinence.' Though Jane Austen's family stressed her conformities, her works, like Wollstonecraft's, back independence of mind. In *Mansfield Park*, Fanny Price, refusing triviality, is rightly said to be a 'legatee of *The Vindication*', as is Anne Elliot in *Persuasion* when she protests against men's view of females as shallow in their attachments: 'Men have had every advantage of us in telling their own story. Education has been theirs in so much higher a degree: the pen has been in their hands. I will not allow books to prove anything.'

In Quincy, Massachusetts, Abigail Adams read Wollstonecraft as an ally against her husband's laugh when she had asked him to 'remember the Laidies' in framing the laws for the new American nation. In 1794, John Adams was Vice-President to Washington when Abigail drew his attention to the fact that of the few queens who had been called on to rule as absolute sovereigns, the greater part had excelled: 'Pliny tells us that . . . among the Lacedemonians, the women had a great share in the political government and that it was agreeable to the laws given them by Licurg[us]'; and in Borneo, she added, 'the women reign alone, and their Husbands enjoy no other privilege than that of being their most dignified subjects; but as reigning and ruling is so much out of fashion at the present day, my ambition will extend no further than reigning in the Heart of my Husband'. John Adams conceded more wit and matter in his wife's letters than he heard in Congress, but owned in private to Abigail that he remained unwilling to lose time in women's company. 'Pardon me!' he teased, 'Disciple of Woolstoncraft!'

'*Pupil* of Woolsoncroft,' she retorted, 'confess the Truth that when you are sick of the Ambition the intrigue the duplicity and the Treachery of the aspiring part of your own sex', there is consolation in 'the simplicity the gentleness and tenderness of the Female character.' Where these qualities exist – and Abigail was too civil to point to herself – they are, she argued, 'more beneficial to the human race' than Adams prudence, 'and when conducted with good sense, approach to perfection'.

Abigail's sister, Elizabeth, asked her for a copy of the *Rights of Woman*, while Aaron Burr, a future Vice-President, wrote to his wife from the seat of government in Philadelphia: 'I made haste to procure [the book], and spent the last night, almost the whole of it, in reading it. Be assured that your sex has in *her* an able advocate.' He admired her ability to adapt the liberating style of Rousseau in such a way as to resist him when it came to his demeaning stance on female education. 'It is, in my opinion, a work of genius,' Burr concludes. 'I promise myself much pleasure in reading it to you.'

At the same time, the *Rights of Woman* provoked outrage. Thomas Taylor the Platonist, who had been the Wollstonecrafts' landlord at Walworth when Mary was eighteen, published *A Vindication of the Rights of Brutes* (1792) proposing animal rights with laboured sarcasm, and sneering at her case for sex education as 'beyond all doubt a most striking proof of her uncommon capacity, and the truth of her grand theory, *the equality of the female nature with the male*'. Hannah More clung to her status as a rarity in what she called an 'unstable and capricious' sex: 'I have been pestered to read the "Rights of Women," but am invincibly resolved not to do it,' she assured Horace Walpole. 'There is something fantastic and absurd in the very title.' Another bluestocking, the classicist Elizabeth Carter, was also put out. At Pembroke in Wales, the gentry declared itself 'shocked': 'the most indecent Rhapsody that ever was penned by man or woman,' they said, and discussed which parts 'wounded modesty the most'. When an effigy of Paine was burnt, people spoke of 'immortalizing Miss Wollstonecraft in the like manner'.

One of the mature men who were impressed with Wollstonecraft was lawyer and banker William Roscoe, a Liverpool connection of Johnson. He was a tall, upright man with mild grey eyes and a cheerful expression, who went in for a range of aesthetic and reforming projects, amongst them a botanical garden for Liverpool, a poem against 'The Wrongs of Africa', and a pamphlet against slavery. In 1791 he applauded Wollstonecraft as an Amazon who conquers with her pen. Roscoe commissioned a portrait of her while she was writing *A Vindication of the Rights of Woman.*

If it was not a good likeness, she told him, he would find a more faithful sketch in her newest book: 'a book . . . in which I myself . . . shall certainly appear, head and heart – but this between ourselves – pray respect a woman's secret!'

Likeness was not the prime aim of this portrait. It's not designed to show the day-to-day Mary Wollstonecraft toiling behind her specs, or trudging City streets on her way to the printing house. The portrait is a statement designed to refute the lightweight image of women's nature. Something else stands forth – a presence in sober black, setting off the dignity of her powdered head and long white fingers like a set of sensitive tendrils. Her white fichu is swathed to her chin like a cravat, drawing attention from her breasts to her face. Here is a sight never pictured before: a stateswoman about to speak – as austere and reserved as Mr Pitt. Her reserves have shed all extraneous attributes of femininity: no frill, no smile, no winning ways, and no silence. It's a vindication of the 'new plan of life', a public image of a new genus.

8

RIVAL LIVES

Fanny's sister Caroline Blood was always in trouble. She had remained in London when her parents returned to Dublin. At the end of 1787, just as Mary was completing *Real Life*, having her sisters to stay and planning to fund Everina abroad, Mrs Burgh heard that Caroline had been found 'in a *dreadful* situation' – prostitution – by parish officers in Islington. The parish would give her shelter in its workhouse if Mary were prepared to pay half-a-crown a week for her keep.

'I cannot allow them again to turn her out,' Mary thought, as she took in the girl's need for 'a few clothes to cover her'.

The great are ordinary as well as great: they are friends, sisters, daughters; they lose heart; struggle; lose patience; and always, they are beset by conventional scenarios – in Caroline's case the Hogarthian downfall of the unprotected girl exposed to the predators of the town. Mary did not go to see Caroline, fearing to take on 'a burden I could *not* bear' had Caroline discovered her hideaway. Instead, she sent the sum through Mrs Burgh, and in the meantime informed Caroline's father. Mr Blood did send the £10 Mary had requested, but then conveniently forgot any further responsibility for his daughter. Mary described him as a 'sensualist' who thought of honesty as romance. She projects her resentment into the 'torment' of Mrs Blood who, with no existence in law except as the property of her husband, could only wait to be delivered by his death. In the end it was not Mr

Blood but Caroline's brother George who had to be pressed, year by year, for her support.

It so happened that George himself was not above a dodge when conditions seemed favourable. Mary had offered him some business for Joseph Johnson, so was 'disappointed' when he failed to acknowledge a sum of money he should have received. She had to tell George, in words that left no room for misunderstanding, 'I wish when I transact any business for Mr Johnson to observe a certain degree of punctuality . . . I am convinced it is vain to attempt to teach some people to be punctual; but as I thought you were scrupulously so your silence alarms me.' Words like 'alarm', 'uneasy', 'apprehension' warn George that she's getting a whiff of dishonesty in view of his recent 'embarrassments', associated, she suspected, with 'that artful, selfish man', his father. A rumour of 'lame' schemes had reached her. Was George diverting Mr Johnson's money – temporarily, he would have told himself – into one of them? Could she sustain herself against others' plots of dependence or against diversionary narratives of helplessness, fortune-making and marriage, or, most dangerous, against the distorting narrative of a rival 'genius' in the shape of the artist Fuseli?

Since the summer of 1786, her sister Bess had languished in a series of unsatisfactory posts. Her past dependence on Mary's protection left her the person most affected by Mary's decision to become a writer. It meant that Mary was never going to support Bess or initiate any further scheme to bring them together. If Mary's existence was an experiment, Bess provides the control: subject to the same kind of constricting employment, panting to be free and subject to depression – remarkably articulate depression. More ladylike and less spirited than her sisters, Bess lacked Mary's counter to self-absorption: an interest in others. She also had the bad luck to have to cope with the quarrelsome lodgers at Newington Green; the sticklers for piety at the school in Market Harborough; and now she quailed before Mrs Bregantz, the headmistress in Putney. When Everina returned from Paris and joined Bess intermittently at Putney, Mrs Bregantz deplored Everina's ignorance of fancywork, her 'mere smattering of Botany' and her '*dread*ful indolence'. The head's 'snarling' manner 'unhinged' Bess, who trembled when Mrs Bregantz approached. Mary was hardly soothing when she advised Bess to take some salts, since 'ugly spots sometimes appear on your fair face'.

Of the Wollstonecrafts, Bess was the least adaptable, trapped as she was in the disheartening grind of a young woman without the will to discover in herself the germ of a new genus. This is to say that Bess was like most of us – a measure for the rarer creature who extends herself beyond predictable bounds.

Mary remarked that Bess 'wants activity'. She was active enough when it came to walking and riding, but too often 'on stilts', as Mary put it, to win the hearts of pupils. In Putney, there were impolitic rows with the head's daughter, Miss Bregantz. Once, Mary lost patience with her sister's habit of inflating minor irritations with the strongest expressions of anguish. Bess gave her no peace, she burst out. It would be a 'holiday' to dismiss her from her thoughts.

Bess was still in her twenties in May 1791, when she moved from Putney to Wales, to take up a post as governess at Upton Castle on the Pembrokeshire coast. She travelled by coach to Bath and Bristol; then by sea across the Bristol Channel to Port Lewis in Wales; and finally on horseback via her father's home in Laugharne. At Pembroke, she took charge of dark, bustling Ria, almost grown-up in her plaid flannel skirts; sickly, beautiful Sophia lolling beside the fire; and the youngest, Harriot, clever and spoilt. They were the heirs of a nabob called Tasker who had bought a castle from an old family, retaining a steward called Rees and a housekeeper called Molley. Tasker was a reader, like Bess, and gave her an ancient turret room next to a fine library, but she took against him too much to talk of books. She didn't exert herself to know the girls beyond what were harmless faults: their idleness and addiction to ghost stories told by the servants. Bess swept off to read Robertson's history of America. Her only friends were the dogs and, for a short space, a little boy, Tasker's nephew, shipped from India to remote rainy Wales and received, according to Bess, with chill parsimony – Tasker refused to buy the child the outdoor shoes he needed if he was to have exercise. Bess did love this lonely, homesick boy; loved him as much as she dared, for she knew they would have to part.

Some of her plaints sound justified: snubs, for instance, at a naval party where officers' admiring glances at her black eyes and healthy cheeks faded to audible 'Ohhhs' when they learnt that she worked for her bread. None offered to hand her downstairs as she watched Ria – so very plain, but an heiress – descend on an officer's arm. Those who greeted the governess did so with averted eyes when others weren't looking. Venus herself, Bess thought, would be ignored in the guise of a governess, as would 'Mary's Sappho'.

Later in the evening all changed when she danced with a sailor who introduced her as 'Wollstonecraft's sister' to a messmate of her brother James.

The messmate exclaimed with warmth, 'You are dancing with the *cleverest* Woman in the World.'

She was another sister, Bess had to explain; all the same he insisted on a dance, and she took his hand with unaccustomed pleasure. Her cheeks glowed as she talked of Mary.

Though Bess found Welshmen free from affectation, they had little to say – any man who could speak was sure to be Irish. As for her employer, whom she called 'the miser', he sat down to a dirty tablecloth on which reposed the remains of dead animals. 'Learn,' she wrote to Everina, 'that I can swallow cocks and hens that *die* of *diverse diseases* – nay can digest a *dead mutton* . . .' She was hungry enough to devour the whole flock.

The miser had to be persuaded to provide a fire in the draughty turret. He required Bess to ask permission if she went for a walk. She did like an autumn walk, the melancholy colours, the dogs bounding about her; but on her return, he met her with grumbles. 'It is happy for fine ladies who have nothing else to do – ' he would say. The more Tasker sniped, the more he smiled.

All this, Bess set down, page after page of wailing, raging letters. She often refers to herself in the third person – 'poor Bess' – an ill-used heroine, whose beauty and suffering make her worthy of rescue. A false narrative, of course – Bess could see this – yet the passive scenario was ingrained. Hers wasn't stupid wailing, but grim for her sisters to read because these were cries for release as youth passed – and release was not at hand. Mary could neither take her on nor encourage false hope.

So, in the 1790s, the gap between the sisters widened. While the productive years in Johnson's circle developed Mary's assurance, Bess, with no protector, abandoned by Mary, was slowly crushed. She wrote to Everina: 'I never think of *our sister* but in the light of a *friend* who had been dead some years; and when all here is fast asleep and nought is to be heard but the screech-owl I sigh to think we shall never meet, as such, again – though perhaps in a better world the *Love of Fame cannot corrupt the soul*[.]' She speaks of 'surprise' and 'indignation' at Mary's 'sudden change' of character. The date is 29 January 1792, the month *A Vindication of the Rights of Woman* came out. Six months later Bess was further put out to hear 'that *Mrs Wollstonecraft* is grown quite handsome', adding snidely that 'being conscious that she is on the wrong side of slight Thirty she now endeavors to set of[f] those charmes (*she once* despised) to the best advantage'. Bess did concede that Mary had a Great Soul, and did occasionally show self-insight: 'I blush to own that my temper is now so *soured*; and continually ruffled that I almost despise myself.' Damage to character is the darkest effect of the social exclusion Bess experienced as a governess. A doctor who at first invited her out, stopped doing so, and talked past her to the family, Bess noticed, 'without one word or look at me – yet never did I strive so hard to please'. The harder she tried to please a gentleman, the more suspect she would, of course, appear. Respectability required self-effacement.

Bess languished at Upton Castle from year to year, 'while the sad hours of life were *rolling* away'. It may have been impossible to find a post paying more than £40 a year, especially for a governess without French. Bess still looked to Mary to provide her a turn in Paris, and was jealous when the needs of their brothers came first.

During 1791, James Wollstonecraft left his ship. Mary took him in and sent him to study mathematics for a few months under Mr Bonnycastle at the Military Academy, Woolwich. James was master of fine turns of phrase in pursuit of money – he could tell a plausible and urgent story promising a quick return that soon turned out to be impossible; it could be, a friend to whom he gave the money for safe-keeping passed it on to a banker who, it just so happened at that moment, failed and killed himself. It was never

James's fault. But he knew how to pull out the stops of abjection: ever the well-spoken gentleman, he longed to call on his creditor but trembled to anger one whose good opinion mattered to him beyond that of anyone else in all the world. Mary was undeceived when James 'threw some money away to dance after preferment when the fleet . . . paraded at Portsmouth', and then idled around her while she worked. At length, in September 1791, he 'condescended to take command of a trading vessel'. It was a speculative voyage, bringing Mary further 'vexation'. In the course of the following year, he tried her patience till she no longer wished to see him.

After James left, Mary moved to Store Street, off the Tottenham Court Road, to the north of Bedford Square. Here, she had the fresh air of the nursery gardens that used to stretch behind the British Museum (now the site of University College). At this point, she acquired good furniture, a cat, and a seven-year-old orphan called Ann, an Irish niece of Betty Delane who had married Hugh Skeys (and was said to be none too kind to the child). Again, Mary's hope to be of use came to nought: she could not warm to Ann, who was loud, stole sugar, and lied. No more is known, but to take on so young a child, far from home, and then dismiss her, has to be questionable. Mary passed her on, for a time, to a childless American, new to London, called Mrs Barlow.

Mary put her most sustained effort into her youngest brother and sister. When Everina returned from Paris, Mary continued to protect her: she was encouraged to wait for the right post, and in the meantime had a home with Mary who passed on translating jobs for Johnson, interspersed with teaching spells in Putney.

When Mary set up in London, her brother Charles had been an articled clerk in the law firm of their eldest brother. His connection with Ned made Mary careful not to divulge where she stayed. As he came to meet her at Johnson's print shop, she noticed the way his 'warm youthful blood paints joy on his cheeks, and dances in his eyes'. When Mr Wollstonecraft removed Charles from Ned during a dispute over property, Mary placed Charles in another law firm. Dismissed in April 1789, he went to Ireland to 'eat the bread of idleness', as Mary put it, when he thought to settle with their relations in Cork who had no wish to keep him. Mr Johnson paid Charles's debts, while Mary begged help from George Blood in Dublin:

> Dear George,
>
> I know you will be sorry to hear that I am envolved in numberless difficulties – but Heaven grant me patience – and I will labour to overcome them all – Before you receive this, or very soon after, you will see Charles – he will tell you some of my vexations – those *very* severe ones he has brought on me I suppose he will throw into the shade; but I would not prejudice you against him though he has wound[ed] a heart that was full of anxiety on his account – and disappointed hopes, which my benevolence makes me regret, more than reason can justify . . . I would fain have made him a virtuous character and have improved his understanding at the same time – had I succeed[ed] I should have been amply reward[ed] – but he has disappointed me – disappointed me . . . My hand trembles – I will write again as soon I can calm my mind – I beg you if you have any love for me – try to make him exert himself – try to fix him in a situation or heaven knows into what vices he may sink! you may tell him that I feel more sorrow than resentment – say that I forgive him – yet think he must be devoid of all feeling if he can forgive himself –
>
> Yours affectionately
>
> Mary Wollstonecraft
>
> April 16 [1789]
>
> I know he will plunge into pleasure while he has a farthing left – for God endeavour to save him from ruin by employing him!

No answer came. In September, she wrote again to 'my still dear George'. At length, a letter with protestations and excuses arrived seven months after her first plea. She replied:

> I cannot comprehend what you intended to say to account for (an excuse was out of the question) your strange unfriendly conduct – nor

> can I conceive that my letters, or silence, could affect you, in the manner you mention, after your unaccountable neglect of a friend who placed the most unreserved confidence in your affection and goodness of heart – all this appears inexplicable – I *cannot* understand it – I am not easily offended with those I love – . . . but . . . I cannot in a moment . . . allow vague expressions of sorrow to work on me like a charm . . . As for the pretext of business, I shall not admit it, an hour might have been stolen from sleep without injuring your health – in short, you have obliged me to alter my opinion of you.

Mary now tried in vain to find Charles a post in the East India Company; then one through William Roscoe. When Roscoe saw an opening in the autumn of 1791, Mary thought her brother would not be up to it. By this time, aged twenty-one, he was loitering with his father at Laugharne. Bess had found him almost naked, frustrated with his father's unconcern, and talking of enlistment in the army. Bess advised Charles not to call on Mary when he left for London at the close of the year. Mary was puzzled and hurt, but when they met he found her affectionate and ready once more with help.

The help came in the shape of Joel Barlow, a young American with a steady gaze whose wife looked after Ann. Barlow was fresh from four years in Paris where he had promoted the dream of the moment: post-revolutionary America as a utopian possibility for emigration. He was handsome, with loose, wavy locks about his cheeks; the rest tied back in a tight queue. Even in youth his hair receded slightly at the temples. A square jaw countermanded the delicacy of his upper lip, warmed in Barlow by the charm of the light-hearted. When he laughed, his mouth curled and long lines ran down the sides of his face from cheek to chin. 'He has a sound understanding with great mildness of temper,' Mary observed. His humour was deliberate – 'regulated' – to an English ear.

This was her first American. Here, at Johnson's, she encountered the New World model which Dr Price had upheld in Newington Green in the aftermath of the war. Barlow was immediately congenial; his revisionist view of 'nature' as the basis for social revolution beat pulse for pulse with the central vein of Mary's *Vindication*. Both fought off Rousseau: Mary, his subjection of women to a false construct of their 'nature'; Barlow, his fantasy of a state of nature antecedent to society.

'The only state of nature is a state of society,' Barlow liked to declare, 'the perfect state of society is a perfect state of nature.' Men who acquiesce in a society based on privilege and exclusion, he went on, are in an 'unnatural' state. This 'unnatural' state has shaped our definitions, our very language. The word 'liberty', he argued, would not have been known in any language, had people not felt deprived of it; and some are 'free men' because 'men are not all free'.

Johnson published Barlow's *Advice to the Privileged Orders* in February 1792, a month after *A Vindication of the Rights of Woman*. Its author struck Barlow as 'a woman of great original genius'. She befriended his wife, Ruth, in their lodgings at 18 Great Titchfield Street, off Oxford Street, and Ruth took Mary to meet a fellow-American, a Mrs Leavenworth. Mary warmed to 'the easy, unreserved behaviour' of American women, and their readiness to discuss subjects of interest to both sexes instead of the usual froth. Ruth described herself as 'incapable of disguise'. 'What I feel, I say,' she declared. She shared Mary's trust in private integrity as the basis of public and patriotic virtues. When Mary confided her anxieties over Charles, the Barlows offered to take the boy with them when they returned to America that spring. There was even a sign that they might make him part of their family. Barlow had clapped him on the knee, and said in his dry way that as he and his wife 'could never contrive to make any boys they must try what they could do with one ready brought up to our hands'.

Parented by the Barlows, Charles was to farm. Mary arranged to sell some of their father's Primrose Street property to buy land in Ohio, most likely through Barlow himself who had recently controlled a huge tract. Tom Paine assured Mary that Charles could not go with a better man, and Fuseli gave him £10. Hopes now rose: if a brother prospered, his sisters might prosper too. Mary busied herself with preparations. She sent Charles, at 'heavy expence', to study farming in Leatherhead in Surrey; backed a letter from Charles to Everina (by now, a governess in Ireland), begging £20 (about half her year's salary); consulted lawyer Roscoe on how to ensure that Ned could not 'snap at the last morsel' by means of some quirk of law (that is, wrest Charles's inheritance from his sisters should their father die, and should Charles too happen to die abroad before he received news of their father's death); and, finally, she kitted Charles out for emigration. It worried her, though, to see him wearing out these clothes as he waited for Barlow to return from a visit to Paris.

One day, while Barlow was away, Ruth laid out a private treasure: her husband's love-letters. Mary was taken aback by the passion of this couple who had been married ten years. Though she thought this the ardour of the exploiter, there's a tinge of envy. She confided to Everina: 'Mrs. B. has a very benevolent, affectionate heart, and a tolerable understanding, a little warped it is true by romance; but she is not the less friendly on that account.' When Mary spoke of her sisters' troubles, Ruth went so far as to suggest that Bess and Everina, too, should migrate to America. Bess could keep house for Charles, and Everina was invited to live with the Barlows until she found a husband – she and Bess were sure to find a welcome in good society. Mary passed this on to Everina, urging her to correspond with Ruth.

The Barlows entered Mary's life through her family problems, but their continued significance remains to be explored. Of all the people Mary knew, Joel Barlow has to be the likeliest to lead us into what we might call the American mystery in her life: the nature and mysterious activities of Barlow's friend Gilbert Imlay whom Mary was soon to meet. Vast collections of Barlow Papers in American archives offer a new approach to Imlay and the networks in Europe that connect him with Barlow. American historian John Seelye has described Imlay as 'in so many ways Barlow's shadow . . . for a time the two operated almost as one'.

Their similarities have to do with character in the formative period of national identity. As Barlow and Imlay plant themselves in corrupt old Europe, they appear as new-grown specimens from the New World. Barlow's appeal for Mary Wollstonecraft prepares her for Imlay, whom Barlow resembles: they are well-read Americans of the revolutionary generation; they come from the officer class in the War of Independence; and like many after the war they are opportunists. They turn their hand to writing, but it's not what they want from life, which is to make their fortune. At this stage, both are at a loose end – possibly, stranded in Europe without money – yet they remain keen-eyed, storing information for future use, as they rove from Paris to London, and London to Paris. With women, they offer a new breed of virile respect. Their manners are attractively transparent, though not quite as readable as they appear. The Barlows' friendship with Mary Wollstonecraft has appeared peripheral, but new findings will prove Joel Barlow to have had a larger impact on her future than any other London attachment. His history – his habits of business and ways with Ruth – foreshadows a drama that was to rock Mary Wollstonecraft's life.

Joel Barlow came from a family long settled in Connecticut. At Yale, his teacher Abraham Baldwin introduced his plainly dressed sister Ruth. She had a delicate oval face; light, curly hair; and steady eyes under finely arched brows. During the Revolutionary War, Barlow served as chaplain, the reward of a crash course in divinity; he showed no further interest in faith once the army disbanded. Optimistic, good-tempered, ready to please, he was ambitious to distinguish himself without too clear an idea how this was to be effected. His youthful moment of glory came when he dined with General Washington. 'How do you think I felt when the greatest man on earth placed me at his right hand,' he wrote from camp to Ruth Baldwin.

Ruth agreed to elope, for Barlow was 'destitute' – not the suitor her family approved. Their marriage took place while the troops were in winter quarters in January 1781. 'You are the tenderest & best of *Lovers*,' Ruth told him.

When the war ended, Barlow settled in Hartford, Connecticut, as a journalist and one of 'the Hartford Wits', a conservative set of Yale poets. In 1787 he published an American epic, *The Vision of Columbus*, foretelling God's plan to make 'the spirit of commerce' the refining agent of the world. It's a heavy poem – Barlow was better at light verse (as in his nostalgia for 'Hasty Pudding') – but an epic was timely, and famous names subscribed to it, amongst them Washington, Lafayette (commander of America's French allies) and King Louis XVI. This encouraged Barlow's selection as European agent of the Scioto Land Company, organised under an act of Congress to purchase a large tract of public land in the wilderness of Ohio and sell plots to prospective settlers. Leaders of the scheme, William Duer (Secretary of the Treasury) and Major Winthrop Sargent, gave an indenture for one-sixtieth of the entire Ohio tract to Barlow, who undertook sales in France. He sailed for Europe in the spring of 1788.

In London that summer Barlow visited Dr Price (with an introduction from Jefferson). He had sent Price a copy of his epic, hoping it would pay. Price saw that it fell sadly below Milton, who had earned all of £10 for *Paradise Lost.* He had to advise Barlow that an anti-English poem would not be welcome. Barlow, nothing daunted, was thrilled with London, and particularly, as a keen reader, the 'prodigious' library (fifty thousand volumes) of the late Prime Minister Lord Shelburne, whose house in Berkeley Square was thought the best in Town. One thing shocked him: the blatant fiddle of a British election, when the Whig leader, Fox, replaced Pitt. Barlow's diary notes that one duke sent sixteen members to Parliament and that a merchant bought a borough for £10,000, which gave him a seat in Parliament and the right to sell another seat for £300 at each election.

All through 1789 Barlow worked at selling the Scioto scheme. He was in Paris to see the fall of the Bastille; the early days of the French Revolution convinced him of an inexorable march of human rights issuing from the American Revolution. But upheavals were not good for business, as Barlow tried to alert Winthrop Sargent on 25 August: 'Since I wrote last in July the Revolution in Paris & through France has intervened & somewhat retarded my operations.' He had to conceal 'a total destitution of money . . . from those with whom I was negotiating . . . Nobody knows it here who should not know it.'

It was characteristic to buoy himself up with hope, in the face of the separation Ruth was enduring. After a year he wondered if she might join him, but though dribbles of money came in, there was not enough to pay her passage. When he thought to borrow from the takings of the company, Ruth refused. She stopped writing in the second half of 1789, speaking to her husband through a silence he was not disposed to hear. His incurious reproaches take for granted a way of life where enterprise took precedence over home ties. In this way the gender gap widened even as respect for womankind appeared to close it. He was busy, Barlow wrote, and 'in the highest train of success, but it is not yet in my power to say the thing is finished'. He signs himself 'your constant & unchangeable lover friend adorer & husband' – it does show a sure touch to put 'husband' last.

When news came of further separation, Ruth could bear it no longer, and agreed to sail. Barlow was teasingly relieved to hear from her in March 1790: 'I thot [*sic*] you were dead, and wished myself so,' he replied. 'I am charmed & astonished & restored to happiness to find you well & fat.'

After the charms of persuasion, followed by thirty-seven days at sea, it was disconcerting for Ruth to arrive in London and find that she must wait day by day for her husband to appear. 'Can you forgive your old cruel Husband for playing such tricks with you?' he demanded from Paris. 'To run away from you – then to make you come after him, & still to hide himself from you.' To keep Ruth company, he dispatched the Connecticut wife of Colonel Samuel Blackden from South Carolina, an associate in Paris who was selling Kentucky land – gossip had it that Barlow was enamoured of Mrs Blackden. Meanwhile, he addressed the French National Assembly on 10 July to 'thunderclaps of applause'. This triumph he presents to Ruth in lieu of his person. Ruth suspected a mistress in Paris, for Barlow asked Mrs Blackden to convince his wife, 'I have not slept with any body but God since I slept with her –.' The same day, he urges Ruth once more: 'Be a sweet patient girl. I know your patience is exhausted, but it shall be the business of my life to reward you for all your goodness.'

Ruth's confidence was shaken. She stopped eating and sleeping – so Barlow was advised by Mrs Blackden. And still, he delayed. In the end it took him five weeks to appear.

During this and the following year Barlow failed to raise the necessary loan for the Scioto scheme to proceed, yet glowing promotions continued to circulate. Fellow-agents in America made no provision for receiving the first six hundred immigrants, who were mostly from the cities – professional men, tailors, wig-makers, clock-makers, dancing masters – not exactly the stuff of pioneers. Barlow made two efforts to avert disaster.

One was to ask a Virginia merchant to welcome the arrivals, but this merchant had nothing to gain beyond Barlow's vague hint of future commerce. To write to a stranger, be it never so civilly, was to grasp at a straw.

Barlow's other plan was to persuade Étienne Sulpice Hallet, French architect, engineer and surveyor, to migrate with the settlers and to construct buildings and roads, in concert with the company's agents living on the spot. So far, no company agent had contracted to live on the spot, yet Barlow's ready voice rolls on to Ruth: 'I have taken infinite pains to make every arrangement for [the settlers'] agreeable reception & happiness.'

Only when the settlers landed in America did they discover that the company had not found the money to buy the land, and could therefore give them no title to farms for which they had paid $15 an acre. After the long Atlantic crossing, and a further journey to the frontier, they found nothing but a few scattered log cabins and the fury of native Americans whose raids on settlers the promotions had neglected to mention.*

When a number of the emigrants made their way back to France, there were threats on Barlow's life. He lay low on the top floor of the quadrangle of the Palais d'Orléans (now the Palais-Royal), its entrance concealed by a gambling den on the floor below. In later years, Barlow looked back on 1791 as 'the period of our deepest difficulties', when Ruth's virtues showed themselves superior to his own: 'if I had been alone, or with a partner no better than myself, I should have sunk'. Barlow warned Ruth's influential brother, Abraham Baldwin (a Congressman since 1785; later, Senator for Georgia), of 'misinformation'. He and Ruth were moving to England.

In October 1791, from their new perch in London, he declared his intention to move on to a new role as disseminator of revolution. The date coincides with the stateswoman portrait of Mary Wollstonecraft in the course of writing her more famous *Vindication*. She was a central figure in

*Congress partly repaired the wrong some years later by giving the remnant of the colony a tract of 24,000 acres on the Ohio, known as the 'French grant' or Gallipolis. When Constantin Volney visited in 1796 he found only about eighty sallow, thin, sickly remnants of the six hundred settlers.

Johnson's circle where Barlow made his intellectual home for the following year. He invited her to tea, along with the radical dramatist Thomas Holcroft and the political philosopher Godwin.

His *Advice to the Privileged Orders* recommended commerce as a corrective to 'feudal' privilege, and an end to monarchies. The British government ordered it burnt. Undeterred, Barlow proceeded to publish a volume of populist verse, *The Conspiracy of Kings* (March 1792), applauding the French as they 'shake tyrants from their thrones and cheer the waking world'.

'Be assured,' Jefferson wrote to Barlow, 'that your endeavours to bring the transAtlantic world into the road of reason, are not without their effect here.' This letter was to be carried by Mr Pinckney, the incoming American Minister in London. 'He will arrive at an interesting moment in Europe. God send that all the nations who join in attacking the liberation of France may end in the attainment of their own.'

The worsening situation (the war to which Jefferson refers: the coalition of Prussia and Austria against revolutionary France) offered gains for neutral foreigners like Barlow. An American was persona grata in Paris, a product of a successful revolution and possessor of a passport granting him freedom to move between warring countries: an advantage for spying. Barlow was in France on a mysterious 'mission' when riots broke out at the royal palace of the Tuileries in Paris on 20 June 1792. 'The visit to the king by armed citizens was undoubtedly against the law,' he wrote to Ruth, 'but the existence of a king is contrary to another law of a higher original.' Not so long ago he had dedicated *The Vision of Columbus* to his most gracious majesty Louis XVI.

Fellow-Americans who had known Barlow in his earlier incarnations as chaplain and Hartford conservative were puzzled by the extremity of this shift. A simple explanation has been his conversion to the French Revolution. The completeness of this conversion – going further than others in Johnson's circle – was a passport to their confidence, though its speed in the wake of Scioto suggests that Europe's distance from America (and ignorance of America's scope for nuance) allowed Barlow to assume the identity of a revolutionary American, a national image as cover for a more complex character. Mary Wollstonecraft, with her 'fondness for tracing the passions in all their different forms', picked up a performing note in his voice: phrases to Ruth like 'I find heaven in your arms' were too ready, Mary thought, to be entirely honest. When Benjamin Franklin had played up to a simplified, Parisian construct of American identity, he was a diplomat; Barlow, too, in time, became a diplomat – American Ambassador in Paris under Napoleon – but how should we distinguish between the legitimate shifts of the diplomat and the shadier shifts of the opportunist? If we bring to public life the morality of private life, there may be no distinction to be made.

Fabulous wealth was to be the reward of the Barlow story. Where did it come from? When Ruth had joined Barlow in Paris in 1790, she had found herself 'pent up in a narrow dirty street'; by the end of the decade, they owned a Paris house opposite the Luxembourg Gardens, grand enough to be coveted by Lady Hamilton (wife of the British Ambassador to Naples and mistress of Nelson); and soon after, Barlow, this opponent of 'the privileged orders', is able to buy a 'seat' in Washington, second only to Mount Vernon, home of the first President.

This fortune was not made in America. It came about somewhere in Europe, but nowhere amongst Barlow's account books and careful copies of business letters is there mention of any activity, after Scioto, that could lead to wealth on that scale. We'll return to Barlow and his associate Imlay when this story shifts to France. Suffice to say now that Barlow lingered in Paris week after week in 1792, and eventually Mary decided that Charles must sail on his own for America. Her chief concern from spring to autumn that year was the fate of her favourite brother. For the Barlows stayed on in Europe, and the scheme of adoption faded. Everina made no contact with Ruth Barlow; she and Bess never tried their luck across the Atlantic where Ruth had promised answers to all their needs: money, security, homes, husbands, and the society closed to a governess. America hovered on the Wollstonecrafts' horizon in the course of 1792, and not for the last time a fantasy of the fresh promise of the New World entered their minds. In the case of Charles Wollstonecraft it became a reality. He sailed in October with an introduction from Barlow to a Yale friend, James Watson, in New York: 'This . . . will probably be handed you by Mr Wollstoncraft, a young gentleman of singular merit. I would thank you to notice him &

give such advice as may be necessary to speed him on his way to the Ohio, where he goes to be an American farmer.'

Every one of Mary's plans for her brothers and sisters failed: none of them was satisfied. All kept their beaks open. Her father's beak never ceased to clamour. In Wales, Bess found Mr Wollstonecraft in a poor state: skeletal, with a long beard, dirty, coughing, groaning, but still able to '*drink* very hearty' and ride ten miles a day. When Bess and her horse had a fall, she came round to hear her father calling to know if the horse's knees were hurt (for the horse had been borrowed). It was shaming for a governess to have her unkempt father turn up at the Castle, ostensibly to check whether a trunk had arrived but really after whatever pickings he could glean. When Mary wished him to '*save* in *trifles*', he flew into a rage and threatened to chase her up in London. With the help of Johnson, she took his affairs in hand. Godwin records: 'The exertions she made, and the struggle into which she entered . . . were ultimately fruitless. To the day of her death her father was almost wholly supported by funds which she supplied to him.'

Mary's efforts for her family were more practical than loving. Love had been Fanny's, and the word returns in relation to Margaret King. Their correspondence has not survived – unsurprisingly, since it was secret – but Mary referred several times to their continued solidarity. On 3 March 1788, she enclosed a letter for her 'dear Margaret King' in one to George Blood, who still lodged with his Dublin employer Mr Noble. 'Be very careful to not let any body see it,' Mary warned, ' – and keep it till she *sends* to Mr Noble's for it.' On 26 May she asked George again: 'Have you delivered – or rather has my letter to M:K. been called for? . . . I have received another letter from that dear girl – I scarcely know how much I loved her till I was torn from her,' Margaret said, 'From the time she left me my chief objects were to correct those faults she had pointed out & to cultivate my understanding as much as possible.' This fits the conclusion to *Real Life*, where Mrs Mason takes leave of her charges in this memorable way:

> I now . . . give you a book, in which I have written the subjects we have discussed. Recur frequently to it, [and you will not feel] the want of my personal advice . . . You are now candidates for my friendship, and on your advancement in virtue my regard will depend. Write often to me, I will punctually answer your letters; let me have the genuine sentiments of your hearts . . . Adieu! When you think of your friend, observe her precepts; and let the recollection of my affection give additional weight to the truths I have endeavoured to instill.

Mary had hoped not to lose touch with Margaret's step-grandmother Mrs FitzGerald. While in Dublin she had borrowed ten guineas on behalf of Betty Delane. The money had slipped through 'quicksand'. Two years later she asked George Blood to return £10 (an amount due to Mary for Caroline's care) to Mrs FitzGerald together with a friendly message: 'Tell Mrs Fitz-Gerald that I am well, and enjoy more worldly comfort than I ever did – you may add, that I should have written to her if she had asked me, as it would really give me pleasure to hear some times of her welfare – enquire about my Margaret &c.' George let this slide, and forgot. It was at this point that Mary let George know how he 'disappointed' her. His neglect of her letters and 'inconsiderate' excuses for his lack of punctiliousness in money matters were of a piece. Mr Noble had overworked him, he had whined, and he had again lent money to the painter Mr Home, a persistent borrower who had drained George before.

In the play of Mary's words, the sharp of 'independence' and the plangent note of 'tenderness' have as counterpoint the repeated flat of 'disappointed'. Her brothers and George Blood repeatedly disappointed her efforts to renew their lives. Yet these years developed her own remarkable capacity to rise, and rise again. 'Blessed be that Power who gave me an active mind!' she said in the manner so irritating to her dependants, 'if it does not smooth[,] it enables me to jump over the rough places in life. I had had a number of draw-backs on my spirits and purse; but I still . . . cry avaunt despair – and I push forward.'

Her resilience was reinforced by the kindness of William Roscoe and Joseph Johnson. Once, when she asked the latter to add up her debts so she might settle them, she added, 'do not suppose that I feel any uneasiness on that score. The generality of people in trade would not be much obliged to me for a like civility, *but you were a man* before you were a bookseller.'

Johnson, Mary, the artist Fuseli and his wife planned to visit Paris for six weeks in August 1792. They got as far as Dover, then turned back in fear of bloodshed when Louis XVI, Queen Marie Antoinette and their two children were caught trying to flee France. The flight to Varennes appeared to confirm a suspicion of the royal family as traitors to the Revolution, in league with France's enemy, the Austrian monarchy, the family of the unpopular Queen. The French monarchy fell, revolutionaries slaughtered the Swiss Guard at the Tuileries, the King and his family were sent to prison, and a republic was proclaimed. During these momentous events, Mary stayed a few weeks with Johnson in the country. When they returned to London on 12 September, she heard with amusement that 'the world . . . married m[e] to him whilst we were away'.

She had set herself against marriage from the age of fifteen, but two samples of married happiness forced themselves on her attention: first, the Gabells with whom she had stayed for two or three weeks in the summer of 1790 at Warminster in Wiltshire; and more recently Ruth Barlow, whose husband found 'heaven' in her arms. Neither fitted Mary's view of marriage as legalised rape. The caresses of the new-married Henry Gabell and large, sturdy Ann Gage had been absurdly unrestrained – though 'as pure', Mary assured herself, as those Darwin 'lavishes on his favourites'. She had to concede '*how much* happiness and innocent fondness constantly illumines the eyes of this good couple – so that I am never disgusted by the frequent *bodily* display of it'. They made her think rather crossly of Milton's Adam and Eve in Paradise, and count herself superior. Was this mistaken pride, she wondered, 'whispering me that my soul is immortal & should have a nobler ambition'? Sometimes she felt an intruder, and then she longed for her little London room and a life wholly tuned to 'intellectual pursuits'. She could not surrender those hard-won freedoms. There seemed no return route: 'my die is cast'. Even so, the road not taken did not entirely fade from view.

Mary's independence as a single woman was threatened less by dependants than by her strange fixation on Henry Fuseli. In 1789 she had attended a masquerade at the Opera House with Fuseli and the son of an early object of his ardour, his Zurich friend Johann Caspar Lavater. On that occasion, Fuseli had stirred up trouble with a fancy-dress devil. From the autumn of 1790, and increasingly in 1791 and '92, Mary took up with the artist, who was eighteen years older. It seemed to her that she had never before encountered a man of such 'grandeur of soul' and 'quickness of comprehension'. In October 1791, Bess received a letter from her that was 'brimful' of Fuseli.

After completing the *Rights of Woman* in a mounting burst of energy, Mary fell into inertia – 'palsied', Johnson put it – during 1792. In that state she became dependent on Fuseli to bestir her. A later protégé of his, the artist Benjamin Robert Haydon, describes how Fuseli took hold of a disciple with 'the most grotesque mixture of literature, art, scepticism, indelicacy, profanity, and kindness. He put me in mind of Archimago* in Spenser. Weak minds he destroyed. They mistook his wit for reason, his indelicacy for breeding, his swearing for manliness, and his infidelity for strength of mind; but he was accomplished in elegant literature, and had the art of inspiring young minds with high and grand views.' Mary, always on the lookout for 'genius', found in Fuseli the real thing: a painter who, a century before Freud, explored the psyche. Fuseli was himself obsessed with the divinity of genius (his own above all), a worshipper of Rousseau, Shakespeare and Milton, and given to explosive quotations from Homer, Tasso, Dante, Ovid, Virgil and the *Niebelungenlied*. Between 1786 and '89 he had painted a series of nine scenes from Shakespeare, exhibited with others by leading artists like Sir Joshua Reynolds and Angelica Kauffmann at Boydell's gallery in London. At this time they were the rage. Fuseli's figure drawings have a Michelangelesque muscularity. *The Kiss* is not confined to lips; it carries through arms and thighs to the extremities of fingers and heels.

Fuseli's own body looked a good deal tamer, but he held the eye with

---

*Archimago, in *The Faerie Queene*, 'the falsest man alive', is one of the devil's emissaries: false religion masquerading as truth's champion.

his steady gaze, provoking his listeners without relinquishing his own pre-eminence at the centre of attention. Godwin, who had reason to be jealous, recalled him as a monster of conceit and a great hater: 'he hated a dull fellow, as men of wit and talents naturally do; and he hated a brilliant man, because he could not bear another near the throne'. He was about five foot five, with small eyes between which sprang a fierce nose. A faintly cynical smile turned down the corners of his mouth. There's the look of a neat reptile in a portrait by John Opie: the narrowness of his green coat; the wide, down-turned mouth; and the dapper, darting steadiness of his turning head – turning a cold gaze on the viewer. The backdrop is a hellish dark red. Fuseli retained a strong Swiss accent, with a guttural, energetic diction. During a dinner at Johnson's, an admirer called out to him from the other end of the room, 'Mr Fuseli, I have lately purchased a picture of yours.'

'Did you?' said Fuseli, 'what is the *sob*ject?'

'I don't know what the devil it is.'

'Perhaps it *is* the devil,' Fuseli replied. ' I have often painted him.'

'Perhaps it is.'

'Well!' said Fuseli, 'you have him now; take care he does not one day have *you*!'

He seemed always in a rage, on the point of a rage, or recovering from one – useful for keeping people at a distance. His habits were bisexual, if we are to believe his boasts – and he did like to boast and jolt civil company with sudden obscenities. It was certainly his habit to write erotic letters to both men and women. Though married, he did not father a child. The smallness of his frame matched that of 'Little Johnson': both had a tight look, evident in Johnson's compressed lips and Fuseli's tight, attenuated body. In character, they were opposites: Fuseli's spitfire manner set off by Johnson's calm. Their attachment – maybe love – was lifelong. Mary's willingness to stay with Johnson in the summer of 1787 and again in the summer of 1792 shows he was no sexual threat, either too modest or uninterested in women's bodies. The way his lower lip seals the upper, pressing hard upon it, the two tight rolls of hair on either side of his cheekbones, and the stiffness of elbows and clasped fingers all signal reserve. When Johnson shook hands, a mixed reserve and affability swung his trunk from side to side; Fuseli was reserved in a different way: he met the inward eye and reserved what he saw.

To Mary, who longed to travel, Fuseli was a cosmopolitan with a worldliness beyond her range. He was born Johann Heinrich Füssli in Zurich in 1741, the son of an artist who brought him up to be a Protestant pastor. When he found Zurich too rule-bound, at the age of twenty-three he moved to London. There, he met Joseph Johnson who was three years older. He shared Johnson's quarters until the fire in Paternoster Row (in 1770) destroyed all Fuseli's paintings, manuscripts, clothes and savings. Soon after, he left for Rome, and when he returned to London near the end of the decade, Johnson welcomed him back. Their closeness lasted another forty years, with vacations on the coast at Margate or Ramsgate or, after 1804, at Johnson's rented house at Fulham on the Thames.

When Mary met Fuseli, he was at the height of his fame, following the success of *The Nightmare* when it was exhibited at the Royal Academy. In this painting a young woman in a white shift lies on her back. Her curved, exposed throat, head and hair fall back in terror, as a little, curled, grinning demon (with a look of Fuseli himself in the first sketch) plants himself atop her helpless body. An early version of this painting hung in the dining-room of Johnson's house. Later, Fuseli became pornographer to the Prince Regent. He could draw women with intent empathy, like his Swiss love, Martha Hess, and the graceful figure of her sister Magdalena ('Madeleine') Schweizer as she sits at work, but he could also obliterate character (on Pope's principle that 'women have no character at all'). He thought women were emotionally rampant and irresponsible, and that to believe in their virtue could only be an act of charity. When he uses his exquisite graphic skills to draw women flaunting their breasts, the viewer is invited to a voyeurism uncomfortably close to pornography – uncomfortable, because the breasts are delicately erotic, yet the woman's desires are not involved. The appeal of Fuseli as witness of character makes him all the more disturbing when he wipes it out.

Mary visited Fuseli at home at 72 Queen Anne Street East (later, Foley Street) to see his work, and he came to Store Street. Johnson encouraged their friendship, and they were often together at his house. Mary declared that she 'loved the man and admired his art' when she was seeking subscriptions for a volume of Milton, to be edited by Cowper, with forty-seven illustrations by Fuseli, engraved by Blake. Fuseli toiled slowly, subsidised by Johnson: the Milton project didn't come out till 1799, and then was a commercial failure. Wollstonecraft foresaw problems in the first sketches.

'Like Milton he seems quite at home in hell,' she told Roscoe (who had exhibited Fuseli in Liverpool), 'his Devil will be the hero of the poetic series; for, *entre nous*, I rather doubt whether he will produce an Eve to please me in any of the situations, which he has selected, unless it be after the fall.' This was the day she sent her final page of the *Rights of Woman* to the printer: 3 January 1792. She knew, then, that Fuseli had little sympathy with her cause, for he believed that women's genius was too 'fugitive' and 'intangible' to be communicated to posterity. Nor did she hesitate to alert Fuseli to his prime fault. 'I hate to see that reptile Vanity sliming over the noble qualities of your heart.' She also knew he was not to be trusted as a confidant.

Despite these reservations, Mary was gripped. In 1788, at the age of forty-seven, Fuseli had married an artist's model from Bath, Sophia Rawlins. His many sketches of her show off a stylish temptress, but she's no mere type. She's there, alive, disseminating sexual aplomb. Though she was never part of the company at Johnson's, she and Mary were friends and remained so. Late in 1792 Mary asked if she might move in with the Fuselis. Much has been made of this, but the evidence of what she actually had in mind is poor. Was she in love with Fuseli, as most assume? Or was she needy like a child who has lacked paternal protection and will seek out alternative fathers? Johnson was ready to oblige; his friend Fuseli was not. Or was Mary simply a kind of disciple? It's reported that she told Sophia that her idea arose 'from the sincere affection which I have for your husband, for I find that I cannot live without the satisfaction of seeing and conversing with him daily'. The statement is wooden – unlike Wollstonecraft's voice – and unreliable, transmitted through one source only, Fuseli's biographer, writing forty years after the event.

Even if she did speak in this peculiar way, she denied sexual feeling. '[I] hope to unite [myself] with your mind . . . [I] was designed to rise superior to [my] earthly habitation . . . [I] always thought, with some degree of horror, of falling a sacrifice to a passion which may have a mixture of dross in it . . . If I thought my passion criminal, I would conquer it, or die in the attempt. For immodesty, in my eyes, is ugliness.' These sole surviving snips taken out of context from Mary's letters to Fuseli are too insubstantial not to be, as William St Clair put it, 'biographically misleading'. Whatever it was that she had in mind, it was a plan easy for a man of Fuseli's vanity to misrepresent. He fancied that she had acquired elegant furniture and better clothes to impress him. One of his boasts was that he'd leave the letters of this attractive woman unopened in his pocket for several days. The sexual fantasies in his own letters show him as a master of innuendo channelled through a third party, suggesting, say, the complicity of the newly married Mrs Schweizer. This, then, was the man who turned on Mary as a temptress threatening his marriage with a *ménage à trois*, even though there is ample evidence that she opposed any form of promiscuity. In fact, violence at home had made her unusually self-protective. The presence of so eroticised a wife might have seemed to offer the security that Mary would *not* be a candidate for sex. Her daughterly attachment to the Blood family, her missing the caresses of her pupil, her bungled attempt to replace Margaret with a convenient orphan, her recent homemaking in Store Street, followed by this misguided impulse to attach herself to the Fuselis, add up to a frustrated craving to reconstitute a family.

In her own terms, Mary's proposal was innocent; the leer came from Fuseli. It should not be forgotten that 1792 is the year Wollstonecraft's fame surpassed that of Fuseli, a man who could not bear to be outdone and thought no woman could do lasting work. Conceivably, his rejection of Mary's proposal was not to protect his wife, but to score in the simplest way: to take the moral high ground and cast this woman in the banal role of sexual intruder. For Mary what hurt was the rejection, repeating the rejections of her early life. Whatever the truth, this is a complex relationship, but there's enough to question Fuseli's insinuations.

When Mary found she had been intrusive, she backed off and apologised. To Johnson, who could not console her (given his loyalty to Fuseli), she said with a winterly smile: 'We must each of us wear a fool's cap; but mine, alas! has lost its bells, and is grown so heavy, I find it intolerably troublesome.' She wondered if she had 'intruded' on Johnson too. Crying and laughing at once, she acknowledged 'that life is but a jest – and often a frightful dream – yet I catch myself every day searching for something serious – and feel misery from the disappointment'.

A remaining puzzle is whether Fuseli's rejection preceded, succeeded or coincided with Mary's decision to move to Paris – in effect, to press on with a story of her own. Ahead lay a risky move carried forward with jokes: 'I am still a Spinster on the wing,' she told Roscoe. 'At Paris, indeed, I might take a husband for the time being, and get divorced when my truant heart longed again to nestle with its old friends; but this speculation has not yet entered into my plan.'

Fuseli's biographer, John Knowles, propagated the myth that Mary Wollstonecraft was loose: 'Having a face and person which had some pretensions to beauty and comeliness', she 'fancied' possibilities with Fuseli, and had 'the temerity to go to Mrs. Fuseli' with a 'frank avowal' that 'opened the eyes of Mrs. Fuseli, who . . . not only refused her solicitation, but she instantly forbade her the house'. Knowles goes on in the same melodramatic vein: Wollstonecraft had 'to fly from the object which she regarded', and wrote a letter to Fuseli begging pardon 'for having disturbed the quiet tenour of his life'. Only the last rings true. The legend was sealed when Knowles let it be known that Mary's letters were too 'amatory' for public consumption. This claim is impossible to prove or disprove, for the letters have vanished.

Many have believed that Mary ran away to Paris because of Fuseli. Knowles stated: 'the attachment on her part . . . would be the cause of her leaving this country'. Adherents of this version are convinced that when Mary declared on 12 November 1792, 'I intend no longer to struggle with a rational desire, so have determined to set out for Paris in the course of a fortnight or three weeks', she was referring to Fuseli as the object of desire. Yet the word 'desire' in the eighteenth century merely meant 'wish'. Mary's 'desire' could just as well refer to a wish to witness the Revolution at first hand, a continuation of the June–August plan – a 'rational' plan formulated *before* Fuseli's rejection. Revolutionary leaders had talked of restoring 'natural' rights to women: property rights and an equal voice in family matters. In September 1792 a humane divorce law had been passed, the kind of legislation that Mary Wollstonecraft wished to see in England. She said later that to go to Paris was the only way to form 'a just opinion of the most extraordinary event that has ever been recorded'. Given a commission from Johnson to write a series of 'Letters from the Revolution', she resolved, as winter approached, to brave the guillotine and go on her own – 'neck or nothing', she joked as she set out.

9

INTO THE TERROR

In Calais, words flew past her straining ears. She caught 'a violent cold' as she sat three days in the winter *diligence*, a young woman travelling, as always, alone. She had intent, rather dreamy eyes (the 'most meaning eyes I ever saw', said the poet Southey) with a hint of vulnerability in her air of 'the princess'. As the vehicle, carrying eight to twelve passengers, rolled heavily over the wide flintstone road towards the capital, she wondered how she was to report on a revolution without the language to unlock it.

So it was that Mary arrived 'awkward' and coughing at her destination in the third *arrondissement*, 22 rue Meslée, a six-storey house opening straight on to the narrow street. She had arranged to stay with a married daughter of the Putney headmistress Mrs Bregantz, but it turned out that Aline and her husband, M. Fillietaz, were away. Here, knocking at the door and mounting the stair, is this venturesome woman we've come some way to know. She moves through deserted rooms, folding doors opening one beyond another, till the servant leaves the stranger – 'almost stunned by the flying sounds' of the spoken language and 'unable to utter a word' – in a remote room at the back of the house.

Mary holed up here, alone, for the two weeks before Christmas, shrinking from dirty streets (she excused herself to Everina), but as yet unready 'to form a just opinion of public affairs'. At home in Store Street, in the company of her cat and breathing the clean, damp air of market gardens, Mary had said tartly, 'children who meddle with edged tools are bound to cut themselves'. So she'd remarked after the September Massacre of fourteen hundred prisoners whose screams were heard with indifference by fellow-Parisians until the mob bashed the Queen's friend the Princesse de Lamballe, disembowelled her, and waved her dressed head on a pike. Blood had seemed to Mary an aberration from the rational progress of the Revolution. Now, on 26 December, she witnessed the renewal of the King's trial: the emptied streets, the eyes behind the shutters, and Louis himself, upright in the close-guarded coach, rolling along the boulevard Saint-Martin on his way from the Temple prison to his tribunal. A spatter of drumbeats seemed to deepen the silence surrounding the King. As Mary relived the scene some hours later, writing in the shadows of her room, eyes glared through a glass door. Then hands approached, dark with blood. 'For the first time in my life,' she wrote, 'I cannot put out my candle.' That night, she came face to face with the oncoming Terror.

Until this night, she had delayed her letter to Johnson, in the hope of being able to say that more blood would not flow. Tom Paine, as honorary French citizen, member of the National Convention (replacing the Assembly), and the first (as in America) to propose a republic, argued for the King's life. All he achieved was to cast doubt on his own loyalty to the Revolution. On 21 January 1793, the King was guillotined before a crowd of eighty thousand. Ruth Barlow reported from London that the friends of the regime, as well as its enemies, condemned it. 'The National Convention have most certainly disgraced themselves . . . they have shewed themselves too cruel & blood thirsty . . . .their friends are ashamed of them.' The English, Ruth said, were venting their outrage on the French Ambassador with a view to provoking a declaration of war: 'M Chauvelin is infamously treated.' On 1 February he was ordered to leave the country, the English Ambassador left Paris, and English residents there, fearing to be trapped in the walled city, rushed homewards in a second wave of panic. The first, in December, had carried the young Wordsworth away from Annette Vallon, his French teacher, who had given birth to his child. Wordworth's desertion is said to be sensible under these circumstances. For Paris, he saw, had become a 'place of fear/ Unfit for repose of night/ Defenceless as a wood where tigers roam'.

Thinkers, not tigers, had initiated the Revolution, from *philosophes* Voltaire and Rousseau to Mary's hero Mirabeau, an orator whose resolution had been born during forty-one months in a dungeon when he was twenty-eight. He had come into his own when the Estates-General were summoned to consider the failing economy in the fateful year of 1789. This body consisted of the three 'estates': the first was the nobility; the second, the higher clergy; and the third, men of property, professionals and the middle-class intelligentsia. Six months before, at the time Joel Barlow had arrived in Paris, Jefferson, the American Ambassador, and Lafayette had introduced Barlow to prospective leaders of the Third Estate. At a Lafayette dinner, he was impressed by a conviction that the whole of France was so united behind the idea of regime change, and the old regime so susceptible to public opinion – even 'soft' – that the leaders expected to bring off a bloodless revolution. All the same, Barlow picked up what sounded for him a warning note: a feeling in Paris that the Americans had thrown liberty away, with their recent acceptance of a federal constitution with controls on every side. 'They say we [Americans] have given up all that we contended for.' Why did these Frenchmen speak as though liberty were inconsistent with control? To Barlow such criticism showed the bravado of novices: 'they are as intemperate in their idea of liberty as we were in the year seventy-five – '.

The first real threat to the monarchy had come on 5–6 October 1789, when six thousand market women marched on Versailles. Two of them rode astride their cannon. The others walked the twelve miles in the rain, bearing pikes and muskets, armed for the first time. They were protesting about the shortage of bread and the price of it. The King agreed to help and to accept the August Declaration of the Rights of Man, yet the crowd at the gates of Versailles remained unsatisfied. Next morning at dawn they broke into the palace and swept through the Hall of Mirrors towards the royal apartments. Narrowly escaping death, the royal family was brought to the Tuileries in Paris. Here the King would be more accessible to the pressure of the street. The 'Austrian' Queen, Marie Antoinette, was particularly unpopular. In 1792 when the Austrians and Prussians declared war on revolutionary France, initial French defeats and a climate of threat and fear began to foment suspicions of internal traitors. On 10 August that year the mob – joined by the National Guard – slaughtered the King's Swiss Guard. This marked the fall of the monarchy (that anarchic moment when Mary Wollstonecraft, Fuseli and Johnson had to abandon their journey to France). From now on France became a police state. About a thousand arbitrary arrests followed a systematic massacre of helpless prisoners, including nuns and priests who had not sworn allegiance to the Revolution. The Princesse de Lamballe was prepared to offer allegiance to the Revolution, but was doomed when she refused to swear an oath of hatred against the Queen.

This atrocity was the turning-point for public opinion in England. Up till then the Revolution had stirred the freedom to rethink every aspect of life anew. What, for instance, does it mean to be a human being? Wordsworth famously wrote, 'Bliss was it in that dawn to be alive.' But after the bloodshed of 2–3 September, England, including Wordsworth, turned its back on the Revolution.

Some Englishmen who remained in Paris were refugees, amongst them Paine and John Hurford Stone, an abruptly enthusiastic man of thirty who had been a member of Dr Price's congregation in Newington Green. In November 1792 they and other foreigners gathered at White's Hotel at no. 7 passage des Petits Pères, not far from the clash of factions, moderates (Girondins) versus radicals (Jacobins), in the geometric arcades of the Palais d'Orléans (where Barlow had once lain low). The company at White's was celebrating the anniversary of the revolutionary sermon Dr Price had delivered in November 1789. Back in England, this event was reported as treason, and the same month Barlow was accused of sedition. Instead of sailing for America, as Ruth wished, he fled to Paris where he was made an honorary citizen of the new French republic. At the time Mary arrived he was away, taking immediate advantage of his citizenship to stand for election in Savoy.

Mary did not share the plight of the refugees, and as war broke out her obvious course was to go home. Soon, France would be out of bounds to

English travellers. The mockery of the King's trial and the look on faces under red woollen caps decorated with the tricolour did not inspire confidence in the will of the people. Vehicles leaving Paris were filled to capacity when an Englishman offered Mary a place in his carriage.

Should she go or stay?

It was typical of her rationality that, at this moment of decision, she set herself to examine the French character. An uneasy article, dated 15 February, questions the amiability of French manners in the lull after the King's execution. It did prove a deceptive lull. The Revolution was moving too fast, she thought, wishing it would release its energies through a slower, more durable process of education. Withdrawing her foot from the carriage bound for England, she turned to formulating 'a plan of education' for the Committee for Public Instruction. It's not known if this contribution was an initiative of her own or invited by a member of the committee: Paine or perhaps Condorcet, a deputy who concerned himself with the position of women. To Mary Wollstonecraft a new system of education presented an irresistible opportunity to act on the platform of history. So she joined a dwindling circle of foreigners who continued to trust in the ultimate benefit of the Revolution.

Amongst the first of her contacts was another Englishwoman, Helen Maria Williams, whose *Letters From France* had done much to win support for the Revolution in its hopeful phase. Richard Price had been her political mentor, as he'd been for Mary Wollstonecraft: all three warned that it would be useless to level differences of birth, only to make way for that of wealth. Expatriates at White's sang a song by Helen Maria Williams, and toasted the 'Women of Great Britain . . . who have distinguished themselves by writing of the French Revolution'. Bluestockings Mrs Montagu and Anna Seward praised her sensibilities, and one of Wordsworth's first poems was a 'Sonnet on seeing Miss Helen Maria Williams weep at a tale of Distress'. It starts: 'She wept. – Life's purple tide began to flow/ In languid streams through every thrilling vein/ Dim were my swimming eyes . . .' Mary Wollstonecraft was less charmed. 'Authorship is a heavy weight for female shoulders,' she thought, 'especially in the sunshine of prosperity.'

And yet she did 'rather like' Helen Maria Williams. A genuine civility and '*simple* goodness' broke through 'the varnish', drawing Mary to her Sunday evening salon, over English cups of tea. The French company was refreshingly intellectual: leading Girondins like Brissot de Warville and Vergniaud might be heard; Mme Roland (influential wife of the Minister of the Interior) and Mme de Genlis (whose tales Mary had read at Mitchelstown Castle) aired ideas without self-consciousness as women. Tall, dignified Mme Roland was an idealist who saw the Revolution as the purification of the soul of France, and was known to have drafted her husband's petition to the King to accept the people's decrees. She spoke energetically with grace and eloquence. Mme de Genlis had once been governess to the children of the pro-revolutionary duc d'Orléans, who called himself Philippe Égalité. Frenchwomen, Mary saw, had 'acquired a portion of taste and knowledge rarely to be found in the women of other countries'. It delighted her to meet women whose 'affectionate urbanity' seemed to demonstrate 'the true art of living'. These were 'the rational few', she mused. They were less romantic than Englishwomen, less prey to unsatisfied desires, and married couples seemed 'the civilest of friends'. Her *Rights of Woman*, translated as *Une Défense des Droits des Femmes, suivie de quelques Considérations sur des Sujets Politiques et Moraux*, was her entrée to some of the Revolution's leading figures, including Roland himself and Jérôme Pétion, the complaisant mayor of Paris, as they became its victims.

Fuseli had given her an introduction to his Swiss friend Madeleine Schweizer (formerly, Magdalena Hess). During those first weeks in Paris, Mme Schweizer was Mary's 'favorite'. Eight years older than Mary and dressed in the flowing drapery of classical figures, she and her banker husband entertained a cosmopolitan circle in the rue de la Chaussée-d'Antin. In that luxurious setting, Mary was amused and not displeased to find herself attracting men's attention. Once, when her mind was on England, she replied '*oui, oui, oui*' to a man who teased her that she might chance to say '*oui*' when she did not intend it, '*par habitude*'. At the Schweizers she had to speak French, and though by February she was 'turning a corner', sometimes a 'foolish bashfulness,' she said, 'stops my mouth when I am most desirous to make myself understood', and 'all the fine French phrases, ready cut and dry for use, fly away the Lord knows where'.

These two circles were peripheral to Mary's mainstays during her first months in Paris: her host, M. Fillietaz, who turned out to have the soft manner the Wollstonecraft sisters liked in a man, and Thomas Christie, co-founder of the *Analytical*. Christie was an old pupil of Dr Price who in 1789 had sent him to France with introductions to Mirabeau and Finance Minister Necker. In a book which explained the French constitution to English readers, Christie had hailed the Revolution as the regeneration of mankind. At his house Mary met leading deputies of the Convention like Henri-Maximin Isnard who proposed the fearsome Committee of Public Safety, the prime organ of the Terror, and Bertrand Barère, a lawyer who was shifting with the political wind and soon to be elected to that committee. Both had voted for the execution of the King. At the other end of the political spectrum was Pierre Louis, comte de Roederer, a moderate associated with Lafayette, who had been a member of the Constituent Assembly in 1789 and had tried to warn the King before the onslaught of 10 August. All three visited Miss Wollstonecraft, recalled a wealthy young guest from Derby, I. B. Johnson, a follower of Paine. He observed that she was 'particularly anxious for the success of the revolution[,] & the hideous aspect of the . . . political horizon hurt her exceedingly'. The previous September, Christie had married a London heiress called Rebecca Thomson, granddaughter of a carpet manufacturer in Finsbury Square. This changed Christie's fortunes: he had been bankrupt, and pursued by a French mistress who had borne his child and whom he had promised to marry. The inconvenient mistress he shook off. It's not known what Mary thought of this, but she found a friend in his wife.

Tom Paine, whom Mary had met at her publisher's in November 1791, was part of Christie's circle. She visited him at an *hôtel* with an elegant façade and long windows, 63 Faubourg Saint-Denis. Mme de Pompadour had lived there, outside the city walls, through the Porte Saint-Denis and up a slow rise – a long but manageable walk from the rue Meslée. Once, Mary dined with Paine and a militant female of a kind she had not encountered before.

Twenty-seven-year-old Anne-Joseph Méricourt from Liège was a failed singer turned courtesan. In that role she had been an immaculate beauty, her short dark hair setting off the creaminess of her perfectly oval face, with a crown of ribbons, a delicate ruffle about her neck and a rose at her breast. Come the Revolution, she donned Grecian robes or played the role of 'Théroigne the Amazon'. On 5 October 1789, when she joined the march on Versailles, she arrived in a red riding-coat, astride a black horse, sabre in hand and pistols at the ready. She called for a legion of Amazons to defend Paris and the Revolution, and declared that bearing arms made each woman a citizen.

Women must 'compete' with men, Théroigne argued. 'We too wish to gain a civic crown and gain the right to die for liberty, a liberty perhaps dearer to us since our sufferings under despotism have been greater.' To pursue this aim, she founded the Club des Amis de la Loi, yet women did not rush to fight for their country. Mary Wollstonecraft met Théroigne de Méricourt when her fortunes were about to turn.

On 15 May1793 she was attacked by market-women in red pantaloons and caps who were terrorising the streets. These women, who had marched on Versailles, now supported a mob from a rival club, the militant Société des Républicaines-Révolutionnaires led by a chocolate-maker's daughter, Pauline Léon – the only club with any chance of winning a popular base amongst the women of Paris. This Jacobin club had recently formed to overthrow the ruling Girondins, who favoured a free market. Girondin supporters like Théroigne were the particular *bête noire* of the market-women, who fought for fixed prices. They advanced on Théroigne while she was making a speech on the Terrasse des Feuillants, stripped her, and bashed her head with stones, leaving her deranged. In her cell at the Salpêtrière asylum, she sat naked, refusing clothes and muttering against her betrayers. Théroigne's mistake was to think that women's rights had ever mattered to the *poissonnières*, whose prime concern was the price of bread. Wollstonecraft judged the lot as 'the lowest refuse of the streets, women who had thrown off the virtues of one sex without having the power to assume more than the vices of the other'. She had no truck with violence and never participated in women's revolutionary clubs.

As winter advanced into spring and France faced defeat on its borders, the moderates in the Convention were losing ground to the mob rule deployed by Jacobin leaders, Robespierre, Marat, Danton and Saint-Just. Roland was forced out of office; Mme Roland, whom Mary is said to have visited, was blamed for undue political influence; her salon in the rue de la Harpe was fading, as was that of Mme Condorcet, wife of the deputy who supported the rights of women. Two milieux Mary had joined, those of Christie and Helen Maria Williams, by virtue of being English, were now composed of 'enemy aliens'. Early in March, she shifted towards a set of Americans who sustained a fluid existence between revolutionists in France and counter-revolutionists in England. Sliding between opposed worlds, they held to the enterprising edge of their own independence. Mary Wollstonecraft's entrée to this milieu came about through the Barlows.

In London, Ruth Barlow was torn between fear of France and English blame of her husband: ' – here you cannot return at present,' she warned him, ' – every thing evil is said of you, & I am obliged to avoid company not to hear you abused. I hope you may be provided for in some eligible way in Paris – or what is to become of us.'

Ruth argued once more for what seemed to her the only reasonable course of action: return to America. Barlow's characteristic answer was to pull out all the stops: his wife was good, she was constant, she was his 'tender friend', and he vowed to love her all his life. No, he could not consent to Ruth's sailing home separately. He relied on fellow-Americans to meet her needs in his absence: Benjamin Jarvis, John Browne Cutting (whose brother, Nathaniel, was consul in Le Havre), and Mark Leavenworth (another Yale graduate whose wife Ruth had introduced to Mary), all part of a network linking the warring cities.

Ruth asked her husband to enclose her letters in those to Mr Leavenworth, 'as all letters with my name on them are opened'; Barlow preferred to address 'Mrs Brownlow' at her lodgings at 10 Great Titchfield Street. The war had placed Ruth in a difficult situation: ports were blockaded, frontiers surrounded with armies, and sailing to America increasingly unsafe. It was rare for her to protest, but now she did: 'I fear my love you did wrong in going to Paris'.

In answer, Barlow outdid himself in his twelfth-anniversary verse for 26 January:

> . . . Those charms that still, with ever new delight,
> Assuage and feed the flame of young desire;
> Whose magic powers can temper & unite
> The husband's friendship with the lover's fire.

Ruth stood by him. 'I know your conduct will always be directed by humanity integrity & a desire to promote the good of your fellow creatures.'

'*Tu m'as enchanté, ma charmante épouse, par ta lettre du 28 janvier*,' Barlow replied. He claimed to fall back on French or Italian for the language of love, but his verse, always in English, is equal to his extravagance, and truer to his playful humour. On 5 March he returned to Paris from Savoy, blithe as ever after failing to be elected to the National Convention.

'I am not at all disappointed or mortified,' he assured Ruth. 'Another thing which is better I believe will succeed soon.'

This unnamed 'thing' was nothing less than to wrest Louisiana from Spain. It was then a vast territory stretching from the southern frontier of Tennessee and the western frontier of the Mississippi across the central plains as far as New Spain in the west. France had ceded this territory to Spain in 1762, and for some years had eyed it regretfully – particularly Brissot, an ex-spy, now a leading deputy of the Convention, who had visited the States in 1788 and involved himself in the Scioto scheme. Barlow, who had recently translated Brissot's *New Travels in the United States*, now offered a plan of action. He was not the first to do so. Ahead of him was a tall, handsome American, verging on forty. His name was Gilbert Imlay and he was an authority on the frontier.

Mary Wollstonecraft met this frontiersman in March–April 1793. At the time, she was in constant touch with Barlow, whose plans for a coup were identical with those of Imlay. Barlow was the first to predict a romance for Mary when, on 19 April, he told Ruth: 'Between you and me – you must not hint it to her nor to J[ohnso]n nor to any one else – I believe she has got a sweetheart – and that she will finish by going with him to A[meric]a a wife. He is of Kentucky and a very sensible man.'

*

Gilbert Imlay was one of the most enigmatic figures to emerge on the frontier in the 1780s. He had a solid background. Within a generation of arriving in America, the Imlays were rich enough to be gentlemen, and by the time Mary Wollstonecraft crossed Gilbert's path, leading members of his family were living in style in the new-built Imlay Mansion in Allen's Town, New Jersey.

In the 1690s Peter Imlay, a settler from Scotland, had acquired tracts of land in eastern New Jersey. His son sold these lands, and in 1710 a deed records that 480 acres of flat farmland in western New Jersey were sold to 'Patrick Imlay, Gent', for £330. This became Imlaystown, where a great-grandson, Gilbert, was born in about 1754, the last of three children. Their mother must have died when they were small, as their father took another wife, Mary Holmes, when Gilbert was eight. His elder brother, Robert, settled in Philadelphia as a merchant in the firm of Imlay & Potts. Philadelphia was America's busiest port, with the topgallants of merchantmen looming over Water and Front Streets, and wharves stretching nearly two miles along the river. Here, a cousin, John Imlay, five years older than Gilbert, made his fortune in shipping. His trade lay with the West Indies where he owned a plantation and slaves on the island of St Thomas. It was John Imlay who, in 1787, built the fifteen-room mansion on South Main Street, Allen's Town, a white house with shutters, fanlight, and hand-carved winding stair, in a white-washed village close to Imlaystown. Had Mary Wollstonecraft accompanied Imlay to America, as he and Barlow proposed, the family of her frontiersman might have surprised her, together with their Queen Anne chairs, Chippendale dining-table and high post bed with a canopy to match the wallpaper. It was not the grandeur of the Earl's house in Henrietta Street, but its clean lines and balanced wings have the quiet grace of the domestic architecture of the Early Republic.

By the late eighteenth century the Imlay clan stretched from western New Jersey across the Pennsylvania border to the newly independent capital of Philadelphia. The papers the Imlays preserved are almost all accounts. No deal was too minute to be recorded; no bit of ribbon unpriced, along with great cargoes of beef and sugar. The odd surviving letter is purely functional. No flicker of character comes through these transactions, apart from pride in commerce and the orderliness of settled enterprise.

During the American Revolution, New Jersey was the site of more battles than any other state: three of the most decisive, Trenton, Princeton and Monmouth, were fought near Gilbert's home. The area was largely Loyalist (loyal to the Crown), including an Imlay cousin who was imprisoned in New York by American forces. Eight other Imlays, including Gilbert's father (imprisoned by the British), fought on the American side. In January 1777 Gilbert, aged twenty-two, joined up as 1st lieutenant and paymaster in Colonel David Forman's Regiment of Continental Troops. Four months later, he persuaded seven out of eleven Loyalist prisoners to change sides and join his regiment. Surprise attacks leading to American victories at Trenton and Princeton were helped by secret agents paid by Washington to cross the lines carrying goods for sale. The spy as salesman: this is how some daring young men of Imlay's generation developed a commercial face.

Imlay had signed up for the duration of the war, so it may seem odd that after only a year and a half, in July 1778, he was 'omitted' from his regiment. A common suggestion is that he was wounded in the battle of Monmouth at the end of June, but a wounded man would have returned to his regiment after his recovery. Imlay bore no subsequent signs of injury. A more likely explanation may lie in the fact that the month of Imlay's 'omission' coincides with Washington's formation of a new intelligence service. It's conceivable that Imlay's initiative and success in converting Loyalists to the revolutionary cause could have led to an appointment as secret agent moving between the lines close to home. This would explain his disappearance for the remainder of the war (apart from one sighting of him in the capital, fashionably dressed). Washington acted as his own effective spymaster. His spies were sworn to permanent secrecy so that some remained unrecognised. If Imlay did serve his country in this way it would explain his disappearance, and also a quality Mary warmed to in his character, an easy dignity. After the war he reappeared as 'Captain' Imlay. It's thought the rank was bogus, yet not one of the high-ranking officers who were his postwar associates ever questioned it.*

Imlay surfaced in 1783. The peace of 1783 between the United States and Britain left the frontier open to American settlement, and Imlay was speculating in Kentucky land as early as April of that year. His earliest deals were with Isaac Hite, who never recovered what Imlay owed him. On one day, 11 November, Imlay acquired vast tracts amounting to 17,400 acres, on the Licking River in Fayette County. At the time there were expanding settlements at Lexington and Louisville, but Kentucky was not yet a state; it was largely wilderness reached via a narrow, rocky track passable only by packhorse. In 1775 Daniel Boone, prime pathfinder of the frontier, had hacked this track through a gap in the north–south chain of the Allegheny Mountains. Boone was a legendary frontiersman, shaking off the trammels of civilisation, a fighter and one-time prisoner of the Native Americans who learnt their lore. Nine or so years after Boone broke through the Cumberland Gap, Imlay appropriated ten thousand of Boone's loveliest acres, near Limestone, worth at the time £2000. This was the bond Imlay offered Boone, entranced, he said, by the balmy climate on the far side of the Gap, the azure sky, the flowers, the gentle breezes, the birds singing 'in unison with love and nature'. His paean to a new Eden was like Barlow's promotion of Scioto, except that Imlay had inspected it in person.

He took an alternative route, travelling from Pittsburgh in March 1784, a journey of five days on a flatboat down the Ohio River as it flowed towards the Mississippi. His arrival and quick plunge into the ferment of

**The Imlay Family* by Hugh and Nella Imlay (1953) reports (p. 188) that 'Captain' Imlay 'was a British officer in the Revolution'. It could mean that he crossed sides in the course of the war when he was 'omitted', but this is unlikely, given the high-ranking veterans of the American army with whom he associated after the war. Could he have become a spy who, as a double agent, *appeared* to spy for the British (like Harvey Birch in James Cooper's *The Spy*)? It would explain why the listings of revolutionary soldiers from Monmouth County (in the Van Kirk Historical Collection, Allentown Public Library) register other members of the Imlay family but not Gilbert. Coming as he did from a Tory town, men known to him were fighting for George III, so he could have joined them in the guise of a Loyalist convert. If he did so, it would seem to have been a secret kept from his family, who give no source for their identification of Gilbert as a 'British officer', a fact tucked away without comment, as though it were shaming, in an appendix of the Imlays' book.

land speculation are registered on a scrap of rough paper recently acquired by the Beinecke Library. It's a scrawled note in a looped hand, sent from the furthest outpost of the frontier, on the western edge of Kentucky, to the New Jersey commander Colonel Henry ('Light Foot Harry') Lee who was now settled in Lexington:

> Dear Sir,
> I omit[t]ed mentioning in my last by M[r] [Alexander Scot] Bullet that Cap[t] Martin would survey the 2000 of [Col. John] Holder's opposite the little Miami [river] also that M[r] Hite will survey the 2000 of Hichman[.] You can know from M[r] Friplet what he has done[.] Success attend
>
> Adieu
> G. Imlay
> Falls of Ohio 21st Apl *1784*. . .

Imlay bought the Boone legend together with the land. The legend was set out by John Filson in 1784, and published on Boone's fiftieth birthday at the time Imlay met him. Until then America had turned east, to Europe, the Atlantic, and fortunes, like that of John Imlay, made by sea; Boone, and those who followed him, saw a future in the West. In October 1784, Imlay, aged thirty, took off into the pathless woods. He did not intend to return till Christmas, which means that he would have been without shelter for nights on end in winter temperatures.

The more he saw of the frontier, the hungrier he grew for lands stretching to untold horizons. That vast obscurity rolling westward beyond the Mississippi filled his imagination. Its vastness was commensurate with the new-found possession of America for which he and his friends had fought. It felt to them as though it were theirs for the taking. In February 1785 he entered into another bout of buying, from John May and his partner Mr Beall. These two rather cautious men, aware that Imlay could offer no satisfactory securities, did nevertheless sell him numerous tracts of land, reassured by his wilderness heroics and the prestige of his associates, heroes of the late war, Colonels Henry Lee and James Wilkinson. Imlay looked to what the French politician and traveller Brissot de Warville called 'the Empire of the West'; Imlay envisioned 'the vast extent of empire . . . only to be equalled by the [commercial] objects of its aggrandizement'. Seeing thus far, he neglected sometimes to 'see' the native inhabitants of the lands he and others had to have: the 'Indians', he said, must acculturate to the ways of the settlers, or end in reserves. Yet he got close enough to the land's inhabitants to admire them as 'very understanding people, quick of apprehension', industrious and hardy. Though squaws, he saw, were 'very slaves', the men treated one another as equals, esteeming 'personal qualities'. He particularly admired their eloquence, public confabs and long remembrance of the dead.

In the spring of 1785 Congress passed an ordinance that western lands be surveyed for townships or 'ranges' six miles square. Imlay now became a deputy surveyor, with the backing of a surveyor who had been in the same battalion as John Imlay. A week later he received a court order to appear for debt. This didn't stop him acquiring more and more land, outdoing old Peter Imlay; only, he couldn't pay for these lands. By December 1786 Gilbert was writing to Boone that he was 'sincerely sorry it is not in my power to pay [what he owed him], for Such is the embarrassing State of affairs in this Country that I have not been able to recover a pound from all the engagements that have been made me'. It was common for speculations to fail, said William Cooper (father of James Fenimore Cooper), who established Cooperstown at this time on the New York frontier. His own practice was to give settlers a genuine deal to aid their survival. Success would quicken their efforts with dreams for their posterity if the speculator put something back, if he settled with his settlers and saw to their needs.

Involvement at that level did not occur to Imlay, any more than it occurred to Barlow and the promoters of Scioto. Yet Imlay was not cold to the wilderness; it was not solely a source of gain. He saw it as the paradise of a man finding a simplicity at odds with 'the distorted and unnatural habits of the Europeans'. This projects a quintessential American drama, from the simplicity of Boone and Cooper's fictional frontiersman Leatherstocking, to Twain's innocents abroad, and to the innocence of *The American*, *The Portrait of a Lady*, and other American encounters with the corrupt Old World designed by Henry James. Undeniably, it was in Imlay's interest to present the virtue of American simplicity to the settlers he hoped to attract, but he was not a cynic. It was part of the ambiguity of Imlay that he did partake of the wilderness and feel its transforming power. If he lived out the American myth during his years on the frontier, if afterwards he presented himself as 'of Kentucky', embodying for Mary Wollstonecraft the natural man of Rousseau, it was not mere performance. There was a readiness, even innocence, as Imlay plunged into the chaos of land speculation. His temper was sanguine. When he lost money, he hoped a bigger deal or the introduction of ironworks would bring the fortune he was determined to make. So, little was said of wild buffalo, biting insects, and the determination of the Shawnee chief Tecumseh to check settlers' encroachments, especially after surveyors like Imlay began to act. The Shawnees and Cherokees had been deserted by their British allies, who handed over their lands without consulting them. Imlay underrated their rage: the 'Indians', he told prospective settlers, had now ceded their lands 'as a consideration for former massacres', and 'people in the interior settlements pursued their business in as much quiet and safety as they could have done in any part of Europe'. He did not mention armed ex-soldiers shooting at will, and Shawnees in the process of killing or carrying off fifteen hundred settlers in Kentucky.

Further south, beyond the Tennessee River, lay Spanish territory with its French population and disaffected Americans, including a number of outlaws for whom the 'trace' (as buffalo tracks were called before they became roads) to the Spanish border post of Natchez offered a route of escape. It was a refuge for Rachel Robards, who broke the law when she ran away from a brutal husband. In Natchez she lived with future President Andrew Jackson, man and wife in their own eyes but beyond the law. Separatism was a mood Imlay played on, together with the sharpest operator in Kentucky, Colonel Wilkinson, Imlay's partner in some land deals: one letter from Wilkinson warns a buyer not to 'expose' and 'ruin' Imlay by pressing for titles to his land unless the buyer wished to lose his claims. When Imlay had to vanish, his properties in Kentucky and New Jersey were administered by top veterans Wilkinson and Lee.

Wilkinson himself provides a clue to Imlay's secret life. He was a known secret agent who was daring, plausible, and made loads of money. As a fighter he had proved his worth at the battles of Princeton and Trenton, and Washington – acting as his own spymaster and commandeering a massive 13 per cent of the federal budget for spying – continued to trust him despite rumours that Wilkinson was in the pay of Spain. Wilkinson and Imlay arrived on the frontier in 1784, as alarm spread when the Spanish governor of West Florida and the Louisiana Territory, Esteban Miró, closed the Spanish port of New Orleans, the outlet of the Mississippi, to American boats. Wilkinson, posing as a sympathiser of Spain, warned Miró of an American attack. He appears as number 13 on a roster of the Spanish espionage system with the code name 'El Brigadier Americano'. Number 24 is Imlay's associate Colonel Bullet of Kentucky, and number 37's name, for what it's worth, is given as 'Gilberto'. The name is too common to draw conclusions, but it opens up a possible line of enquiry.

Gilbert Imlay disappears from the start of 1787 until 1792. He lets it be known that he has left for Europe at the end of 1786, but this remains unproved. The only sign of him is his grant of a deed for two thousand acres of Kentucky land on 27 December 1789, and a further deal in 1791, but unlike other grantors, Imlay gives no address. One clue lurks in a batch of spy-letters: a letter from Wilkinson asks the Spanish governor in West Florida to remember him affectionately to 'Gilberto'. So, did Gilbert Imlay slip across the south-west and follow the trace through 'Indian' territory to Spanish Florida, operating there as a spy, as well as hiding to evade prosecution for debt (like Barlow after the collapse of Scioto)? At a guess, he played a double game, posing as a Spanish spy to reinforce Wilkinson's successful efforts to soothe the Spanish governor with an illusion of influence, while at the same time acting as an agent provocateur on behalf of the US to foment separatism from Spain during a phase when the vast Louisiana Territory was prey to contesting imperial powers.

Wilkinson now put together a band of confidential agents who would sell Louisiana land to western Americans (encouraged by handouts from Spain) who would then, he promised Miró, 'constitute a Barrier against the encroachments of Great Britain or the United States'. Yet the upshot was, in effect, the encroachment of the US on Spanish lands – at Spanish expense. In all, Wilkinson extracted over thirty thousand dollars from the weak governor of what Spain regarded as a minor colonial outpost.

Since 1785 a French spy, one D'Argès, had been planted at the Falls of Ohio. French interest in Louisiana raised its head in 1787, reinforced by Brissot after his visit to the United States. Wilkinson danced his way into a minuet with France who wished to regain the Louisiana Territory from Spain. These contesting powers had their eye also on the uncertain fate of the adjacent region of Kentucky. During the 1780s there were no fewer than nine US conventions over the issue of Kentucky's independence from external control.

Great Britain, yet another of the contesting powers, had dispatched its own spy, Colonel Conelly, to Louisville in 1788. In addition to the dance Wilkinson led Spain and France, he also flirted with a Britain keen to invite trade agreements that could detach Kentucky from the United States, and lure it into its sphere of influence – though less inclined to offer Wilkinson the regular pension he received from Spain. Early in 1792, the British Foreign Secretary was warned that Wilkinson (newly appointed by Washington as second-in-command of the American army) was moving frontiersmen to advance on Spanish territory. During this year America won Kentucky when it entered the Union. In this same year, Imlay re-emerged as an author in London.

He approached Debrett, publisher of the guide to the peerage. The social prestige of Debrett suggests that Imlay – of whom no portrait exists – could blend his frontier image with the manners of the Imlay Mansion. His generation of army officers modelled themselves on Washington: the calm dignity, the sturdy conviction, with no look of the revolutionary.

Imlay claimed to have written *A Topographical Description of the Western Territory of North America* (1792) in the backwoods which is its subject. Though this claim is usually discounted (because of the smokescreen of his supposed departure for Europe five years earlier), we don't have to read far into the book to see that Imlay had to be telling the truth. Its local detail is so abundant, its measurements and figures so precise, that it would be impossible to write such a book from memory alone. The book went into three editions during the 1790s. A publisher, Kegan Paul, reading it nearly a century later, saw 'a model of what a monograph on a new country should be. It is at once clear, full, and condensed . . . Its language throughout shows an educated accomplished man.' Imlay projects a vision of two Americas: the genuine America in the West as distinct from European colonisation of eastern States, a belief in the West as '*the centre of the earth . . . at once the emporium and protectors of the world*'.

As much as it was in the interest of the States to acquire the central plains, it was not in their interest to tussle with Spain, nor to infringe their neutrality. Following the failure of Wilkinson's attempt to advance on Spanish territory in January 1792, the next best move would be for an ally – France – to wrest these lands from Spain. This, conceivably, was the situation when, in December 1792, Imlay set out two secret plans for the capture of the Louisiana Territory, and registered them with Louis Otto (son-in-law of his acquaintance St Jean de Crèvecoeur) in the French Foreign Office.

There were four thousand men, Imlay claimed, burning with the Fire of Liberty and incensed with Spain's violation of their rights, who would leap to arms against the oppressor. On 25 January 1793 the Comité de Défense Générale (precursor to the Committee of Public Safety) asked Brissot to present a report on the feasibility of the proposed expedition against Spain's colony. Imlay, who knew the terrain, was to have the help of General Francisco di Miranda, a Venezuelan who had fought in the War of Independence.

In February a new French Minister to the United States, Citizen Edmond Charles Genêt, was dispatched with instructions to pursue Imlay's plan and extend the Revolution to far-flung subjects of Spain who were said to be panting for its principles. His instructions were to stir up frontiersmen, *secrètement*, to bring independence to the *Louisianais*. They were to sweep down the Mississippi into Spanish territory and take New Orleans at the mouth of the river, where a small French fleet would lie in wait to back the invasion. In the meantime, on 7 March, France declared war on Spain.

Later that month, Imlay pressed his plan with a letter of introduction to

Brissot from Thomas Cooper. Imlay was, as always, well connected. Cooper, an Oxford graduate, was a radical cotton manufacturer in Manchester, a small man with a large wedge of a head, brilliant in conversation and terrifying in controversy, who published with Joseph Johnson and supported Mary Wollstonecraft. 'Let the defenders of male despotism answer (if they can) the Rights of Woman,' he had challenged Burke. As emissary from the democratic clubs of England to affiliated clubs in France, he enjoyed the status of French citizen. Brissot used Cooper's support of Imlay to pressure the Minister of Foreign Affairs, Le Brun, that if *ce Capitaine* (Imlay) and his cohorts did not leave within fifteen days, they would have to give up the enterprise.

A committee was formed to forward the expedition, with none other than Joel Barlow top of the list – chosen to lend his lustre as a thinker devoted to the cause of liberty in both the American and the French Revolutions. It can't be a coincidence that Barlow returned to Paris just two days before France declared war on Spain. He was not merely a figurehead: a document in the archives of the French foreign ministry, dated March 1793, describes him as a person 'to whom one might entrust the general direction [of the coup] under Genêt and the handling of the groundwork'. He did undertake some hush-hush work, for on 21 March he refers to 'a bundle' which, if not collected by an unnamed secret agent, Ruth was to place in their baggage for America.

Barlow and Imlay wanted land as the spoil of war, but this private motive blends with public-minded intentions to be of use to France and the Revolution, and of use, too, to their own country by keeping the outlet of the Mississippi open to American trade. There was a third motive. If the Mississippi expedition went through, Barlow would be able to offer Ruth an advantageous return to America with expenses paid by France. Of course, so secret a plot could not be communicated in letters. He said, it 'will suit you my love much better & me too, because it will carry us both home upon a good mission'. That much he could tell her. He also dwells on the friendship of Mary Wollstonecraft.

'W[ollstonecraft] expresses the greatest love for my dearest,' he had said on 6 March. He saw her within a day of his arrival in Paris, and again on the 18th: 'Mrs W. speaks of you with more affection than you can imagine. I never knew her praise any person so much. What is the reason? It is not to flatter me, for she never flatters any one. Why is it that she loves you? Is it because she has found out that every body else loves you? It is well to be in the mode, but I thought her of a more independent spirit.'

Barlow put further pressure on Ruth by travelling to collect her at Boulogne, but still, she did not come. Frustrated, because he didn't dare to enter England, he wrote to her on 5 April: 'Mrs W[ollstonecraft] was exceedingly disappointed to see me return without my dearest. I told you before how she loves you. She never loved any body so well.' Ten days later, he reproached Ruth once more, making light of the rising Terror: the obsession with security, the establishment of the Revolutionary Tribunal, which sent its victims to the guillotine, the curfews, the bread riots in the market, the casual violence in the streets. Far away in Wales, Bess Wollstonecraft was taunted 'that Miss W[ollstonecraft] is massacred before this'. At this very time Barlow reassures Ruth: 'you would have been perfectly safe here . . . Mrs W[ollstonecraft] is exceedingly affectionate to you.' On 19 April, 'Mary is exceedingly distressed at your last letter to think you are not to come here. She writes you today, she wrote a long letter before which it seems you have not got.'

In this way, from March to June, Barlow repeats Mary's attachment with an insistence that co-opts her for his plans. Her obvious role was to lure Ruth to Paris; the more novel part was to take off with them, and Imlay, for America.

As the Louisiana scheme gained momentum in late March and early April, with Ruth packing up in London in preparation for crossing the Atlantic, Mary Wollstonecraft was drawn into the little group as Imlay's 'sweetheart'. She had no inkling of the coup in the offing. Imlay pictured for her the fresh green breast of the New World. On the frontier, he liked to say, 'we feel that dignity nature bestowed upon us at the creation'. He promised a society in which 'sympathy was regarded as the essence of the human soul', and envisioned, a hundred years on, the entire continent peopled by republicans. A federal government, Imlay was sure, would be able to introduce the change that must take place for man 'to resume his pristine dignity'.

Mary caught fire. As Barlow had predicted, she stood ready to leave for America. There, they would farm, and she hoped to bring her sisters over. In mid-June, Mary hints of her American plan in a letter to Bess: 'I cannot explain myself excepting just to tell you that I have a plan in my head, it may prove abortive, in which you and Everina are included, if you find it good, that I contemplate with pleasure as a way of bringing us all together again.'

Mary was dreaming of a frontier home for herself and her sisters. With Charles already there, it would unite four favourite members of her family. The utopian dream of the New World was at its height in the early 1790s when Coleridge planned a community on the banks of the Susquehanna, and persuaded Joseph Priestley and Thomas Cooper to cross the sea, with thousands of others. To a reformer like Mary Wollstonecraft, bent on a new social system, revolutionised America presented a hopeful alternative to the violence of revolutionary France. Imlay epitomised this hope, averse (he said) to violent men who trample on laws and civil authority, to standing armies, and to 'the contumely and ignorance of men educated with none but military ideas'. To Mary, here was a 'most worthy man', confident, cheerful, indestructible. Depression vanished in his presence.

'Whilst you love me,' she told him, 'I cannot again fall into the miserable state, which rendered life a burden almost too heavy to be borne.'

At thirty-four she was a virgin, apprehensive of situations that shaded into the wide meaning she gave to prostitution. Her judgements of people had been astute: she had detected vanity in Fuseli and excessive protestation in Barlow's love-letters. She had warmed to older men of sturdy morals, Dr Price and Joseph Johnson. Here, now, was a man of her own generation who stated morals she could share, especially his abhorrence of slavery as 'a traffic which disgraced human nature'.

'As to whites being more elegantly formed, as asserted by Mr Jefferson, I must confess that it has never appeared so to me,' he said. 'Indeed my admiration has often been arrested in examining [blacks'] proportion, muscular strength, and athletic powers.' No race is better than another, he added. We are 'essentially the same in shape and intellect'. He praised the American slave poet Phillis Wheatley; urged women's right to own and inherit property; and the revised edition of his *Topographical Description* of the frontier in 1797 – showing the Wollstonecraft influence – insists that women should not have to answer to laws they can't promulgate. He wished for women's sake to reform divorce law.

The rights of women are central to Imlay's novel, *The Emigrants*, published in London three or four months after he began to pursue Mary Wollstonecraft. The novel joins the issue of injured women to Imlay's subject of frontier heroics. An English family braves dangers and setbacks on their way from Philadelphia via Pittsburgh to settle in the Ohio country, a sentimental adventure complete with 'Indians', capture, rescue and camps in the wilderness – anticipating the frontier fiction of Fenimore Cooper. Unfortunately, Imlay's is dead on the page with flat characters and inflated emotions; and though the hero's sexual appetite is unusually blatant for American writing of the period, it comes over as comic – half-reverent, half-gloating voyeurism as a beauty's breasts heave into view through torn clothes. Unfortunately, the tosh overshadows the moral debates where Imlay comes into his own as a clever, well-read man who concurs with Wollstonecraft's call for the education of women: 'few women have had strength of mind equal to burst the bands of prejudice' by 'soaring into the regions of science and nature'.

Imlay also confronts rape in marriage. The fictional Lord B— 'considered women merely as a domestic machine, necessary only as they are an embellishment to their house, and the only means by which their family can be perpetuated'. His wife resolves 'never to enter the bed of my Lord B— again; for his conduct to her that morning, after coming to her two hours after midnight in a state of intoxication, was too gross for a woman of spirit and delicacy to forget . . . [To comply again would be,] after the treatment she had received, a most ignominious prostitution.' Lady B's 'prostitution' dramatises the case for divorce reform. The 'man of honor' doesn't just talk of morals; he puts them into practice. Whether Imlay himself followed his principles remains to be seen, but as principles go, he was sure to please.

When Genêt reached America, he set about enlisting enthusiastic frontiersmen for an assault on Spanish Florida. But he did it so flagrantly that

Washington was forced, on 19 April, to put an end to his activities with a reassertion of American neutrality. Jefferson, as Secretary of State, concluded (or, for diplomatic reasons, pretended to conclude) that Genêt was not representing the policy of his government, and asked for his recall. Since Genêt was seen to be an appointment of Brissot, this weakened Brissot's position. Robespierre was to use the Genêt Affair to justify Brissot's execution. It could be argued that Genêt and Brissot were puppets of Imlay and Barlow, whose names do not appear in the diplomatic crossfire that cut off their Louisiana scheme.

Washington soothed relations with Spain when he spoke publicly of prosecuting American participants in this plot. Vice-President Adams sent his wife Abigail a rhyme about the unidentified secret agents behind it all:

> At home dissensions seem to rend
> Or threat, our Infant State
> 'Bout Treaties made; yet unexplain'd
> With Citizen Genêt.

The conspirators in Paris were undismayed, since they were already on to their next scheme. Colonel Benjamin Hichborn of Massachusetts had proposed they use American neutrality in order to ship goods between warring countries. Scores of ships were doing this. The American Minister in London, Mr Pinckney, protested repeatedly over the British navy's retaliation against American shipping. So these were risky ventures – useful, though, for spying.

Barlow continued to use Mary Wollstonecraft as a draw for his wife. 'Mary writes to have you come here & take lodgings with her at Meudon 5 miles from town. She really loves you very much. Her sweetheart affair goes on well. Don't say a word of it to any creature.' Mary herself wrote four letters to Ruth in quick succession, the last on 7 May (none has survived). Barlow sent his final letter to Ruth on 10 June: 'Mary is much disappointed & grieved at your not having come . . . I think she loves [you above] all other creatures.' Ruth then crossed the Channel.

Despite her husband's assurance that Paris was safe and quiet, Ruth arrived just as the Terror took over. On 2 June the moderates had been detained in their lodgings; on 13 June they were imprisoned; and on 28 July declared traitors. Brissot was amongst them, as was leading feminist Olympe de Gouges, who had set out her *Rights of Woman* to match, point for point, the French Declaration of the Rights of Man. 'Men!' she exclaims. 'Are you capable of justice? It is a woman who asks the question.' She wished to be 'a man of state' and demanded women's suffrage as well as the abolition of the slave trade, workshops for the unemployed and a national theatre for women. The immediate cause of her arrest was disseminating anti-Jacobin pamphlets, which she continued to manage from prison until her execution.

During this time, the names of enemy aliens had to be chalked on the doors where they lived. Mary could not endanger the Fillietaz family by her continued presence. In May she had found a post for Bess in Geneva, and applied for a pass in order to join her. A pass was not forthcoming, and Bess decided against Switzerland as too expensive, but significant here is Mary's willingness to leave France. It seems that her romance was at this point no more than romance. There is no sign of Imlay's intervention. If they were sleeping together, Imlay would have made a plan to keep her safe – he was a great instigator of plans. In any case, for Mary to leap into bed with an admirer without a serious understanding would have been out of character, inconsistent with the modesty and sexual caution she expressed in the *Rights of Woman*. It had expanded on her earlier warning against women's susceptibility to flirts and rakes, the residue of her own encounters with Joshua Waterhouse and Neptune Blood. If her first plan had worked, she would have left the country that May, and possibly not seen Imlay again.

As it was, her alternative was to find lodgings outside Paris. Plan number two, to live with Ruth at Meudon, was once again not designed to promote an affair with Imlay. Meudon is likely to have been the subject of her four urgent letters to Ruth in London. It was only when Ruth's intentions remained too uncertain that Mary accepted a third solution, offered by her hosts: she took refuge in a cottage at Neuilly-sur-Seine, north of the Bois de Boulogne beyond the city wall. It was a cottage belonging to or tended by an elderly gardener who worked for the Fillietaz family. He plied Mary with home-grown fruit and warned against her taste for solitary walks in the wooded lanes leading to the river.

Here, Mary lived quietly and happily from June to August. She had begun to write a 'great book' on the French Revolution, 'great' because she resolved 'to trace the hidden springs and secret mechanism, which have put in motion a revolution, the most important that has ever been recorded in the annals of man'. That summer of 1793 she explored the deserted Versailles, with its fading phantoms. Nothing testifies more hauntingly to the evanescence of the past in the present context of the Terror:

> How silent is now Versailles! . . . The train of the Louises, like the posterity of the Banquos, pass in solemn sadness, pointing at the nothingness of grandeur, fading away on the cold canvas, which covers the nakedness of the spacious wall – whilst the gloominess of the atmosphere gives a deeper shade to the gigantic figures, that seem to be sinking into the embraces of death.

So Mary's days were filled with the history of the present, enlivened by visits from Imlay, his face flushed with expectation as he came to meet her through the *barrière* at Longchamps (one of the guarded exits from the walled Paris). Long afterwards, she would remember the approach of his 'barrier face' as she waited on the far side, holding up a basket of grapes. 'Dear girl,' he called her, with lips 'softer than soft'.

'I am confident my heart has found peace in your bosom,' she said. 'Cherish me with that dignified tenderness, which I have only found in you.' So she recalls him by night in the light of her candle; seals him in the folds of her paper; repeats his parting phrase, 'God bless you.' She was often alone, even 'quite lonely'. Imlay's visits may have been fewer than we have come to imagine. Business came first.

The first shipping ventures were disastrous. Cargo that Barlow sent to New York on a ship called the *Hannah* turned out to be illegal. A court case ensued, and judgement went against the captain, a man called Parrot. Sixteen years later Parrot finally tracked down Barlow, who had to arrange a reimbursement of $20,000. It's hard to fit Barlow's humorous intimacy, his warmth towards Mary, his love for Ruth, with this irresponsible distance from the small man who's caught and tried and pays the price. Yet it fits the irresponsibility of the Scioto scam.

Barlow and Hichborn chartered an English ship called the *Cumberland*, and loaded it with flour and rice to relieve the food shortage in France. Their associates were Mark Leavenworth who often acted as Imlay's agent; Colonel Blackden, another speculator in Kentucky land, whose wife had welcomed Ruth to London; and a banker called Daniel Parker who had involved himself in the Scioto scheme when Barlow arrived in Paris, then had the prudence to withdraw. Since the British Parliament had just then passed a bill prohibiting trade between Americans and France passing through England, the cargo was said to be bound for Spain. When this ship came to land at Bordeaux, suspicious Frenchmen embargoed a vessel whose papers indicated traffic with the enemy. The *Cumberland* was one of a hundred foreign ships, most of them American, trapped at Bordeaux during the second half of 1793. The American Minister, upholding Washington's policy of neutrality, ignored their pleas for rescue. The balance of power between England and France was being fought out at the French coastal ports, and for the Americans to engage in shipping was to enter this war zone.

Meanwhile, Mary worked away at her 'great book' in the woods of Neuilly. On 13 July she received £20 from Joseph Johnson, probably an advance for her proposal to attempt a history at a time when her status as enemy alien put an end to earning a living as a foreign correspondent. 'I am now hard at work,' she reported to Bess, 'for I could not return to England without proofs that I have not been idle.'

To write a history of the Revolution during the Terror was 'almost impossible' from so close a perspective. Her proximity to events must be coloured by the 'prejudice' of current political sentiments (evident in her virulence towards Marie Antoinette as a type of Lady Kingsborough). Yet, for all her disclaimers, the closeness of her perspective grants her a unique advantage. When she approached the barrier to meet Imlay, she observed through its towering, stately frame the prison that Paris had become. Its expansive approaches, its people lounging 'with an easy gaity peculiar to the nation', and the beauty of its buildings were all, she saw, locked and confined as in a cage. The barriers, she wrote, 'have fatally assisted to render anarchy more violent by concentration, cutting off the possibility of innocent victims escaping from the fury, or the mistake of the moment'. Her eye takes in the Terror not as a number of heads to be counted, but as architecture, the design of a city conceived in exquisite taste by 'miscreants' of the past – a continuum of powerbrokers, uninterrupted by revolution. 'Thus miscreants have had sufficient influence to guard these barriers, and caging the objects of their fear or vengeance, have slaughtered them.' She bears witness to 'the effect of the enclosure of Paris' on her own observant eye, 'disenchanting the senses' so that 'the elegant structures, which served as gates to this great prison, no longer appear magnificent porticoes'.

A tear starts to her eye, blocking the 'inlets of joy' to her heart as Imlay approaches. For that eye looks beyond private excitement, towards the anguish of the city where a 'cavalcade of death moves along, shedding mildew over all the beauties of the scene, and blasting every joy! The elegance of the palaces and buildings is revolting, when they are viewed as prisons, and the sprightliness of the people disgusting, when they are hastening to view the operations of the guillotine, or carelessly passing over the earth stained with blood.' Only education 'will prevent those baneful excesses of passion which poison the heart'.

She writes from where she stands, and the tense is the present – this is her strength: she feels as well as sees through her welling tear beauty dissolve in terror. Another advantage is her use of her outsider-insider position to make a balanced statement designed to outlive its time:

> . . . The french had undertaken to support a cause, which they had neither sufficient purity of heart, nor maturity of judgement, to conduct with moderation and prudence . . . Malevolence has been gratified by the errours they have committed, attributing that imperfection to the theory they adopted, which was applicable only to the folly of their practice.
>
> However, frenchmen have reason to rejoice, and posterity will be grateful, for what was done by the assembly.

John Adams, as second President of the United States, would read the book twice and, though he often disagreed, think her 'a Lady of a masculine masterly Understanding' with a 'clear often elegant' style. He was struck by her critique of revolutionary France in favour of the American Revolution: 'She seems to have half a mind to be an English woman; yet more inclined to be an American,' he remarked.

Critical judgements would have been dangerous for anyone to make at the height of the Terror, but especially for an Englishwoman whose country was at war with France. Mary Wollstonecraft therefore took the precaution of confining her history to the run-up and first three months of the Revolution, the hopeful period before it turned on its leaders. Her chief hero, Mirabeau, had died in 1791. His passionate speeches had denounced inherited wealth, required the rich to pay the national debt, demanded equal rights for Jews and Protestants, and proposed the abolition of the slave trade. More recent events, the turn to bloodshed, were reserved for a sequel Wollstonecraft meant to write when it was safer.

So she lay low at Neuilly while the Terror took hold, and Imlay came and went in that city where the guillotine was in motion, the Committee of Public Safety tightened its grip on the populace, its Law of Suspects spun around pointing its arbitrary talon, and thousands eventually lost their heads. Imlay was a man of secrets who worked through others. The only way to reach him is through the activities of his associates, who were careful what they put on paper. In this, Mary was his opposite: she made no secret of her feelings.

'I obey an emotion of my heart, which made me think of wishing thee, my love, goodnight! Before I go to rest,' she writes at midnight to Imlay from Neuilly. 'You can scarcely imagine with what pleasure I anticipate the day, when we are to begin almost to live together.'

The Revolution was not only an event that had happened outside Mary Wollstonecraft; it was active in her own blood. She had been in revolt all her life – against tyranny, against the prostitution of women by law and custom, against misshapen femininity. The storming of the Bastille on 14 July 1789 had been the potent image of release: what was locked up in women's minds, feelings, desires. Imlay's frontiering freedom encouraged her interior revolution. Then, too, he had a tenderness missing in Fuseli, in her violent father, and in her sister's obtuse husband. The question – one only time could answer – was whether this was going to add up to a new genus of manhood with whom to invent a new kind of union. Wollstonecraft had written that the marriage tie should not bind a couple if love should die. Yet once she fell in love with Imlay, she craved permanence.

'I like the word affection,' she told him, 'because it signifies something habitual.' She saw a test ahead: 'We are soon to meet, to try whether we have mind enough to keep our hearts warm.'

Imlay was no sexual innocent. In Kentucky there had been a 'girl' whom he sent away; more recently, a mistress whom Mary called a 'cunning' woman.* This past was not an issue at Neuilly: in a period of coming and going, the present drama was at the barrier: partings, meetings, and a flush on Imlay's face as he drew near.

She did occasionally pass through the barrier herself. 'Why cannot we meet and breakfast together *quite alone*, as in days of yore?' she wrote to Ruth from Neuilly, recalling their breakfasts in Titchfield or Store Streets when they had lived close by in London. 'I will tell you how – will you meet me at the Bath about 8 o'clock . . . I will come on Monday unless it should rain . . . We may then breakfast in your favourite place and chat as long as we please.'

Ruth's favourite place was the new Chinese Baths – so called because of the twin pagodas of its design. Its aromatic water, warmed robes and restaurant would remain popular with Americans in Paris for the next sixty years. The Baths were in the boulevard des Italiens, not far from the guillotine. One day Mary entered Paris on foot through the Place de Louis Quinze soon after an execution. Her foot slipped, and looking down, she saw that she had stepped on blood left by the last batch of victims. Shocked, she burst into protest, when a passer-by warned her in low tones not to put her own life at risk.

---

*There was a rumour that Imlay had spent a night with Helen Maria Williams. This has never been substantiated.

After some four or five months she and Imlay still lived apart. We don't know when, exactly, they became lovers, but it's worth noting that Barlow, the only witness to leave a record, saw Mary in 'sweetheart' mode. We know, while Imlay was absent, she longed to kiss him, 'glowing with gratitude to Heaven'; we know the 'barrier' roused their desires; but nothing more until Mary left Neuilly to live with Imlay on the Left Bank in the Faubourg Saint-Germain. A likely place could be the Barlows' lodgings in the Maison de Bretagne, 22 rue Jacob, small, only two storeys with an attic, along the same street as the statelier York Hotel where Benjamin Franklin, John Adams and John Jay had signed the peace treaty with Britain in 1783.

By August, Imlay realised that the French defeat when the British captured Toulon meant new danger for the dwindling number of British citizens in Paris. That month Thomas Christie, another honorary citizen of France, was arrested, and when he and his wife were released they fled to Switzerland. To protect Mary, Imlay gave her the cover of his name and nationality by certifying her as his wife with the American Ambassador. As such, she carried a certificate of her status as an American. It was vital at that time, for Americans continued to have standing in France under Robespierre. This move was to keep Mary Wollstonecraft safe when the time came for fresh threats against enemy aliens.

Imlay may have saved Mary's life. It was certainly a generous act, for to declare this woman his wife meant taking on responsibility for her support. All the evidence suggests that he undertook this obligation. Other aspects are less certain. How did Imlay convince his ambassador, Gouverneur Morris, to certify a marriage? Was there some kind of republican ceremony? Certainly, whatever they signed was a document of doubtful legality.

Imlay's action impressed Mary as a commitment. This was a novel solution to the problem of marriage: she would not bind herself in a legal contract unjust to women. Imlay concurred. He called the laws of matrimony 'barbarous', laws set up for the 'aggrandizement of families'. Mary meant to pursue a purer union with a 'worthy' man, consistent with the promise of perfectibility that radical intellectuals of her generation read into revolution. Once she had the certificate, it was safe to leave Neuilly and move into the city. It was a major move – she hired a carriage for her books alone. 'I have so many books, that I immediately want,' she said. So Mary returned to Paris, lived with Imlay, and soon was pregnant. To make love was, for her, a sacred as well as passionate act.

'My friend,' she told Imlay, 'I feel my fate united to yours by the most sacred principles of my soul.'

New emotions stirred desire, 'sacred emotions' she called them, 'that are the sure harbingers of the delights I was formed to enjoy, for nothing can extinguish the heavenly spark'. Once, when a Frenchwoman boasted her lack of passion, Mary was heard to reply: '*Tant pis pour vous, madame. C'est un défaut de la nature.*' (The worse for you, madame. That's a defect of nature.)

From now on, she called herself 'Mary Imlay' – that is, when she chose to do so. Mary Imlay's identity eludes the usual categorisations of women as virgin, wife, mother or mistress, for she never discarded her independent and celebrated character as 'Mary Wollstonecraft'. The ambiguity comes from the peculiar situation of an Englishwoman who chose to stay in Paris during the Terror; at the same time, her needs for commitment as well as freedom rehearsed the conflicts of women in future generations.

10

RISKS IN LOVE

Events soon bore out Imlay's precaution in lending Mary his name and nationality. On 10 September 1793, it was announced that all foreigners 'born within the territory of Powers with which the French republic is at war' would be imprisoned.

On the night of 9–10 October, in one relentlessly efficient swoop followed by mop-up operations over the next four days, some two hundred and fifty Britons in Paris were arrested as spies or counter-revolutionaries. They were taken to the Luxembourg, once a palace, now a prison. The prisoners included John Hurford Stone, his lover Helen Maria Williams, and her mother and sisters, who were eventually released through the intercession of a Frenchman attached to Helen's sister Cecilia. General Miranda's association with Brissot brought him before the Revolutionary Tribunal. Barlow defended him, and he was fortunate to be deported. Condorcet, condemned and in hiding, was hunted down eventually, and cheated the guillotine by taking poison. Marie Antoinette was executed on 16 October, and Brissot on the 31st together with twenty former deputies of the Convention. Brissot looked each of his colleagues in the eye before laying down his head. On the scaffold the efficient executioner Sanson took thirty-six minutes to cut off twenty-one heads. Imlay broke this news to Mary.

'I guess you have not heard the sad news of to-day?'

'What is it? Is Brissot guillotined?'

'Not only Brissot, but *les vingt et un*.'

As faces rushed towards her – Brissot, Vergniaud, Brûlart de Genlis (husband of Mme de Genlis) – Mary fainted.

Robespierre's party denounced women who meddled in politics, whether it be the behind-the-scenes influence of Mme Roland or Charlotte Corday, who stabbed the bloodthirsty Marat in his bath. Helen Maria Williams thought the Jacobins masculinised the Revolution, breaking up families and destroying domestic affections (persuading the dauphin, for instance, to turn on his mother Marie Antoinette, with accusations of sexual abuse). On 17 August 1792 a proposal at the Jacobin Club that women be given the vote had not found favour. The Jacobins followed Rousseau's sentimental picture of women's nature: only mothers and helpmates were to have a place in an ideal society. No more revolutionary-republican *citoyennes*, no more 'Amazons'. At this time the Jacobins expelled women from the public arena when they closed their political clubs and banned them from forming political associations. On 10 November they executed Mme Roland. The Convention declared that 'a woman's honour . . . precludes her from a struggle with men'. She must be 'confined' to the home.

Imlay was not a man to settle for domesticity – or not for long. While he sat at their fireside, schemes fired in his head. A bar of soap that had cost twelve *sous* in 1790 had risen to as much as twenty-eight by 1793. The hunger of the populace was barely relieved by fixing food prices, '*le maximum*', as it was called. Parisians were crying out for *du pain et du savon*, and Imlay sent out ships to bring in cargoes of wheat and soap. Shipping took him to the port of Le Havre (renamed Havre-Marat), a hundred and twenty-six miles north-east of Paris. Although all goods from Britain and its colonies were barred in France, Imlay did do business with London, in league with an English soap trader, John Wheatcroft, a resident of Le Havre where he owned properties and remained in good standing with the Committee of Public Safety. Imlay traded with the respectable London firms of Turnbull, Forbes & Co. and Chalmers & Cowie, but these are unlikely to represent the full extent of his interests. Did he pursue a career as a spy with commerce as his cover, validated by the name of his shipping family? Such activities are difficult to prove, but unquestionably Le Havre, then London, became over the next year his centres of operations.

As Mary's pregnancy advanced, he was mostly away. 'But, my love, to the old story – am I to see you this week, or this month?' she writes to Imlay on 29 December, some six to eight weeks after his last visit. 'I do not know what you are about – for you did not tell me.'

Strange that so bright a woman had so dim a notion of what this was. Eventually, he told her it was trade in soap and alum – a reply she found disconcerting. Soap did not fit her idea of the New World thinker she had come to love. Nor did he tell her of his New Jersey origins and the commercial interests of the Imlays. She did not know of the Imlay Mansion in Allen's Town tended by slaves brought back from the West Indies. As late as March 1794 – having known Imlay a whole year – she still talks of his growing up in 'the interior regions' of America. It was as though he had come to her from some far region – some edge of existence like that of her secret self, far out and alone. Yet he was changing before her eyes: the backwoods simplicity was vanishing, and in its place was a hard-eyed man preoccupied with schemes he presented as wholly – and banally – commercial. When his talk turned on what seemed to her a gambler's hope for riches, he appeared a stranger.

'Recollection now makes my heart bound to thee,' she told him, 'but it is not to thy money-getting face.'

Imlay, in his New World character, had presented an undivided figure. It puzzled her to come upon his ambiguities, the backwoodsman who required the comforts of the affluent classes; the patriot who stayed in France; the outsider who was in some part insider; the lover-evader, dreaming of their future fireside, who hardly came home. Possibly he had more than one identity, and this fits Mary's sense of his duality. She was in love with one of these men, the gentle, philosophic frontiersman, who appeared to her blocked by the contrived identity of a piddling soap merchant, one of the 'square-headed money-getters' who are 'stupidly useful to the stupid'. She knocked at this image – softly, then loudly – convinced the real man was there, 'concealing' himself and increasingly inaccessible to her.

Her letters call up the passionate Imlay who had been her counterpart, using the sacred 'thy' she reserves for him alone. 'I have thy honest countenance before me . . . relaxed by tenderness; a little – little wounded by my whims; and thy eyes glistening with sympathy. – Thy lips then feel softer than soft – and I rest my cheek on thine, forgetting all the world. – I have not left the hue of love out of the picture – the rosy glow; and fancy has spread it over my own cheeks, I believe, for I feel them burning . . .' Her heart beats through her pen, directing her gratitude to 'the Father of nature, who has made me thus alive to happiness'.

As she approached the fifth month of pregnancy, maternal feeling woke to the first 'gentle twitches'. She thought: I am nourishing a creature who will soon be sensible of my care. To preserve this creature she resolved to exercise her body and calm her mind. As she did so, she relived desires heightened by the coming and going at Neuilly, and now began to fret that this was not, as it had seemed, an anticipatory phase; it was a pattern, the way Imlay conducted a relationship. She learnt to know the crack of his whip in the air as his horse galloped off into the distance. And, with that, the 'we' on Mary's lips withdraws to 'I'. One thought in particular plagued her, and eventually she put it to Imlay.

'These continual separations were necessary to warm your affection. – Of late, we are always separating. – Crack! – crack! – and away you go.'

This was a man with a purpose that seemed to her debased by commerce.

'I hate commerce,' she said.

He could not agree. Commerce, he argued, tended 'to civilize and embellish the human mind'.

Many call him rascal, scoundrel or cad, as though he were unusual and Mary Wollstonecraft a sex-starved dupe. These have always been reductive myths. Imlay was no different from other men on the make in an age of smuggling, piracy and colonisation. Men like Clive of India wrested private fortunes from peoples on the far side of the globe; privateers raided the high seas; and navies took 'prizes' in the form of enemy vessels – piracy legalised in games of war. Nothing was more respectable than impecunious Captain Wentworth taking prizes at sea in Jane Austen's *Persuasion* – as the youngest Austen brothers would have done as they rose through the ranks of the navy during the Napoleonic Wars. What is ugly in such acts takes place across the horizon, unseen by society. Captain Imlay was another gentleman who hoped to make his fortune far from home. A game called 'Speculation' entered British drawing-rooms in 1804: players in *Mansfield Park* are advised to sharpen their avarice and harden their hearts, while the master of the 'turns' in the game shows 'impudence' and 'quick resources'. In *Pride and Prejudice* an early sign of Jane's and Bingley's decency is that they prefer a different game to one called 'Commerce'.

The Wollstonecrafts came from a class structure where to be 'in trade' was low. Money should not be new – not like that of the nabob of Upton Castle, whom Bess Wollstonecraft despised for his incivilities and dirty tablecloths. The Wollstonecraft sisters positioned themselves as thinking individuals, defined by serious books, as by their aspiration to know French, for them (as for the upper classes in Europe throughout the wars with France) a sign of cultivation. The Imlays, on the other hand, came from a society where making money was admired. The power attained by politics or arms is brief, Imlay believed; 'wealth is the source of power; and the attainment of wealth can only be brought about by a wise and happy attention to commerce'. This was what a man did if he was manly. Mary questioned Imlay's 'struggles to be manly', and welcomed the counter-struggle of sensibility, she told him, 'striving to master your features'. Mary saw manliness as a construct that could be remade in the light of Rousseau's proposition: 'We know not what Nature allows us to be.'

He wrote often, every two or three days. Mary found his letters 'a cordial': he was 'sweet', 'cheerful' and 'considerate', sustaining a kindness that invites return.

'Write to me my best love, and bid me be patient – kindly – and the expressions of kindness will again beguile the time, as sweetly as they have done to-night,' she wrote. 'I am going to rest very happy, and you have made me so. – this is the kindest good-night I can utter.' And again: 'I have just received your kind and rational letter.'

Imlay's attentiveness encouraged Mary to express her desires. 'The way to my senses is through my heart; but, forgive me! I think there is sometimes a shorter cut to yours,' she said. 'I shall not . . . be content with only a kiss of DUTY – you *must* be glad to see me – because you are glad – or I will make love to the *shade* of Mirabeau.' Imlay was habituated to the 'cunning' or '*piquante*' woman who plays male games, deflecting her own desires, Mary argued, to the detriment of her own sex. Mirabeau, she fancied, would have been more amenable. She thought the run of men should change habits based on 'casual ebullitions of sympathy', and adapt to a more imaginative form of desire 'by fostering a passion in their hearts'. While Imlay was away, she found her own imagination 'as lively, as if my senses had never been gratified by [your] presence – I was going to say caresses – and why should I not?'

Absence intensified her sense of the man she felt him to be. In her thoughts and desires she lived with the Imlay who wrote to the woman she felt herself to be. She called him up in her imagination, making him in this way more her own. This was easier when she could not see his commercial face.

'I do not know why,' she confessed, 'but I have more confidence in your affection when absent, than present; nay, I think that you must love me, for, in the sincerity of my heart let me say it, I believe I deserve your tenderness, because I am true, and have a degree of sensibility that you can see and relish.'

She doesn't talk of her physical attractions, though she had them; she doesn't see her own face as she sees his, but exists (in her own terms) as a speaker of 'truth' who can enter into the life of feeling. Facelessness deepens her presence in her writings, where inward truth takes precedence over face and form. 'Be not too anxious to get money!' she counselled, ' – for nothing worth having is to be purchased.' The *Rights of Woman* had called for fidelity on the part of men, and for women to refuse maintenance where there is no love. Mary was determined to put this kind of union into practice, based on what Imlay called the soul's sympathy.

As December crept by there were moments when impatience could not be contained: 'You seem to have taken up your abode at H[avre]. Pray sir! When do you think of coming home? Or, to write very considerately, when will business permit you?' She tried to make light of it. 'The creature!' she would exclaim as she passed his tatty slippers at her door. If he did not return soon, she told him on New Year's day 1794, 'I will throw your slippers out at [the] window, and be off – nobody knows where.'

Imlay justified his absence as provider for her and their unborn child. 'Exertions are necessary,' he reminded her. Why then was she cold to him? Why had she been 'three days without writing'? Did she not know that he revered her?

'I do not want to be loved like a goddess,' she replied, 'but I wish to be necessary to you.'

The kinder his voice, the guiltier she felt for her mistrust. 'Your own dear girl will try to keep under a quickness of feeling, that has sometimes given you pain,' she apologised. 'Yes, I will be *good*, that I may deserve to be happy.' How badly she wished to retrieve trust – badly enough to take the blame. 'Quickness,' she said, was her flaw. Lonely for Imlay, and wanting reassurance, she begged: 'bear with me a *little* longer.'

Imlay did not as yet reveal to Mary his biggest commercial scheme, though eventually she was drawn into a venture that would uncover its character only by degrees. Fashionable society continued to look to France as the arbiter of taste; the cultivated spoke French, drank French wines, and filled their houses with French porcelain. Amongst Imlay's exports were glassware and other portable objects from abandoned chateaux. No sooner did Robespierre come to power in 1793 than he passed a law (10 June) by which the contents of royal palaces, 'the vast possessions which the last tyrants of France reserved for their pleasure', were to be sold off to aid 'the defence of liberty'. A series of public sales of undervalued items began on 25 August at Versailles, where a poster announced that objects 'may be taken abroad free of all duty'. This was a time of ideological purity when despised luxuries were to be discarded; storehouses were piled high with unwanted treasure. Americans like Imlay were able to buy costly objects with devaluing paper money.

One way to approach the elusive Imlay is through American associates like Richard Codman who made fortunes in France during this decade. Imlay and Barlow had failed in massive schemes without losing their incorrigible self-confidence. It was common in their milieu for fortunes to fluctuate and to have their shady aspect. The most prominent figure amongst them was James Swan, another who had speculated in Kentucky lands. Mired in debt in 1787, he had come to France to recuperate his fortunes. He was unprincipled (according to future President James Monroe), and was to spend his last twenty-two years in a debtors' prison. But in 1793–4 he flourished as a shipper, banker and dealer in art objects. His Boston firm was the official dealer in French objets d'art and furniture confiscated from the royal palaces after the fall of the monarchy and exported to the United States. Dallarde & Swan, his Paris branch at 63 rue Montmorency, was Barlow's business address.

Imlay, meanwhile, was developing trade with Hamburg, Gothenburg, and Denmark, particularly with the mighty Copenhagen firm of Niels Ryberg who traded with the US and France. His ships had to reckon with the British blockade of French ports on the one hand, and on the other, the arbitrariness of French embargoes. So was born a series of high-risk schemes requiring secrecy and vigilance.

At the same time Imlay and Barlow did not give up on the capture of Louisiana. A member of the Paris conspiracy (formed in March 1793 with Barlow at its head) alerted Citizen Otto (in the French Foreign Office) on 22 May that Imlay was still game. Having failed in the Louisiana scheme, Genêt was denounced as an instrument of the late Brissot when Robespierre addressed the National Convention on 15 November. Only eight days later Barlow and Leavenworth (writing from their base at the Maison de Bretagne in the rue Jacob) put forward a renewed plan for a coup. Its selling point was the advantage Louisiana would bring to France in terms of needed goods, while costing France nothing ('*sans coûter rien*', as it was headed). This longer, more closely argued plan, set out point for point, was designed to appeal to Robespierre with its blend of logic and moral loftiness: to take Louisiana would be '*une action d'humanité*' for a people who '*soupirent pour la liberté, et que leurs coeurs réclament l'identité politique avec leur mère patrie*' ('yearn for freedom, and where hearts cry out for political identity with their mother country'). Together, Barlow and Leavenworth asked the Committee of Public Safety to authorise their raising a force of two thousand. The names of their officers were to be withheld. In other words, control of the expedition was to remain in private hands, and its leaders protected by secrecy. The French Foreign Office endorsed the Barlow–Leavenworth scheme on 7 December. Leavenworth may have acted for Imlay, whose association with the guillotined Brissot would have made it impolitic for his name to appear.

Imlay expressed contempt for the military; when he planned to take over Louisiana, he had a mind to expansion and land, not the cut-throat acts of a coup. That he left to the generals like Wilkinson, who was promoted to lead the frontier wars of the 1790s. Imlay drew an uncritical portrait of Wilkinson (as General W—), 'expatiating, in his usual way, upon what would be the brilliancy and extent of the empire which is forming in this part of the world [the American frontier]; which he said would eclipse the grandeur of the Roman dominion in the zenith of their glory'. Mary Wollstonecraft could not contemplate that glory game without loathing its violence. She deplored 'the hard-hearted savage romans' who sacrificed lives with a *sang froid* 'from the bare idea of which the mind turns, disgusted with the whole empire'. The clarity of her intelligence saw through false fantasies of militarised omnipotence, which disempowers individual judgement – the alternative form of power open to our species, which education, she saw, must promote. If Imlay never saw the monster in his mentor, neither did Washington or Jefferson. Imlay's duality as idealist-entrepreneur, a quasi-innocent who partakes in the underworld, makes him a precursor of Gatsby in certain ways – even to his taste in fine shirts.

Many of Imlay's shifts remain secret, but one fact is suggestive of his standing with both French and American authorities, and that is Mary Wollstonecraft's change of status as soon as she lives with him. It's remarkable that Imlay is able to pass Mary off as his wife with the collusion of Gouverneur Morris, the cautious, patrician American Minister who, incidentally or not, is well placed to spy (with Washington's secret concurrence) for the British and Austrian Cabinets after their ambassadors leave Paris. It's even more remarkable that Mary, whose name had been chalked on her host's door to mark the presence of an enemy alien during the dangerous spring of 1793, is suddenly safe in Paris a few months later when the full Terror is unleashed.

The most significant fact is often something that should have happened and didn't. What should have happened – and what people back in England assumed *must* have happened – was for Mary Wollstonecraft to be arrested along with other British aliens in Paris. Yet she wasn't. Neither she nor Imlay thought it unsafe for her to remain alone in Paris during the blood-soaked autumn of 1793, while Paine's naturalisation and long years in America could not protect him from arrest on grounds of English birth. It did become clear that even the status of honorary French citizen was no protection when agents of the Terror came for Paine after Christmas.

At dawn they banged on the door of the Maison Philadelphie (the new, more politic name for White's Hotel) in the passage des Petits-Pères, where he had spent the night. Orders were to search for incriminating manuscripts. Paine, anxious about his manuscript of *The Age of Reason*, invited the party to sit down to breakfast. The demands of this meal lasted at least three hours. Commissioner Doilé's wordy report explains how a '*fatigue*' descending on the captors between seven and eight had obliged them to take a bit of nourishment and call off the search until eleven ('*exténués de fatigue nous nous sommes trouvés forcés de prendre quelque nourriture*'). While they were champing, Paine secreted his manuscript on his person. It was vital to pass it on, so he directed the searchers to the Maison de Bretagne on the Left Bank, where Joel Barlow lived. While Barlow's *armoires* were thrown open and policemen's heads lost inside, Paine must have managed to slip Barlow his manuscript.

All this time Paine pretended that he could not speak French (though he had made himself perfectly – dangerously – comprehensible in his speech at the Convention, a year before, when he had made a case for sparing the King's life). Slowly, through the simulated fog of Paine's English-French, it was borne in on Doilé that the Bretagne was not the home of his captive. Paine, he realised, had simply wanted to be with his countryman ('*ami natal*', says Doilé, a little put out, but respectful of sentiment). So the party, with the addition of Barlow, trundled back to the Right Bank and out beyond the city wall to the place Doilé triumphantly calls 'the true domicile of the said "PEINE"'. The afternoon search yielded no incriminating matter. It's like a comic scene from *The Scarlet Pimpernel*: the dimwit guard; the resourceful foreigners. But it is, in fact, a darker tale. For when the search ended eventually at four o'clock, Doilé reports, 'we requested citizen Payne to come with us to be transferred to prison which request he obeyed without difficulty'. There, in the Luxembourg, Paine was to languish for almost a year.

It seems extraordinary that Mary Wollstonecraft should have been protected by her far more dubious status as an American. Even to be a genuine American was no guarantee of safety: the papers of Gouverneur Morris are full of petitions from his countrymen for rescue. Yet Gilbert Imlay came and went as though he had a special immunity, and could extend it at will to shield Mary Wollstonecraft completely in a situation where no one else felt safe for very long. The arbitrariness of the Terror, particularly the ease of denunciations under the Law of Suspects, did make Paris a place of danger, especially for anyone who did not lie low – and Mary Wollstonecraft did not lie low. A Silesian count, Gustav von Schlabrendorf, who was betrothed for a while to Christie's sister Jane, was arrested with the English and condemned to the guillotine, missing death narrowly by mislaying his boots when the guard came, and again the following day when the guard forgot to call his name. Later, he recalled how Mary 'often' visited him, Helen Maria Williams and other friends. There can't have been many visitors to the Luxembourg, revealing a glimpse of the quality in Wollstonecraft some describe as rash, others fearless. She must have been the only Englishwoman not behind bars. In fact, the British press assumed her imprisonment, a news item relayed to Bess Wollstonecraft by 'every brute' in her vicinity. Bess instinctively knew the report was untrue, but it left her 'haunted' by the image of her sister. She regretted 'having been too severe on a heart capable of all that commands respect and Love'. For a few weeks in November–December 1793 Bess reproached herself – so long as she feared Mary was 'in greater danger than a more insignificant character'.

This fear was justified. Mary Wollstonecraft was pressing on with her history of the French Revolution at a time when the Revolutionary Tribunal was obsessed with incriminating papers. Helen Maria Williams had burnt her own papers as well as those entrusted to her by Mme Roland and Mme de Genlis. Paine was currently imprisoned to stop him writing what he saw. Mary Wollstonecraft was likewise too honest not to write as an eyewitness to the Terror. 'I am grieved – sorely grieved when I think of the blood that has stained the cause of freedom in Paris . . . Alas!' she cries. 'Justice had never been known in France. Retaliation and vengeance had been its fatal substitutes.' It would have been simple for the Public Prosecutor, Fouquier-Tinville, to pick out such passages. Her opposition to bloodshed aligns her with those now judged traitors, so that to continue to set down what she saw risked death.

When Mary was four and a half months pregnant, at the start of 1794, she noticed eyes on her belly. 'Finding that I was observed, I told the good women . . . simply that I was with child: and let them stare! and – , and – , nay, all the world, may know it for aught I care!'

Her words were defiant, but she did shrink from 'coarse jokes'. It took some pluck to sustain nonchalance with people who questioned a pregnant woman on her own, and this may be one reason for her decision, a few days later, to leave Paris. During that first week of January she was visited by a member of the American network who (she reports to Imlay) 'incautiously let fall' something she hadn't known – clearly, a secret matter, because she skirts it on paper.

Mary accused Imlay of 'a want of confidence, and consequently affection'.

Given his long habit of secrecy, he took a risk when he involved himself with a celebrity, engaging in intensive correspondence with a woman whose voice proclaimed a morality of clarity and candour. Imlay acted swiftly. He asked her to respect his motives, and fastened their tie with an idyllic picture of domestic bliss with six children grouped about their fireside. Best of all, he wished her to join him.

Soothed and '*lightsome*' (in her Yorkshire idiom), she agreed to leave as soon as she was fit to travel. This letter of 9 January 1794 is the first signed 'Mary Imlay'. During the visible stage of her pregnancy, she was going to be 'Madame Imlay' living with her husband in Le Havre where her history was unknown. Body and mood healed at the prospect: 'I look forward to a rational prospect of as much felicity as the earth affords – with a little dash of rapture into the bargain.' A little shamefaced, she tries out the language of dependence – her 'tendrils' clinging to his 'elm' for support – but soon shifts to her own candid manner when she dreams of their reunion: 'Knowing I am not a parasite-plant, I am willing to receive the proofs of affection, that every pulse replies to, when I think of being once more in the same house with you.'

When she applied to leave Paris, she had none of the difficulty others experienced with officials of the Terror. A pass was hers for the asking, as though Imlay was either influential at the American Embassy or shielded by a link with the Terror itself – the much-needed supplies shipped through a British blockade. If Mary's English identity were to present a problem, this would have surfaced over the pass (as it did before her association with Imlay, when she had failed to get a pass the previous spring). On Thursday 16 January 1794, she set off for Le Havre. As she produced her pass for the guards at the exit from Paris, her real danger lay in her luggage: her manuscript with its talk of 'butchery' and critique of cold eloquence – unmistakably Robespierre. To write '*merde à la république*' was a crime and Mary had still not burnt this incriminating document. Shaken by all she had witnessed of 'death and misery, in every shape of terrour' for 'the unfortunate beings cut off' around her, she was not to be deterred. She carried that sheaf of paper out of Paris, though her life, she well knew, 'would not have been worth much, had it been *found*'.

Mary's spirits rose as she journeyed towards Imlay. Her notes to him touch on her feelings from moment to moment: 'I am driving towards you in person! My mind, unfettered, has flown to you long since, or rather has never left you.' 'I hope to tell you soon (on your lips) how glad I shall be to see you' and to 'bid you goodnight . . . in my new apartment where I am to meet you and love, in spite of care, to smile me to sleep'.

Havre-Marat, at the mouth of the Seine, was then a port with a population of twenty-five thousand. A wall, fifteen feet wide at the top, held back the tide about a mile in front of the town with its tall, close-packed houses. Imlay had prepared good lodgings, rented from John Wheatcroft, on the rue de Corderie, Section des Sans-Culottes, near the harbour. They dined in a big room; a *gigot* smoked on the sideboard. Mary was given the means to hire a servant and buy fine linen for Imlay's shirts. Though in the course of business he was troubled by sudden embargoes and ships that did not arrive, they lived well – as Imlay always did. Their second period together was like marriage in everything but law. After two months Mary reassured Everina, 'I am safe, through the protection of an American . . . who joins to uncommon tenderness of heart and quickness of feeling, a soundness of understanding, and reasonableness of temper, rarely to be met with – having been brought up in the interiour parts of America, he is a most natural, unaffected creature.' With Imlay she hid the 'shades' in their relationship, and contrived to dissolve these blots of 'darkness' in companionable squabbles over pillows.

'I could not sleep. – I turned to your side of the bed,' she wrote to him when he was away, 'and tried to make the most of the comfort of a pillow, which you used to tell me I was churlish about; but all would not do.'

Mary asked Ruth Barlow to send her *Le Journal des débats et des décrets* (the Debates and Decrees in the National Assembly) and, if possible, her books. When Richard Codman passed through Le Havre en route to London, she gave him a package for Johnson: part of her *Historical and Moral View of the Origin and Progress of the French Revolution.* A 'View' is what it is, not a full-scale history obedient to the institutional language of the dominant order. Soon after, Jane Austen likewise questions the supposed objectivity of institutional history ('the quarrels of popes and kings, the men all so good for nothing and hardly any women at all'). More specifically, Wollstonecraft questions the spectator habit induced by court government. Monarchy is entertainment. It's theatre for those it rules. Drama, the great cultural achievement of the reign of Louis XIV, was so pervasive that even the wars of that long reign seemed like spectacles. If the Terror, too, was theatre – and here Wollstonecraft concurs with Burke – this taste for theatre was carried over from the *ancien régime*.

In her critique of theatricality as a mode of existence, Wollstonecraft authenticates a character who appeared ten years before she was born. Clarissa (in Richardson's novel) resists the theatricality of Lord Lovelace, who expects to dazzle her with verbal flourishes, ingenious plots and amusing disguises. When these fail he rapes her. Clarissa's vindication of her integrity made her the most popular heroine of the century. Her story exposes the stale props of a rake who looks back to plots of woman's complicity, the lusty, conniving image of woman promoted by the Restoration comedies of the previous century. Lovelace proves an anachronism in 1748, beside the new independence of a literate woman. Clarissa's pen confounds Lovelace; her rational language takes us to conclusions that close his show. For Wollstonecraft's generation of the 1770s and '80s, the anachronism was no longer the rake, but the theatre of royalty propped by the parade of courtiers. Mary Wollstonecraft's Clarissa voice offers reason in place of stupid awe at the glamour of privilege, for reason is 'the image of God planted in our nature'.

A View discerns the workings of character at the heart of public events. 'The lively effusions of mind', characteristic of the French, provoke 'sudden gusts of sympathy' which evaporate as new impressions come from another quarter. It's an English critique of susceptibility and brilliance in favour of solidity: 'Freedom is a solid good, that requires to be treated with reverence and respect.' Freedom is also female – a mother bird who had shown her 'sober matron graces' in America, promising 'to shelter all mankind' beneath her 'maternal wing'. Wollstonecraft was not content to expound the history of the old order; she explores the psychology of servility 'destroying the natural energy of man' while it 'stifles the noblest sentiments of the soul'. The result is 'a set of cannibals [revolutionaries], who have gloried in their crimes; and tearing out the hearts that did not feel for them, have proved, that they themselves had iron bowels'. Her cool conclusion is that the retaliation of slaves is always terrible, and that the ferocity of Parisians – 'barbarous beyond the tiger's cruelty' – was a consequence of bent laws, nothing but 'cobwebs to catch small flies'.

Wollstonecraft worked on this *View* throughout her pregnancy, and finished it during her ninth month. Soon after, the *Grande Terreur* set in, as Robespierre pushed through the Law of 22 Prairial, depriving the accused of counsel or witness, with verdicts limited to acquittal or death. The acknowledged aim was to encourage denunciation. For the next forty-seven days, citizens of Paris went to the guillotine at a rate of thirty a day; as Wordsworth put it: 'never heads enough for those that bade them fall'.

With daily raids on the Luxembourg, Paine fell into a semiconscious fever – jail fever or typhus – and was reduced to a skeleton unable to speak. It was only a matter of time for his name to appear on the list of the condemned; one dawn a turnkey chalked the door of his cell. Since Paine was perspiring with fever, the door had been allowed to remain open. It opened outwards, so that the chalk mark was on the *inside* of the door which Paine's cellmates, claiming a change of temperature, got permission (from another turnkey) to close. The mark now faced inwards. At eleven that night they heard the guards with lanterns trained on each door, approach and stop, approach and pass, and the fading screams down the passage as others were hauled to execution. 'My God, how many victims fall beneath the sword and Guillotine!' Mary groaned to Ruth. 'My blood runs cold, and I sicken at thoughts of a Revolution which costs so much blood and bitter tears.'

As Mary expanded into her last month of pregnancy, she asked Ruth to send her dress fabric: white dimity or calico or printed cotton – the new simplicity in reaction against brocade and satin. Just after she turned thirty-five, Mary gave birth, on 10 May 1794. She was attended by a midwife, a sensible choice in France where midwives were properly trained. Mary had taken against the displacement of midwives by doctors plying their forceps (designed in the seventeenth century, lost sight of, then reintroduced in the 1730s). Doctors tended to treat women's bodies like ill-designed machines that required their intervention. The danger of forceps to the foetus was satirised in *Tristram Shandy*, where Dr Slop sets a series of disasters in train when he manages to crush the nose of the emerging hero. Mary, by contrast, practised what we call 'natural' childbirth: an informed, matter-of-fact attitude; home surroundings; support from the father; and an experienced midwife. The pains, though fierce enough, were not prolonged; and, against the custom of 'lying in' for weeks, Mary was up the following day. The baby was registered as 'Françoise', born in the legitimate marriage of 'Guilbert' Imlay and 'Marie' Wollstonecraft his wife. She was to be known as Fanny, after Fanny Blood. Mary described her labour to Ruth:

> Havre May 20th [17]94
>
> Here I am, my Dear Friend, and so well, that were it not for the inundation of milk, which for the moment incommodes me, I could forget the pain I endured six days ago. – Yet nothing could be more natural or easy than my labour – still it is not smooth work – I dwell on these circumstances not only as I know it will give you pleasure; but to prove that this struggle of nature is rendered much more cruel by the ignorance and affectation of women. My nurse has been twenty years in this employment, and she tells me, she never knew a woman so well – adding, Frenchwoman like, that I ought to make children for the Republic, since I treat [childbirth] so slightly – It is true, at first, she was convinced that I should kill myself and the child; but since we are alive and so astonishingly well, she begins to think that the *Bon Dieu* takes care of those who take no care of themselves. But, while I think of it . . . let me tell you that I have got a vigorous little Girl, and you were so out in your calculation respecting the quantity of brains she was to have, and the skull it would require to contain them, that you made almost all the caps so small I cannot use them; but it is of little consequence for she will soon have hair enough to do without any. – I feel great pleasure at being a mother – and the constant tenderness of my most affectionate companion makes me regard a fresh tie as a blessing.

When the carrier of this letter was delayed three days, she added a post-script: 'I am now, the 10th day, as well as I ever was in my life – In defiance of the dangers of the ninth day, I know not what they are, entre nous, I took a little walk out on the eighth – and intend to lengthen it today. – My little Girl begins to suck so *manfully* that her father reckons saucily on her writing the second part of the Rts of Woman.'

Mary Wollstonecraft had always been an advocate of breast-feeding. 'The suckling of a child . . . excites the warmest glow of tenderness,' she had said after watching Fanny Blood. 'I have even felt it, when I have seen a mother perform that office.' During the 1780s only about thirty per cent of Parisian mothers breast-fed their babies; the rest, mainly those who could afford it, packed them off to a wet-nurse. For centuries the Greek physician Galen was believed when he said that sex and nursing were incompatible: 'carnal copulation troubleth the blood, and so by consequence the milk'. In the *Rights of Woman*, Mary Wollstonecraft took the baby's part: 'There are many husbands so devoid of sense and parental affection that, during the first effervescence of voluptuous fondness, they refuse to let their wives suckle their children.' Wet-nursing was often fatal: in France one in two babies died. For those who survived, the custom of suspending babies from the rafters of cottages was hardly beneficial to their emotional and mental development: infants hung there unchanged, sluggish with inactivity, their mouths crammed with rotting rags. Mary Wollstonecraft insisted on the free exercise of the child's body, and its need to be held and touched. She played with Fanny, fed, kissed and hugged her, and confided to Imlay that her initial sense of duty was giving way to love, warmed as much by Fanny's ruddy cheeks, vigour and intelligence as by nursing. Motherhood did bear out her belief that the vital education in tenderness begins at the breast.

Her unorthodox methods caused talk amongst the women of Le Havre, who called her the 'raven mother'. When asked what this meant, she replied: 'They all thought I was not worthy of having such a child.'

At the age of three months, Fanny went down with the dread disease of smallpox. Mary's common sense told her to distrust doctors with their bleeding, purging and other potentially fatal interventions. She always put hygiene first, in contrast with the superstition of certain reputable members of the Royal Society like Robert Boyle who had advised blowing dried and powdered human excrement into the eye as a remedy for cataract, and Robert Hooke who had taken medicines made up of powdered human skull. Mary Wollstonecraft could have picked up Dr Haygarth's sounder advice on cleanliness in a book on smallpox published by Joseph Johnson. She tended Fanny herself and washed the child's pustules in warm baths twice a day. Within a few weeks Fanny recovered her bloom and vivacity – 'our little Hercules', her mother crowed.

11

THE SILVER SHIP

In the summer of 1794 Imlay launched a treasure ship. Its mysterious course would eventually send Mary to Scandinavia in its wake. The need for that journey starts in this quiet domestic period when Mary was nursing her newborn baby and apparently inactive. At this early stage of motherhood she accepted her dependence on Imlay – a little rueful about the clinging tendrils she was growing. But it turns out that the spring and summer of 1794 were not quite as quiet as they appear. It's now clear that Mary played an active part in a sequence of events *before* the treasure left Le Havre, and that buried in this run-up time lie obscure, disparate, but crucial clues – old shipping records, communications sent in cipher, and, most telling, a newly discovered letter from Mary Wollstonecraft herself – to questions long unanswered. Why should she make her way to a remote, craggy region of Norway where there were no roads, only small towns reached by boat? How much of the mystery surrounding this treasure did she know? And more important for her future peace of mind: what *didn't* she know?

This is no fiction: there is no certainty where this story began. We might say it began with Gilbert Imlay's frontier character, or with French needs in time of war. A convenient starting-point could be when Barlow, the only American in Europe with the standing of honorary French citizen, began dealings with a new three-man Commission des Substances, in charge of provisions. In need of grain, beset with enemies and a British blockade, France was sending out commissioners to neutral ports – Copenhagen, Gothenburg, Hamburg – with sixty million livres to spend on supplies. The commissioners saw the export of frowned-on luxuries as one means of feeding the army and Paris: suppliers could be paid from storehouses crammed with exquisite silverware, glassware and other goods seized from émigrés and those condemned to death. Here is an unacknowledged link between the Terror and American profiteers. Naturally, men like Barlow and Imlay would have deplored the Terror, as Tom Paine did, as Mary Wollstonecraft did, yet commercially they benefited. This may be a rational source for Mary's dislike of what she called Imlay's 'commercial face': she caught sight of something she could not condone, without knowing quite what it was that he did.

Imlay's contact in Gothenburg was Elias Backman, a Finn of Swedish extraction who exported grain, gunpowder and alum to France. Imlay conceived a plan of paying for such commodities with French silver and bullion – a trove that would multiply his shipping profits – and he may have approached Backman to receive a first consignment. For, in March 1794, Backman sent a petition to the Swedish Crown, asking leave to import gold and silver worth 172 million *riksdaler*, to be carried by an English cutter called the *Rambler*, lighter and swifter than any Swedish vessel. The French trove was to sail under the cover of a Swedish owner (Backman) and neutral Swedish flag. Permission to go ahead was granted.

'The successful outcome of this valuable transport depends on its secrecy,' Backman warned.

Secrecy was not always preserved. The British already had intelligence of traffic between Sweden and Le Havre, and intercepted a letter from Le Havre to a Swedish merchant: 'Whatever merchandise you send to this place will be gratefully received, but particularly provisions and military stores; name your own price, it will be paid – mention any house of Amsterdam, where you choose that the money should be placed . . .' This is the way Imlay spoke, the persuasive imperatives very like those in a letter to Wollstonecraft. She, too, slipped up, even when she tried to be secret, and clues lurk in her letters.

One to Ruth Barlow refers to a 'disappointment' for the Barlows early in February 1794: this is when France rejected the Louisiana coup that Barlow and Leavenworth had tried to relaunch the previous November. It left the Barlows with no means to escape the Terror and return home, as Ruth had begged. An alternative plan – in league with Imlay – was their move to Hamburg. France's rejection of the coup was formally announced on 6 March. By the 10th the Barlows were ready to leave, a date that coincides with Swedish permission to import the *Rambler*'s bullion. In Hamburg and its environs – a free port and an increasingly lucrative commercial centre – Joel was to handle ships and cargoes in transit. The following month Mary Wollstonecraft, writing to Ruth from Le Havre, mentions the joint enterprise of Imlay and Barlow. This letter slips through the silence preserved by their menfolk:

> My Dear Friend,
>
> I wrote to Mr B[arlow] by post the other day telling him that I indulge the expectation of success, in which you are included, with great pleasure – and I do hope that he will not suffer his sore mind to be hurt, sufficiently to damp his exertions, by any impediments or disappointments, which may, at first cloud his views or darken his prospect – Teasing hindrance of one kind or other continually occur to *us* here – you perceive that I am acquiring the matrimonial phraseology without having clogged my soul by promising obedience &c &c – Still we do not despair – Let but the first ground be secured – and in the course of the summer we may, perhaps celebrate our good luck, not forgetting good management, together. – There has been some plague about the shipping of the goods, which Mr Imlay will doubtless fully explain – but the delay is not of much consequence as I hope to hear that Mr B[arlow] enters fully into the whole interest . . .
>
> Believe me yours
>
> affectionately
>
> Mary

Worse plague and delay followed, only three days after Mary gave birth: the British capture of the armed *Rambler* on 13 May as it sailed from Sweden to pick up its cargo in France. A court case in London challenged Swedish neutrality, and Imlay's London agent, Mr Cowie, reported to the Swedish authorities. Nothing was proved because the captain had taken the precaution of keeping no papers on board. Supported by a cover story from Sweden – denying the French destination – the *Rambler* could not be convicted, but its release took a while. It was therefore not until the autumn that the ship finally carried that treasure of gold and silver out of France. On the last lap of this voyage, it sailed from Glückstadt (a Holstein port on the Elbe, downstream from Hamburg) for Gothenburg, where it finally arrived on 25 October 1794.

The capture and delay of the *Rambler* is crucial to Imlay's plan for an interim treasure ship, to sail in August. It has come to light that Mary involved herself in this scheme to the extent of dispatching the treasure herself on its precarious journey through the British blockade.

Thirty-six silver platters, rumoured to carry the Bourbon crest, came somehow into Imlay's hands. The plate, together with thirty-two silver bars, was said to be worth £3500. With the *Rambler* held up for an indeterminate time, Imlay meant to ship this hoard separately. He approached a Norwegian captain, a young man of twenty-five called Peder Ellefsen, who journeyed two or three times to Paris to collect the silver and bring it to Imlay in Le Havre. These journeys took place during the first half of June 1795, at the height of the Great Terror when Paris was a blood-soaked fortress. To arouse suspicion was death. Though Ellefsen would have had passes secured by Imlay (with the same know-how, we can assume, as ensured Mary Wollstonecraft's safety when she had passed out of Paris six months earlier), it took some courage for a stranger to enter that ever more suspicious fortress and exit with secret goods. Ellefsen would have known how to keep quiet, and Imlay must have had reason to trust his nerve and honesty. At Le Havre, Imlay lodged him and the silver with a merchant, probably the Englishman Wheatcroft from whom Imlay and Mary rented their house. Mary's long-lost letter reveals that she herself inspected the treasure – the forbidden fleur-de-lis gleaming on royal plates – in Ellefsen's room.

It was Ellefsen who pointed out a ship for their purpose. On 18 June (when Fanny was a month old), Imlay bought this oak three-master named the *Liberté*. The question was how to transport the treasure through a British blockade raring to teach the Swedes a lesson. Imlay had to take the precaution of dissociating his new ship from Backman and all things Swedish. This interim treasure was to be more secret than the first: the cargo small enough to hide; the crew unaware of its presence; and its destination disguised. The *Liberté* was to sail as a Norwegian ship. Its papers named Ellefsen as the owner, carrying only ballast. His mate was a New Englander called Thomas Coleman who took down the *Liberté*'s tricolour flag and draped it like a scarf about his waist, in the fashion of seamen. He was the only other man aboard to know of the silver.

Their plan appears to have been twofold. First, to evade the British navy as well as Algerian pirates along the north German and Danish coasts, it would have made sense for Ellefsen to sail for his native shore, taking a northerly route along the coast of Norway. The ship could dodge its way in and out of fjords familiar to Ellefsen, and then follow the Swedish coast south to the trading port of Gothenburg.

The second part of the plan had to be a private agreement (signed on 12 August) between Imlay and Ellefsen: that though Ellefsen was officially the owner of the ship, the true owner was Imlay. Ellefsen renamed the ship *Maria and Margrethe*, after Marie de Fine Fasting who became his wife, and after his mother Margrethe, a wealthy woman of statuesque beauty whose seven sons would row her from her estate to the local town of Arendal on the southern shore of Norway.

Ellefsen and the mate, Coleman, quietly loaded the silver at the back of the hold. Ellefsen signed a receipt for it, which Mary and the merchant Mr Wheatcroft checked. The latter witnessed Ellefsen's signature. The receipt was then enclosed in a letter, dated 13 August, to Elias Backman. It appeared to be a letter of introduction, written in English. Should the ship and its treasure, like the *Rambler*, fall into English hands, this letter was designed to confirm that it was not carrying a cargo destined for

Sweden, and that Imlay and the Swedish merchant were strangers who might find it convenient to do business at some later date. Backman was not informed of the silver on its way, and one reason for this could be the danger that such a communication could be intercepted.*

Since Imlay had to set out for Paris, it was Mary Wollstonecraft who gave Ellefsen his last orders . On 14 or 15 August the silver ship sailed out to sea, bearing Mary's hope that it would bring Imlay the fortune he had always wanted – in her mind, a fortune that would free him to stop scheming and stay with her. She did put probity before money. 'The fulfilling of engagements appears to me of more importance than the making of a fortune,' she had said to Imlay that summer when he was stressed (presumably by the capture of the *Rambler*) to the point of sickness. Yet events will show that, for all her professed aversion to dealing, Mary did have some investment in the silver ship. Money itself meant little to her beyond a means to pursue a new kind of life, but this silver had come to stand for a future with Imlay, who had become an inextricable part of that life, and all the more so since the birth of their child, now three months old and dearer to her mother every day.

The *Maria and Margrethe* reached Norway after nine days at sea. Then it vanished. The captain was rumoured to be back home. It was said that there had been a plan to sink the ship near Arendal, the coastal base of the Ellefsen family, only twenty miles west of Peder's own coastal home at Risør. He denied a cargo of silver. Had Ellefsen betrayed Mary's and Imlay's trust? This mystery would take months – in a fuller sense, centuries – to solve.

Meanwhile, a political upheaval took place in Paris. On 27 July 1794 Robespierre fell. Next day he was guillotined with Saint-Just, followed in the next two days by eighty-three members of the Commune of Paris. As each head fell, the mob roared '*À bas le Maximum!*', hailing the end of price

---

*Another possibility – hard to prove but one that can't be ruled out – is that, unknown to Mary, this particular cache of silver was always due to end up elsewhere. The ship's papers, obtained from the Danish Consul in Rouen on 20 July, give a Danish destination, but it's unclear whether this was genuine or a blind, or whether there was a change of plan.

controls. Jean-Lambert Tallien, a political moderate, came to power, and the Terror was over. A fortnight later, Imlay left for Paris.

Barlow, too, visited Paris from Hamburg at this time. It's inconceivable that the two did not converge to review their joint enterprise. As the silver ship sailed northwards, and before it disappeared, Mary adds a coda to a letter she sent to Imlay from Le Havre: 'I will not mix any comments on the inclosed (it roused my indignation) with the effusion of tenderness, with which I assure you, that you are the friend of my bosom, and the prop of my heart.' Her indignation over what appears to be a business enclosure is easy to overlook, buried in a love-letter, but 'indignation' reappears in the opening of her next letter the following day, 20 August, together with intimations of betrayal by a person whose name is left as a blank.

'I want to know what steps you have taken respecting ——. Knavery always rouses my indignation – I should be gratified to hear that the law had chastised —— severely; but I do not wish you to see him, because the business does not now admit of peaceful discussion, and I do not exactly know how you would express your contempt.'

Had this 'knavery' to do with the silver ship? If so, it is worth noting that the knave was in Paris, while Ellefsen – the apparent culprit – was at this date still at sea. A little over a month later, on 28 September, when Mary writes from Paris to Imlay in London, she again encloses business letters from ——. 'I want to hear how that affair finishes,' she adds, 'though I do not know whether I have most contempt for his folly or knavery.' We can assume from the pattern of repeated words – 'knavery', 'contempt' – that she refers to the same betrayal. Clues to the vanished treasure may have lain all along in Mary Wollstonecraft's letters. What can we deduce from them?

First of all, she was no longer entirely in the dark as to Imlay's schemes; she was privy to his letters, and declared later that she was 'fully acquainted with all the circumstances' of the silver ship. Second, whoever was in Paris in mid-August was not there at the end of September. Presumably, the knave was not at this time resident in Paris. There is yet another, apparently incidental, fact which could cast an unexpected light on the mystery: from the time Mary left Paris in January 1794, she was writing every month or two to Ruth Barlow who kept her letters. These are intimate letters, showing that Ruth to some extent replaced Fanny Blood as a friend. Ruth was confiding and affectionate, and Mary confiding in return when it came to her experience of childbirth. Her final letter to Ruth (about Imlay's stress) is dated 8 July 1794, and, given their habits, a letter should have followed in August or September. But Mary never, it seems, wrote to Ruth again. What could have ended their friendship?

That September, Imlay took off for London. Mary was to wait for him in Paris. She returned to the place where she and Imray had stayed: 'I slept at St. Germain's,' she wrote to him, 'in the very room (if you have not forgot) in which you pressed me very tenderly to your heart. – I did not forget to fold my darling [Fanny] to mine, with sensations almost too sacred to be alluded to.'

She had been away for nine months. Paris in the meantime had turned against the Terror: the Jacobin Club was closed on 8 September; salons reopened; theatres filled; and the uniform of baggy trousers and red wool hats was now outré. Men changed into plain cloth coats and stout boots – no ruffles at the wrist, no silk stockings – copying the clothes of an English countryman who had never fluttered around a court. Women floated in white, unstructured gowns in flimsy fabrics like muslin – high-waisted *robes en chemise* – with heelless slippers, imitating the simplicity of ancient Greece. The naturalness of the naked body replaced panniers, stomachers, loops, layers and theatrical makeup. Tallien's wife slit her gowns to the thigh. Mary's nursing breasts and rounded figure would have been in fashion, as were the flowing curves of Venus; instead of tight lacing, women in 1795 adopted 'corselettes' only six inches long. It was done to stand out again: the author Mme de Staël (daughter of Necker and wife to the Swedish Ambassador) wrapped herself in stoles and turban. It was also fashionable to have suffered: there were balls for victims; widows cropped their hair and twined thin red ribbons around their necks to signify execution. Newspapers had been freed, Mary informed Imlay. But women's protest was still suppressed, France was still at war with the monarchies of Europe, and foreigners, apart from Americans, were still the enemy – if anything, the laws against them were stricter.

So Mary returned to find all the English gone. The Christies were back in London; I. B. Johnson, who had associated with Mary during her first months in Paris, and Thomas Cooper, Imlay's backer, had escaped at the start of the Terror; and Helen Maria Williams had departed in April 1794 when the Committee of Public Safety had ordered foreigners to leave Paris and not return on pain of death. The Williams women had gone to ground near the village of Marly, sixteen miles away, and early in July had fled briefly to Switzerland. Paine still languished in prison after eight months. His plea to the new order shows the continued danger of being English, however revolutionary such a person might be, however insistent his adoption of American nationality:

> Citizen Representatives
> . . . When I left the United States of America in the year 1787, I promised to all my friends, that I would return to them the next year; but the hope of seeing a Revolution happily established in France, that might serve as a model to the rest of Europe . . . induced me to defer my return to that country, and to the society of my friends, for more than seven years. This long sacrifice of private tranquillity, especially after having gone through the fatigues and dangers of the American revolution, which continued almost eight years, deserve[s] a better fate than the long imprisonment I have silently suffered . . .
>
> It is perhaps proper that I inform you of the cause assigned in the order for my imprisonment. It is that *I am a foreigner*; where-as the *foreigner*, thus in prison, was invited into France by a Décret of the late national assembly, and that in the hour of her greatest danger when invaded by Austrians and Prussians. He was moreover, a citizen of the United States of America, an ally of France . . .
>
> Thomas Paine
>
> Luxembourg thermidor 19th 2nd year
> of the French Republic one and indivisible

This plea failed. Paine was kept in prison until November 1794, when a new American Minister, James Monroe, came to his rescue. Paine had deteriorated during his year in prison, drowned his sorrows in drink, and become offensive to his friends. Though Monroe took him into his home to aid his recovery, he was less than delighted when Paine stayed two years.

The Swiss Schweizers were still in Paris. Once, while Mary was composing a letter to Imlay, Madeleine Schweizer sat beside her reading his *Topography* of the frontier. 'She desires me to give her love to you, on account of what you say about the negroes,' Mary relayed.

Count von Schlabrendorf, too, remained in Paris. He was charmed to meet again the woman who had come to comfort him in prison. He saw a face 'so full of expression' and a grace beyond any ordinary standard of beauty. Her glance, her voice, her movement, he said, 'enthralled me more and more'. '*Sinnvoll*', he called her, the most perceptive women he had met. It's not certain what he meant when he said that she was 'not to be judged as an Englishwoman', but he liked her for being neither prudish nor licentious: 'She was of an opinion that chastity consisted in fidelity.' He was in love with her, something he realised only later. 'Her unhappy union with Imlay prevented any closer bond.'

Imlay wrote regularly from London, though sometimes in haste. He saw his conduct as responsible; Mary, on the contrary, saw in his long departures a denial of their love. This disturbed her the more because, unlike her father, he was not cruel. He was amiable and well intentioned: he tried to console her with 'permanent views and future comfort' and declared, 'our being together is paramount to every other consideration!' She could never quite relinquish her belief in his honesty, even though she was led to expect him each week, and each week disappointed. Of course, if he was, as suggested, some sort of secret agent – his *Topography* is the work of a collector and spinner of information who gives nothing of himself away – he would not have been as free to join her as she assumed, nor at liberty to explain his movements.

Since the War of Independence, London was tense with the United States, keen to restore dominance, and seething with secret agents. The idea was to prevent new settlements in the American interior becoming dependent on France or Spain. Might the West separate off as an independent republic of Louisiana, stretching from the Mississippi to the Rockies, under the protection of Britain? Treaties of commerce and friendship would be welcomed. During the mid-1790s, General Wilkinson (calling himself 'the Washington of the West') devised a plot to bring together a band of settlers and 'Indians' who would capture Florida and those vast plains of the Louisiana Territory, with British support. Then there was talk of a joint US-British coup to be led by General Miranda in 1798. This British interest, replacing that of the French, stirs at a time when Imlay turns from Paris to London. He appeared the leading expert on the frontier – an expanded edition of his *Topography* came out in 1797. Could he have been an agent promoting interest at the London end? If so, it explains his prolonged stay in London – the repeated postponement of his return to Paris. At this time, Mary was still reverting to their plan for a farm on the frontier. It may sound deluded, given Imlay's zest for commerce, but if he was secretly involved, once more, in a frontier venture, she was not far off from his dream of land. Only, where she thought in modest terms of pastoral retirement, he planned on a vast scale.

When he postponed their departure for America, his excuse was the need to amass more funds.

'The secondary pleasures of life are very necessary to my comfort,' he said.

'It may be so,' Mary retorted, 'but I have ever considered them as secondary.'

Her doubts about his schemes and associates continued to rankle. 'I would share poverty with you,' she said, 'but I turn with affright from the sea of trouble on which you are entering.'

Imlay's material support added to Mary's unease: if he didn't love her or was unfaithful, it relegated her to the mistress position she always feared. For a year following Fanny's birth – so long as she breast-fed – she did not earn. Johnson published her *French Revolution* late in 1794, with a second printing the next year. It was his custom to pay her an advance, and this may have been the money he is known to have sent her in the spring and summer of 1793. By the winter of 1794–5 she had no funds of her own. Imlay had instructed an American contact in Paris to provide for her whenever she asked, but Mary shrank from doing so, for the contact would taunt her with talk of unexplained commitments that would keep Imlay in London 'indefinitely'. This unnamed man 'inconceivably depressed' her during the last days of December as he flexed a closer connection than that of the woman who called herself 'Mrs Imlay'.

'Yesterday,' she writes again to Imlay on 10 February, 'he very unmanlily exulted over me, on account of your determination to stay [in London]. I had provoked it, it is true by some asperities against commerce . . .' Her sense of the two in league exploded in sudden disillusion with the nation she had most admired. She burst out: 'I shall entirely give up the acquaintance of Americans.'

Imlay tried to assure her their futures were knit. Mary brushed this aside with the urgency of a 'fever that nightly devours me'. Breast-feeding seemed to deplete her further. Still, he did not come. Some see in this a confirmation of a familiar scenario. First scene: ambitious female shows true colours as sex-starved dupe. Second scene: fallen woman gives birth. Third scene: abandoned mother becomes clinging bore, to be shaken off with repeated excuses. But if Imlay *was* in the pay of US or British Intelligence, his commitment to Mary Wollstonecraft would have been an aberration in a secret life that didn't admit of serious intimacy, least of all with a celebrity. What little can be discerned through the screen of the unnamed Paris agent – disapproval – may not be disapproval of Mary herself, but realistic caution. For if the Paris agent was in some sense a secret agent, there would have been a danger in Imlay's involvement with a clever, probing partner. The shadows on the other side of the screen look (and declare themselves) more purposeful than what is so far conceived. Mary thought herself party to Imlay's silver scheme, but there was much she didn't know – including the fact that secret agents used inside knowledge to speculate. The two activities were linked.

The winter of 1794–5 was one of the coldest on record. The Seine froze; so did the fountains. The Convention had abolished more price controls, the cost of bread rose, and the freezing of harbours like Le Havre on the Normandy coast prevented the import of emergency grain. People died of starvation; wolves howled at the gates of Paris; coal was scarce and queues lengthened. That February, Mary moved in with a German family who had a child Fanny's age and shared Mary's ideas on childcare. She began to think that if she died, her housemate would be a better person to rear

Fanny than would her father. During those dark days she often thought of death, and almost regretted the intensity of her love for Fanny that tied her to life. Unable to bring herself to approach Imlay's contact, she had to chop wood to keep them warm. At last she was driven by continued night sweats to reveal her reluctance to Imlay.

'I have gone half a dozen times to the house to ask for [money], and come away without speaking – you may guess why.'

She had so far had 3000 livres, and proposed to take a thousand more to pay for wood; and there her dependence must end. Her milk had always been abundant; now, her longing for Imlay seemed a curse, she told him, 'that burns up the vital stream I am imparting'. At this, Imlay asked her to London. Mary was reluctant, for the invitation was complicated by a warning that he might find it necessary to return afterwards to Paris.

'Is our relationship then to be made up of separations?' she put it to him, 'and am I only to return to a country . . . for which I feel a repugnance that almost amounts to horror, only to be left there a prey to it!'

Her country had changed in the two years she had been away. The British government was fighting on two fronts: with its revolutionary neighbour across the Channel and with reformers at home who had roused the workers all over the country. There had been an influx of French aristocrats whose tales terrified the landed classes, while in 1794 the English army (made up of mercenaries and 'pressed' men) had been chased from Europe by the new French army of citizen soldiers. Defeat filled Prime Minister Pitt with fear of sedition. Talk of electoral reform and rights for the working class now took on the taint of treason. Late in 1794 Pitt had instigated Treason Trials for twelve members of the London Corresponding Society, though their aim was reform, not revolution or regicide. The association of reform with the excesses of the French Revolution led England into a phase of extreme reaction. One of Johnson's radical authors was Joseph Priestley, the discoverer of oxygen. When the cartoonist Gillray pictured Priestley toasting the King's head, a mob burned down his laboratory, papers and library in Birmingham. He emigrated to America, telling Vice-President Adams that all he wanted was refuge from his countrymen with ears bent to the lies of 'illiterate' informers. Pitt's informers were indeed everywhere. One sent later to spy on Wordsworth and Coleridge famously overheard suspicious talk of 'Spy Nozy' (Spinoza).

Women who asked for rights were now 'Amazons', in contrast to virtuous homebodies; to be 'daring' and 'restless' was to be 'unprincipled'. The result was to polarise women in the old way as saints or sinners, wives or wantons, negating Wollstonecraft's fusion of rights and domesticity. Burke aligned her with Mme Roland, Helen Maria Williams, Mme de Staël and Mme de Genlis, as 'that Clan of desperate, Wicked, and mischievously ingenious Women, who have brought, or are likely to bring Ruin and shame upon all those that listen to them'. He implored mothers to 'make their very names odious to your Children'. Women had to conform to the swing against equal rights if they were to survive as writers: 'Do not . . . call me "a champion for the rights of women",' Maria Edgeworth says in her 1795 *Letters for Literary Ladies*. 'Prodigies are scarcely less offensive to my taste than monsters.'

A third cause for caution on Mary's part was Imlay's maxim: to 'live in the present moment'. When he had not returned at Christmas, it had occurred to her that another woman – 'a wandering of the heart, or even a caprice of the imagination' – detained him. If so, she told him, 'there is an end of all my hopes of happiness'. She begged for 'the truth'. If she was wrong, she said, 'tell me when I may expect to see you, and let me not be always vainly looking for you, till I grow sick at heart'.

Next day she demanded his fidelity against the prevailing double standard: 'such a degree of respect do I think due to myself'. Only a slave would open her arms to a sultan 'polluted by half a hundred promiscuous amours during his absence'. Infidelity had to be an unacceptable betrayal of her belief that their extra-legal partnership would not grant him the customary advantage of master over mistress. 'I could not forgive it, if I would.' If he came back only to prove his rectitude, she made it clear he was not to come at all.

Imlay tried to convince her that her suspicion was the result of her poor mental state.

Though Imlay and Barlow had appeared as new men from the New

World, respectful of women and tender as lovers, they practised the double standard as a matter of course.* Ruth came to accept that her husband would have 'amours' during their long separations. 'I will be indulgent & only require you to love me best when with me,' she assured him, promising on his return to 'press you to my heart with as much ardour as the first day of our marriage – for you are every day more dear to me'. Such complicity was unthinkable for Mary Wollstonecraft.

The Barlows remained in Hamburg over that winter of 1794–5. They lived in the adjoining area of Altona, rising above the River Elbe, the chief town of the Duchy of Holstein, then under the jurisdiction of Denmark. Ostensibly, Barlow worked as a shipping agent, but that particular winter the port froze, and shipping came to a standstill. Curiously, at this very time, he confides to his brother-in-law Senator Abraham Baldwin (the increasingly distinguished member of Ruth's family who had not welcomed Barlow) that 'pecuniaries' are prospering. Some of his gains could have come from his membership of a secret network provisioning the French war effort and from ships like the *Rambler*, but there remains the question: what happened to the silver in the hold of the *Maria and Margrethe*?

When the Barlows had left Paris with Joel and Imlay in cahoots in the spring of 1794, they had been in their usual financial straits. By the end of 1796, Joel Barlow had millions in today's terms – an estate valued at $126,000. No one so far knows how he made his fortune. Biographers who talk of 'part-cargoes' and the 'percents' he could levy as a shipping agent, forget the ice. Northern ports were impassable from the end of November 1794 until 11 March 1795. Gouverneur Morris, who wintered in Altona that same year, records in his diary that on 22 and 23 January 1795

*A telling exchange between Dr Johnson and Boswell in 1779

Johnson: 'A husband's infidelity is nothing.' [A couple is connected by children and fortune.] 'Wise married women don't trouble themselves about infidelity.'

Boswell: 'To be sure there is a great difference between the offence of infidelity in a man and that of his wife.'

Johnson: 'The difference is boundless. The man imposes no bastards upon his wife.'

many people who ventured out froze to death and frozen horses waiting in the street crashed over. Dealing did go on. Morris, for instance, managed to dispose of some silver on behalf of a French aristocrat, acting quickly, he explains, because it was dangerous to hold on to silver. Joel and Ruth spent that winter learning German, so they said; yet from the depths of the freeze, on 10 February 1795, Barlow informs Ruth's brother that 'pecuniaries' will be bettered by this time in Altona. In the same breath he says that the Elbe has been shut to the mouth for near two months. Even after the port opened in March a traveller standing at the mouth of the Elbe saw ships 'beat about by the Ice which still floats down the River in huge masses in a most frightful manner . . . I can almost from the appearance of the Sea conceive myself at Greenland.' If Barlow did grow rich over that winter, whatever shipping made his fortune had to have happened *before* the freeze – before, that is, November 1794. Can it be that the Barlow fortune began in the hold of the treasure ship that sailed north in August?

Mary made a new friend in Paris, Archibald Hamilton Rowan, aged forty-two, a commanding figure and refugee from County Kildare in Ireland. His first political act had been to challenge the clemency of the Viceroy in a case of an accessory to a rapist in high station. Though himself a wealthy landowner – a descendant of a sixteenth-century aristocrat – Rowan had joined the United Irishmen, a reforming coalition of radicals, Catholics, a few liberal aristocrats, and the politically sidelined professional and business classes. Rowan became Secretary, and then on 29 January 1794 was prosecuted for handing out seditious leaflets. Not only was he sentenced to two years in prison – he might have hanged when a government spy discovered that France had approached him to inform on Ireland. In fear for his life, he had escaped from Dublin Prison and fled to Paris. On arrival, he had been sickened by the blood that had literally washed his feet where he stood a hundred yards from the guillotine, witness to the orgy of executions that ended the Terror.

A letter to his wife in Ireland describes a post-Terror fête to celebrate the reburial of Mirabeau (for Wollstonecraft the pre-eminent leader of the Revolution, in contrast to his murderous successors) in the

Panthéon. Rowan's attention was caught by 'a lady who spoke English, and who was followed by a maid with an infant in her arms, which I found belonged to the lady. Her manners were interesting, and her conversation spirited, yet not out of the sex.' A friend whispered that she was the author of the *Rights of Woman*.

> I started! 'what!' said I within myself, 'this is Miss Mary Wollstonecraft, parading about with a child at her heels, with as little ceremony as if it were a watch she had just bought at the jeweller's. So much for the rights of women,' thought I. But upon further inquiry, I found that she had, very fortunately for her, married an American gentleman a short time before the passing of that decree which indiscriminately incarcerated all the British subjects who were at that moment in this country. My society, which before this time was entirely male, was now most agreeably increased, and I got a dish of tea and an hour's rational conversation, whenever I called on her. The relative duties of man and wife was frequently the topic of our conversation; and here I found myself deeply wounded; because if my dearest thought as Mrs. Imlay did, and many of their sentiments seemed to coincide, my happiness was at an end. I have sometimes told her so; but there must be something about me of deep deception, for I never seemed to have persuaded her that I had merited, or that you would treat me with the neglect which I then thought was my portion. Her account of Mr. Imlay made me wish for his acquaintance; and my description of my love made her desirous of your acquaintance, which it is possible may happen; and until you can decide for yourself, repay her, my dearest friend, some of those kind attentions which I received from her when my heart was ill at ease.

Exiled from all he held dear, Rowan began to 'croak' his dark reveries to Mary, and passed messages between her and his level-headed wife. In doing so, he felt himself a connoisseur of female accomplishment. Since everyone addressed Mary as Mme Imlay, he assumed she had submitted to a 'republican marriage' for the sake of security, but whatever the legal situation, her connection with Imlay 'had with her all the sanctity and devotedness of a matrimonial engagement'. At the same time she persisted in her opinion 'that no motive upon earth ought to make a man and wife live together a moment after mutual love and regard were gone'.

This made Rowan reflect on his treatment of his wife, and wonder if her attachment to him had been impaired.

Mary assured him he 'had no reason to be alarmed; for when a person whom we love is absent, all the faults he might have are diminished, and his virtues augmented in proportion'.

Imlay continued to summon Mary to London – somewhat against her will. For she still had hopes of the Revolution, and wished Fanny to grow up in France.

Finally, in April, she agreed, and Imlay sent a servant to ease her journey. Passing through Le Havre, she prepared their house for Archie Rowan, who was to stay there en route to America. France was too beset by war to interest itself in the fate of Ireland, and Rowan left for Philadelphia in July. He carried an introduction to Mary's brother Charles Wollstonecraft, and went on to went on to lose money in a calico printing factory they set up together in Wilmington, Delaware.

Ten or so days before Mary was due to sail for England, she weaned Fanny, hoping to endear the child to her father. She glanced across the Channel with a half-ray of hope hardly daring to light her eyes, but ready to return at a signal from Imlay. He did seem to give this signal when he wrote to her: 'Business alone has kept me from you. – Come to any port, and I will fly down to my two dear girls with a heart all their own.'

'I do not see any necessity for your coming to me,' she replied on 7 April, wary, suddenly, of lending herself to his endearments. 'I cannot indulge the very affectionate tenderness which glows in my bosom, without trembling, till I see, by your eyes it is mutual.' On the brink of her crossing, her almost extinguished hope held to only the thinnest edge of life:

> I sit, lost in thought, looking at the sea – and tears rush into my eyes when I find that I am cherishing any fond expectations. – I have indeed been so unhappy this winter, I find it as difficult to acquire fresh hopes,

as to regain tranquillity. – Enough of this – lie still, foolish heart! – But for the little girl, I could almost wish that it should cease to beat, to be no more alive to the anguish of disappointment.

When Mary docked at Brighthelmstone (Brighton) on Saturday 11 April, she let Imlay know she would meet him at his London hotel the following evening: 'I hope you will take care to be there to receive us.' By 'us' she meant the child of a friend whom she had undertaken to escort from France; Fanny's French nursemaid, Marguerite Fournée; and 'our little darling' of eleven months who, done with the breast, was 'eating away at the white bread'. 'But why do I write about trifles?' she adds. 'Are we not to meet soon? – What does your heart say!'

Imlay had a furnished house waiting for her at 26 Charlotte Street, Rathbone Place (off Oxford Street, running north), only a block away from her old flat in Store Street. Though they shared the house, Imlay was evasive, claiming the demands of business. Fuseli, to whom she turned, snubbed her. Mary's wish to hide her distress made her lie to Bess. Already, her sisters had wondered at Mary's staying in France, and wondered too at Imlay's avoidance. (So far, Imlay had distanced himself from Bess, while appearing the devoted husband: 'I am in but indifferent spirits occasioned by my long absence from Mrs Imlay, and our little girl, while I am deprived of a chance of hearing from them.') The sisters' needs, as so often, centred on money – Bess, in particular, looked for rescue from her drudgery as governess - and here there was scope for misunderstanding: a husband was expected to help his wife's family; one reason Mary had not married Imlay was to protect him from that obligation. She had to disabuse her sisters of a rumour (from their brother James) that Imlay had made £100,000. At the same time she was aware that Bess expected to join her married sister in Charlotte Street. Mary now made a bad blunder: in order to disguise the facts that she was not married and that the relationship was just then falling apart, she told Bess it would mar her happiness to have a third person with them. This blow to Bess led to permanent estrangement.

Towards 20 May it was revealed that Imlay had taken up with an actress from a strolling company. Mary had long feared something of this sort, but the truth overwhelmed her. The fiction of living as a family ended when Imlay left the house, and she found herself cast in the mistress character she most despised: a discarded mistress whom a dutiful man continues to support despite his pressing debts. Support of this kind was insupportable. It was more than the loss of a lover. His explanations were 'cruel' because they locked her into the routine, age-old story of sexual surrender and betrayal. At the deepest level, it was a denial of the new genus, a threat to its continued existence. Explanations, she said, might convince the reason, 'whilst they carry death to the heart'.

Imlay protested his attachment to her.

'My friend – my dear friend,' she replied, 'examine yourself well – I am out of the question; for, alas! I am nothing – and discover what you wish to do – what will render you most comfortable – or, to be more explicit – whether you desire to live with me, or part forever? When you can once ascertain it, tell me frankly, I conjure you!'

Imlay's unwillingness to hurt her with a straight answer prolonged the torment. During the last days of May she not only lost hope but seemed to have lost her very self: 'My soul has been shook.'

There were efforts to be calm. She even had Imlay to dine, promising to greet him with a cheerful face and avoid contention – yet at the centre of her calm was a 'whirl' of grief she could not subdue.

At the end of May, she swallowed an overdose of laudanum. Later, her novel *The Wrongs of Woman* recreates what happened: 'Maria' feels her head swim and then faintness. Mrs Wollstonecraft's dying words echo in her fading consciousness – 'have a little patience and all will be over' – before she allows herself to sink with the thought: 'what is a little bodily pain to the pangs I've endured?' Her final thought is for her child. At this, she vomits and resolves to live.

Mary too did not die. Imlay, advised in time of her intentions, came to her rescue, and his promptness must have saved her life. With roused feelings, he urged her to go on. She had to live for her child; and she had to live because something indomitable in her could not accept Imlay's conduct. She could not accept that this man of many gifts should not be educable, like Clarissa with Lovelace. Like Lovelace, Imlay had the attractions of a confident talker whose intelligence invites a woman to believe he is educable if only she can convey her character; but both Imlay and the fictional Lovelace are fixed in sexual habit. Lovelace is a practised deceiver of women, who at times can admit that the rake's code betrays his own best interest. Imlay was even tougher to counter because libertinism was at home in the rhetoric of liberty. And then, too, he really did believe himself a friend to women. His rescue of Mary Wollstonecraft during the Terror appeared the act of a 'generous soul'; only slowly did she find herself trapped in the fatal plot of the fallen woman – as Clarissa finds herself locked, literally, with whores by the very man to whom she had looked for rescue.

Mary's act stung Imlay's conscience. His reputation, more than hers, was at stake: he risked the scandal of a celebrity whose attempted suicide declared ill use, even though she never spoke ill of him. He could have defended himself by denying her claim to be 'Mrs Imlay'. He could have cast her off with the backing of society's righteous virulence towards a wanton with a bastard. That Imlay did not do so tells us he was not that sort of scoundrel, and this is one reason why Mary held on as long as she did.

Imlay's answer to her rejection of his support was to offer her work in Scandinavia. For two centuries it has not been known what exactly this entailed, but it turns out that to discover the fate of the silver ship was not the purpose of her voyage. New archival evidence shows that, as the ice thawed in March, 'Gilbert Inckay' (as Swedish record calls him) had sold a ship called the *Maria and Margrethe* to Elias Backman of Gothenburg. Divers off the coast of Norway have looked for sunken treasure in vain. It's now beyond doubt that the ship did not sink, and that Mary Wollstonecraft knew where it was before she arrived on the scene.

Why did Imlay not undertake this journey himself? Some priority that 'alarmed' Mary Wollstonecraft directed him not to Norway but to *France*. It was more urgent to settle some difficult business there, most likely, awkward questions from Barlow's contacts on the Food Commission, about the loss of the expected return of grain. Imlay had some explaining to do: he would have to convince the Commission that the silver had been stolen. He, the benevolent, trusting Imlay had been tricked by a thieving captain. Since Mary now warned him against his associates on both sides of the Channel, it may be worth noting that Imlay would again coincide with Joel Barlow, who was due to leave Hamburg for Paris.

If Imlay's part in this should prove 'unlucky', Mary told him, she would regret it less if it sent him home to her convinced that a true friend was another form of 'treasure'.

Mary Wollstonecraft was indeed an asset, known throughout Europe, and particularly in Paris, for her high-mindedness. It would vindicate Imlay's case to have a celebrity of this calibre state his position, with her customary eloquence, to the prime movers in the case: Ellefsen's lawyers, his judge, and most powerful of all, the Prime Minister of Denmark as ruler of Norway. So, while Imlay did his part in Paris, Mary Wollstonecraft was to back his case against the accused in Norway. There were two aspects to her projected journey: the more important, perhaps, from Imlay's point of view was that a celebrity of moral standing would be seen to be on his side; and at the same time there were the practical matters: she was to carry through the repairs to the ship and confront Ellefsen.

What is certain is that Imlay led Mary to believe that by pressing the rich Ellefsens for restitution, she could again make him, Imlay, her own. She agreed, she told him, 'to extricate you out of your pecuniary difficulties', while he, in turn, heartened her purpose by declaring her to be his wife in a letter of attorney addressed to Mrs Mary Imlay. Here were her orders:

> Know all men by these presents that I Gilbert Imlay citizen of the United States of America residing at present in London do nominate institute and appoint Mary Imlay my best friend and wife to take the sole management and direction of all my affairs and business which I had placed in the hands of Mr Elias Backman negotiant of Gothenburg or those of Messrs Ryberg & Co Copenhagen, desiring that she will manage and direct such concerns in such manner as she may deem most wise & prudent. For which this letter shall be a sufficient power enabling her to receive all the money that may be recovered from Peter Ellefsen or his connections whenever the issue of the tryal now

> carrying on against him, instigated by Mr Elias Backman as my agent for the violation of the trust which I had reposed in his integrity.

There follows a tight, cryptic direction* concerning Norway, which could be unravelled in this way: Mary Imlay was to fix in her mind Imlay's distress over the loss of the silver, 'aggravated' by further distress and loss resulting from 'Ellefsen's disobedience of . . . instructions' to do with the ship. On the basis of that accusation (not, it appears here, an accusation that Ellefsen himself stole the silver), Mrs Imlay was to extract an unspecified sum as 'damages' due to her husband. The actual amount would depend on what she would find out on the spot: whether Ellefsen's family found themselves 'implicated in his guilt'; whether they had the 'means' to make a substantial 'resistitution'; and whether they could be persuaded to settle out of court. The next paragraph, also cryptic, turns to Mary Imlay's task in Denmark:

> Respecting the cargo or goods in the hands of Messrs Ryberg and Co Mrs Imlay has only to consult the most experienced persons engaged in the disposition of such articles, and then[,] placing them at their disposal[,] act as she may deem right and proper[,] always I trust governing herself according to the best of her judgment in which I have no doubt but that the opinions of Messrs Ryberg & Co will have a considerable and due influence.
>
> Thus, confiding in the talents zeal and compassion of my dearly beloved friend and companion[,] I submit the management of these affairs intirely and implicitly to her direction[,] remaining most sincerely & affectionately hers truly
>
> May 19th 1795 G. Imlay.

*'Considering the aggravated distress is the accumulated losses and damages sustained in consequence of the said Ellefsen's disobedience of my instructions I desire the said Mary Imlay will clearly ascertain the amount of such damages, taking first the advice of persons qualified to judge of the probability of obtaining satisfaction or the means the said Ellefsen or his connection, who may be proved to be implicated in his guilt, may have, or power of being able to make restitution and thus commence a new prosecution for the same accordingly . . .'

The date shows this plan to have been in place two weeks before Mary took the overdose. It revived immediately after, encouraged by Imlay's suggestion they meet up when their missions were done. The journey ahead also opened up the possibility of a travel book. An advance from Johnson would pay Mary's debts and renew her confidence as a writer who could support herself through an uncertain future. Though hopes of Imlay were not dead, her doubts, understandably, had more hold.

Ten years earlier, after Fanny Blood's death, she had fought depression. Now, she had to pull herself free of the doomed plot of the fallen woman. Scandinavia, as a destination, was way off course from the usual Grand Tour centred on the Mediterranean. Between June and September 1795, Mary Wollstonecraft would show her mettle far north in lands no one visited at the time.

12

FAR NORTH

Men often travelled with swords or pistols along roads infested with highwaymen. On 9 June, Mary, Marguerite and Fanny, three unprotected females, set off on an overnight coach for the north of England. Mary had little sleep with a one-year-old on her lap. When they arrived at the port of Hull, they found themselves in a damp room in a house like a tomb. It was two in the afternoon. The hoot of the post-horn broke the silence, Fanny echoed it, and tired as Mary was, she dashed off a letter to Imlay to catch the next post:

> I will not distress you by talking of the depression of spirits, or the struggle I had to keep alive my dying heart . . . Imlay, dear Imlay, – am I always to be tossed about thus? . . . How can you love to fly about continually – dropping down, as it were, in a new world – cold and strange! – every other day? Why do you not attach those tender emotions round the idea of home, which even now dim my eyes?

His determination to save her, and signs of regret when she departed, had raised 'involuntary hopes' he yet might change.

Casual sex had atrophied his heart, she warned him. 'You have a heart, my friend, yet . . . you have sought in vulgar excesses, for that gratification only the heart can bestow . . . – Ah! my friend, you know not the ineffable delight, the exquisite pleasure, which arises from a unison of affection and desire, when the whole soul and senses are abandoned to a lively imagination, and renders every emotion delicate and rapturous. Yes; these are emotions . . . of which the common herd of eaters and drinkers and *child-begeters*, certainly have no idea.'

She could hear Imlay demand to know the purpose of all this, and had her answer ready: 'I cannot help thinking that it is possible for you, having great strength of mind, to return to . . . purity of feeling – which would open your heart to me.'

During her stay in Hull, Mary visited her childhood town of Beverley. Her nostalgia had a jolt. She had changed, while Beverley had remained the same; only, eclipsed by the growth of Hull, it had shrunk into 'sullen narrowness', more class-bound than ever in obedience to the 'fanaticism' of counter-revolution.

A cargo vessel was due to sail for Copenhagen in six days. It was not fitted out for passengers, but would have to do since the captain agreed to take her to Arendal (the Norwegian base of the Ellefsens) or Gothenburg (the official destination of the silver ship). So it was that an Englishwoman and a Frenchwoman with a toddler between them went aboard on 16 June. At the last minute, departure was delayed. Marguerite became seasick as the vessel rode at anchor. Fanny, 'gay as a lark', began to play with the cabin boy, then became restless when rain kept her below deck. Whenever the ship prepared to sail, the wind changed or dropped. Mary used the delay to write almost daily to Imlay, who sent as many replies, urging her on. For she remained depressed, and was telling him in every way short of explicitness that she felt unfit for so uncertain a journey.

'Now I am going towards the North in search of sunbeams!' she put it wryly.

It was a mere fortnight since she had taken the overdose. Though 'a determination to live' had revived, she did let mention 'the secret wish' that the sea might become her tomb, 'and that the heart, still so alive to anguish, might there be quieted by death'. Another would have fetched her back, and this is what she hints when she observes, with apparent surprise, how she had improved in London (near Imlay) but was now losing ground at the prospect of leaving her country (Imlay) behind. She lived for his words, and at length persuaded the kindly captain to take her ashore to see if an extra letter had arrived. Her disappointment was unreasonable, for Imlay thought her at sea, but so long as she lingered in England, hope still flickered that he'd come after all. She could not ask it, and Imlay could not – or would not – decode the language of depression.

Eventually, the ship sailed on 21 June. Winds prevented a landing at Arendal, so they made for Sweden. Not far from Gothenburg, the vessel was becalmed. Mary persuaded the captain, against rules of the sea, to have her rowed to a lighthouse on what was probably the Onsala peninsula. There, Marguerite's apprehensions were realised when, casting about for inhabitants, they came upon two shaggy men covered in coal. Poor, cowering Marguerite, so far from Paris, implored their return to the ship. Mary insisted they must go on, and Marguerite had no alternative but to follow her back into the boat where Mary begged the sailors – against orders – to row them six miles to the mainland. They did so sturdily, watching their vessel in case it should sail without them. After eight days at sea, Mary stumbled ashore and collapsed on a rock that seemed to heave under her. She struck her head, and was unconscious for a quarter of an hour. The sensation of liquid – blood – running over her eyes brought her round.

'I am not well,' Mary informed Imlay, 'and yet you see I cannot die.'

Set in a land of firs and lakes, and opposite rocky, wooded hillocks in a wavy line against the pale summer sky, eighteenth-century Gothenburg was laid out in the Dutch style with canals and merchants' houses lining the streets near the harbour. Elias Backman, a good-natured man of thirty-five, invited Mary's party to stay with his family in the country. Mrs Backman was French, which must have eased Marguerite. Fanny could be with their four little boys, especially the youngest, only a few months older. The pure northern air, the long light of summer nights when Mary could write without a candle, the balm of sleep outdoors, and the kindness of the Backmans, began to restore her. Pink crept back into her cheeks and her body rounded with returning health. Since Backman had lived in France, French must have been their common language over the next twelve days as they reviewed the misfortunes of the silver ship.

How bad was the damage, Mary would have asked. And what was Backman's understanding of Ellefsen the thief?

Captain Peder Ellefsen of the *Maria and Margrethe* came from a family of ironmasters and shipowners whose wealth had been established in the seventeenth century. When his father died at the age of forty-three, after fathering fifteen children, the entire fortune came into the hands of Peder's proud and beautiful mother. At the time of this story Margrethe was the dominant figure in a leading family who were amongst the founders of the seafaring town of Arendal.

On about 24 August 1794 Peder had disembarked at an obscure spot called Groos on the southern coast of Norway. He planned to go ahead to Arendal, while the mate, Thomas Coleman, was told to delay a few days, then bring the ship there. With no roads as yet in that remote part of Norway, Peder galloped away through the woods in the long-trailing light. When he reached Arendal, he sold the ship to his mother and stepfather. Then, on 1 September, Coleman duly docked the *Maria and Margrethe* at Sandviga on the island of Hisøy in the channel leading towards the town.

What happened next is murky: even at the time, it proved difficult to verify conflicting stories. There was Coleman's story that it was at Sandviga that the crew was ordered ashore, and while the ship was deserted, Captain Ellefsen returned in a boat with four men to carry off the silver. There would have been many convenient hideaways, for the buildings on the shore were propped on poles and hung over the water, serving as warehouses as well as homes.

Another story, later investigated but unsolved in court, said the silver was ferried by a smuggler called Søren Ploug to Flensburg (on the Schleswig-Holstein coast). Rumours of lost treasure will always fly about, but there's something convincing in the specificity of Flensburg. In the late eighteenth century Schleswig-Holstein specialised in silverware for the flourishing burghers of Hamburg. Wealth was displayed in silver necklaces cascading across the bosom, candelabra, spoons with ornamental handles, tea-sets, silver hymn books, great clasps for Bibles and waists, and even a silver model of a ship with three masts and sails, all combining an almost filigree craft and opulence.

Finally, there was the story – lingering in Norwegian folklore – of shipwreck on a rock at Skurvene, near Arendal, of treasure lost overboard, of one silver bar remaining. What emerged eventually from judicial inquiries is that there had indeed been a plan to sink the ship.

Something in Ellefsen's plan went wrong, because on 10 September his sale was revoked, followed by fresh moves on his part to detach himself from the ship. On 20 September he signed it over to Coleman, with three witnesses, including Peder's lawyer. The crew – kept in the dark – were not present at the handover. They were paid part of their wages for a further voyage, and promised the rest if they proved loyal.

Coleman hung on in Arendal, with talk of contrary winds, for three weeks after he took command. If there was an attempt to sink the ship, this would have been the most likely time. At last Coleman set sail, for Gothenburg he said, on the morning of 10 October, but the ship was damaged by a storm – so the crew said. Planks were damaged, they said, and the ship sprang a leak. Two days later it was back in Norway, slipping into Oksefjorden, near Tvedestrand, halfway between Arendal and Captain Ellefsen's homeport of Risør. It was a long inlet and deep enough for a sailing ship, though it had to pass through foul water. Then, on the 17th, the *Maria and Margrethe* shifted once more, slipping by rocky, fretted inlets whose entrance was guarded by skerries, small islands, a tightly wooded land shadowed by the serrated edges of firs. The ship lingered a further month at Risør with its chain of outlying islands, dense with brushwood. As the home of Coleman's former captain, this can't be a coincidence. It was here the trouble started for the Ellefsens.

At the end of October, Imlay at last revealed to Backman that the delayed ship had carried a cargo of silver. Backman leapt to action. First, he dispatched trusty Captain Waak, who had finally sailed the other treasure ship, the *Rambler*, safely to port at Gothenburg at just this time. Waak talked quietly to Coleman as one seaman to another, and this is when

Coleman owned that silver – whether some or all, he didn't say – had been taken ashore in Norway without the crew's knowledge. As tensions rose, Ellefsen seems to have pressed Coleman to back him, which Coleman – having told his story – could no longer do.

The magistrate of Risør cross-questioned Ellefsen about the letter Imlay had given him in Le Havre to convey to Backman. Ellefsen opened the letter, extracted and destroyed the enclosed receipt for the silver, and denied the silver's existence. He also tried to retrieve a responsible image as captain. On 14 November, he advertised for a lost ship, claiming that the mate had that day 'escaped with the ship from East Risør harbour'.

Where did the *Maria and Margrethe* go? This is where Per Nyström lost sight of the ship when he did his pioneering archival searches in the 1970s. Norwegian historian Gunnar Molden has carried these searches a stage further with a series of remarkable discoveries, including the crew's testimony. It reports two efforts to reach Sweden, coastal pilots not responding to signals, and the ship battling with a torn sail through successive storms. A man is washed overboard and swept away by heaving seas. The hold carries two feet of water, and pumping must go on continuously. Twice, the ship ventures in as close as it dares to the rock-bound land, and twice it goes back out to sea – a half-wrecked vessel divested of its treasure, adrift in the Skagerrak.

When Mary asked questions, she learnt that 'Swedish harbours [are] very dangerous . . . and the help of experience is not often at hand, to enable strange vessels to steer clear of the rocks, which lurk below the water, close to the shore'. It was not uncommon, she heard, that 'boats are driven far out and lost'.

There was another danger: the oncoming winter, turning into that worst of winters. Hardening ice in Swedish inlets could have made entry difficult if not impossible, especially where fresh water from rivers enters the ocean. Coleman retreated to south-west Norway where harbours were still open. A badly damaged ship came to land, way off course, at Nye-Hellesund in the parish of Sogne. The crew refused to put to sea again, and the ship was moved further into Hellesund harbour for nearly a month, until mid-December, when it sailed east to Kristiansand. There the *Byfogd*

(Town Magistrate) and the Chief of Police interrogated Coleman. The case at once attracted high-level attention. The magistrate gave his opinion that guilt lay with Ellefsen, not Coleman. Talk in Norway was shocked to find a son of the grand Ellefsens accused of 'crimes very awful and dishonourable to the Nation, while in charge of the ship'. A document was found in the captain's quarters. It was in English and vital to the case against Ellefsen: his declaration of 12 August 1794 that 'the ship belongs to Gilbert Imlay and is his absolute property'.

'How I hate this crooked business!' Mary Wollstonecraft exclaimed to Imlay on 29 December. At that time he was turning to legal redress. The following month a Royal Commission was set up by Norway's rulers in Copenhagen to investigate the affair. Peder Ellefsen was arrested, and only released when his mother paid an enormous bail of ten thousand *riksdaler* (equivalent to more than half the value of the silver).

In March, when the ice broke up, Backman bought the ship from Imlay. Its imminent departure for Sweden led to a flurry of further judicial inquiries in the town hall of Arendal in the spring of 1795. Without Ellefsen's receipt for the silver it was impossible to prove it had ever been aboard; then, too, Ellefsen claimed to have 'mislaid' Imlay's instructions to him. In all, sixty-nine witnesses took the stand and their testimonies now cast doubt on Ellefsen's guilt. Defence lawyers also cast aspersions on one of the two judges appointed by the Royal Commission, a Norwegian called Jacob Wulfsberg. Since he had acted in the past for Backman, Wulfsberg was too biased in Backman's favour, lawyers said, to be an impartial judge. 'Who were fooled by whom?' Gunnar Molden asks. After two centuries that question is suddenly hot with a sense that answers can be found.

Early in June, at the same time as Mary Wollstonecraft was dispatched to Scandinavia, Captain Waak arrived again in Norway, now to take charge of the *Maria and Margrethe* and sail it to the nearest Swedish port of Strömstad for repairs. This was Mary 's first destination when she left Gothenburg in July for points north.

Mary's journey, ending on the River Elbe near Hamburg, tracks back along the ship's and treasure's path. It was their mysterious course that dictated the course of her travels. On 1 July 1795 she suggested to Imlay that he meet her at her final destination at the end of August. This was a critical letter for their future, where Mary asks Imlay to decide what that future was to be. Imlay avoided a direct answer, but urged her on with the prospect of reunion in Hamburg. As usual, he wrote often: Mary's replies indicate there were at least ten letters from the time she took off for Strömstad. He understood how much her far-flung tasks depended on the comfort and promise of his words.

From now on the journey would be too arduous for a child. Fanny remained in the Backman home, while Elias accompanied Mary to see his ship at Strömstad. Their journey north took them to inns with feather beds so deep in their wooden frames they seemed to Mary like graves from which she would never emerge. Windows were shut even in summer, as though Swedes could never feel warm. An abundance of wild flowers in the vale of Kvistrum made her reflect that Sweden had been the right country to initiate the study of botany. Linnaeus's system of binomial classification according to genera and species was the source of the 'genus' terms of her own self-making.

Flowers, alas, failed to compete with the stench of herrings, spread as manure near bleak log homes without paths or gardens, as though all energy had drained into the rudiments of survival through harsh winters. Female servants had to crack the ice in the streams and wash laundry with red and bleeding hands.

Would male servants ever help? Mary asked.

The answer was no.

Approaching Strömstad under heavy skies, she felt a little oppressed by its overhanging rock. This small town, a spa in the eighteenth century, is still a springboard for journeys to Norway, sails bouncing off the hills about the harbour. Rocky islets stretch far out to sea on every side. Here, in the harbour, Mary and Backman would have examined the damage to the ship and assessed the costs of repair. Records of extensive repairs have survived. These were going on from 16 June until 1 September, which is to say they were carried out over Mary's period in Scandinavia. One of the tasks of 'Mary Imlay' was to oversee the sale of a seaworthy ship. It's on record that 'Marin Inclay' was able to finalise the sale of the ship for £250. The repairs cost three times as much, which, if Imlay was liable, explains his need to seek restitution from Ellefsen. Norway, then, was Mary's next destination.

Mary and Backman parted on 14 July. Leaving the harbour to cross the Oslo fjord in an open boat, she observed the difficulty of steering amongst a myriad islands. There were no beaches as far as she could see: the waves beat against bare rock. Mary wrapped her cloak about her and lay down on the bottom of the boat when a 'discourteous wave' broke her sleep. Larvik, where she landed at three in the afternoon, proved a clean town with a wealth of ironwork. Travellers were so few that townsfolk gathered to stare. The questions in their eyes tempted Mary to adopt Benjamin Franklin's solution when he had travelled in America: to carry a placard with her name, place of origin and business. Her revolutionised French dress, light, unstructured, yielding to the curves of the body, appeared of particular interest to female starers. The only available transport was a rude one-horse vehicle with a half-drunk driver. As a sailor lumbered up beside her, Mary spied a gentleman emerging from an inn and eyeing her assorted party. She was further disconcerted when the driver cracked his whip for attention, but seeing the gentleman smile, she began to laugh. Off they dashed, at full gallop, with Mary musing 'whether I can, or cannot obtain the money I am come here about'.

She had come to negotiate an out-of-court settlement, so her first move was to confer with Judge Wulfsberg, in Tønsberg to the north of Larvik. He had a reputation for solving cases, sometimes with the aid of a disguise, and later became Chief of Police for Christiania (Oslo). He was known too for his efforts to calm disputes – the kind of man Mary respected. As ever, she was drawn to benevolent men unlike her thuggish father.

Wulfsberg advised her in English. Matters must be allowed to move slowly, he warned; she should prepare to make Tønsberg her base for at least a month. Originally, she had intended to go to Arendal; now, Wulfsberg shifted her sights to Risør, where he had set up a further hearing for mid-August. Accordingly, she settled down at a pleasant inn with a view of the sea.

When Mary came to Tønsberg she still suffered some residue of the night sweats that had laid her low the previous winter. Now, energy surged back as she rode and climbed. Breezes fanned her sleep. Nature was her balm, while 'employment' restored her agency. In the course of her rambles, she came upon a stream said to be rich in iron, and walked there each morning, seeking 'health from the nymph of the fountain', though her improvement, she believed, came more from air and exercise. She took up rowing to a place where she could swim, the rhythm of the oars keeping time with her memories.

'You have often wondered, my dear friend, at the extreme affection of my nature – but such is the temperature of my soul,' she reminded Imlay. 'I must love and admire with warmth or I sink into sadness. Tokens of love which I have received have rapt me in elysium – purifying the heart they enchanted. – My bosom still glows . . . and if I blush at recollecting past enjoyment, it is the rosy hue of pleasure heightened by modesty.'

Reason warned that hopes of Imlay could be delusive. Yet without hope, there was nothing to sustain life, and life seemed precious again.

'I cannot bear to think of being no more . . . nay, it appears to me impossible that I should cease to exist, or that this active, restless spirit, equally alive to joy and sorrow, should only be organised dust,' she ruminated in the emerging voice of her travel book. 'Surely something resides . . . that is not perishable.' This was the impetus to start what was to be her finest work. Here, at Tønsberg, she 'turned over a new page in the history of my own heart'.

Again, she took on a well-tried genre, and again transformed it, blending episodes of travel with reverie, a new, Romantic voice, attuned to nature. As she dozed under a rock, she felt herself 'sovereign of the waste', lulled by the 'prattling' of the sea amongst the pebbles. Her soul seemed diffused through her senses until she felt a child again, resting – she liked to think – on the footstool of her Creator. Yet her allegiance to reason, and what she had seen of Parisian salons, led her to question a sentimental embrace of rural solitude. Even as it healed body and spirit, she derides Rousseau's ideal state of nature as 'a golden age of stupidity'. An intelligent being wants the arts and sciences of civilisation. She asked herself whether happiness lies in unconscious ignorance or the educated mind, and imagined Imlay with similar doubts in the wilds of America. Boldly, her *Travels** concede that 'nothing so soon wearies out the feelings as unmarked simplicity'. The woods of Norway taught her that she would be as much an exile on the American frontier as she had been at Mitchelstown Castle. For a while, she might bury herself in the woods, but she would 'find it necessary to emerge again'.

Wollstonecraft's *Travels* take the form of twenty-five letters to a friend, Imlay himself, unnamed but identified as the father of the traveller's child. Her private drama erupts through the lulling sea, like volcanic heaves as the traveller calls up a look in her correspondent's eyes. Can she revive his flush of ardour – through words, loyalty and far-flung business on his behalf, unprecedented for a woman in the eighteenth century to undertake alone? The 'Mary' who signs publishable letters to her one-time lover has obvious links with the Mary Wollstonecraft who wrote more challenging private letters to Gilbert Imlay. The lifeline of her private journey was Imlay's promise to meet in Hamburg. Its immediate effect was to aid her recovery, but it held some danger. For Imlay broke promises casually as new schemes opened up. Mary, in contrast, had the tenacity to carry through plans conceived at a higher level as plans for existence itself. Her Fuseli plan – to live with the painter and his wife – had been unworkable. But, soon after, Imlay's novel in defence of women's rights had seemed to present a viable narrative: a chance to integrate desire with domesticity without the destructive scenes of seduction or marriage. When Mary found herself trapped in a traditional plot after all, it had struck at the root of her existence. In his frontier character Imlay had presented himself as her natural mate, a man of high-minded simplicity, and though she soon discovered her mistake, she did not give up. There was this extraordinarily protracted effort to draw out the American promise he had seemed to embody and which seemed to her still concealed in his core. For she believed in the perfectibility of human nature. It was her form of faith.

---

*Published as *A Short Residence in Sweden, Norway and Denmark.* William Godwin called it *Travels in Norway*, and shortened it to 'Travels' in his Diary on 6 Sept 1796. His briefer title is adopted here.

Norway was the high point of her journey: she felt kinship with the sturdy independence of its people, and their warmth towards a woman 'dropping down from the clouds in a strange land'. She could not speak to them in Danish (the dominant language in Norway at the time), but they did manage to communicate sympathy for the French Revolution and reluctance to hear her call Robespierre 'a monster'. At a Tønsberg party, yellow-haired women gathered round her, sang to her, pitied her singleness at thirty-six. They enquired after her needs, 'as if they were afraid to hurt, and wished to protect' her, perhaps sensing the melancholy that hung about her whenever she contemplated the fading promise of her union with Imlay and the lost bond with Fanny Blood whose soft voice she still could hear. She felt more than a mother's anxiety for the future of Fanny her child. When she thought of 'the dependent and oppressed state of her sex', she feared to unfold Fanny's mind 'lest it should render her unfit for the world she is to inhabit – Hapless woman! What a fate is thine!'

Fears for Fanny were the obverse of retrieved command of her own fate. There is a decided contrast between the pleading letters to Imlay and her purposeful travel book. Although the *Travels* are presented as flowing instantaneously from her experience, the first six episodes were, in fact, composed retrospectively – the art lay in apparent artlessness – 'my desultory manner', as Wollstonecraft termed it. Impressions are to be tossed off 'without . . . endeavouring to arrange them'.

The best place to seek the character of a country is in its provinces, she thought, not its cities where the inhabitants tend to sameness. It was therefore an advantage that her travels so far had taken her from a Swedish pilot-house with fresh white curtains and fragrant with juniper berries, to the poor log-shelters in the 'fields of rock' near Strömstad, and then on to energetic towns along the greener Norwegian coast. Her comparisons of primitive with polished societies foreshadow anthropological travel, lending her intelligence to what she observes, for 'the art of travelling is only a branch of the art of thinking'. She takes in women weaving and knitting to keep warm during the deep winter; the smell of children's bodies seeping through layers of linen; the hospitable warmth of peasants; and the communicative smiles exchanged with women who share no language. She is detached enough to resist the vanity that might be excited by the attention she gets as an attractive woman travelling alone. Her glance at the rosy calm of her sleeping child, her wonder at the unextinguished shades of a summer night, and hints of unexplained estrangement from the man to whom she writes, give her a quickening presence, a traveller with a finger on her pulse at the very moment it beats.

From her Tønsberg base, Mary made forays along the south coast of Norway. Her first business was with lawyers back in Larvik. Since she thought them corrupt, we can assume they represented the Ellefsens and resisted her claim for compensation. To the indignation of Judge Wulfsberg, one lawyer named Lars Lind was 'not even ashamed to say that as long as he is offered the right amount of money, the case will never be solved'. From Larvik, Mary then pursued Ellefsen himself along a 'wild coast' to Risør, a westward journey by sea.

Risør turned out to be two hundred houses, red and ochre, huddled together, with planks for passages from house to house, to be mounted like steps, under a high rock which looked to her like a Bastille, a place 'shut out from all that opens the understanding, or enlarges the heart'. To be born here was to be 'bastilled by nature'. This constriction had to do with commerce, men immured in contraband and trickery, infecting the whole town. Wollstonecraft pictures these tricksters drinking and smoking in rooms whose windows are never opened; by evening their breath, teeth and clothes are foul. The obvious subtext is distaste for Captain Ellefsen and his supporters, for the actual landscape of Risør is not as claustrophobic as Wollstonecraft imagined. There is a steep rise to be sure, but further inland were oak woods, with logs for export floating down the rivers to the town at the outlet of three fjords. At the time, it was the third-largest shipping centre in Norway, and it's hard to believe Mary Wollstonecraft that all goods were contraband. She dined with the British Consul, who would hardly have been there were Risør merely a smugglers' haven. She pictures herself writing in preference to pacing up and down her room, as though imprisoned, but she has come of her own accord.

What she doesn't tell us is that she's at hand for another judicial hearing and that she herself confronted the accused, Peder Ellefsen.

At first, Peder behaved humbly with Mary when she reproached him for the damage she had assessed during her visit to Strömstad, and demanded compensation.

He said he would think it over.

After discussing the matter with family and lawyers, he returned. His manner, at this second meeting, had changed. He refused her demand with the assurance of a man in the right.

Mary, in turn, refused to accept this. She resolved to go to Copenhagen, and lay the case in person before the highest official to whom she could appeal, no less than the Prime Minister of Denmark, Andreas Peter Bernstorff. Wulfsberg immediately backed her with a letter of introduction. He presents her as a talented woman, well known for her writings as Mary Wollstonecraft, but now as 'Madame Imlay' seeking compensation for wrongs done to her husband. He presents Imlay as a Benefactor betrayed by a man he had tried to help. Wulfsberg's letter recapitulates the case against Ellefsen, stressing his conviction that the Ellefsens had leaned on witnesses to produce false testimonies, so that the truth of the matter was impossible to prove. He complains that Peder Ellefsen had somehow managed to instigate a High Court appeal before a judgement had been made – 'which seems to be contrary to all laws'.

Then, just as Mary won this gesture of support with a sense of 'self-applause', Imlay withdrew his lifeline. The change came in a letter he sent her on 20 August.

Mary's attempted suicide had affected his standing, he said. He suspected that she had not talked of him 'with respect'.

Her suicide note had been in his praise, she protested, 'to prevent any odium being thrown on you'.

For Mary what rankled was Imlay's infidelity. 'I will not torment you,' she resolved, 'I will not complain', but she could not stop. Imlay had to hear that there 'are wounds that can never be healed – but they may be allowed to fester in silence without wincing'. Hardly an inducement to live with her. At this moment, Mary exchanged her Hamlet drama of flatness for that of King Lear: 'What is the war of the elements to the pangs of disappointed affection, and the horror arising from a discovery of a breach of confidence, that snaps every social tie!' She declared to Imlay that to tame her feeling would wither what she has come to be. 'Love is a want of my heart . . . To deaden is not to calm the mind – Aiming at tranquillity, I have almost destroyed all the energy of my soul . . . Despair, since the birth of my child, has rendered me stupid – soul and body seemed to be fading away before the withering touch of disappointment.'

On her way south from Christiania to the Swedish border, she contemplated the torrential waterfall from the 'dark cavities' of Frederikstad. Her Romantic bravura lingered in Coleridge's mind when he read Wollstonecraft's *Travels* a year later: 'Kubla Khan', his most famous poem, has a waterfall that seems to fountain from the centre of the earth through 'dark caverns measureless to man'. As Wollstonecraft gazed at the torrent, she asked herself 'why I was chained to life and its misery', and then her soul 'rose with renewed dignity above its cares – grasping at immortality . . . I stretched out my hand to eternity, bounding over the dark speck of life to come.'

She reached Gothenburg on 25 August, and after a seven-week absence clasped her 'Fannykin', her 'little frolicker', delighted to find the child venting sounds on the brink of words. But three waiting letters from Imlay reversed her joy. Her reply next day is filled with the sound of his voice.

'You tell me that my letters torture you . . . Certainly you are right; our minds are not congenial . . . You need not continually tell me that our fortune is inseparable, *that you will try to cherish tenderness* for me. Do no violence to yourself!'

Mary's next stop was Copenhagen. There, 'fresh proofs' of Imlay's 'indifference' arrived on 6 September. A quarter of Copenhagen had been gutted by fire: amidst its 'heaps of ruins', she felt 'strangely cast off'. From now on she moves, suicidal, through alien landscapes. Imlay's evasion of her questions about the future – he was a master of the devout wish – leaves her panting.

'I do not understand you. It is necessary for you to write more explicitly – and determine on some mode of conduct.–I cannot endure this suspense – Decide – Do you fear to strike another blow?'

Stretched to the limits of endurance, she remained thoroughly professional. She wrote to Imlay, 'I . . . only converse with people immersed in trade'; presumably, she was carrying out his instruction to dispose of 'goods' in the hands of the Danish traders Ryberg & Co. She also sent a statement on Imlay as the injured party to Prime Minister Bernstorff (together with supportive letters from Wulfsberg and other Norwegian and Danish public figures). Mary Wollstonecraft's letter to the Prime Minister, buried for more than two centuries, is a new discovery by Gunnar Molden:

> Impressed with a respect for your character, I venture, Sir, to expostulate with you relative to an affair which Mr Wulfsberg has already in some measure explained to you, in the letter which accompanies this brief statement.
>
> Previously allow me to introduce myself to you by [my] own name, Mary Wollstonecraft, and I think I may be permitted, in a strange country, without any breach of modesty, to assert that my character as a moral writer is too well established for any one to suspect that I would condescend to gloss over the truth, or to anything like subterfuge, even in my own cause.
>
> Mr Imlay, my husband, being very much engaged in business could not, at this juncture, leave England to pursue, according to law, Peter Elefsen, who had fraudulently deprived him, and his Partner, of a considerable property. I, therefore, wishing to have an opportunity of writing an account of the present state of Sweden, Norway and Denmark, determined to undertake the business, being fully acquainted with all the circumstances.
>
> Will you, Sir, spare a moment to peruse the following narrative.
>
> In the spring of 94 Mr Imlay bought a ship of an American captain, who had previously engaged Peter Elefsen to be his flag master. The transfer of the vessel deprived him of his employment, and his distress introduced him to our notice. For some time, without having any first plan of rendering him useful, Mr Imlay let him have the money necessary to support him – and at last sent him to Paris, two or three

times, to bring down some silver to Havre de Grace. During the intervals between these journeys Mr Elefsen pointed out a vessel that could be bought cheap and Mr Imlay purchased it; and giving the command to Elefsen[,] preparations were made for a voyage to Gothenburg. Mr Elefsen mean time lodged in the house of a merchant at Havre[,] Mr Imlay paying for it, as well as supplying his other wants. In his room the silver was deposited. I saw it there and the mate, Thomas Coleman, an American, assisted Elefsen to carry it on board the ship.

Previous to his sailing he signed a bill of parcels, in which the articles he took were not specified, as well as a receipt for the silver, both of which Mr Wheatcroft and I read over, Mr Wheatcroft (a merchant at Havre) witnessing that Elefsen signed them. The receipt was enclosed in a letter to Mr Backman of Gothenburg with other instructions for clearing and loading the vessel. I, Sir, gave Elefsen his last orders[,] Mr Imlay having set out for Paris the day before.

I have now to inform you that Elefsen took the silver privately on shore at Arendall, as the mate, Thos. Coleman, [h]as fully proved, and opened the letter addressed to Mr Backman, taking out the receipt. The cover of the letter has been brought into court. Many corroborating testimonies have supported the evidence of the mate to the conviction of the judges and every impartial person; still the atrocities carried on during the time the trial has been pending[,] to retard the march of justice[,] have even been more flagrant than the breach of trust. Many of the inhabitants, particularly the post master, who has but one character in the country, having shared in the spoil, bribery has produced prevarication and perjury. His [Ellefsen's] father-in-law, a major in the army, offered five hundred dollars to the wife of the first judge. When I arrived at East Risoer[,]* Elefsen waited on me and, as we were alone, behaved in the humblest manner, wished that the affair had never happened, though he assured me that I never

*Risør was called East Risoer at this time.

should be able to bring the proofs forward sufficient to convict him. He enlarged on the expense we must run into – appealed to my humanity and assured me that he could not now return the money. Willing to settle the business I desired him to inquire of his relations, who are people of property, what they would advance, and come to me in the evening when I would endeavour to compromise the matter. – He came and was almost impertinent. He had been spurred on by his attorneys, the pest of the country. Their plan, I plainly perceive, was to weary us out by procrastination. The suit has already been pending a twelvemonth, and the want of such a considerable sum in trade, as well as the expenses incurred by the detention of the vessel, which it has been proved he endeavoured to sink, is a very serious injury to us, not to dwell on the vexatious circumstances attending the failure of a commercial plan. I am very well convinced that an English jury would long ago have decided in our favour[,] not suffering justice to be insulted in the manner it has been with impunity; but the judges are timid.

To you, Sir, as a known lover of justice, I appeal, and I am supported by the most worthy Norwegians who wished by the respect they paid me to disavow the conduct of their countryman.

I am Sir yours
Respectfully

Copenhagen
Sepr 5th 1795

Mary Wollstonecraft
femme Imlay

Mary's irritability with the Prime Minister's polite caution tells us that neither her nor Wulfsberg's letter succeeded in convincing Bernstorff to squeeze the Ellefsens on Imlay's behalf.

Mary pressed on in the face of depression. Why did she and Fanny make for Hamburg when, by now, she knew that Imlay had gone back on the plan to meet them? If the source of 'knavery' was not on the remote Norwegian coast but in the get-rich city of Hamburg, then it makes sense that Hamburg was always going to be her final stop on her business journey.

And this would explain, too, the loathing she felt, as though Hamburg were to blame for what went wrong in her life. For the silver ship as bearer of her hope that Imlay would make his fortune and settle down was foiled, it would seem, at Hamburg. To Mary, the city stank of commerce. It seemed to contaminate the air she breathed.

One of her contacts was Imlay's acquaintance, the Franco-American writer St Jean de Crèvecoeur, who earlier in 1795 had moved to Altona, the Holstein town within walking distance of Hamburg. As a tolerant place – Danish, not German – Altona welcomed Jews, fleeing French émigrés (among them Mme de Lafayette), and many Americans including of course Joel and Ruth Barlow who had returned to Paris two months before Mary arrived. She too stayed in Altona, a little put out by the lack of paths and steps to ease Fanny's climb down the bluffs to the beach on the edge of the Elbe. Mary's only pleasure was to dine with Crèvecoeur daily, grateful for a companion who shared her distaste for commerce. She echoed what Dr Price had warned American leaders in 1784, that addiction to commerce could debase the new national character. 'England and America owe their liberty to commerce,' she said, 'which created a new species of power to undermine the feudal system. But . . . the tyranny of wealth is still more galling and debasing than that of rank.' Talk, in Hamburg, ran in 'muddy channels'.

Imlay hit back her volleys and continued to believe he had not wronged Mary Wollstonecraft. Had she not wanted to earn her keep? Had he not funded her journey in Scandinavia? Had she not travelled; met leading men; regained health; prolonged her stay by an extra month; and gathered matter for a travel book? He could not see that all this was secondary to her mission to earn the standing he had seemed to confer on her as his 'best friend' and 'wife'. Every letter of hers sets out this claim: 'I cannot live without some particular affection – I am afraid[,] not without a passion' (Norway, 5 August); 'I am weary of travelling – yet seem to have no home – no resting place to look to' (Copenhagen, 6 September); 'I am content to be wretched; but I will not be contemptible' (Hamburg, 27 September). Her pleas cry out in every form of love, plaint and logic she could devise, but Imlay refused to respond on any but his own terms.

'I am going to be the first of a new genus'. This portrait of Mary Wollstonecraft by John Opie has been lost to sight for seventy years.

Abigail Adams asked her husband to 'remember the Laidies' in framing laws for the new American nation.

John Adams teased his wife as a 'Disciple of Woolstoncraft!'

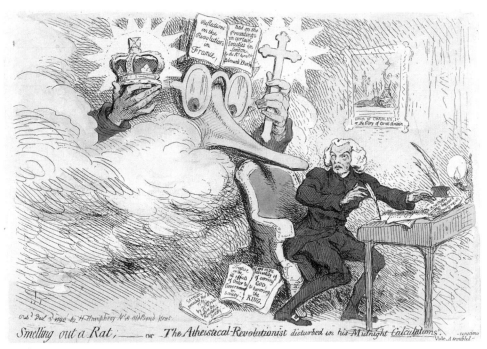

Gillray's cartoon of Wollstonecraft's political mentor Dr Price as regicide, caught by the counter-revolutionary Burke. Wollstonecraft defended Dr Price in *A Vindication of the Rights of Men*.

Mary Wollstonecraft's favourite pupil, Margaret King.

George Ogle befriended Wollstonecraft while she was a governess.

Joseph Johnson, Wollstonecraft's benevolent publisher.

Johnson's lifelong friend, the artist Henry Fuseli in 1794.

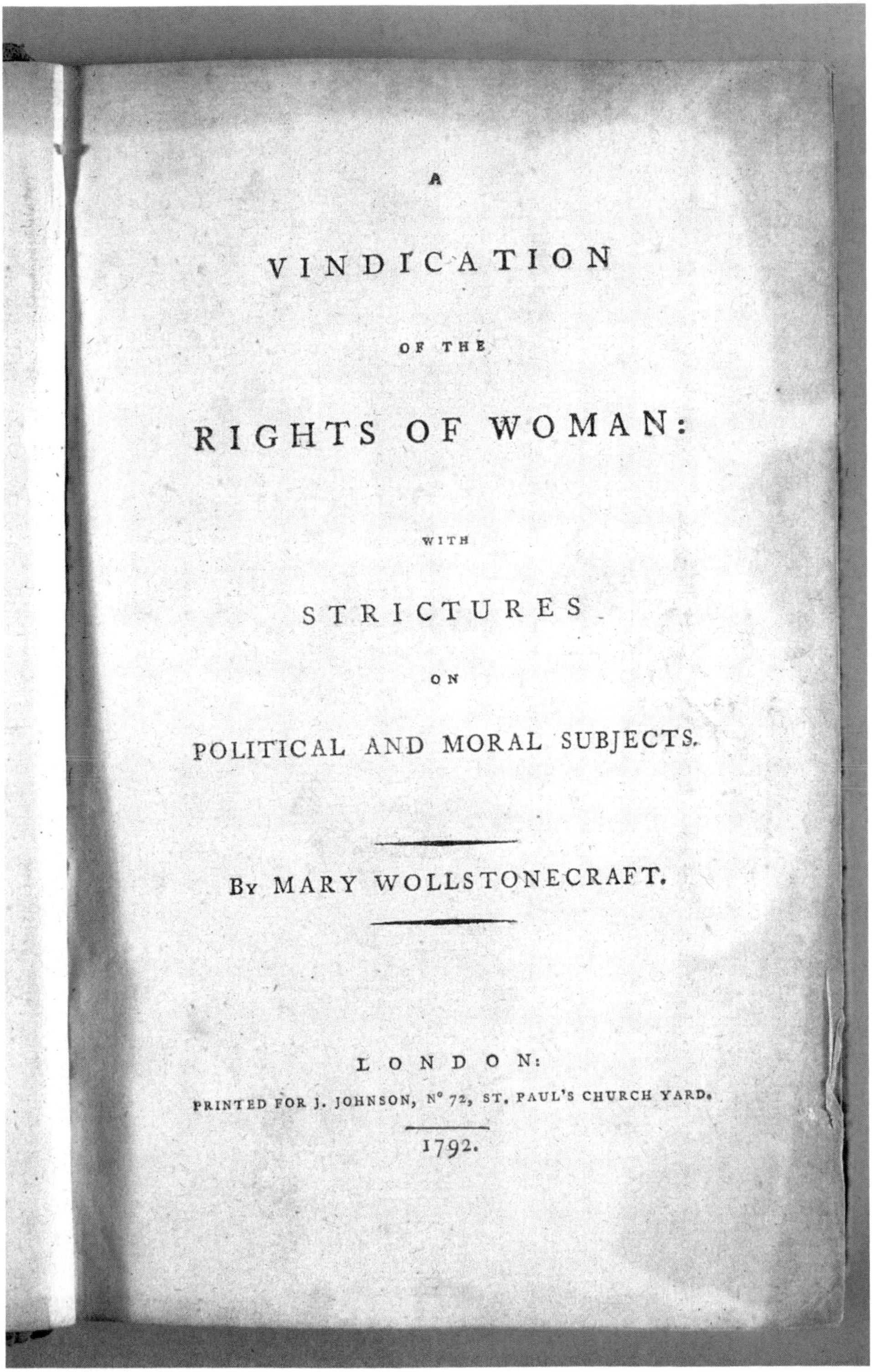

A

VINDICATION

OF THE

RIGHTS OF WOMAN:

WITH

STRICTURES

ON

POLITICAL AND MORAL SUBJECTS.

BY MARY WOLLSTONECRAFT.

LONDON:

PRINTED FOR J. JOHNSON, Nº 72, ST. PAUL'S CHURCH YARD.

1792.

The first edition of *A Vindication of the Rights of Woman.*

Ruth Barlow, who had a confiding friendship with Wollstonecraft.

Joel Barlow, American poet and diplomat.

Tom Paine, whom Wollstonecraft visited in prison.

Helen Maria Williams, an English writer in Paris.

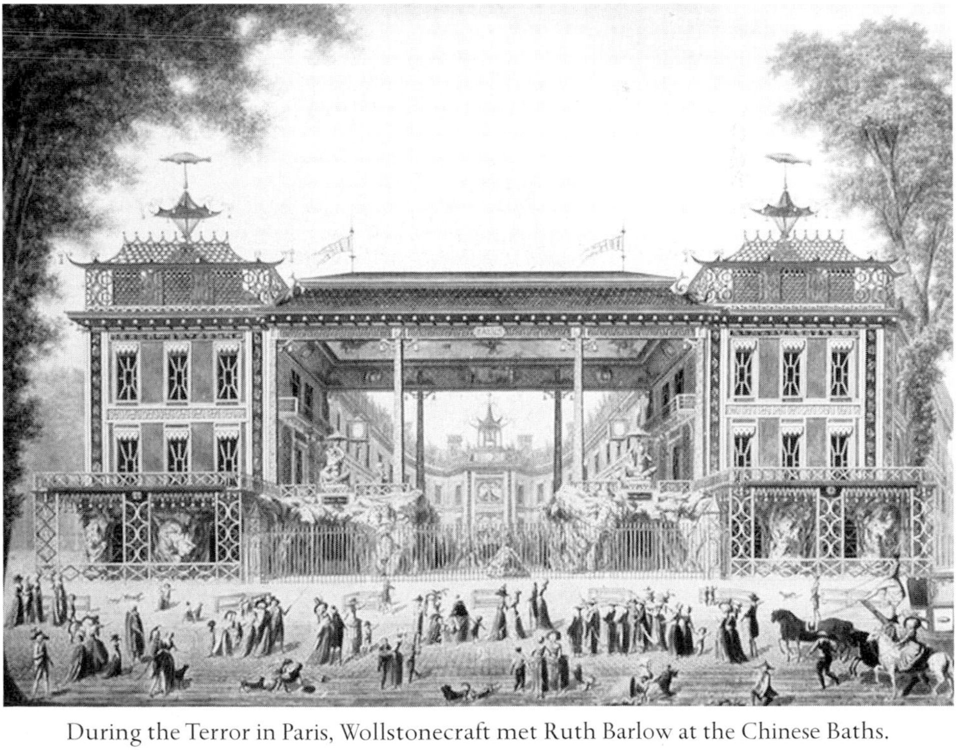

During the Terror in Paris, Wollstonecraft met Ruth Barlow at the Chinese Baths.

Gilbert Imlay modelled himself on the frontiersman Daniel Boone opening up the West for settlers. (Detail from a painting by George Caleb Bingham.)

William Godwin in 1795. Hazlitt said of him, 'He blazed as a sun in the firmament of his reputation'.

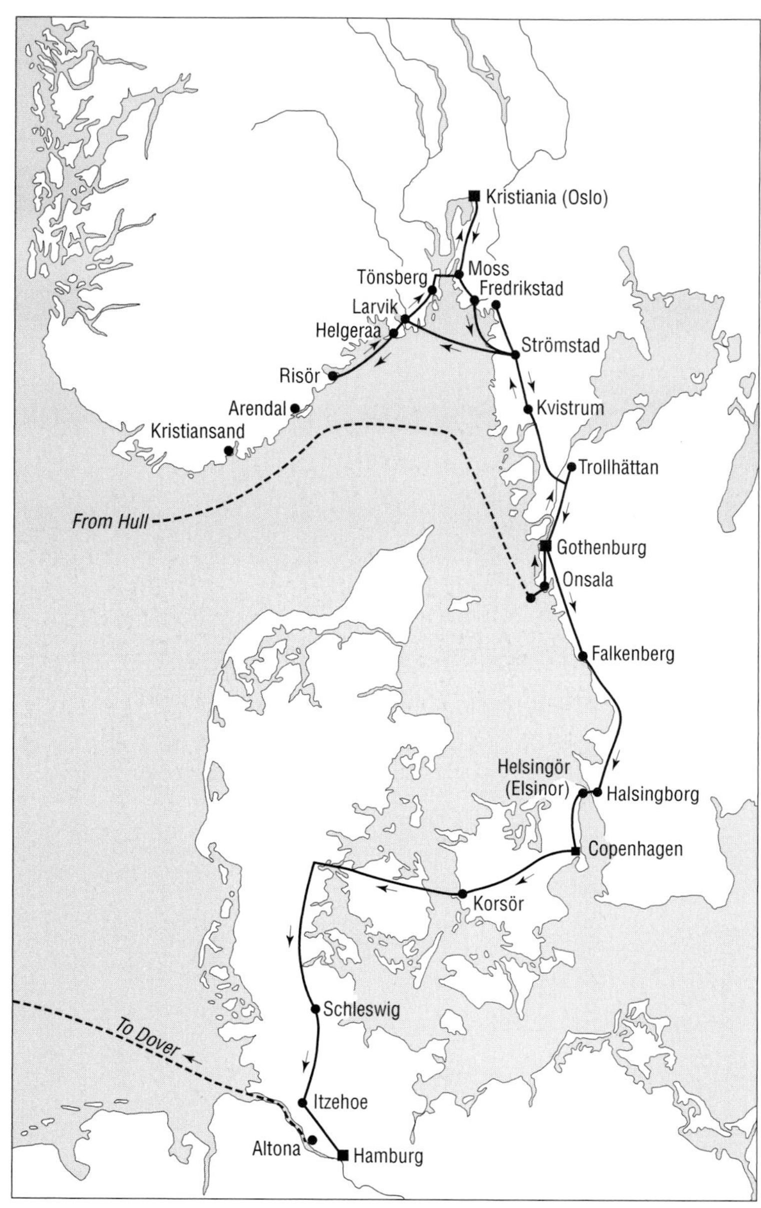

Wollstonecraft's travels to Scandinavia and Hamburg.

Details of the silver ship in a Swedish log.

Risør, in Norway, where Mary Wollstonecraft came to confront Ellefsen.

The river Elbe at Altona, near Hamburg: 'the silvery expanse . . . bearing on its bosom so much treasure'.

'Mrs Perfection': the writer Elizabeth Inchbald.

Amelia Opie, née Alderson.

Mary Wollstonecraft in 1797, married to Godwin and pregnant with Mary.

Mary Wollstonecraft Godwin (Mary Shelley) in 1815. This miniature shows her resemblance, at the age of eighteen, to her mother.

'I love to cherish melancholy'. Godwin mourning Mary Wollstonecraft in 1798.

FABLES
ANCIENT AND MODERN.
Adapted for the Use of Children from Three to Eight Years of Age.
BY EDWARD BALDWIN, ESQ.
VOL. I.
ADORNED WITH THIRTY-SIX COPPER-PLATES.

Published by Thos. Hodgkins Hanway Street; Oct. 6th 1805.

Illustration for Godwin's retelling of Aesop's *Fables.* He tried them out on his own girls, and discussed Wollstonecraft's methods of childcare with her follower, below.

An 1802 profile of Margaret Mount Cashell (née King) in Paris, where she met Wollstonecraft's circle.

A sentimental Victorian image of Mary Godwin and Percy Bysshe Shelley at Wollstonecraft's grave.

Claire Clairmont.

Byron in 1818. Sketched in Venice, where he took over the care of Allegra from Claire.

Writers who took up 'the great problem of the true nature of woman':

Elizabeth Barrett Browning

Charlotte Brontë

Emily Dickinson

George Eliot

Henry James

Virginia Woolf

Their exchange became a fight to the death. Mary was like the knight who performs a series of ordeals for the love of his lady – only, in this scenario, a woman replaced the hero of romance – and for so unlikely a phenomenon there was no reward. Imlay's drama was about gain; from his point of view, it should have satisfied both sides. For Mary it was unacceptable unless commerce bought a domestic reward.

During most of her fortnight or so in Hamburg, she did not hear from Imlay; then, a letter arrived on 27 September. Something told Mary he had a new mistress, even as he confirmed 'the ties which bind me to you and the child'. At this point she rebelled: she left Hamburg that same day.

The lawyers in the case did not solve the mystery of who fooled whom. Four unanswered questions remain. If we go back to the question why, given the value of the silver, did Imlay not go to Scandinavia himself, the answer now seems plain: because the silver *wasn't there*. We must separate the shady issue of the ship from the shady issue of the silver. If the bulk of the silver was *not* dispersed to the Ellefsens in Arendal, we can understand why Mary Wollstonecraft saw no necessity to go there once she had seen Ellefsen in Risør.

The second question is what took Imlay to France in the summer of 1795 while Mary Wollstonecraft was in Scandinavia? What business was more pressing than the fortune he had supposedly lost far north? Here is another piece of circumstantial evidence that the silver, or the bulk of the silver, was not with the Ellefsens.

Mary sided with Imlay's vilification of Ellefsen: she went to Scandinavia to 'get the money' from his family. But in view of the 'knavery' that happened while Ellefsen was at sea, is the focus too narrowed? Should Ellefsen take the whole brunt of the wrong, or had he only a share in it?

These three questions are interrelated, but the fourth and most important stands alone. No one asks why Wollstonecraft's last destination was Hamburg. Why did she not sail back to England from Copenhagen? It was a long, wearisome journey from Copenhagen to Hamburg in those days (becalmed as she was on the stagnant water of the Little Belt, with Fanny crying for food), and she was, by then, in a bad way. For two hundred years, the love-drama has acted as a partial blind, obscuring a further purpose: a switch from business to do with the ship (in Scandinavia) to the business of the silver. 'Cast off' as she was by Imlay's refusal to join her, business alone could have taken her to Hamburg, where she entered into enquiries with reluctance. She had to talk to sharks whose dealings she reported 'fully' in statements to Imlay that were not allowed to survive. The only certain fact is that she could not find the 'information' Imlay needed. There was something to do with the silver that Imlay did not know, and needed to know – conceivably a source of the 'knavery'. Hamburg was Europe's spy capital, called by her Irish friend Rowan an 'emporium of mischief', for the spies were a venal lot. Carriers of cargoes swarmed through dark 'lanes' (the *Gängeviertel*) around the port with its thicket of sailing ships and barges anchored along the mighty river. Ships brought epidemics along with trade, and twisted phantoms begged outside the Pesthof, the fever hospital. Mary's concentrated investigations allowed no time for the cultured inhabitants of Hamburg. Detection led her into the over-stuffed homes of merchants or forced her, possibly, to penetrate the murkier 'lanes'. Her hatred of Hamburg is projected on its commerce in general, in the same way as her hatred of Risør had been projected on its landscape: in both places she came too close for comfort to the 'crooked business' at the core of this mystery.

This is a Conradian tale. Barlow, as Imlay's business partner, and Coleman, the American mate, emerge and fade. At times, they hover into view as pivotal figures. Mostly, they vanish off-stage while the court lights up a single figure, a captain who blunders from story to story, tricky, impertinent – but not, in the end, a proven criminal. Imlay's side claimed that the Ellefsens bribed witnesses to tell false stories. But what if those witnesses were telling the truth? Could Ellefsen himself have told the truth, or some part of the truth, when he claimed the ship as his?

At this distance in time, it may be impossible to solve a case unsolved in its day, but two things are worth noting. The historian Gunnar Molden, listening, as it were, in Norwegian, had the impression that Ellefsen's was not entirely a guilty voice. This, in itself, has no status as evidence, but Molden has also dispelled the legend that Ellefsen, disgraced at home, went into exile. In fact, he went to live on Gjessoy, an island at the mouth of Arendal; married Marie; had five children (one named Margrethe) whose descendants live in Arendal today; and pursued the captain's life he would have lived had there been no silver. The silver existed, of course, but not in Ellefsen's pocket. He did not retire; he did not, in other words, get rich. Imlay, on the other hand, did benefit. New evidence shows that he did gain £1000 from his 'goods', a bit less than a third of the silver's value. Huge as this amount was in the eighteenth century, it could explain Imlay's need to know more of what happened to the other two-thirds of the silver.

To open up the mystery we must start not with Captain Ellefsen, but the unidentified man whom Mary advised Imlay *not* to visit in Paris on 20 August 1794 because what he had done was too wrong, literally, for words. Who was he? There is no proof, but one person does fit the case at every point: a man who visited Paris just then and whose wife's intimate correspondence with Mary Wollstonecraft came just then to an end. Another way of approaching the mystery is to ask who benefited. Who got rich in 1794–5? Not Ellefsen. Only one person emerged from that stillest of winters with a fortune, and that person was of course Joel Barlow.

Imagine Barlow, quick-eyed as ever, spotting an opportunity in Hamburg: some unforeseen chance to dispose of silver. This seems too good to be missed. A man with his Scioto history will not be afraid to seize the initiative. It's likely to have been a high-handed change of plan that Mary Wollstonecraft calls 'knavery' in her letter sent soon after she saw the silver ship out to sea; but what part of it is 'folly' – a judgement she adds to 'knavery' a month later, at the end of September?

Between the 'knavery' in mid-August and the 'folly' of the following month comes the loss of the ship when Ellefsen hands it over to the mate on 20 September. If, at Hamburg or at a Danish-Holstein port along the Elbe – Glückstadt (the stopover for the *Rambler*) or Altona – a substantial part of the treasure is taken off the vessel, Barlow will have to satisfy the captain. Since Barlow is always impecunious until this time, what has he to offer but the ship itself? In such an event, Ellefsen does not sail along the

Norwegian coast because Imlay planned that route for security, but because he is simply on his way home, content with his gains.

Another way to start is with the treasure hidden in the hold. Officially, the silver is to pay for a great shipment of Swedish grain at the end of summer. This is the high-level export deal Barlow would have made with French commissioners. Let's say again that, en route, the ship docks on the Elbe. Its Holsteiner crew is drawn from that region. The Norwegian captain and the American mate – hired by Imlay – are set apart as keepers of the secret, secret sharers amongst an unknowing crew who will remain on board throughout the ship's subsequent adventures. At Hamburg, where Barlow is active as Imlay's agent, he hears of a more lucrative way to dispose of the silver than to purchase grain. Barlow writes to advise Imlay that he has taken matters into his own hands, and devised a new scheme. For this he needs Ellefsen's compliance. Ellefsen's new instruction is to stage a shipwreck on his home coast where shipwrecks are common. This is designed to satisfy French authorities that the silver has been lost with the ship, since the expected grain will fail to appear.

An alternative scenario, with a similar result, might be that Ellefsen departs from Le Havre with secret instructions – unknown to Mary or the mate – to offload the silver, or some part of it, in Arendal. This Ellefsen duly does, sending it on via Danish channels (Ryberg, say, or Flensburg) to the silver centre of Schleswig-Holstein. In such a case, Imlay has to be party to the scheme together with Barlow, and Backman at first irrelevant. This would explain two puzzling facts: Imlay's making no plan with Backman to receive so precious a cargo, and his informing him of the silver only after some part of the plan goes publicly wrong (in October 1794). It's inconceivable that, back in August, Imlay should have sent off treasure to a complete stranger who has no idea what will arrive – and will therefore have no precautions in place. The very casualness of the letter Ellefsen carried from Imlay to Backman, its air of strangers who may contemplate future business, must be a cover for a different, more calculated plan.

Meanwhile, back in Le Havre, Mary Wollstonecraft opens a letter to Imlay, and stumbles on a plan unknown to her. Indignant, she warns

Imlay to have nothing to do with a knave who will be in Paris to discuss the deal in person. I think we can safely assume that Imlay ignores Mary's advice – he's dealt with knaves before. We can't know what is said in Paris but we can guess that he and Barlow come to some private agreement as to how the silver will be handled. When Imlay returns to Le Havre, Mary is still not as fully informed as she believes. For, in her eyes, Imlay remains a victim of 'folly' when, in late September, she tells him bad news from the 'knave', probably the failed plan to sink the ship, with consequent damage. Something definitely goes awry at Arendal. The mate's protest, for instance, can't be contained. Coleman has seen too much: seen silver disappear; seen and perhaps stopped the attempt to sink the ship. So, to shut him up and get him off the scene, he's allowed to take the ship to Gothenburg, where he thinks it has to go. Peder Ellefsen is not an accomplished liar; he gets deeper into trouble as he stumbles from story to story.

Then, too, the plan does not allow for a trader of Backman's character. This is a stable family man who has spent his youth in France and married a Frenchwoman. Backman, attached to France and habitually a good citizen, feels responsible for the shipment of grain to starving Paris. His emissary, Waak, bears witness to the ship in Risør – an unsunken ship for which its captain, Ellefsen, must be held to account. So Ellefsen, shaken, makes a private confession to his lawyer, who invokes a law of 19 July 1793 which protects its confidentiality. Norwegian law has been quick to follow the Fifth Amendment to the American Bill of Rights (1791), which protects the accused from self-incrimination. Tantalisingly, Ellefsen's act of confession is noted in surviving records, together with an acknowledgement that it cannot be used as evidence in court. All the same, suspicions about Ellefsen, shaming to Norwegian shipping, reach Norway's rulers in Copenhagen. The shady locus of wrongdoing becomes from this point an international incident – the prestige, wealth and pride of a Norwegian clan versus Swedish insistence on rectitude – played out at influential levels on both sides.

Meanwhile, Coleman's authority is challenged by his Holsteiner crew, who refuse to stay at sea in a leaking ship as fogs and darkness descend and winter storms close in. The ship is sent to the nearest port of Kristiansand. Its illegitimate use of the national flag exercises local officials, who interrogate Coleman and turn against Ellefsen. The ship lies at anchor through that ice-bound winter; then Backman buys it when the ice breaks up in March. The ship proves almost a wreck, and Imlay finds himself liable for massive repairs that will cost a good deal more than the ship is worth. So, on 4 June, when the ship finally leaves Norway for the repair dock at Strömstad, Imlay is bent on pressing for compensation from the Ellefsens, a course of action Mary Wollstonecraft agrees to pursue in person. Five days later she is hastening north. So far, the criminal case can't convict Peder Ellefsen, but it has not absolved him – an advantage Imlay is quick to seize with a view to an out-of-court settlement. It's always assumed that the compensation Mary seeks is for stolen silver, but she can make a surer case against a captain who has abandoned his command in mid-voyage, with resultant damage. Can it be entirely coincidental that it's after her unsuccessful confrontation with the Ellefsens that Imlay turns cold? Mary is still hopeful of her plea to the Danish Prime Minister, but Imlay foresees he will have to fork out, after all, for the repairs.

While she's far north, Imlay travels to Paris, where he intercepts the flourishing Barlow who returns to Paris in July 1795 and starts investing his gains there. A year later Barlow records that the greater part of his property is 'now lying in Paris'. In his small leather-bound notebook the single word 'silver' shines amongst cargoes of soap, alum, candles, tallow, gloves, salt and hides. It's a matter of urgency that Imlay review his business with Barlow because it now appears that Barlow had defrauded him in Hamburg – pocketing a larger share of their mutual gains. Their deal has been secret, and so it would have remained had it not been for a reference to Imlay's visit to Paris amongst Mary Wollstonecraft's pleas to meet her. She refers to the precariousness of what he must 'settle' in Paris – to her, Imlay still appears in the light of victim. Later, when Godwin obeys Imlay's order to strike out all business dealings in Wollstonecraft's letters, he retains this Paris reference, having no idea of its significance. Once the letters are published, any protest from Imlay would only draw attention to it.

He appears culpable for not meeting Mary as promised, but Mary's negotiations take a month longer than predicted. She had thought to meet Imlay in August, but only reaches Hamburg in September. Imlay is too active to loll around Paris once his business there is settled, so returns to London – this may be what disappoints Mary in the three letters from him that reach her in Gothenburg on 26 August. The month in Tønsberg held her up – not against her will. Norway and Sweden are healing to Mary, and so long as she is in the company of Backman and Wulfsberg, right and wrong seem clear. But something else has long troubled her: the 'knavery' that came to her knowledge in August–September 1794; the 'crooked business' she deplored in December. At the end of September 1795, she finally comes face to face with crookedness. In Risør she preserved her distance from small trickster minds closed off in their rocky fastness; in Hamburg, it's different. Nothing here is small, and crookedness can't be distanced. It's under her skin. She uncovers things in Hamburg she has not expected; evidence, perhaps, that Imlay is less the victim than he has led her to believe. Though the nature of his business has been eliminated from Mary's letters, they seethe with accusation: she speaks '*entre nous*' of 'fraud'. Hamburg sucks her into the slime that gives rise to 'mushroom fortunes'. It's her private heart of darkness.

A day came when Mary had endured enough. It coincided with the arrival of Imlay's 'unkind' letter on 27 September, refusing her repeated pleas to set out his position on their future.

'Extraordinary and unnecessary,' he said.

'I have leant on you for support, and been pierced by a spear,' Mary groaned.

At once, without awaiting further instructions, she boarded a vessel bound for England. The ship sped before the wind. By Sunday 4 October she had landed at Dover with no one to meet her and nowhere to go. She would never again be 'humbled' by accepting Imlay's support, yet did advise him of her return. He found her lodgings in London, but was, again, distant.

On 10 October Mary pressed his cook, who admitted that Imlay did have a new mistress. Stunned, once again, by confirmation of a half-known truth, Mary went to see for herself the furnished house he had provided for the mistress. There she confronted Imlay's lies.

'Nothing but my extreme stupidity could have rendered me blind for so long,' she said. 'Yet, while you assured me that you had no attachment, I thought we might still have lived together.'

The pain of the night that followed was too unbearable to prolong what she called 'my hated existence'. Imlay's indifference had left her 'outcast' in Copenhagen; his prevarication had 'pierced' her in Hamburg; but what he said to her in London was worse, for it left her helpless – a state alien to her active nature. After her three-month journey on Imlay's behalf, there was no more she could do. Rage at her betrayal gave way to the calm of surrender – an acceptance that struggle was useless.

'Your treatment has thrown my mind into a state of chaos,' she told him. Yet she is clear as to his wrongdoing, and composed enough to assure him he has nothing to fear: 'Let my wrongs sleep with me!' She is also capable of plans: begs him to put Fanny in the care of her German friend in Paris, and leaves her clothes to Marguerite. When she had taken an overdose in June, she had notified Imlay in time to be saved; this time she meant to die. In Hull she had looked at the sea as a possible 'tomb'; her eye now fell on the Thames.

It was raining hard that Saturday night when she took a boat from the steps at the Strand and was rowed up the Thames as far as Battersea Bridge in the village of Chelsea. Too many people were about. She had to find a deserted spot, so offered the waterman six shillings to row her further upstream as far as Putney Bridge (near Walham Green where she'd lived long ago with Fanny Blood). No one was there. Trusting to sink more quickly if she weighted her clothes, she trod back and forth on the bridge for half an hour till shoes and dress were soaked with rain. Then, she climbed on to the frame of the central arch, and plunged into the dark water. Suffocation was slow and infinitely more horrible than she had expected. She pressed her skirts to her sides, trying to hasten death.

About two hundred yards from the bridge, a boatman found her unconscious, and rowed her to the Duke's Head Tavern at Fulham. There, she had the good fortune to be resuscitated by a member of the Royal Humane Society (set up to teach artificial respiration in an age of canal and river transport). Her recovery was quick, though drowning seemed less painful than coming back to 'life and misery'. *The Times* reported the rescue of an elegantly dressed lady who explained 'that the cause of this, which was the second act of desperation she had attempted on her life, was the brutal behaviour of her husband'. This time, Imlay did not come in person – the rescue was too public – but within two hours he sent a physician, dry clothes and a coach to take Mary to their friends the Christies in Finsbury Square.

It was a spacious square, a little to the west of Mary's birthplace in Spitalfields. Rebecca Christie nursed her, and for some time she saw no one else.

'I know not how to extricate ourselves out of the wretchedness into which we have been plunged,' Imlay wrote.

Mary replied with her mother's dying words – 'Have but a little patience' – promising him to 'remove' herself beyond his reach.

Imlay behaved with respectful concern. He visited Mary, and once more offered to do her service if she would allow it. Though service of course fell short of love, he was shocked enough to offer the prospect of living together when his present affair came to an end. Mary realised the danger of continued waiting and uncertainty. She gave Imlay an ultimatum.

'If we are ever to live together again, it must be now. We meet now, or we part for ever. You say, you cannot abruptly break off the connection you have formed. It is unworthy of my courage and character, to wait the uncertain issue of that connection. I am determined to come to a decision. I consent, then, for the present, to live with you, and the woman to whom you have associated yourself. I think it important that you should learn habitually to feel for your child the affection of a father.'

Mary's extraordinary proposal may seem similar to the threesome she had urged on Fuseli, but was fraught with parental bonds. Imlay's awareness of his responsibility towards Mary is nowhere more evident than in his agreement, at first, to her plan. He even took her to view a house he was on the point of renting, to see if it would suit her. It's tempting to clothe the mistress in resentment, as Mrs Fuseli has been clothed. Yet, according to Mary, she and the mistress were to some extent in sympathy: 'I never blamed the woman for whom I was abandoned. I offered to see her, nay, even to live with her, and I should have tried to improve her.' It's doubtful if the mistress longed to be improved, but she did probably hear Mary's version of Imlay's treatment. Not surprisingly, Imlay changed his mind. Mary was not one to play a tame third in a household; the situation was bound to be explosive.

Defeated, Mary took lodgings around the corner from the Christies, at 16 Finsbury Place, and asked Imlay to send her things, including her letters. For a while she cast the letters aside, unable 'to look over a register of sorrow'. She therefore overlooked a new letter Imlay had enclosed. From November to January he was again in Paris, accompanied by his mistress. Thinking he had left without a word, Mary reproached him, and on 27 November received a lofty retort. He stood by 'the most refined' feelings. He demonstrated this with an offer of 'friendship'.

'You will judge more coolly of your mode of acting, some time hence,' Imlay tried to comfort them both.

'*Do you judge coolly*', she flashed back, flinging aside his flattering self-portrait, and, instead holding up the mirror to a man of caprice who had failed as a father. 'If your theory of morals is the most "exalted", it is certainly the most easy. – It does not require much magnanimity, to determine to please ourselves for the moment, let others suffer what they will!' She interpreted the offer of friendship as no more than the 'pecuniary support' she had to refuse – together with his '*ingenious* arguments'.

It was wounding on both sides. Mary knew she tormented the man she still loved, but could not contain her 'thirst for justice'.

Imlay accused her of 'decided conduct, which appeared to me so unfeeling'.

Mary replied, 'my mind is injured – I scarcely know where I am, or what I do'.

Even now she could not curtail her 'grief' – not only over losing Imlay, but losing the frontier promise of the man she believed he was. 'My affection for you is rooted in my heart. – I know you are not what you now seem,' she went on. 'I have loved with my whole soul, only to discover that I had no chance of a return – and that existence is a burthen without it.' In

Finsbury Place, her life seemed 'but an exercise of fortitude, continually on the stretch – and hope never gleams in this tomb, where I am buried alive'.

It did not help that gossip was circulating. Women whispered to one another that Mary Wollstonecraft's connection with Imlay had no legal sanction. The novelist Mary Hays told William Godwin that some 'amiable, sensible & worthy women . . . especially lamented that it would no longer be proper for them to visit Mrs W'. Hays herself visited Wollstonecraft, and promised to come again.

One day, to Mary's astonishment, she had a proposal from a rich man of fifty. It was someone she had met through Johnson, to whom she had spoken of her attachment to Imlay and recent sufferings. The unnamed suitor fantasised a woman of reason by day who, by night, became 'the playful and passionate child of love . . . One in whose arms I should encounter all that playful luxuriance, those warm balmy kisses, and that soft yet eager and extatic assaulting and yielding known only to beings . . . that breath[e] and imbibe nothing but soul. Yes: you are this being. Yet paradoxical to say, you never yet were this being. If you have been, I am unjust: toward him [Imlay] whom I estimate, not from my personal knowledge for I never saw him, but from your own affectionate descriptions. Well, well: I never touched your lips; yet I have felt them, sleeping and waking, present and absent. I feel them now . . .'

To Mary, it was an insult, 'the bare supposition that I could for a moment think of *prostituting* my person for a maintenance'. He had 'grossly' mistaken her character. 'I am, sir, poor and destitute. – Yet I have a spirit that will never bend.'

Another source of income was now urgent. Mary approached Imlay's business associate, Mr Cowie of the firm of Chalmers & Cowie, to lend her a sum. Another associate who lent her a small amount is called a 'long-tried friend' in a new-found letter. It reveals that Cowie was to repay himself out of a 'venture or cargo of Mr Imlay's that would come into his hands'. Some of the lost silver had apparently made its way into London in the course of the preceding spring. If Imlay's gains produced more than £50, Mr Cowie felt 'bound to pay the surplus into her [Mary's] hands'. She remained confident of this return, according to Godwin, writing on

2 January 1798: 'she understood that the goods produced more than they were estimated at, I think about £1,000', and counted this 'amongst her future resources'. The £1000 must have come from the silver ship, since this was the only Imlay venture in which Mary took part.

Mr Cowie's motive lies in shadow: Mary Wollstonecraft perceived a 'friendly intention'; also that he felt 'bound' to hand over the profit. But why did he not do so, and why was Mary content to receive a loan from Mr Cowie but somewhat reluctant to receive profits from Imlay, as the letter makes plain? A likely answer could be that until she reached Altona and Hamburg, she had not realised the fraudulent element in Imlay's transactions. Was the nature of his gains on Mr Cowie's conscience? Was the loan something of a silencer? And was Mary unhappy to accept tainted money – or was it simply that it was Imlay's money? Two facts are certain: a large sum for Imlay's 'goods' did materialise, and Mary did not receive what his associate considered to be her due when she returned to London in October 1795.

That autumn the Ellefsen trial was still going on. A new witness was John Wheatcroft, the English trader in Le Havre who had been Imlay's landlord and probably the one to harbour Ellefsen and the silver in his house. He is likely to have testified in Imlay's favour. But the defence lawyers' objection to Judge Wulfsberg was eventually upheld by the Danish Supreme Court in 1796. Wulfsberg had to vacate his seat on the commission. The case dragged on till November 1797, without resolution, though, conceivably, a private settlement was reached. In any event, the payment due to Mary continued to be deferred. Imlay could have argued that her 'surplus' was unearned because she left Hamburg without the information he wanted, yet he was not vindictive, more the sort of person who would have meant to pay her eventually after diverting the sum – temporarily, of course – into some other venture.

At Mary's most destitute moment in the autumn of 1795, Mr Cowie agreed to fund her on the basis of future writings and the returns due to her from Imlay. Accordingly, she began to prepare her travel letters for publication, improvising an artless manner designed to convince the reader of their truth. However keenly these letters appear to come off the pulse in the course of her journey, they actually took their final cast in the aftermath of suicidal despair. The linear plot of an episodic journey carries a counterpoint of the traveller's up-down interior life: first, the restorative process during the northwards journey to Norway, aided by the healing effects of the northern summer – the finest summer she had ever experienced; then, on the traveller's return to Gothenburg, a reconnection with her lover plunges her back into a situation that darkens as she journeys in the opposite direction. Hamburg immerses her in a 'whirlpool of gain'. She is exposed to a 'swarm of locusts who have battened on the pestilence they spread abroad'.

When Godwin later came to write a life of Mary Wollstonecraft, he refers the reader to her *Travels* as though they were an unmediated report of what took place. Yet her Hamburg is surreal, not documentary truth, like a precursor to the 'Unreal City' of *The Waste Land.* The horror inflates as the physical journey turns into an interior journey towards a collapse that will be all the greater for the intimations of recovery that precede it.

Wollstonecraft reinvents her traveller as a victim sent by her lover into hell. He is the type of all who sell their souls to commerce. The letter pretends to be a private whisper, but is designed for publication: it will be read by contemporaries, while the permanence of print will damn Imlay for all time. Instead of choosing to pair with a new genus, he proves to be 'of the species of the fungus' aligned with the 'mushroom' fortunes of Hamburg.

The traveller's dissociation of her own species from that of the fungus, omits two crucial facts: first, the real-life complexities of Imlay's attitude to his 'dearly beloved friend and companion', and second, the fact that Imlay had co-opted Wollstonecraft as his accomplice. To earn her 'surplus' she did wade into those 'muddy channels' – before she wrenched herself away. Now, abandoned in Finsbury Place in the autumn of 1795, Wollstonecraft omits particulars of the sordid dealings to which she was party, and so conceals the full story of her darkening view of Imlay. In short, she conceals her complicity – perhaps no more than knowledge. She can't relay what she knows of the silver ship, yet we glimpse its unmarked passage, treasure still intact, through her sight of ships stilled on the Elbe: 'the silvery expanse, scarcely gliding though bearing on its bosom so much treasure'. She can't relay the deals that contaminate the muddier channels, but the facelessness of the passive voice – the artful grammar of power – resonates with her knowledge of 'particular' gamblers: 'Immense fortunes have been acquired by the *per cents* arising from commissions, nominally only two and a half; but mounted to eight or ten at least, by the secret *manoeuvres* of trade.' She does not mention the Swedish grain that the *Maria and Margrethe* failed to deliver to the starving of Paris before the nightmare winter she experienced there; she does say, though, that the 'interests of nations are bartered by speculating merchants. My God! With what *sang froid* artful trains of corruption bring lucrative commissions into particular hands . . . and can much common honesty be expected in the discharge of trusts obtained by fraud?' Again, fraud.

So Imlay is fixed in his enduring image as rogue, while his arguments have vanished with his letters. The drama that survives is the one sustained by Mary Wollstonecraft's fame: the drama of a new kind of creature who struggled against a persistent refusal of the way of life she'd conceived. The interior drama of the Imlay years was to see destroyed the independent and desiring creature she had become. Her two suicide attempts were temporary capitulations to that denial. Yet, reviving, she continued to reject Imlay's determination to construct her as a grateful dependant of his largesse. Neither her reasoning nor her fondness had much impact; and though barbs did get through, he always resealed the carapace of the model man.

'My conduct was unequivocal,' he insisted; his principles were 'exalted'; he was 'magnanimous'.

Mary had to point out that for all this, he had not seen to the needs of her father and sisters, as she had asked before she sailed; nor had he paid her small debts which now weighed on her. 'Will you not grant you have forgotten yourself?'

There was also the question of reputation: Mary pressed Imlay to admit he was abandoning two who bore his name. 'The negative was to come from you.' He refused to give it. The vehemence of this struggle for the high moral ground was subsumed in the passive sadness of the fictive traveller. This traveller combines the rarity of a lone woman in unvisited lands with a traditional icon of female vulnerability, bound to win readers of both sexes. Her story borrows its pathos from women's laments of attachment (like Anne Hunter's 'My Heart Is Fixed on Thee', set to music by Haydn in 1794–5). The *Travels* end with the journey home – only, there is no home, no finality. The narrative breaks off at Dover, and though the formal journey is over, the inward journey still moves towards the lover whose rejection the traveller foresees.

The month after her plunge from the bridge, memory cast its beam back on this approach to death. It is the context for a narrative moving towards an end about to happen, which is just, only just, off-stage as the traveller returns, sadder and more shaken than ever, to the London she had left. It's not an ending so much as a shutting off. Art shuts off when life is to end. In the immediate after-time of the narrative, an actual attempt at suicide sealed the pathos of this work with a tenacious 'truth'. Yet, to prepare a book for the press was, in itself, an act of renewal. And as it turned out, the book was a huge success, with translations in Dutch, German, Swedish, and later, in 1806, extracts in Portuguese. Mary's friend Rowan brought out an American edition in Wilmington, Delaware, and admiring comments came from contemporaries like Robert Southey who said: 'She has made me in love with a cold climate . . . with a northern moonlight.' At the same time, the book proved a lifeline to a further phase of its author's existence. Addressed to a faithless lover, it touched a kinder man.

Early in January 1796 – the month the *Travels* appeared in London – William Godwin answered an invitation from Miss Hays to renew acquaintance with Mary Wollstonecraft, whose intrusive talk had put him off at the Tom Paine dinner four years before:

> *Tuesday* [5 January], *11 o'clock.*
>
> I will do myself the pleasure of waiting on you on Friday, and shall be happy to meet Mrs. Wolstencraft, of whom I know not that I ever said a word of harm, & who has frequently amused herself with

> depreciating me. But I trust you acknowledge in me the reality of a habit upon which I pique myself, that I speak of the qualities of others uninfluenced by personal considerations, & am as prompt to do justice to an enemy as to a friend.

So this cool philosopher met Mary Wollstonecraft once more, at 30 Kirby Street in Holborn. She saw a man with a head too big for his body and a nose too long for his face, holding forth amongst the teacups. His voice was that of a judging minister who confronts human flaws in a level tone of unabashed but not unforgiving truth. He was, in fact, a lapsed clergyman; a convert to rationalism. While Mary had lived in France, Godwin had come to fame with *Political Justice* (1793), the most radical of all English exposés of power. He was also a bachelor of forty waking up to the want of a woman in his life. Godwin's diary records the Hays tea on the 8th and dining out in a party including Mrs Christie and Mary Wollstonecraft on the 14th. The next day he read the first seventy-eight pages of the *Travels*, and then as he continued day by day from 25 January till 3 February, his dislike of the 'harsh' feminist dissolved 'in tenderness' for her sorrows; at the same time, he recognised 'a genius which commands all our imagination'. Perhaps there had never been a book of travels 'that so irresistibly seizes on the heart'. It was 'calculated to make a man fall in love with the author'.

13

WOMAN'S WORDS

Godwin called on Mrs Christie on 13 February, hoping to find Mary Wollstonecraft. He discovered that she had left town to stay with a friend, Mrs Cotton, in the village of Sonning, on the Thames near Reading. There, in February–March 1796, Mary received the attentions of Mrs Cotton's near-neighbours, Sir William and Lady East of Hall Place in Hurley, Berkshire. This couple was on visiting terms with Jane Austen's aunt Mrs Leigh-Perrot, and their son had attended the school run by Jane Austen's father at Steventon Rectory in the adjacent county of Hampshire. The connection was close enough for Sir William to have graciously sent the Austens a portrait of himself. Shortly after Mary's stay near the Easts, Jane Austen, at twenty-one, proposed to turn doctor, lawyer or Guard at St James's, should she find herself alone in London. It's unlikely that so well known a visitor to the area as Mary Wollstonecraft went unnoticed. No scandal had touched her beyond some unproven gossip. Most pitied her as a serious-minded wife who had been deserted by her 'infamous' husband.

Now, it was thought, Mary must get over Imlay. Godwin believed that her mood improved. Letters tell a different story. As the months passed, Mary continued to relive her incredulity that 'the heart on which I leaned has peirced mine to the quick' – this cry of pain in Hamburg she now repeated to Archie Rowan in far-off America. Her voice seems to come in panting bursts, yet her hand is even. Each line begins in the margin and flows steadily across the page, as though she were a habituée of despair. Letters to Rowan in January 1796 and to von Schlabrendorf in May show no change. She is 'weary' of herself. She is 'broken'. 'I live but for my child,' she told them both.

The question of support for Fanny brought her back into contact with Imlay. One evening early in February she had visited her neighbour Mrs Christie, as she often did, when she caught sight of Imlay, returned from France, as she entered with Fanny, now aged twenty-one months. Mrs Christie tried to intercept Mary and persuade her to leave. But she swept past with Fanny in her arms, and – in front of an assembled company – dropped her at Imlay's feet. It was a public rebuke for his failure to support his child.

If Imlay was still in league with Leavenworth, the going was rough, for in January 1796 Leavenworth was ruined. It was a massive crash, a loss of £40,000 he told Ruth Barlow – 'he is quite in the horrors about it', she said. On the 28th everything he had was seized, and he went to prison. If Imlay had been trying to avert that crash, he would have had sufficient reason to be in Paris for the previous three months. It's possible, then, that in February 1796 his affairs were on the edge. In his gentle way, he promised to see Mary the following day to discuss what he could do. This interview must have felt futile, for it was the very next day that Mary left town.

A month or two later, soon after her return to London, she spied Imlay riding along the New Road (as the Euston Road was called). When he saw her he reined and dismounted, and they walked some way together, with Imlay defending his 'forbearance'. To Mary, words were irrelevant beside the facts of betrayal. Once more, they fought for the moral high ground. Mary's part was to 'disdain' to reproach Imlay; then retorts – great sprays of indignant eloquence – would fountain from her opening throat. Imlay's part was not to budge, as a man of 'principle'. He was not about to shed his winning blend of high-mindedness and frontier vigour. His presence made it all the harder for Mary to forget a passion that was poisoning her residue of life. They agreed that she would continue to be 'Mrs Imlay', and Imlay offered to take out a bond for Fanny, the interest of which would contribute to her support.

'You must do as you please in respect to the child,' she returned in the last of their letters:

> I could wish that might be done soon, that my name may be no more mentioned to you . . . I am glad that you are satisfied with your own conduct . . .
>
> Your understanding or mine must be strangely warped – for what you term 'delicacy,' appears to me to be exactly the contrary. I have no criterion for morality, and have thought in vain, if the sensations which lead you to follow an ancle or step, be the sacred foundation of principle and affection. Mine has been of a very different nature, or it would not have stood the brunt of your sarcasms. The sentiment in me is still sacred. If there be any part of me that will survive the sense of my misfortunes, it is the purity of my affections. The impetuosity of your senses, may have led you to term mere animal desire, the source of principle; and it may give zest to some years to come. – Whether you will always think so, I shall never know.
>
> It is strange that, in spite of all you do, something like conviction forces me to believe, that you are not what you appear to be.
>
> I part with you in peace.

The New Road saw the finale to what had been ending since 1794, with numerous curtain scenes. At this time the Christie household broke up, with Mr Christie's departure for Surinam where he had business interests. There was no further reason for Mary to remain in Finsbury Place, and she moved to lodgings at 1 Cumming Street, off the Pentonville Road. She was now not too far from Godwin's lodgings in Somers Town (a new-built suburb, still incomplete, near the present site of King's Cross), and on 14 April she knocked at his door. In her soft voice she greeted the pale thinker with the prominent forehead and sharp edges who had been put out, but coldly just when she had fired up against him at Joseph Johnson's table. His face was fine, thoughtful, likened by one contemporary to portraits of Locke with a long, elegant nose. He saw an intelligent face at once dreamy and resolute, the left eye very slightly veiled by its lid. There was attentiveness in each glance, and, he was now aware, a reserve of extraordinary passion. She was tall and well proportioned with a fullness of form that was thought 'voluptuous'. Some of her auburn hair escaped from the modest cap worn by married women. It was one of the curiosities of English manners that though Mary never denied her history, and though whispers there were, the façade of a married name and dress did carry her day. Deceptive as this may seem, the deference to custom was obligatory if Mary was to survive with her child in a society which made female sexual conduct (more than violence, exploitation or fraud) central to its moral system.

In appearing at Godwin's door, Mary ignored a convention that forbade a woman to go alone to a man's rooms. He lived at 25 Chalton Street on the wrong (northern) side – the workers' side – of the New Road. Godwin did not take this visit amiss; he admired a woman who 'trusted to the clearness of her spirit for the direction of her conduct'. Next day they met again for tea. Godwin was not drawn to women's rights, but his response to her *Travels*, combined with a habit of enquiry, opened his mind to the novelty of Mary's character. Here was a woman worth knowing, and he not only a competent judge but with a mind so fearless that there was nothing to block her disclosure of what she was. Hardly anything could be rarer than this conjunction.

Godwin's name goes back to Anglo-Saxon England. The Anglo-Saxons called their deity 'God', their word for 'good', as William Godwin did not fail to note, together with the fact that the founder of Saxon policy was called Wodin or Goden. His rooted Englishness contrasts with the shifting, half-Irish Wollstonecrafts. Godwin's traceable forebears surface in the region of Newbury in Berkshire from the late sixteenth century. They were professional men of words, lawyers and divines. When William's father, a Dissenting minister, moved northwards to Wisbech in Cambridgeshire, he retained a relic from the seventeenth century, the barrister's wig belonging to his grandfather's brother. His son, William, would sport it when he acted the Roman statesman Cato in the old barn – his image of Cato in a wig came from the frontispiece to Addison's play of that name. William was born in 1756, the seventh child of thirteen in a family with little in the way of money. His grandfather, still alive during William's childhood, was a minister and scholar; his father, less of a scholar, was devout in a strict, acerbic way – he once rebuked William for picking up the family cat on a Sunday. From the earliest age, William felt a call to follow in the line of ministers, and even before he could write he was delivering sermons from a kitchen chair. Written sermons were his first form of literary composition. His mother took pride in the speed of her retorts and knack for telling a story. Livelier than her husband, and rebuked by his congregation for unsober dress, she was the uneducated granddaughter of Northumberland landowners, called Hull, whose male line had died out.

As a child, William was sickly. From the time the newborn was packed off to a wet-nurse till he caught smallpox at the age of twelve, he was not expected to live, and this may be one reason why his parents conserved their emotions where he was concerned. There was a tragedy two or three years after his birth: a younger brother was drowned while his two eldest brothers were absorbed in flying their kite. Though William was aware he was not a favourite with either parent, it did not trouble him unduly because he was the chosen bedfellow of his father's cousin. Miss Godwin had been a schoolmistress and now lived with the family, probably since the death of William's grandfather, when she came into an inheritance of £40 a year, £16 of which she gave to the Godwins for board and lodging. Godwin's autobiography calls her Mrs Sothren, but she married only later, in 1772. While he was growing up she was a single woman who virtually adopted him as her only child – to the happiness of both. Gentle and loving, she kept the boy close, taught him to read, praised his brains, and did not encourage outdoor activity apart from her enthusiasm for gardening. At a young age William read the whole of the Bible, *The Pilgrim's Progress* and the first children's book, James Janeway's *A Token for Children,* with its image of the dying Godly child surrounded by marvelling spectators whom the child rebukes and exhorts. His words are treasured; his immortality assured. Godwin recalled, 'I felt as if I were willing to die with them if I could with equal success, engage the admiration of my friends and mankind.' The high-minded writer, with a reserve of feeling that women sensed and were drawn to, was what the schoolmistress had backed in Godwin as a boy.

In his teens, he was sent to board with a tutor called Newton in Norfolk. He was beaten, as all boys were, and bore with the vindictive caprice of an inferior mind. He was aware of Newton's limitations, but like many a sensible child of poor parents, resolved to gain what he could from the education on offer. When Newton was away – a lot of the time – his pupil would slip into his library and read of the Greeks' struggle 'for independence against the assaults of the Persian despot'. Words like 'independence' and 'despot' fired him long before the American and French Revolutions. That volume of ancient history, written in the 1730s, awakened, he said, 'a passion in my soul, which will never cease but with life'. The boy would sit close to the shelf so that he could slip the book back into place, should the master return. It did occur to him that he might ask to borrow it, but decided not to put himself in a position of being refused. It was a matter of not allowing a 'despot' to exercise his will, whenever he could prevent it. He would promise himself, 'my mind, my mind shall be the master of me!' – one of those adult children who want to leave childhood behind as soon as they can.

Newton promoted the harsh Calvinist doctrine of the Sandeman sect. Where most Calvinists believed that ninety per cent of people would be damned, Sandemans believed the number was closer to ninety-nine per cent. Later, in reaction, Godwin took up the liberal dogma of Unitarianism. Eventually, he discounted dogma itself. Though he trained at the Hoxton Academy and served as a minister for a few years, he lost his faith and resigned in 1783. At about that date, at the age of twenty-eight, he had a fit. The fits, with one minute's notice – probably epilepsy – were to come back seventeen years later, without impairing his general health.

Godwin began his literary career with biography: he proposed a biographical series for a magazine, but the first, on the former Prime Minister William Pitt, grew into a full-length study. Published in 1783, it was particularly well timed, as the Younger Pitt came to power. Godwin's eventual biography of Mary Wollstonecraft in 1798 has been called his best work; and he turned out a third biography, on Chaucer, in 1803. He's rarely thought of in terms of this genre, but his alertness to the inner life combined with narrative momentum is new, taking biography closer to the inwardness and readability of fiction without sacrificing the advantage of authenticity. Philosophically, Godwin defended individual over general histories as a way to promote the private transformation that, he believed, must precede a change in society: 'the contemplation of illustrious men . . . kindles into flame the hidden fire within us'. For it is necessary, he argues, 'to scrutinise the nature of man, before we can pronounce what it is of which social man is capable'. He is called an 'Anarchist' in politics and 'Jacobin' in fiction, but, like Mary Wollstonecraft, he's an innovator across genres, defying classification. As an educator he backed the unpretentious language that Wollstonecraft upholds in her *Thoughts on the Education of Daughters*, what he termed 'real English' as opposed to rhetorical flourishes. He advised a correspondent to trust 'to the strength & energy of your ideas . . . & do not think they stand in need of an embroidery of fine words to set them off'.

In London his connections multiplied, including some who had crossed Mary Wollstonecraft's track – Dr Price, Paine, Fuseli, Joseph Johnson – but Godwin's circle was wider and more diverse. Where she was grateful for her publisher's support, and loyal to him alone, Godwin worked for an array of publishers, amongst them the firm of the first John Murray which, under his son, would become pre-eminent for writers of the next generation like Scott, Byron and Jane Austen. In 1787, before the young Helen Maria Williams moved to France, she invited Godwin to her salon in London. There, he met the poet Samuel Rogers from Newington Green, and the diarist and letter-writer Mrs Thrale who had taken care of Dr Johnson and was now – to her family's disgust (for marrying an impecunious musician) – Mrs Piozzi. Godwin knew the future American President, John Adams, during his tour of duty in London from 1785 to 1788, his son John Quincy Adams (another future American President), and Wilberforce (leader of the anti-slavery movement). In 1791 Godwin was living in Great Titchfield Street, Marylebone, when Joel Barlow settled there, and they discussed questions of justice. Godwin's autobiographical fragments suggest how small educated society was, and how accessible to a penniless but able man. His leaning towards the Whigs led him into contact with certain members of the aristocracy, including Fox.

Godwin was stirred to his depths by politics. From 1780 he had been a republican, and when the French Revolution came nine years later, his heart, he said, 'beat high with great swelling sentiments of Liberty'. Yet, from the start, he could not condone mob government and violence.

Overnight, it seemed, Godwin became a cult figure with the publication of his *Enquiry Concerning Political Justice* in February 1793. At this time, the poor, who were victims of the Industrial Revolution, were perceived as dangerous rabble. Godwin exposed the self-interest of the ruling class in its deployment of labour, property, law and punishment, and advanced in their place the voluntary redistribution of property and the free exercise of private judgement, especially in condemning the use of force. 'Who shall say how far the whole species might be improved, were they accustomed to despise force in others, and did they refuse to employ it for themselves.' War was justifiable only to repel invasion, not to prevent it, and he believed there would be less talk about a 'justifiable' cause for war if we trained our imaginations to call up the unfeeling carnage which 'justifiable' intended. It's a fallacy, he warned, that our war may be ended by making it more and more terrible: 'a most mistaken way of teaching men to feel that they are brothers by imbuing their minds with perpetual hatred.'

Mary Wollstonecraft's feminist friend Mary Hays was an early convert. Godwin suffered Hays to turn him into a confidant, and she deluged him with letters in which she analysed her unrequited feelings for a Dissenting minister called Frend. 'I am sorry . . . that the nature of my avocations restrain me from entering into regular discussions,' Godwin at length protested. He advised fiction as a consolation, with the result that she used their correspondence when she wrote *The Memoirs of Emma Courtney* (1796), an autobiographical novel in which Godwin appears as the 'understanding' Mr Francis.

Three weeks after he completed *Political Justice*, Godwin began a novel. *Things as They Are; or the Adventures of Caleb Williams* remains a classic, dramatising the miscarriage of justice in a society where moral and psychological issues are too complex for the law, which in any case is the instrument of the squires and the nobility. Williams stumbles on the fact that his refined master has committed a murder. To get Williams out of the way, the master uses his power to trump up a false charge. When Williams is on the run, he finds refuge with thieves, who justify their lives in this way: 'We who are thieves without a licence, are at open war with another set of men, who are thieves according to law.' Godwin was the first to use courtroom drama in fiction.

In the same month as this novel was published, May 1794, twelve leading members of the Society for Constitutional Information and the artisan-based London Corresponding Society were arrested for high treason. If convicted, the punishment was death. Amongst the accused was Godwin's best friend, the dramatist Thomas Holcroft. The government believed that the men on trial presented a republican threat, whereas in fact their aim was the electoral reform that began in the course of the next century. Since Parliament was grossly unrepresentative of the population (including the accused), they were indeed questioning its legitimacy, but there was no infringement of Edward III's Statute 25, which limited high treason to an intent to kill the King or the use of armed force against him. Judge Eyre's opening address to the jury went beyond this in attempting to construct a 'conspiracy to subvert the Monarchy', in effect a new crime for which there was no known statute or precedent. It was, then, an arbitrary attempt on the part of the judiciary to establish a law without recourse to the proper procedures of Parliament.

During the Treason Trials that October, Godwin published – anonymously, for his own safety – a long piece in the *Morning Chronicle*. 'Cursory Strictures on the Charge Delivered by Lord Chief Justice Eyre to the Grand Jury' argued that a wish to reform institutions could not be classed as a crime, for this could not be construed a conspiracy to kill George III. Godwin's method is to invoke the conservatism of law on its own turf, praising Statute 25 as a 'wise and moderate law' that had stood the test of four centuries. Repeatedly attacked by the encroachments of 'tyrannical princes' and the decisions of 'profligate judges', Englishmen had always found it necessary to restore the statute to its original simplicity. Godwin accordingly nails Judge Eyre for exploiting the ambiguity of the word 'force' to imply an armed force, and underpins legal conservatism with Judge Blackstone's commentary on the law of treason as 'a great security to the public' that 'leaves a weighty *momento* to judges to be careful, and not overhasty in letting in treasons by construction or interpretation'.

'Did these [men] plan the murder of the King, and the assassination of the royal family?' Godwin demands. ' Where are the proofs of it?'

Like a lawyer for the defence, he switches from logical rigour to withering eloquence: 'It may be doubted whether, in the whole records of the legal proceedings of England, another instance is to be found, of such wild conjecture . . . and dreams so full of sanguinary and tremendous prophecy.' What he means by sanguinary dreams is reserved for the end, where he turns to address 'you', the accused, in the name of the Lord Chief Justice, were he to speak plainly: You had no warning that your attempts at reform were treason; you went to your beds in the happy conviction that you had acted in accordance with your country's legal code. And for this, 'the Sentence of the Court [but not of the law] is "*That you, and each of you, shall be . . . hanged by the neck, but not until you are dead; you shall be taken down alive, your privy members shall be cut off, and your bowels shall be taken out and burnt before your faces . . .*".' It was largely due to the power of Godwin's pen that the case collapsed.

Afterwards, at a dinner in London, the acquitted philologist Horne Tooke pressed Godwin repeatedly whether it was he who had written 'Cursory Strictures'.

Godwin at last said carelessly, 'I believe it was.'

'Give me your hand,' Tooke said, and when Godwin rose from the table to do so, Tooke put that hand to his lips, saying, 'I can do no less for the hand that saved my life.'

*

A week after Godwin had tea with 'Wolstencraft', his diary notes that 'Imlay calls' on Friday 22 April 1796 and that the following day he, in turn, called on 'Imlay'. It's always assumed that 'Imlay' meant Mary, but since this is the only occasion Godwin uses the name in his diary, it's worth considering whether this visitor could be Gilbert Imlay. It would mean that within a week of Mary's approach to him, Godwin saw Imlay on her behalf, with the support of her friends Mary Hays and Rebecca Christie. Over consecutive days there was intensive contact between these four. Godwin also includes 'Imlay' amongst twelve friends at a dinner at his lodgings, with food brought in from a nearby coffee-house. He later remarked that Mary had come to him in trouble, and that he had not hesitated to help her. Mary's final discussion with Imlay had to do with the practical matter of maintenance for Fanny. She had been too proud to go back on her word and become his dependant, even if, as 'Mrs Imlay', she was entitled to support. On 22 and 23 April 1796 her friends, representing social opinion, may have taken it upon themselves to press Imlay to help her after all. If they did act in this way, it would have been a step towards easing her mind.

Amongst those present at Godwin's dinner-party was the actress, novelist and dramatist Mrs Inchbald, remembered now for *Lovers' Vows*, the play that rouses the wrong heartbeats in *Mansfield Park*. Mrs Inchbald had been a widow from the age of twenty-six. A speech impediment had been a bar to stardom on the stage. She had lived in mean lodgings, worn a shabby gown in the midst of finery, and controlled her attraction to worldly men who would not take an actress for a wife. She did want to marry again, and in the meantime made her way with a combination of charm and prudence, cultivating the innocent air of a milkmaid – the modish form of femininity in the 1780s. She was a beauty skilled at wars of words who chose to smile on Godwin. He liked clever women, and was incapable of consorting with anyone he could not respect.* Godwin did not blame Mary for her unmarried plight. He believed (as she did) that

*In contrast with Mr Bennet in *Pride and Prejudice* or Mr Palmer in *Sense and Sensibility*, sensible men who, in their youth (and to their later discomfort) had succumbed to female silliness.

marriage is 'law and the worst of all laws . . . Marriage is an affair of property, and the worst of all properties.'

To Mary, returning to her country in the repressive aftermath of the Treason Trials, the English seemed 'to have lost the common sense which used to distinguish them'. This was a country at war, cut off from travel, filled with soldiers, and draining the poor who were near starvation. When George III rode through London after opening Parliament on 29 October 1795, watchers hissed and threw stones at his coach with cries of 'Bread!' and 'No war! No war!' An Act of Parliament suspended the law of habeas corpus, and Pitt's Combination Acts outlawed trade unionism. Two Whigs, the playwright Sheridan and Lord Holland, were the only important politicians to oppose these Acts. Fox and fellow-Whigs did move motions in Parliament for the reform of rotten boroughs, but were voted down by great majorities who looked on them as eccentric seditionists in sympathy with France – saved only by the respect the English feel for the well-connected. Spies were rife, men against whom there was no evidence were kept in prison for years, and public meetings were not allowed.

Though Mary tried to bestir herself with thoughts of a return to France or a fresh start in Italy or Switzerland, she remained still locked in depression. Disillusion with Imlay infected her with misanthropy: 'ceasing to esteem him', she realised, 'I have almost learned to hate mankind'. She gives out this dark thought as late as 13 May 1796.

The minister in Godwin responded with measured counsel. Injustice had set Mary, as she put it, 'adrift into an ocean of painful conjectures. I ask impatiently what – and where is truth?' It seemed to her that she had 'ceased to expect kindness or affection' and wished to tear from her heart 'its treacherous sympathies'. Godwin talked directly to the soul without fudging: here are your strengths; here your flaws. There was more to this eloquent creature than to any of the lovelier women in his milieu. Hers was no transient feminine beauty; she carried the permanent stamp of nature. As he drew nearer to this nature he was coming to know, he wanted to know it better. Reason ensured that anything could be communicated so long as it was perfectly true – both lived by the Enlightenment ideal. When Godwin admonished, he did so without rancour – that absence of rancour is extraordinary, above the smiting gods, because the feeling's pure. He told Mary later, 'I found a wounded heart, &, as that heart cast itself upon me, it was my ambition to heal it.' We can't know what else he said, but a letter survives where he counsels another woman in a similar state, waking her to the damage she does herself by trying to break through to a man as dense as rock. He pointed to the 'morbid madness' of the persistently lovelorn Hays when Mary's infection returned – as it did in her low times.

She had meant to slough off women's weakness, and had been confounded by her failure. For all her resilience, she could not recover a sense of purpose that had been undermined by the very person she had chosen to promote it. Godwin was sure that no one, especially the great of soul, finds it easy to evade castle-builders of the Imlay sort. 'The whole scene of human life may at least be pronounced a delusion!' he said philosophically. Mary must accept that she had made the mistake of 'imputing to [Imlay] qualities which, in the trial, proved to be imaginary'. In this clarifying way, he addressed that part of her that still clung to delusion, hoping against hope that Imlay might be restored to the man she had believed him to be.

Godwin took an altogether tougher line on suicide: he wanted her to see how 'by insensible degrees' she could come to stake her life upon the consequences of her error. Error, he said. Not love. If love can be recategorised as 'error', it will shrink to nonentity 'when touched by the wand of truth'. It was irrational, Godwin argued, to consign herself 'to premature destruction' for the sake of a man 'so foreign to the true end' of her cultivation – a cultivation 'so pregnant . . . with pleasure' to herself 'and gratification to others'. She was 'formed to adorn society' and, through her books, 'to delight, instruct, and reform mankind'. Godwin's respect reflected her able self, while his cool (though not cynical) acceptance of human nature relieved her of the self-hatred that is the most intractable part of depression. He said that no one would kill herself if she could believe, as it often proved, that years of enjoyment lay ahead. A disappointed woman should try to construct happiness 'out of a set of materials within your reach'.

Over the next few months their friendship grew, in Godwin's words, 'by almost imperceptible degrees'. At first, they met about once a fortnight. 'Dined at H[olcroft]'s with Wolstencraft,' Godwin's diary records on 15 May, and again, 'sup at Wolstencraft's' on the 28th. He was ready to offer a response that could meet the risk she took in being true, and so restore her trust in mankind. 'Nor was she deceived,' he said with justifiable pride. She was not the only one to gain his help, but with Mary alone Godwin did something utterly uncharacteristic. This matter-of-fact man wrote a poem.

Her tart reply is the first sign of release from depression: it marks a return of the humour that had failed since she began to suspect Imlay's infidelities. The poem, which does not survive, must have been a set-piece, for it reminded Mary of a couplet from Samuel Butler: 'Shee that with *Poetry* is won/ Is but a *Desk* to write upon.' In such love poems a woman serves as a prop for rhetorical extravagance. Mary had a cure of her own to propose: Godwin should sensitise himself to the play of character in genuine passion, as in Rousseau's novel *Julie, ou La Nouvelle Héloïse* (1761):

> July 1, 1796
>
> I send you the last volume of 'Heloïse,' because, if you have it not, you may chance to wish for it. You may perceive by this remark that I do not give you credit for as much philosophy as our friend [Rousseau], and I want besides to remind you, when you write to me in *verse*, not to choose the easiest task, my perfections, but to dwell on your own feelings – that is to say, give me a bird's-eye view of your heart. Do not make me a desk 'to write upon,' I humbly pray – unless you honestly acknowledge yourself *bewitch'd*.
>
> Of that I shall judge by the style in which the eulogiums flow, for I think I have observed that you compliment without rhyme or reason, when you are almost at a loss what to say.

On the same day as Mary laughed off the poem, Godwin arrived to say goodbye before leaving for three weeks to visit his mother and his friend Dr Alderson in Norwich. From there, he affects to have a go at heartfelt words: 'Now, I take all my Gods to witness . . . – but I obtest & obsecrate them all – that your company infinitely delights me, that I love your imagination, your delicate epicurism, the malicious leer of your eye [with slight paralysis of the lid], in short every thing that constitutes the bewitching tout ensemble of the celebrated Mary.' Her spontaneity seems to draw out a playfulness, untapped till now by Godwin's ministerial *gravitas*. He continues to send up romance: 'Shall I write a love letter? May Lucifer fly away with me if I do! No, when I make love, it shall be with the eloquent tones of my voice, with dying accents, with speaking glances (through the glass of my spectacles) . . .'

He also challenged her attachment to an increasingly militarised France. 'Shall I write to citizenness Wolstencraft a congratulatory epistle upon the victories of Buonaparti?' If it would rejoice the cockles of her heart he was prepared to pass in review before her the art treasures, like Raphael's *St Cecilia*, which 'that ferocious freebooter' had recently stolen from Italy.

By now, Mary had given up all thought of returning to the Continent. She took her furniture out of storage and moved to 16 Judd Place West (situated at what is now the front gate to the British Library, and around the corner from Godwin's lodgings). No mean inducement was his offer to look at her work. Since Mary chose to write 'for independence', and since she had to support Fanny when maintenance from Imlay failed to appear, she did have bits and pieces of work in progress. In January she had put together a stage play based on her experiences in Paris – a comedy, of all things – which she had offered to managers without success. Godwin had read it on 2 June – without encouragement. As summer came on, she was making stabs at a second novel, on the wrongs of women. During July, she reworked her manuscript and began to seek out friendships in Godwin's circle, a rather belated encounter with sophisticated women of the capital. She dined with Godwin's recent acquaintance Mary ('Perdita') Robinson, once an actress and mistress of the Prince of Wales, now a writer whom Mary had reviewed. She dined also with Sarah Siddons, the foremost actress of the day (famous for her performance as Lady Macbeth), who declared that no one could have read Wollstonecraft's *Travels* with 'more reciprocity of feeling, or more deeply impressed with admiration of the writer's extraordinary powers'. Mary, in turn, admired the 'dignified delicacy' of Mrs

Siddons in Nicholas Rowe's tragedy, *The Fair Penitent.* Calista's line about hearts that were 'joined not matched' spoke to her own sufferings with Imlay.*

Then there was Dr Alderson's daughter, Amelia. Before they met, Amelia had let her know that 'as soon as I read your letters from Norway, the cold awe which the philosopher has excited, was lost in the tender sympathy called forth by the woman. I saw nothing but the interesting creature of feeling and imagination.' After they met, Amelia declared that whatever she had seen had always disappointed her 'except Mrs Imlay and the Cumberland lakes'. Mrs Imlay appeared to her another Cleopatra, reminiscent of her 'Princess' character with George Blood. Until this time, Mary's ties with intelligent women had been intense but few: her sisters; Jane Arden in Yorkshire; Fanny Blood; her pupil Margaret King; and more recently Ruth Barlow.

Though Mary's friendship with the Barlows had appeared to end during their period in Hamburg, they, at least, did not forget her. In the summer of 1796 Joel Barlow was acting as American agent in Algiers – in his words, an 'abominable sink of wickedness, pestilence and folly' – when the plague broke out. With victims dropping on every side, on 8 July he warned Ruth, far away in rue du Bac in Paris, that he had to expose himself to the disease if he was to rescue American citizens and board as many as possible on ships leaving the country. In case of his death, she was to know that he was leaving her what he considered their joint property of about 120,000 dollars (equivalent to about three and a half million today), with bequests to be made at her discretion. The most interesting part of this letter is his concern for Mary Wollstonecraft, who was the only non-American and the only non-member of the Barlow family (apart from the loyal Blackdens) whom Barlow thought of when death came close:

---

*Mrs Siddons (1755–1831) was the daughter of the stage manager Roger Kemble. Her brother was John Kemble who took the lead in the dramatisation of *Caleb Williams* which opened on 12 March 1796. Mrs Siddons performed in *The Fair Penitent* on 22 and 23 November 1796. Mary Wollstonecraft recalls Calista's line in *The Wrongs of Woman*, ch. 9.

> Mary Woolstonecraft, – poor girl! You know her worth, her virtues and her talents; and I am sure you will not fail to keep yourself informed of her circumstances. She has friends, or at least *had* them, more able than you will be to yield her assistance in case of need. But they may forsake her for reasons which to your enlightened and benevolent mind would rather be an additional inducement to contribute to her happiness.

We have seen a Barlow whom Mary had sized up in London as one 'devoured by ambition'; and we have seen the confidence man of the Scioto scam; and possibly Mary's unidentified 'knave' in the business of the silver ship. Yet Joel Barlow can rise to a hero's action, and risk his life when rescue is needed. His duality mirrors Imlay who can save Mary Wollstonecraft from the Terror and yet be the same person to dodge inconvenient obligations. Barlow, though, remains visible, while Imlay manages to vanish, leaving no trail – only creditors. Frustration and rage are left to fill up the place where he has been. His voice leaves no sound; his name is rarely on paper. His associates – including Barlow and Mary Wollstonecraft – sustain his secrecy, as when Barlow tells Ruth that Mary has a sweetheart. Both times when Barlow puts this on paper, he gives no name: the man is 'of Kentucky'. The frontier cover must suffice. Despite their business partnership and network of the same contacts, there is no mention of Imlay in Barlow's papers – no more than a note at the end of 1795 to the effect that Imlay had marked out seven Parisian cafés. For what purpose? This still remains unknown, but much about Imlay, especially his unshakable sense of himself as a man of 'principle', could fall into place were he part of a new secret service, run by the President himself.

Barlow, too, may have played a part for the man who sent him to Algiers and his immediate boss during his tour of duty there was Colonel David Humphreys, a Yale buddy, once a fellow-member of the Connecticut Wits, then appointed as special secret agent to Europe in 1790. Outwardly he served as a US representative in Portugal, then as American Minister in Spain. Can it be relevant that September 1795 sees him operating in Le Havre? While Wollstonecraft was advancing into the heart of the 'fraud' she uncovered in Hamburg and refusing Imlay's further instructions, Barlow, in Paris, was conducting a cryptic correspondence with Humphreys. They dwell on an 'object' that, throughout, remains unnamed, and Humphreys is reminded to 'instruct' Barlow before leaving Le Havre. Barlow jots down a memo about Imlay's seven cafés in Paris – a rare mention of Imlay by name. A sea captain called O'Brien roves between Algiers, Lisbon, London and the seat of government in Philadelphia. There, O'Brien conveys the situation in Algiers to no less a person than the President.

Washington was prepared to draw on his massive Intelligence fund in order to ransom American captives in Algiers, while Barlow negotiated with the capricious Dey, Hassan Bashaw, who maintained Turkish rule over Moors and Jews and harboured pirates to whom nations paid annual tribute – bar Britain with its formidable navy. So long as the American states had been colonies, they had come under the protection of the British navy, but in the 1790s US trade in the Mediterranean was disrupted, and numbers of American sailors enslaved. Barlow had the help of a Jew called Baccri who understood how to proceed – the tactic was to avert the Dey's threats, before they were made, by tickling his greed with gifts and promises. Barlow promised an American frigate, and then had to cope with a gathering storm when delivery was – deliberately – delayed. Master-spy Humphreys let Barlow know that Washington wished him to extend his stay in North Africa. Barlow, who had left Paris and Ruth at the end of December 1795, would not return until October 1797.

He saw this as a 'disinterested' sacrifice for his country. Once he had a fortune on the go, he followed a wish to rise into public office out of the murk of commerce. This he did by way of Algiers, writing grandly of 'the interest of the United States'. All the same, he was shifting cargoes of his own in the Mediterranean, and it was convenient to have his country pay eighty-three thousand dollars a year to the Pasha of Tripoli for US merchant ships to be protected from the pirates of Algiers. This treaty Barlow secured in 1796. A darker deed was to deliver arms – two thousand bombshells and a thousand quintals of gunpowder – to the unprincipled bully of a Dey. Thanks to Barlow, US trade was to flourish on the backs of subject peoples. In one letter to the Secretary of State, he was shameless enough to quote a common saying in the Levant that 'no honest man goes to Algiers'. He can quote this with humorous ease because, at this moment, Barlow moves from dark tricks to the good deeds on his mind in his will-letter to Ruth. If he does not die in the epidemic, he will put his fortune to the service of his country and others in need. It's the standard sequel to sordid gains to purify them with charity, and I suspect that Barlow's recollection of Mary Wollstonecraft at this time in his life – when he draws close to mortality – has to do with his own need for moral gesture.

Barlow did not die in the foetid summer of 1796. But it was an occasion for trying out a farewell aria to Ruth. 'I have the wife that my youth has chosen and my advancing age has cherished . . . from whom all my joys have risen, in whom all my hopes are centred . . . If you should see me no more, my dearest friend, you will not forget I loved you.' Even more gracefully, he accepts that another man could take his place.

> It is for the living, not the dead, to be rendered happy by the sweetness of your temper, the purity of your heart, your exalted sentiments, your cultivated spirit, your undivided love. Happy man of your choice! should he know and prize the treasure of such a wife! Oh, treat her tenderly, my dear Sir; she is used to nothing but unbounded love and confidence. She is all that any reasonable man could desire; she is more than I merited, or perhaps you can merit. My resigning her to your charge, though but the result of uncontrollable necessity, is done with a degree of cheerfulness, – a cheerfulness inspired by the hope that her happiness will be the object of your care, and the long continued fruit of your affection.
>
> Farewell, my wife; and though I am not used to subscribe my letters addressed to you . . . it seems proper that the last character that this hand shall trace for your perusal should compose the name of your most faithful, most affectionate and most grateful husband,
>
> *Joel Barlow*

Barlow always projects this image of married love, but as we know, his wife, unlike Mary Wollstonecraft, forced herself to condone infidelities. She wondered on one occasion if he had a '*black* sweetheart'. Ruth declined into invalidism, retreating for long spells to spas, while Barlow sent regular, ardent letters – being (like Imlay) too busy to visit the dear sufferer. Once Barlow had a nightmare where he stands at Ruth's bedside gazing at her wasted form – yet still he does not come. Instead, he offers her his dream. His concern steams off the page. It's not surprising that Ruth stayed wilting in those heated epistolary arms. Imlay too was a master of assurance: his affirmations left Mary Wollstonecraft sick in mind for a year and a half. Only Godwin helped her.

What pleased Mary most in Godwin's letter from Norwich was an announcement that he was coming home, 'to depart no more'. He was expected on 20 July, and when he did not appear, Mary found herself 'out of humour'. The following day she left her manuscript ('as requested') at his lodgings, together with a note, signed familiarly with her first name. 'I mean to bottle up my kindness, unless something in your countenance, when I do see you, should make the cork fly out – whether I will or not.'

As it happened, he wasn't back. When he did return, on the 24th, he came ready to renew what was turning into something he had not known before. For his placid face and coldness of manner didn't exclude a lurking responsiveness, keen of eye behind his spectacles. Hazlitt said that Godwin had 'less of the appearance of a man of genius, than any one who has given such decided and ample proofs of it'. Though he was angular and near-sighted, discerning women were drawn to him. A cosmopolitan beauty, Mrs Reveley, daughter of an architect, an unconventional wanderer who had brought her up on his own in Constantinople, fell in love with Godwin and let him know it. One day at Greenwich in 1795 she and Godwin overstepped the mark. It was Godwin who stopped. He was too responsible to deviate from the laws of society in a way that might injure a wife. Mary, unaware of this particular attraction, could not hide her antagonism to the hovering Mrs Inchbald – 'Mrs Perfection', as Godwin called her after she suggested amendments to his novel, saying, 'I have not patience that anything so near perfection should not be perfection.' Godwin enjoyed his role as her favoured escort.

'I suppose you mean to drink tea with me, *one* of these day[s],' Mary grumbled on 2 August. 'How can you find in your heart to let me pass so many evenings alone . . . I did not wish to see you this evening, because you have been dining, I suppose, with Mrs Perfection, and comparison[s] are odious.'

Amelia Alderson was yet another. Mary spent the next evening with her, and was tantalised to hear that Godwin had made overtures to this lovely writer ten years younger than herself. In this case, however, there was no jealousy, for Amelia had turned him off.

'You, I'm told, were ready to devour her – in your little parlor,' Mary accosted Godwin next day. '*Elle est très jolie – n'est[-ce] pas?*'

It was a jolt, yet in a way reassuring. Godwin was not, then, impervious to women. She and Amelia put their heads together over the question of whether he had ever 'kissed a maiden fair'. Mary asked him to answer only after an hour-long visit from Amelia, intended to try his defences. Could a reserved man of middle age, whose beliefs had distanced him from sexual bonds, change his habits, as in Boccaccio's tale of Cymon, an uncouth man transformed by love?

She put this to Godwin . 'I was making a question yesterday, as I talked to myself, whether Cymons of forty could be *informed* – Perhaps, after last night's electrical shock, you can resolve me – .'

The clear-cut Clarissa voice of her relations with Imlay is no more. A more nuanced voice falls into hiatus, a listening silence or invitation – not to sex in the usual sense, but to some preliminary '*informed*' state in a wordless space between the sexes. Sentences are incomplete, and silence is a sort of language – the meaning silence of a creature in the process of evolving new modes of communication with a man selected for alertness.

In order that Godwin might know her more fully, she loaned him *Mary; A Fiction*, and teased him to discern her secret self. Godwin read it on 3 and 4 August. Could he recognise her as a philosopher in her own cause, in this sense his counterpart? Could he *see* a woman so near to his eminence? 'I called on you yesterday, in my way to dinner, not for Mary [the novel] – but to *bring* Mary – Is it necessary to tell your sapient Philosophership that I mean MYSELF –'

Two days later, on Saturday 13 August, they discussed their position. 'It was friendship melting into love', Godwin said later. 'When, in the course of things, the disclosure came, there was nothing in a manner for either party to disclose to the other.' That day there appears to have been some rational agreement to be patient and 'considerate' with temperaments so at variance: Mary active and eager; Godwin sedentary and cautious. When they met again the following Monday in Godwin's lodgings, Mary laid her head on his shoulder. She recreates this scene in her new novel: 'there was a sacredness in her dignified manner of reclining her glowing face on his shoulder, that powerfully impressed him. Desire was lost in more ineffable emotions . . .' Darnford, the lover, feels at this moment that to make Maria happy could be the foremost duty of his life. Maria is scarred by 'recollected disappointment'. Her fear of outrunning this new affection gives way, now and then, to 'the playful emotions of a heart just loosened from the frozen bond of grief'. All we can say for certain is that this was Mary Wollstonecraft's version of what seems to have been an emotional exchange. Next morning she sends to ask if Godwin had felt '*very* lonely' in the night, and invites him to kiss Fanny when she and Marguerite deliver this letter. Mary's tone is confidently hopeful.

His reply strikes an unexpected note: mentor and humorist have turned away. He is wary, even prickly. He had been unwell, he claims, and she had mortified him – unintentionally, he concedes. Godwin was clearly not ready as yet to act as lovers, as they tried that Tuesday night, 16 August 1796, recorded in his diary: 'chez moi'. It went badly wrong.

Afterwards, Godwin admitted to Mary that during the last few days he had been wrapped up in a private fantasy. 'I longed inexpressibly to have you in my arms,' he confessed, but this fantasy had been self-contained. So her offer to stay the night had taken him by surprise. To men who would expect to choose the moment, and particularly to one of Godwin's biblical training, no respectable woman would take the initiative. Yet reason told him that Mary was more than respectable; desire, in her terms, was 'pure', even 'sacred'. That they could love, yet be so awkward, left them shaken. For once, Mary could not speak. She departed in silence, and sat in Judd Place West, brooding over her breakfast, unable to eat. She even shut the door on Fanny.

Could she have been responsible for an act against Godwin's better judgement? He had appeared emotionally withdrawn, unaware of the risk she had taken in trying to make love in the aftermath of Imlay's desertion. Godwin's theory of sex could not have been further from her experience: he thought we should be 'wise enough to consider the sensual intercourse a very trivial object', and cohabitation a mistake – bound to blunt what feelings there were. Unlike Mary, he was anti-domestic, with the self-protectiveness of the bachelor. All the same, he was not quite as logical in practice as in thought. Logic dictated that if sex were as trivial as he believed, then partners might change with little trouble. In practice there was fidelity in all Godwin's relationships and, given his respect for women, it would be natural for him to extend fidelity to 'the sensual intercourse'. There was another discrepancy between theory and actuality: although, theoretically, sex was trivial and marriage out of the question, in actuality Godwin was shaken to find himself in the grip of fantasy, as Mary's warm cheeks and rounded arms came into focus. He was not a man of impulse; his mind led him, and it worked with a measured deliberation. Mary, who was spontaneous, did not fathom his uneasiness when feelings threatened his control.

'Full of your own feelings, little as I comprehend them, you forgot mine – or do not understand my character,' she wrote to him that Wednesday morning. What she feared most was to have lost Godwin's respect: 'Mortified and humbled, I scarcely know why – still, despising false delicacy I almost fear that I have lost sight of the true. Could a wish have transported me to France or Italy, last night, I should have caught up my Fanny and been off in a twinkle, though convinced it is my mind, not the place, which requires changing. My imagination is for ever betraying me into fresh misery . . . I am hurt – But I mean not to hurt you. Consider what has passed as a fever of your imagination . . .' For her own part, she would follow Rousseau's *Solitary Walker*, and take her way through the world alone.

'Do not hate me, Godwin asked. 'Do not cast me off. Do not become again a *solitary walker* . . . You have the feelings of nature, and you have the honesty to avow them.'

He reproached himself that during the many hours he had longed to hold her in his arms, he had made no move. 'Why did I not come to you? I am a fool. I feared still that I might be deceiving myself as to your feelings.' Then, too, he had fantasised a different drama, possibly closer to the conquest he had spoofed in his letter from Norwich: 'When I make love, it shall be in a storm, as Jupiter made love to Semele, & turned her at once to a cinder.' His wish to 'terrify' was misplaced. Mary looked for tenderness and Godwin was, thankfully, a sensitive man.

'At one point we sympathize,' he wrote, 'I had rather at this moment talk to you on paper . . . I should feel ashamed in seeing you.'

Shame did not prevent a brilliant answer to sexual failure that turns it around. He reveals his own character; sees into hers; reassures her; lays out the ground for compatibility; and commands her recovery:

> You don't know how honest I am. I swear to you that I told you nothing but the strict & literal truth, when I described to you the manner in which you set my imagination on fire on Saturday. For six & thirty hours I could think of nothing else . . .
>
> Like any other man, I can speak only of what I know. But this I can boldly affirm, that nothing that I have seen in you would in the slightest degree authorise the opinion, that, *in despising the false delicacy, you have lost sight of the true.* I see nothing in you but what I respect & adore.
>
> I know the acuteness of your feelings, & there is perhaps nothing upon earth that would give me so pungent a remorse, as to add to your unhappiness.
>
> Do not hate me. Indeed I do not deserve it. Do not cast me off. Do not become again a *solitary walker.* Be just to me, & then, though you will discover in me much that is foolish and censurable, yet a woman of your understanding will still regard me with partiality.
>
> Upon consideration I find in you one fault, & but one. You have the

feelings of nature, & you have the honesty to avow them. In all this you do well. I am sure you do. But do not let them tyrannise over you. Estimate everything at its just value. It is best that we should be friends in every sense of the word; but in the mean time let us be friends.

Suffer me to see you. Let us leave every thing else to its own course. My imagination is not dead, I suppose, though it sleeps. But, be it as it will, I will torment you no more. I will be your friend, the friend of your mind, the admirer of your excellencies. All else I commit to the disposition of futurity, glad, if completely happy; passive & silent in this respect, while I am not so.

Be happy. Resolve to be happy. You deserve to be so. Every thing that interferes with it, is weakness & wandering; & a woman, like you, can, must, shall, shake it off . . . Call up, with firmness, the energies, which, I am sure, you so eminently possess.

Send me word that I may call on you in a day or two. Do you not see, while I exhort you to be a philosopher, how painfully acute are my own feelings? I need some soothing, though I cannot ask it from you.

Godwin left this letter at one o'clock. Mary replied within an hour:

I like your last – may I call it a *love* letter? . . . It has calmed my mind – a mind that had been painfully active all the morning, haunted by old sorrows that seemed to come forward with new force to sharpen the present anguish – Well! well – it is almost gone – I mean all my unreasonable fears – and a whole train of tormentors, which you have routed – I can scarcely describe to you their ugly shapes so quickly do they vanish – and let them go, we will not bring them back by talking of them . . .

One word of my ONLY fault – our imaginations have been rather differently employed . . . My affections have been more exercised than yours, I believe, and my senses are quick, without the aid of fancy – yet tenderness always prevails, which inclines me to be angry with myself, when I do not animate and please those I love.

Though she finished the letter at two, she sat on it for another hour or two. Meanwhile, at three, Godwin 'suddenly became awake' to the mistake he had made in response to her offer: 'I was altogether stupid & without intelligence as to your plan of staying, which it was morally impossible should not have given life to the dead.' It was not, he explained, proof of indifference to her, but of thoughts 'obstinately occupied' and intent upon 'an idea I had formed in my own mind of furtive pleasure'. He was struck by the absurdity of his oblivion, and wondered if it was 'too late to repair it'.

He was sinking into 'self-abhorrence' when Mary suddenly appeared at his door. In response to his 'Suffer me to see you', she had come in person to deliver her letter, which closed with an invitation to dine that day at half past four. It proved now possible to face each other. After she left, Godwin decided to pass on his analysis of their sexual differences because he wanted her to know him for what he was; on the other hand, he wished to secure her calmed state of mind. 'Take no notice of it for the present,' he added at the bottom.

When he arrived to dine, Hays was visiting but left, as she always did, before dinner. Godwin and Mary Wollstonecraft dined alone after the day's exchanges, and parted for the night.

By the following night, Thursday, they were emboldened to try again, this time at Mary's lodgings – 'chez elle', in Godwin's code – and again, sex was not a success. Godwin was 'tortured' by Marguerite's presence in the next room. He was unused to assignations. What would the nurse make of this? Would she give him away? Mary, for her part, was disconcerted again by the 'hoar frost' of Godwin's reserve – the phrase came to her with a fable she made up the morning after, as she walked with Fanny before breakfast. She pictures herself as a Sycamore tree, denuded of leaves, amongst a cluster of evergreens. The Sycamore envies the foliage that shelters other trees from the blasts of winter. She craves sun, and at every sign, pesters her neighbours to know if spring has come. One day in February, the sun appears and the Sycamore's sap mounts. She holds out one more day, and the sun darts forth again. Feeling it is spring, the tree's buds 'immediately come forth to revel in existence'. But alas, next day 'a hoar frost . . . shriveled up [the Sycamore's] unfolding leaves, changing in a moment the colour of the living green – a brown, melancholy hue succeeded – and the Sycamore drooped, abashed', for she had mistaken February for April. The fable ends with the dashes common in Mary's communications to Godwin: 'whether the buds recovered, and expanded, when the spring actually arrived – The Fable sayeth not – .'

She sent this to Godwin that same day, Friday19 August. It was a simple plea for warmth, but Godwin could not or would not 'read' it. His bristles rose.

'I have no answer to make to your fable, which I acknowledge to be uncommonly ingenious & well composed. I see not however its application,' he told her, and it 'puts an end to all my hopes. I needed soothing, & you threaten me. Oppressed with a diffidence & uncertainty which I hate, you . . . annihilate me. Use your pleasure. For every pain I have undesignedly given you, I am most sincerely grieved; for the good qualities I discern in you, you shall live for ever embalmed in my memory.'

As it happened, they had grown too necessary to each other to stop now. After they made up, they tried again 'chez elle', and yet again on the 21st, more effectively for Godwin – 'chez elle toute' – though less so for Mary. This time, it was her turn to lose confidence: 'I am sometimes painfully humble,' she told him next day. 'Write me a line, just to assure me, that you have been thinking of me with affection . . .'

'Humble!' Godwin exploded, 'for heaven's sake, be proud, be arrogant! You are – but I cannot tell what you are. I cannot yet find the circumstance about you that allies you to the frailty of our nature. I will hunt it out.'

When he was to dine with Mrs Perfection, she reminded him where his 'fealty' was due. Should there be 'a possible *accident* with the most delightful woman in the world', she warns, 'take care not to look over your left shoulder – I shall be there –'

'I shall report my fealty this evening', Godwin promised.

In uncertain weather, when the wind still whistled through Mary's branches, there was the shared haven of work. She passed Godwin her manuscript – her next chapter of *The Wrongs of Woman* – or asked him to bring a revision of *The Iron Chest* (a dramatisation of *Caleb Williams*) to read aloud at her fire: ' – and we shall be so snug – yet, you are such a kind creature, that I am afraid to express a preference, lest you should think of pleasing me rather than yourself – and is it not the same thing?– for I am never so well pleased with myself, as when I please you – I am not sure please is the exact word to explain my sentiments – May I trust you to search in your own heart for the proper one?'

'Your proposal meets with the wish of my heart,' Godwin returned. It's not quite the language of love, but at least he picks up the words. How delicately she moves from sure ground to that frontier of language. And how disconcerting that a lover like Imlay should cross that frontier with ease, while an honest man like Godwin could not find it in him to speak his heart. Mary repeated her failed message to Imlay: 'I would describe one of those moments when the senses are exactly tuned by the rising tenderness of the heart.'

So she searched for words, making Godwin her co-searcher in the space between the sexes. Her tenderness for him might vent itself in a 'voluptuous sigh', yet 'voluptuous' was often expressive of a meaning she did not intend. 'I am overflowing with the kindest sympathy – I wish I may find you at home when I carry this letter to drop it in the box, – that I may drop a kiss with it into your heart, to be embalmed, till we meet, *closer* – .' She draws a line through 'closer', and adds: 'Don't read the last word – I charge you!' Words are at the centre of what was happening: her efforts to find words that could quicken desire without narrowing its capacity to generate 'closer' affections.

'I shall come to you to night, probably before nine – May I ask you to be at home,' she wrote. 'Should I be later – you will forgive me – It will not be my heart that will loiter – bye the bye – I do not tell *any* body – especially yourself – it is always on my lips at your door – .' Since Godwin is the person she can tell least of all, this is not about secrecy as some suppose; again, it's about what is unsaid between them. Repeatedly, she leads this master of words to the brink of a language he has yet to learn. 'Our *sober* evening was very delicious,' she said once when she was down with a cold, ' – I do believe you love me better than you imagined you should – as for me – judge for yourself – .' There are scenes to be acted; they stir at his door.

Godwin's unpreparedness for her drama went back to the austere Calvinism he had followed as a minister. His father had been an exceptionally silent man who could sit through a social event without uttering; and Godwin himself could scarcely begin a conversation where there was no set topic.

Mary asked if he could be 'gay' without effort.

No, not without effort, Godwin said, reminding her to bring Latin books for her lesson.

'I will tell you why you damped my spirits, last night, in spite of all my efforts,' she said next day. Could 'Man' (as Godwin called himself, mimicking little Fanny) be afraid of her? she wondered, rather hurt.

So effort sometimes replaced spontaneity. And all the while convention would have militated against unacted desires, a convention policed by the small-minded like Francis Twiss, a gaunt Shakespeare scholar (brother-in-law to John Kemble and Sarah Siddons) whose talk, Mary thought, expanded the gender gap, mocking women and bringing out the satyr in men.

Godwin could later own that he did lack 'an intuitive sense of the pleasures of the imagination. Perhaps I feel them as vividly as most men,' he added, 'but it is often rather by an attentive consideration', and 'liable to fail . . . in the first experiment', guarded by caution in case he should be deceived. He saw this as a gender trait, in contrast with Mary's quick warmth. It surprised him that 'her fearless and unstudied veracity' should prove sound. Such a companion, he said, 'excites and animates'. Godwin always remembered 3 September when Mary was present at Kemble's, and conversation became heated on the subject of love. Other celebrities took part, Sheridan (the playwright), Curran (the Dublin lawyer) and the ubiquitous Mrs Inchbald.

Truncated nights called for a two-way passage of notes and letters. If she had written intensively to Imlay (seventy-six letters) because they lived mostly apart, she wrote more intensively to Godwin (146 letters over a briefer span) even though they lodged around the corner. Their very proximity allowed for up to three exchanges a day dropping through their doors, responses coming off the pulse with the speed of emails. It's a remarkable record of intimate conversation two hundred years ago, allowing us to eavesdrop on the past. A jubilant communication from Mary to Godwin on 13 September indicates when their real union took place:

> Let me assure you that you are not only in my heart, but my veins, this morning. I turn from you half abashed – yet you haunt me, and some look, word or touch thrills through my whole frame – yes, at the very moment when I am labouring to think of something, if not somebody, else. Get ye gone Intruder! Though I am forced to add dear – which is a call back –
>
> When the heart and reason accord there is no flying from voluptuous sensations, I find, do what a woman can – Can a philosopher do more?

Godwin records their making love every night. Though Mary had brought him 'a wounded and sick heart', it was not, he insists, 'a heart querulous, and ruined in its taste for pleasure. No; her whole character seemed to change with a change of fortune. Her sorrows, the depression of her spirits, were forgotten, and she assumed all the simplicity and vivacity of a youthful mind.' His words recall the start of Mary's attachment to Imlay, and the first readers of his *Memoirs* were shocked by the keenness with which Godwin appears to register the transforming effect of her love for another man. Those first readers ridiculed Godwin as a sort of cuckold. More astute readers of the present day have realised that he was actually recording what he himself had witnessed as she came back to life. 'She was like a serpent upon a rock, that cast its slough, and appears again with brilliancy, the sleekness, and the elastic activity of its happiest age. She was playful, full of confidence, kindness and sympathy. Her eyes assumed a new lustre, and her cheeks new colour and smoothness. Her voice became chearful, her temper overflowing with universal kindness, and that smile of bewitching tenderness from day to day illuminated her countenance, which all who knew her will so well recollect, and which won, both heart and soul, the affection of almost everyone that beheld it.'

This glow of health can be seen in the portrait of Mary Wollstonecraft painted during her association with Godwin, and now in the National Portrait Gallery. It's a strong rather than pretty face, with a bloom on cheeks and lips. They are well cut, dented lips, the lower lip a little fuller. Her long eyebrows disappear beneath her hair, short and wavy about her face; and her shapely nose is too long for beauty. Her expression is calm and contained.

It's usual for historians to confirm the happiness of Mary Wollstonecraft and Godwin. They are right in general, but day-to-day living can't sustain the finality of romance. For one thing, these were two people with the capacity to go on developing. For another, they were committed to candour. Since they lived apart, their intimate mutters to each other open up the differences that pass unnoted – like conversation – in the lives of most pairs.

Once, when she is at the theatre in a poor seat Godwin has obtained for her, she resents the sight of him in a better seat next to Mrs Inchbald. 'You and Mrs. I were at your ease enjoying yourselves – while, poor I! – .' Ruefully, she asks if she might 'spend my spite'.

One Friday, Godwin complains to Mary, 'You spoil little attentions by anticipating them.'

'Yet to have attention, I find, that it is necessary to demand it.' Mary broods over this, when he doesn't turn up on the Sunday as arranged. 'Your coming would not have been worth anything, if it must be requested.'

Another time, when she is ill in bed, she resents his wish to go out instead of staying with her – and blames herself, too, for not letting him go. We overhear them on an ordinary day – a quiet Wednesday, 21 September 1796.

'Though I am not quite satisfied with myself, for acting like such a mere Girl yesterday – yet I am better,' Mary says. 'Say only that we are friends; and, within an hour or two, the hour when I may expect to see you – I shall be wise and demure – never fear – and you must not leave the philosopher behind – .'

'Friends? Why not? If I thought otherwise, I should be miserable,' Godwin replies. ' In the evening expect me at nine, or a little before.'

There are times when she is put out by his 'false interpretations', and especially his defensiveness when she broaches women's needs. If she opposed him, she wouldn't jest, she says. Could Godwin the 'profound Grammarian', master of the comparative – as in good, better, best – allow her a private *bill of rights* to a *comparative* freedom to be herself, to be playful and even 'frolicksome', once a year – or when the whim seized her 'of skipping out of bounds'?

Godwin agrees by return: 'I can send you a bill of rights . . . *carte blanche* . . . But to fulfil the terms of your note, you must send me a bill of understanding. How can I always distinguish between your jest & earnest, & know when your satire means too much & when it means nothing but I will try.'

His role as her mentor was changing. He felt freer to be severe in the manner of an educated man reproving an untrained woman. At the same time he allowed Mary to see drafts of his essays on education and literature (published in 1797 as *The Enquirer*). She wished him to recommend day schools in preference to boarding-schools, which deprived children of domestic affections – to her, 'the foundation of virtue'. Godwin was not won over – domesticity had never figured on his agenda – but he made a rather grudging effort to oblige her by raising the issue in a noncommittal paragraph at the end of the essay on 'Public and Private Education'.

Her interest was fanned by reading him, she found, 'while other recollections were all alive in my heart – .' Her willingness to lend herself to his writing contrasts with Godwin's criticisms of her own manuscript – *The Wrongs of Woman*. She had sent it to him six days earlier, in need of 'encouragement'. His opinion reduced her to 'painful diffidence'. On this occasion, Godwin could not disguise his irritability with something in her grammar that appeared to him a fundamental flaw. He offered her a grammar lesson. Mary, restraining an impulse to throw down her pen, defended herself.

'I am compelled to think that there is some thing in my writings more valuable, than in the productions of some people on whom you bestow warm eulogiums – I mean more mind – denominate it as you will – more of the observations of my own senses, more of the combining of my own imagination – the effusions of my own feelings and passions than the cold workings of the brain . . .'

Insistence was not her usual way with Godwin. She preferred to 'woo philosophy'. The following year, when Coleridge and Hazlitt were walking in the West Country, Coleridge asked Hazlitt if he had ever seen Mary Wollstonecraft. 'I said I had once for a few moments,' Hazlitt recalled, 'and that she seemed to me to turn off Godwin's objections to something she advanced with quite a playful, easy air.' Her playfulness softens Godwin up in advance of the lesson he was to give her. She fancies how the Grammarian will pause, as Milton's Adam paused in his speeches to kiss Eve in Paradise:

> You are to give me a lesson this evening – And, a word in your ear, I shall not be very angry if you sweeten grammatical disquisitions after the Miltonic mode – Fancy, at this moment, has turned a conjunction into a kiss; and the sensation steals o'er my senses. N'oublierez pas, I pray thee, the graceful pauses, I am alluding to; nay, anticipating – yet now you have led me to discover that I write worse, than I thought I did, there is no stopping short – I must improve, or be dissatisfied with myself –

Sometimes, a figure in recent memory can reflect light on someone further in the past. Mary Wollstonecraft had to cope with a harshness in Godwin, a critical kind of stubbornness that can remind us of Leonard Woolf, and of Virginia Woolf's way of coping through play and jokes that weren't only jokes, as when she named his room 'Hedgehog Hall'. The curtness of Leonard's factual diary – like Godwin's – manifests a respect for fact that is almost fierce. Then, there is his staunchness and honesty in relationships, and above all a biblical sense of justice that Woolf absorbed as a Jew who was also the son of a barrister, and that Godwin absorbed from reading the Bible with his surrogate mother, Miss Godwin.

It is a peculiarity of the English climate that while the month of August is often rainy, summer can make a belated and almost magical appearance in September. So it was in 1796. The fine weather roused Mary to '*vagabondize*' in the country, and introduce Godwin to the landscape of her Essex childhood. 'I love the country,' she told him, 'and like to leave certain associations in my memory, which seem, as it were, the land marks of affections – am I very obscure?' Accordingly, they set out for the Wollstonecraft house near the bend in the road known as the Whalebone, where Mary had lived from the age of six till she was nine. The house was uninhabited and the garden a wilderness. They visited the market at Barking, and the old wharf, crowded with barges. Here, where the Thames, nearing the sea, winds through the Essex marshes, she had lain on her back and gazed at the sky, awed by a sense of creation – a counter-force to an unloved child who was helpless against her father's violence. It mattered to revisit this place in her potent character as loved woman of thirty-seven in the company of a man who was a 'kind creature'. Godwin found that 'no one knew better than Mary how to extract sentiments of exquisite delight from trifles'. A little ride in the country could open her heart to offer 'a sort of infantile, yet dignified endearment'. She found a 'magic in affection' when Godwin clasped his hands round her arm in company. Anguish visited her more seldom, he noticed. She was now more 'tranquil'.

Mary told Godwin that she experienced 'not rapture' but 'sublime tranquillity' in his arms. 'Hush! Let not the light see . . . These confessions should only be uttered – you know where, when the curtains are up – and all the world shut out . . .' However reassuring this was for her, he may have wondered at the absence of 'rapture'. It was something she had known with her frontiersman.

Another uncharacteristic act on Godwin's part – a ploy out of keeping with his straightforwardness – does suggest a wish to provoke Mary's jealousy. To maintain their secret, it was their habit to go separate ways. Mary, for instance, dined often at this time with Fuseli and his wife, every Tuesday and other days as well. For all that, she was disconcerted by Godwin's air of indifference one evening when, accompanied by Mrs Perfection, he left her supping with Mrs Robinson.

'I thought, after you left me, last night, that it was a pity we were obliged to part – just then,' Mary protested the following day. 'I was even vext with myself for staying to supper with Mrs R. There is a manner of leaving a person free to follow their own will that looks so like indifference, I do not like it. Your *tone* would have decided me – but to tell you the truth, I thought, by your voice and manner, that you wished to remain in society – and pride made me *wish* to gratify you.'

'I like the note before me better than six preceding ones,' Godwin admitted. 'I own I had the premeditated malice of making you part with me last night unwillingly. I feared Cupid had taken his final farewel.' Relenting, he offered to call on her in the course of the day, and his diary records that they spent the night together in his rooms.

One Saturday night Godwin let go an idea that it would not be good for his health to make love just then. Next morning, their exchanges flew:

> If the felicity of last night has had the same effect on your health as on my countenance [Mary wrote], you have no cause to lament your failure of resolution: for I have seldom seen so much live fire running about my features as this morning when recollections – very dear, called forth the blush of pleasure, as I adjusted my hair . . .
>
> Return me a line – and I pray thee put this note under lock and key – and unless you love me *very much* do not read it again.

This prompt was either lost on Godwin, or else he avoided lovers' language. 'What can I say?' he scratched while Marguerite waited for his reply. 'What can I write with Marguerite perched in a corner by my side? I know not . . . I do not lament my failure of resolution: I wish I had been a spectator of the live fire you speak of.' Then, as he turns to a practical matter – she must return his bottle of ink – the right words come. 'Fill it as high, as your image at this moment fills my mind.' This noncommittal truth was as far as he would go.

Both were wary of commitment. For seven months they retained their own households, dining at one house or the other in the late afternoon, going separate ways of an evening, and then sleeping together in the darkest hours of the night. Their secret remained secure.

London saw a Godwin still taken with Amelia Alderson.

'You have no idea how gallant [Godwin] has become,' she boasted to a friend on 1 November, 'but indeed he is much more amiable than ever he was. Mrs Inchbald says, the report of the world is that Mr Holcroft is in love with her, *she* with Mr Godwin, Mr Godwin with *me*, and I am in love with Mr Holcroft!'

A further rumour was that John Opie, a divorced artist, was Mary's suitor. He was two years younger, a self-taught son of a Truro carpenter who was hailed in London as 'the Cornish wonder'.

How can you pursue this woman, one Joseph Farington pressed Opie in November. 'She is already married to Mr Imlay an American.'

No obstacle there, Opie replied. 'Mrs Wollstonecraft herself informed me that she never was married to Imlay.'

'Opie called this morning – But you are the man,' Mary reassured Godwin that same month. Another suitor was a Dr Slop* who extolled her in a sonnet, and threatened their secret when he took to visiting Godwin in the morning, with Mary sometimes there. 'I treated him so unceremoniously,' Godwin confided to a friend, 'that he at length ceased to call.'

Continued secrecy depended on contraception. Condoms (called 'armour' by Boswell, and 'English overcoats' by Casanova), made of sheep's guts and tied on with pink ribbons, were on sale in London at prohibitive expense. Aristocrats alone could afford them, and used such armour less for contraception than to protect themselves from venereal disease. Godwin adopted what he called the 'chance-medley system', based on abstinence during what was believed to be three fertile days following the end of menstruation, and frequent sex at other times, for it was widely held that frequency, as in the case of prostitutes, diminished the chance of conception. A code in Godwin's diary (a dash for sexual contact; a dash followed by a dot for full intercourse) shows how responsibly he and Mary followed this system. Unfortunately, though, it was wrong. Godwin might have done better to consult a rabbi in preference to common opinion in late eighteenth-century England. For Jewish practice is based on an accurate observation – going back to biblical times – that women are most fertile a fortnight into their cycle. Godwin and Mary Wollstonecraft,

---

*A joke? Dr Slop is the male midwife who is disastrously clumsy with forceps in *Tristram Shandy*, a novel MW admired.

unknowingly, were following a scheme for procreation. The inevitable happened. By 20 December, Mary began to fear she might be pregnant.

Godwin was due to call that day. She wrote thankfully: 'Fanny says, *perhaps* Man come to day – I am glad that there is no perhaps in the case. – As to the other perhaps – they must rest in the womb of time.'

Three days later there was still no sign: 'I am still at a loss what to say – .' One unexpected consolation was Godwin himself. 'There was a tenderness in your manner,' she noticed, 'as you seemed to be opening your heart, to a new born affection, that rendered you very dear to me. There are other pleasures in the world, you perceive, besides those know[n] to your philosophy.'

By 28 December she gave up hope: 'I am not well to day. A lowness of spirits, which I cannot conquer, leaves me at the mercy of my imagination, and only painful recollections and expectations assail me . . . I dare say you are out of patience with me.' They spent a fraught evening on the 30th, and Godwin was displeased when Mary's irritability went on.

'You do not, I think, make sufficient allowance for the peculiarity of my situation,' she muttered. 'But women are born to suffer.'

She spoke repeatedly of sickness, 'the inelegant complaint, which no novelist has yet ventured to mention as one of the consequences of sentimental distress'. Nausea and pre-natal depression were exacerbated by her debts. Mary had incurred certain debts on the strength of Imlay's promise of maintenance for Fanny. Marguerite, for a start, must be paid. Fanny must be clothed. There were food and other bills. But no sum from Imlay appeared. In mid-November she had approached Joseph Johnson, who thought the time had come for her brother Charles to repay her help.

'She has been deserted by Mr. Imlay whose affairs are in a very deranged state,' Johnson wrote to Charles in America, 'she has herself and child to support by her literary exertion and you will not be surprised to learn has occasion to apply to me long before her productions can be made productive.' Mary, Johnson added, 'deserves more than most women, and cannot live upon a trifle; she has suffered much from the infamous behaviour of Imlay, both in her health and spirits, but she has a strong mind and has in great measure got the better of it.'

Charles did not answer. Godwin himself could not help, but offered to write to his friend Thomas Wedgwood (son of Josiah, founder of the potteries) for a loan of £50.

Mary replied on a wet Sunday morning. 'I cannot bear that you should do violence to your feelings, by writing to Mr Wedgewood. No; you shall not write – I must think of some way of extricating myself. You must have patience with me,' she added, 'for I am sick at heart – dissatisfied with every body and every thing.'

Between them, unspoken, loomed the spectre of marriage and Mary's renewed fear of abandonment. There were further lapses into irritability, requiring apology, self-questioning and veiled anger at Godwin for taking time to deliberate his part in her plight.

'I believe I ought to beg your pardon for talking at you, last night,' she said, 'because there was nobody else worth attacking . . . But, be assured, when I find a man that has any thing in him, I shall let my every day dish alone.'

Her struggle to make no demands succeeded only in an air of grim control. Godwin reproached her for treating him 'with extreme unkindness: the more so because it was calm, melancholy, equable unkindness. You wished we had never met; you wished you could cancel all that had passed between us. Is this, – ask your own heart, – Is this compatible with the passion of love? . . . You wished all the kind things you had ever written me destroyed.'

'This does not appear to me just the moment to have written me such a note.' She was plaintive; then stiffened. Her tones become ominously close to those with Imlay: 'I am, however, prepared for any thing. I can abide by the consequence of my own conduct, and do not wish to envolve any one in my difficulties.' Gloom resounds through her notes to Godwin:

> *1 January*: I have a fever of my spirits that has tormented me these two nights past.
>
> *5 January*: I was very glad that you were not with me last night, for I could not rouse myself . . .
>
> *12 January*: . . . You have no petticoats to dangle in the snow. Poor Women how they are beset with plagues – within – and without.

Her landlady laid a trap one Friday night at the end of January. The plot was to catch Mary in bed with a lover who, it turned out, was not there. Mary brushed off the pry with the same impatience she had felt long ago for the Smallweeds of Newington Green: she had no time for a 'foolish woman' who wished to 'plague' her, and 'put an end to this nonsense'. She tried to keep up her walks, but it was hard to hold back the nausea. It rained, and she came home so drenched she had to undress. When a fine day dawned at last on Friday 3 February, she set out to see Dr George Fordyce, the physician at St Thomas's Hospital, in the early afternoon, before it grew dark. He was the son of James Fordyce who wrote the pre-eminent conduct book that Wollstonecraft had dismissed in her *Rights of Woman*. She knew the doctor as a member of Johnson's circle, a learned eccentric with few private patients – patients did not greatly interest him. His portrait shows a scholar fingering the pages of an open book, a sedentary, rather hunched figure with balloon cheeks and a chin receding into a fur collar. He wears a loose coat like a dressing-gown. The doctor would have confirmed Mary's pregnancy.

At two months gone, she was facing an impossible choice. If she chose to remain the lone Mrs Imlay, she would be stigmatised as soon as her belly began to swell. If she married Godwin, she would have to repudiate the fiction of an Imlay marriage, and Fanny would become illegitimate. And even if she chose that second course, would Godwin, who had publicly repudiated marriage, go through with it? Either way, she would be ostracised.

14

'THE MOST FRUITFUL EXPERIMENT'

Mary no longer opposed marriage. Well before she was pregnant there are signs of a shift. The first – so tenuous as to be almost unstated – appears three weeks after she began to sleep with Godwin. 'You are a tender considerate creature, she told him, 'but, entre nous, do not make too many philosophical experiments, for when a philosopher is put on his metal [mettle], to use your own phrase, there is no knowing where he will stop – and I have not reckoned on having a wild-goose chace after a – wise man – You will ask me what I am writing about – Why, as if you had been listening to my thoughts – .' Then, at the end of September, she asked Godwin for a key to his home, teasing him that a free woman would not observe the punctuality he could expect from a wife. Eventually, on 10 November, the appeal of domesticity became explicit: 'I send you your household linen – I am not sure that I did not feel a sensation of pleasure at thus acting the part of a wife, though you have so little respect for the character.'

Godwin's opposition to marriage was widely known. Nor did his attachment to Mary change his mind. During the period they were lovers he stated 'that two persons of the opposite sexes may be lovers for half their lives, and afterwards a month of unrestrained, domestic, matrimonial intercourse shall bring qualities to light in each, what neither previously suspected'. There could have been an unstated reason for this resistance: earlier signs of latent catalepsy. Though Godwin had no fits in this period, he had fits of sleepiness in the afternoon, even in company. Then, too, he believed his temper unsuited to marriage. 'My temper is of a recluse,' he told himself. His home was no domestic haven; it was a workplace, only partially furnished and shared with an indexer called James Marshall who acted as a kind of factotum. In his early London years Godwin had often pawned his watch so as to eat. He now lived on £110–£130 a year, three times what an educated woman could earn, but not enough to take on the dependants who followed a bride's train. Mary had debts, and a husband would become legally responsible for them. Godwin continued to pity men who were condemned to a lifelong domestic prison, and when circumstance forced him to marry, he made it abundantly plain that he was doing so only to relieve a woman who mattered to him. Calm, deliberate as ever, he reconciled himself to the jeers that would come his way. Mary had reservations of her own, kept in her case under wraps. Only once, in confidence to Amelia Alderson (after a trying day), did she own that she had *not* got over Imlay.

'The wound my unsuspecting heart formerly received is not healed. I found my evenings solitary; and I wished, while fulfilling the duty of a mother, to have some person with similar pursuits, bound to me by affection; and beside, I earnestly desired to resign a name which seemed to disgrace me,' she said. 'Condemned then to toil my hour out, I wish to live as rationally as I can; had fortune or splendour been my aim in life, they have been within my reach [through other suitors], would I have paid the price. Well, enough of the subject; I do not wish to resume it.'

This was a sober version of her remark to Godwin that she found not rapture but tranquillity in his arms. 'Sublime tranquillity,' she had said, for, to her, sex was 'sacred' – in practice (if not in theory) a commitment. They married quietly on 29 March 1797 at Old St Pancras Church, which still stands with its graves on St Pancras Way. In the eighteenth century the setting was pastoral with the River Fleet, as yet unenclosed, running down from the heights of Hampstead and Highgate. After the ceremony, wife and husband walked back to their separate lodgings as if nothing had happened. A note from Godwin's rooms has an air of detachment. No bridegroom, he. 'Non,' he replies coolly to a question as to whether he was vexed over this event, 'je ne veux pas être fâché quant au passé. Au revoir.' His diary ignores the event apart from a laconic 'Panc[ras]'. Two days later, Mary asked her husband to send over some of his dinner, since she was now the sharer of his worldly goods – nonexistent goods, as she well knew.

'But when I press any thing it is always with a true *wifish* submission to your judgment and inclination,' she joked.

Godwin, too, passed off their new status as a joke when he informed Mary Hays. 'My fair neighbour desires me to announce to you a piece of news, which it is consonant to the regard that both she & I entertain for you, you should rather learn from us than from any other quarter. She bids me remind you of the earnest way in which you pressed me to prevail upon her to change her name, & she directs me to add, that it has happened to me, like many other disputants, to be entrapped in my own toils: in short, that we found that there was no way so obvious for her to drop the name of Imlay, as to assume the name of Godwin. Mrs Godwin (who the devil is that?) will be glad to see you at No. 29, Polygon, Somers Town, whenever you are inclined to favour her with a call.'

Four days earlier, on 6 April, they had moved into the Polygon, a circular block of three-storey houses in Somers Town.* There was a parlour near the entrance, a dining room on the first floor, a balcony, and bedrooms on the top floor. Opposite, to the north (now Polygon Street) and west (now Werrington Street), were fields where Fanny played at making hay with her toy pitchfork. Godwin, at forty-one, and Mary, at nearly thirty-eight, each continued to prize independence. Godwin especially, as a long-time bachelor, was alert to the dulling effect of proximity. In *The Enquirer*, published a month before they married, he expressed his concerns in an essay, 'Of Cohabitation': 'It seems to be one of the most important of the arts of life, that men should not come too near each other, or touch in too many points. Excessive familiarity is the bane of social happiness.'

Mary too wished to maintain her space: 'A husband is a convenient part of the furniture of a house, unless he be a clumsy fixture,' she told Godwin. 'I wish you, for my soul, to be rivetted in my heart; but I do not desire to

*A plaque in Werrington Street marks the spot where the Godwins lived.

have you always at my elbow.' So Godwin rented separate quarters for himself, about twenty doors away, at 17 Evesham Buildings in his old haunt of Chalton Street, where he sometimes slept or went in the morning as soon as he woke. It was his habit to wake between seven and eight; read a classic for an hour before breakfast; then to write for the rest of the morning. In the afternoon he took a walk. At dinnertime, between four and five, he would return to the Polygon. Occasionally Mary joined him for his walk, but more often they communicated by letter – Fanny, attended by Marguerite, would trot down the road with a message too hot to wait out the day. They would wait while Godwin (once 'Man', now 'Papa') set down his reply. Evenings were spent, as before, with separate friends. Mary urged her husband to attend Johnson's dinners, declaring, by way of inducement, that she would not be there.

They agreed to disregard the convention that a couple had to move in mixed company as a pair, and that in matters of opinion they would free each other to go their own ways. One potentially divisive area was Mary's faith – her sense of a spirit that rolls through all things.

'I love the country and think with a poor mad woman, I knew, that there is God, or something very consolatory in the air', she defied Godwin who shunned the supernatural.

'I still mean to be independent,' she told Amelia, 'even to cultivating sentiments and principles in my children's minds (should I have more), which he disavows.' To Amelia, she still signed herself 'Mary Wollstonecraft' with 'femme GODWIN' beneath, and Godwin continued to call her 'Wt' in his diary, but she did her husband the courtesy of being 'Mary Godwin' to his friends, and she addressed him in the usual respectful manner by his surname. Before, when they had been lovers, she had called him, at first, 'William' as well as 'Godwin'; marital manners were more formal. Only after two months as man and wife did Godwin invite a return to the familiarity of first-name terms.

'Your William (do you know me by that name?) affectionately salutes [kisses] the trio M[ary], F[anny], & least (in stature at least) little W[illiam, their unborn child],' he wrote when he was away.

Though Godwin often assumed the role of reason, and Mary that of feeling, these roles did not entirely reflect the combinations of thought and feeling both cultivated – Godwin of course with more deliberation. He was astonished at the rightness of her spontaneous judgements and the vividness of her feelings. 'Never,' he said, 'does a man feel himself so much alive as when, bursting the bonds of diffidence, uncertainty and reserve, he pours himself entire into the woman he adores.' The poet Southey thought Mary's face 'the best, infinitely the best' amongst the London literati, 'with a proud look of independence rather than superiority'. This marriage of two independent characters was sustained on the basis of separate quarters. 'We were in no danger of satiety', Godwin reflected later. 'We seemed to combine, in a considerable degree, the novelty of lively sensation of a visit, with the more delicious and heart-felt pleasures of domestic life.'

This was a new man talking of heart and the pleasures of domesticity. He was surprised – but not put out – to find his wife 'a worshipper of domestic life'. She loved to observe the growth of affection between him and Fanny, and his anxiety about the child to come, for pregnancy was 'the source of a thousand endearments'. He was prepared to admit, 'I love these overflowings of the heart', as he came to question that part of the English character that feels a 'sort of mauvaise honte, which prevents men from giving utterance to their sentiments'. The longer he lived with this wife he had never meant to have, the readier he was to regret (though not wholly alter) his custom of treating others 'as if they were so many books'. Virginia Woolf was right to look back on the marriage as the most fruitful experiment of Mary Wollstonecraft's life. 'I think you the most extraordinary married pair in existence,' their friend Thomas Holcroft told them.

Vital to the success of the experiment was repartee without aggression. Both had combined revolutionary principles with a refusal of violence. Godwin had acted against the legal violence of the Treason Trials, while Wollstonecraft had taken a stand against the imitation violence of her own sex on their march to Versailles. Her father's brutality had left her with a horror of all forms of tyranny, which she had early associated with masculinity. Godwin was a rare man who abjured the long habit of power – who abjured power itself. He provided an exact match with the far-sighted womanhood of Wollstonecraft, with its emphasis on inward authority, beyond the scope of public figures playing power games.

Disagreements were laid open by an intelligent and articulate couple who remained separate enough to continue to exchange numerous notes, leaving an unprecedented history of the day-to-day conduct of a new kind of marriage. Some of their misunderstandings seem more common to our egalitarian age than to married life two hundred years ago. Mary regarded her time 'as valuable as that of any other person accustomed to employ themselves', and resented her husband's expectation that she would deal with tradespeople.

'I am, perhaps, as unfit as yourself to do it,' she told him on 11 April, 'and feel, to say the truth, as if I was not treated with respect, owing to your desire not to be disturbed – .'

It was particularly galling when she had no means to pay what she owed. Previously, Johnson had undertaken this for her, but now that she had a husband, Johnson withdrew from his protective role. She could not expect Godwin to become a provider – one of his objections to marriage – but she did want to see him take a firmer line with the manager of the Polygon, who liked to say yes and no at the same time.

The real trial was the scrutiny society turned on the doings of two celebrities. *The Times* announced that 'Mr Godwin, author of a pamphlet against matrimony', had clandestinely wed 'the famous Mrs Wollstonecraft, who wrote in support of the Rights of Woman'. Fuseli was as surprised as strangers – and none too pleased to be surprised. A fortnight into the marriage – on that same 11 April that Mary owned her soberer motives to Amelia – its difficulties came to a head: not only Godwin's expectations of a wife but also flying snubs, as news got round. Mary felt it keenly when Mr and Mrs Twiss let her know that contact must cease – a rebuff to her feelings of gratitude, for she had dined with them every third Sunday and had been sent for whenever they had guests she might like. Worse, Mrs Inchbald, who had been due to share a box at the theatre with Godwin, was forgotten, and soon let him know she had found another escort. This was accompanied by a mock-apology that 'she would not do so the next time he was married'. Nonsense, thought Mary, aware nevertheless that she had played some part in this omission and must act to avert a storm. It would not do to neglect Perfection. She asked Amelia to intervene, to beg Perfection's 'pardon for misunderstanding the business', and to convey Mary's offer to make way and sit, if necessary, in the pit.

The storm was not averted. On the evening of 19 April the Godwins appeared together at the theatre – their first public appearance as a couple. By the fifth month, pregnancy can't be concealed from the sharp-eyed. Mary would have cared too much for the growing foetus to lace tightly. At the sight of the swell under her breast, Perfection unleashed her tongue. Her snub, delivered with the aplomb of a practised actress, was meant to carry – it was certainly heard throughout the box. To Godwin it was 'cruel, base, insulting', and later he let Mrs Inchbald know that others present had shared his opinion. Her ready answer was that she had never wished to meet Wollstonecraft, nor any other unmarried mother, and that Godwin had forced the acquaintance on her. It must be said, in fairness to Mrs Inchbald, that her novel, *Nature and Art* (1796), had received a cool review from Mary Wollstonecraft, who thought it played up to men. Where Mary was content to see off her rival, Godwin was shaken. A favourite of Mrs Inchbald could not dismiss her lightly, and he and Mary had words when they were alone.

'I am pained by the recollection of our conversation last night,' he wrote to her from his rooms next morning. 'The sole principle of conduct of which I am conscious in my behaviour to you, has been in every thing to study your happiness.' He reminded her how hurt she had been when they met, and the care he had taken. Some return was due to his distress. 'Do not let me be wholly disappointed.'

Mrs Inchbald's snub had been predictable, and the quieter withdrawal of Mrs Siddons was accepted with resignation. The very visibility of an actress made it difficult for her to present the acceptable image of retiring womanhood. Those who were mistresses, like Dora Jordan and Mrs Robinson, were thought of as loose. This meant that respectable actresses (like Mrs Siddons and Mrs Inchbald) guarded their reputations with extra caution.

Mary remarked: 'Those who are bold enough to advance before the age they live in . . . must learn to brave censure. We ought not to be too anxious respecting the opinion of others.'

Snubs and jeers from other quarters were not wanting. Godwin patiently explained himself to those who accused him of inconsistency. Here is one such reply to a reader:

Sir,

As your letter consists of little more than a dry question respecting my private conduct it may seem little more than reasonable, that I should be [loath?] to answer this and similar questions to every person who may happen to be a reader of my publications. In another view however, the question may be considered as a testimony of some degree of esteem on the part of the proposer, & as such entitled to an answer.

The first thing I have to say is that you have somewhat mistaken my character, which I conceive to be somewhat of an enquirer, & not a dogmatical deliverer of principles. I certainly desire that my conduct should be found consistent with my general sentiments; but I do not conceive myself bound by my sentiments to-day, not to see a subject in a very different light to-morrow; I should be very sorry to be that oracle to another, which I am very far from being to myself.

But let us examine the amount of difference.

Your reference, I perceive, is to the first edition of *Political Justice*, published in 1793. When that edition appeared, my mind was decided against the European system of marriage, but it was in a state of some doubt as to the question, whether the intercourse of the sexes, in a reasonable state of society, would be wholly promiscuous, or whether each man would select for himself a partner, to whom he would adhere as long as that adherence should continue to be the choice of both parties. In the second edition, page 500, published in 1795, I gave my reasons for determining in favour of the latter part of this alternative . . .

Yes, you will say, I have conformed to the European institution of marriage, an institution which I have long thought of with abhorrence; & probably shall always abhor.

> I find no inconsistency in this. Every day of my life, I comply with institutions & customs which I . . . wish to see abolished. Morality, so far as I understand it, is nothing but a balance between opposite evils. I have to choose between the evils, social & personal, of compliance & non-compliance.
>
> I find the prejudices of the world in arms against the woman who practically opposes herself to the European institution of marriage. I found that the comfort & peace of a woman for whose comfort & peace I interest myself, would be much injured, if I could have prevailed on her to defy these prejudices. I found no evils in conforming to the institution I condemn, that would counter balance this.
>
> That she might not risk [becoming?] the victim of prejudices, I was willing to pass through a certain ceremony. Clear in our own conceptions of the subject, I found little difference in the effect of the ceremony. I will not, I believe, live with her a day longer or treat [her] differently than I should have done, if we had entered into the intercourse without the ceremony.
>
> I am still as ardent an advocate as ever for the abolition of the institution. I should still, for the most part, dissuade any young man for whom I had a regard, from complying with it. I should fear that he would not be able to comply, with so much impunity, as I hope I shall have done; & I continue to think that marriages, in a great majority of instances, are among the most fertile sources of misery to mankind . . .

This rationalisation, on 9 May 1797, conceals the increasingly obvious fact that Godwin's inner and more private feelings were opposed to the supposed gist of his doctrines.

One friend remained staunch. To Mary Hays, Wollstonecraft's manners when she unbent in private conversation 'had a charm that subdued the heart'. Hays had visited her friend almost daily during her bad times, and promoted this marriage, by inviting her to the Godwin tea-party in January 1796 and also by prompting a change of name. Like Wollstonecraft, Hays looked on female desire as a trait in its own right, not as current piety would have it, called out only in answer to male instinct. Hays's passion for Mr Frend was a lone thing. Her recent novel *Memoirs of Emma Courtney*, much discussed early in 1797, scandalised conventional readers when Emma dares to declare her love to the politely indifferent Mr Harley, and though this heroine suffers an excess of sensibility, she never loses her self-respect. Unrequited love, what Emma Courtney calls 'the deceitful poison of hope', bonded Hays and Wollstonecraft more closely than shared ties to Johnson (reinforced by Wollstonecraft's resumption of her assistant-editorship of the *Analytical*, with Hays amongst the reviewers she commissioned). They joined in resisting the marriage market, the mis-education of women and the glorification of war. Soldiers, in Emma's terms, are thugs: 'their trade is *murder*, and their trappings but the gaudy pomp of sacrifice'. Worse than common ruffians and housebreakers, they 'cut down millions of their species, ravage whole towns and cities, and carry devastation through a country'. She echoes Wollstonecraft when she says the Romans 'had in them too much of the destructive spirit'. Godwin was the kind of man Hays upheld; in the character of her 'good' mentor Mr Francis, his eyes pierce the soul.

A longer outcome of the marriage was its release of Imlay from obligation. In legal terms marriage meant economic support. If 'Mrs Imlay' was now Mrs Godwin, she had no further claim on Mr Imlay, and this was to have consequences for blithe little Fanny, aged three. By now, a year had passed since Mary had discussed provision for Fanny as she walked with Imlay along the New Road, and as yet he had sent nothing. Mary was beset with debts and could not expect Godwin to help. He lived in a small way on advances for successive books, with nothing to spare. It would have been painful for a woman as scrupulous as Mary to bring burdens with her into a marriage that was not what Godwin had wanted. Over the last year her power to write had faltered. It may seem strange that this should have happened after the success of her *Travels*, but strong characters like Mary Wollstonecraft are often able to sustain an effort in the heat of crisis, only to sink when the heat is off. She owned something of this to Everina. 'My pecuniary distress arises from myself, from my not having the power to employ my mind and fancy, when my soul was on the rack.'

Since the pregnancy was still a secret, she could not explain herself to Johnson who, understandably, was not disposed to support her indefinitely. Mary was obliged to fund their father, who over the last two years had received only £35 from Ned Wollstonecraft's administration of the Primrose Street property. Ned by now was sunk in debt. Their brother James was equally useless. He had given up his lieutenancy, lost his money, and leant on Mary to open a path to Paris. While there, he was arrested for spying and turned to Joel Barlow to bail him out. Barlow, who obliged for Mary's sake, was not repaid. Eventually, Barlow let loose a bellow of fury with James Wollstonecraft's ingenious excuses. Mary, who had long given up on Ned, blamed Charles for his continued silence on the subject of their father. She was forced to confide in Johnson, on whom her debts fell, and as her pregnancy advanced he took it upon himself to approach Charles once more on her behalf.

> In addition to her own distresses from the rascally Mr Imlay she has had the melancholy condition of your father who has been long left destitute from the imbecility or something worse of Edward [Ned], to relieve, for except the last assistance [of being thrown on the parish] which it would shock her to see him submit to receive, he has had no one to depend upon for nearly three years but her self, & she has no resource but to me . . . She is in a very anxious state of mind and you will give her great relief if you will be good enough to satisfy her that she may look up to you for a regular support for him.

Johnson asked Charles in vain to return a sum of money lent to him when he had sailed for America in 1791, adding with civil irony: 'You will add interest or not at your own pleasure.' Charles preferred to take the line that Johnson had profited so greatly from Mary's publications that he really owed *her* money. It's a measure of what she had to shoulder that mild, benevolent Joseph Johnson eventually lost patience with her family.

During February and early March 1797, Everina had stayed with Mary, en route from Dublin to a temporary post with Josiah Wedgwood II near Stoke-on-Trent. It was a strained visit. The sisters had little to say to each other. After Everina left, her seamstress came to the door, pleading her poverty. Mary had to pay the debt of three pounds and four shillings. Another caller was the brother of Everina's friend Miss Cristall, who prevailed on Mary to give him all her ready cash.

'I am continually getting myself into scrapes of this kind,' she groaned to her sister. 'I must get some [money] in a day or two, Johnson teazes me, and I will then send you a guinea' – for Everina herself was making some claim. 'I did imagine that Mr Imlay would have paid, at least, the first half-year's interest of the bond given to me for Fanny. A year, however, is nearly elapsed, and I hear nothing of it, and have had bills sent to me which, I take it for granted, he *forgot* to pay. Had Mr Imlay been punctual, I should, after the first year . . . have put by the interest for Fanny, never expecting to receive the principal, and not chusing to be under any obligation to him.'

In the end Godwin rescued her by an appeal to Wedgwood for £50 (despite her protest) in January, and then £50 more on 19 April. In the midst of her troubles, Mary took pity on her servant's boy, who had nowhere to stay. She took him in, 'poor fellow', and sent him to school.

'Let my wrongs sleep with me!' Mary Wollstonecraft had declared to Imlay on 10 October 1795 when she had thought to end her life. *The Wrongs of Woman* was to be an ambitious novel uniting the particular wrongs of the Wollstonecraft sisters with a wider spectrum of wrongs inflicted on the defenceless poor, from pressing unwilling men into service in the navy to the sexual abuse of servant-girls. Following *The Rights of Woman*, *The Wrongs* is a sequel on legal wrongs in fictional form. Wollstonecraft began the book with help from Godwin in the second half of 1796, at the time of her delayed recovery from Imlay, but it was not until February 1797 that she explored a setting for her drama: a madhouse. There, a wife called Maria is imprisoned – inexplicably, since her sanity is as clear to the reader as it is to herself. Her challenge is to sustain her sanity as the weeks pass, surrounded as she is by all-too-real gothic scenes: the howls of mental agony and the shrieks of passions run out of control, which she pities but distances by day. By night, however, in her sleep, the horrors penetrate her mind.

These were horrors Mary Wollstonecraft saw for herself. On 6 February, Godwin and the novel's publisher, Johnson, accompanied her to Bedlam Hospital for the insane. Johnson's support for the subject of *Wrongs* went back to the anonymous book he had published long before, in 1777, on *The Laws Respecting Women*, which looks at 'cases upon Women being confined in private Madhouses on the Plea of Insanity'. Before 1700 few were confined in this way, but in the course of the eighteenth century private 'madhouses' became profitable. They were unregulated by law until 1774, and, even after, were as unsupervised as Wollstonecraft reveals. In February 1797 she asked Godwin for the second volume of *Caleb Williams*, where Caleb is unjustly imprisoned and then meets the thieves, who see themselves as the mirror image of their social superiors who thieve within the law.*

Everina's visit that February gave another stir to the brew. Mary complained of her sister: her peevishness, her silence. Embarrassing evenings at home ticked by. Why was Everina Wollstonecraft glum? This was unlike the Everina whose light spirits used to cheer Mary at Newington Green, even if they could not dispel the autumnal moodiness of Bess. It's true that Everina caught a cold in London, but there was another reason for gloom. Everina and Bess had not communicated with Mary for a year and a half, not since May 1795 when Mary had concealed the fact that she and Imlay were unmarried and at breaking point, pretending instead that they were too intimate to take Bess into their home. Bess had copied this humiliating letter to Everina, who shared her dismay. In addition, Mary had promised her sisters that Imlay would take on a husband's obligation to help them, but this was always deferred. In *Wrongs* Maria asks her husband to give £1000 to each of her sisters, and his answer is 'have you lost your senses?'

So it was that Everina sat glum and wordless when she stayed with Mary, who had arranged a post for her with Godwin's friends, the high-minded

*Dickens would later use a prison as a metaphor for Victorians locked in their mental attitudes. In the second half of *Little Dorrit*, the world outside the prison is revealed as a mirror image of the world inside the prison, locked in its absurd dramas and even more absurd rhetoric.

and generous Wedgwoods. The post was too advantageous to turn down, or Everina would not have left Bess (who had joined her in Ireland). As the clock ticked through these tense evenings, the moans of the unfortunate third sister hover in the air of Everina's silence. Bess is present, whether the sisters discuss her or not, for scenes with Bess flood back in Mary's memory – back to 1782 and her escape from the misery of the Wollstonecraft family into the worse misery of marriage; the merchant bridegroom, satisfied to have performed his part in conferring his goods on a wife; her deteriorating state of mind after childbirth; the dark of winter in January 1784, the flight to Hackney, the closed, rocking coach, and Bess bewailing her separation from her baby; Bishop's punitive refusal to send word of the child; then months later, the sad death of little Elizabeth Bishop cut off from her mother in her father's unyielding custody. *Wrongs* conjures up the bereaved mother who 'looked like a spectre', whose 'eyes darted out of her head' as the news is brought, and who asks to be alone, 'hiding her face in her handkerchief, to conceal her anguish'. Her sister's story is central to Wollstonecraft's exploration of psychic damage.

After Maria hears of her daughter's death, she relates her story to her keeper, Jemima, and a fellow-inmate, Darnford. It's a story of marital persecution. English law grants her so little protection that she ends by wondering if England can be her country – if a woman can, in truth, be said to have a country. For the law of the time permits the husband to force himself on his wife in brutal and drunken states, to declare her insane if she resists, to imprison her in a 'madhouse' against her will, and take away their child in his capacity as sole owner. This is what the Wollstonecraft sisters had feared when Bess rejected Meredith Bishop, and this is exactly what the fictional husband, George Venables, does.

Venables is 'gross', staled by prostitutes whose mirth he calls 'nature'. With him, 'imagination was so wholly out of the question, as to render his indulgences of this sort entirely promiscuous'. To his wife his 'brutality was tolerable, compared with his distasteful fondness'. Still, Maria goes on, 'compassion, and the fear of insulting his supposed feelings, by a want of sympathy, made me dissemble, and do violence to my delicacy. What a task!' Wollstonecraft, through Maria, questions the wifely virtue of sexual compliance. 'When novelists or moralists praise as a virtue, a woman's coldness of constitution, and want of passion; and make her yield to the ardour of her lover out of sheer compassion, or to promote a frigid plan of future comfort, I am disgusted.' In *Pride and Prejudice* Elizabeth Bennet is similarly appalled by her friend's decision to marry Mr Collins, who is not a man whom a sensible woman could possibly respect. To marry for an establishment is 'an affair of barter', Wollstonecraft says. Jane Austen, too, could not bring herself to marry her eligible suitor, the well-to-do friend of the family Big-Wither. She accepted his proposal; then, next morning, made her escape. Her most virtuous heroine, Fanny Price, will not marry to please her relations, grateful as she is for the home they have given her. Wollstonecraft reframes marital morality when she says that 'heartless conduct is the contrary of virtuous. Truth is the only basis of virtue; and we cannot, without depraving our minds, endeavour to please a lover or husband, but in proportion as he pleases us.' If a wife is repelled or indifferent is she not 'indelicate'? Wollstonecraft's Maria voices what Hardy's new-married Sue Bridehead dares to voice a century later (to an outcry from Victorian society): 'It is not easy to be pleased, because . . . we are told that it is our duty.'

When Maria can please her husband no longer, she makes an escape. She lives as an outlaw in hiding, as Bess did in Hackney. Eventually, Maria is caught and consigned to an asylum – her 'prison'. There, her solitary confinement again recalls Bess during her years in Wales where she had to plead for permission to take a walk. Shut up and humiliated, it was still preferable to the sexual captivity of her marriage, as is the 'madhouse' for Maria.

'Let me exultingly declare that it is passed,' says Maria, ' – my soul holds fellowship with him no more.' Estrangement is the deep emotion in this story. 'He cut the Gordian knot, which my principles . . . respected; he dissolved the tie, the fetters rather, that ate into my very vitals – and I should rejoice, conscious that my mind is freed.' Mary Wollstonecraft could not have written that without the emotional separation from Imlay that Godwin helped her achieve.

Maria, like Mary the author, is more resilient as well as more fortunate than Bess. She has an alternative in the wings – the congenial if unconvincing Darnford (made up of unblended elements of what had attracted

Wollstonecraft in two quite different men: the intrepid Imlay and the bookish Godwin), whose presence helps Maria assert her freedom. This is not, though, a romance plot where the hero effects the rescue; here a woman is the agent of her fate. Darnford (in one of the provisional conclusions) was to prove unreliable. Maria follows Clarissa, and is followed in turn by Fanny Price, Jane Eyre and Maggie Verver (wife of the adulterous Italian prince in *The Golden Bowl* by Henry Janes), all of whom manage to free themselves from pressure to surrender their bodies to men they can't trust.

'I am vexed and surprised at your not thinking the situation of Maria sufficiently important,' Wollstonecraft complained in mid-May to Godwin's protégé George Dyson, who had read her manuscript. 'Love, in which the imagination mingles its bewitching colouring must be fostered by delicacy – I should despise . . . [a woman] who could endure such a husband as I have sketched – yet you do not seem disgusted with him!!!'

Dyson's 'want of delicacy' goaded Mary to explain her idea. Surprisingly, she does not stress the physical wrong of domestic violence. Her concern is an even more hidden wrong, the power of a husband to 'degrade the mind'. This looks forward to *The Subjection of Women* (1869), where Harriet Taylor Mill and John Stuart Mill recognise that psychic wrongs are more damaging than visible ones; to Wollstonecraft's mid-Victorian admirer George Eliot, who explores the humiliation of Gwendolen Harleth by her politely fearsome husband in *Deronda* (1876); and to Henry James, lighting up more sinister manipulations, licensed by marriage, in *The Portrait of a Lady* (1881).

Wollstonecraft discerns all this almost a century before, when she claims that private pains are as dramatic as 'what are termed the great misfortunes'. Those may have more 'of what might justly be termed *stage effect* but it is the delineation of finer sensations which, in my opinion, constitutes the merit of our best novels, this is what I have in view; and to shew the wrongs of different classes of women equally oppressive'. The parallel history of Jemima, 'witness of many enormities', reveals the wrongs of women of the working class. Like Maria, Jemima has been prostituted. Hers is no sentimental story – she is a thief, criminalised and hardened to some degree by her will to survive, yet still with a residue of compassion.

Wollstonecraft refuses to reduce the fallen woman to sinner or saint. The challenge was to make her entirely believable – she is not as hearty and quick as Moll Flanders, not as enterprising as Defoe's Roxana; not as dazzlingly amoral; she is too ordinary to be a heroine, and her wrongs in proportion more painful. Looming large amongst a working woman's wrongs is the extreme of economic exploitation. Wollstonecraft is accurate when Jemima states that she is paid no more than eighteen pence as a laundress from one in the morning till eight at night.

The wrongs are summed up in a courtroom scene where Maria defends women against the unjust laws of the land – only to be dismissed by the judge. It's a speech that could have been made in the Commons, were women admitted to Parliament. Maria's oratorical voice seems to evoke the pale, austere portrait of Mary Wollstonecraft in the image of Mr Pitt – the epitome of political power – painted while she was writing *The Rights of Woman*. Yet *The Wrongs of Woman*, five years later, presents an alternative image of womanly strength, tried, tested and transformed by the maturity to know 'in what human happiness consists'. This presents an advance on imitation of existing power models, as does a new portrait of the author, painted as she composed this novel.

> Such was the native soundness of [Maria's] constitution, that time had only given to her countenance the character of her mind. Revolving thought, and exercised affections had banished some of the playful graces of innocence, producing insensibly that irregularity of features which the struggles of the understanding to trace or govern the strong emotions of the heart, are wont to imprint . . . Grief and care had mellowed, without obscuring, the bright tints of youth, and the thoughtfulness which resided on her brow did not take from the feminine softness of the features . . . and the activity of her well-proportioned, and even almost voluptuous figure, inspired the idea of strength of mind, rather than of body.

It is as though the living Mary Wollstonecraft is looking at herself as she takes shape on Opie's canvas in the spring of 1797. She wears a green velvet hat and a soft white dress wrapped across full breasts. Sleeves cover her rounded arms to the elbow. Viewers often look for signs of pregnancy, but that's in shadow. Opie's light is on the promise of womanly intelligence.

On 3 June Godwin set off for the Midlands with the keenest of his disciples, a law student called Basil Montagu, an illegitimate son of the Earl of Sandwich, who later became an eminent barrister and author of books on bankruptcy, copyright and the death penalty. At this time, Montagu was engaged to the youngest of the Wedgwood daughters. The plan was to visit the Wedgwoods at Etruria near Stoke-on-Trent (where Everina was employed as governess). On the way the travellers were to take in Oxford, Stratford-upon-Avon, Dr Darwin in Derby, and the novelist Robert Bage at Tamworth. Godwin had to leave open the date for his return, given the uncertainties of travel in the eighteenth century – the rained-on, muddy roads – as well as his wish to follow what chance might throw in their way. His departure for two or three weeks should have presented no problem for an independent woman, but for Mary it brought back Imlay's departures and abandonment.

At first, separation brought them closer. Two days after Godwin left, on 5 June, his first letter struck a sure note, reversing the platitudes of the sentimental journey. He opens with a run of caustic observations, and then, abruptly, throws off distance and plunges into intimacy:

> Stratford-up, Monday
>
> I write at this moment from Hampton Lucy in sight of the house and park of Sir Thomas Lucy, the great benefactor of mankind, who prosecuted William Shakespeare for deer stealing, & obliged him to take refuge in the metropolis . . . [The previous day, they had been in Oxford.] Here we had a grand dinner prepared for us . . . by a Mr Horseman, who says that you & I are the two greatest men in the world. He is very nervous, & thinks he never had a day's health in his life. He intends to return the visit, & eat a grand dinner in the Paragon* but he will find

*Word play on the Polygon. There was, incidentally, a Paragon Place along the Surrey Road where Hays had lived in 1794–5 before moving to the centre of London.

himself mistaken. We saw the buildings, an object that never impresses me with rapture . . . We had also a Mr. Swan, & his two wives or sisters to dinner; but they were no better than geese.

And now, my dear love, what do you think of me? Do not you find solitude infinitely superior to the company of a husband? Will you give me leave to return to you again, when I have finished my pilgrimage, & discharged the penance of absence? Take care of yourself, my love, & take care of William . . . I remember at every moment all the accidents to which your condition subjects you, & wish I knew of some sympathy that could inform me from moment to moment, how you do, & how you feel.

Tell Fanny something about me. Ask her where she thinks I am. Say I am a great way off, & going further, but that I shall turn round & come back again some day. Tell her I have not forgotten her little mug [a promised gift from the potteries] & that I shall chuse a very pretty one. Montagu said this morning about eight o'clock upon the road, Just now little Fanny is going plungity plunge. Was he right? . . .

Farewel.

His speed and unembarrassed baby-talk provoked a quick return (showing off the benefit of Godwin's Latin lessons):

If your heart was in your mouth, as I felt, just now, at the sight of your hand, you may kiss or shake hands with the letter and imagine with what affection it was written – If not – stand off, profane one!*

. . . I am well and tranquil, excepting the disturbance produced by Master William's joy, who took it into his head to frisk a little at being informed of your remembrance. I begin to love this little creature, and to anticipate his birth as a fresh twist to a knot, which I do not wish to untie. Men are spoilt by frankness, I believe, yet I must tell you that I

*Virgil, *Aeneid*, VI: '*Procul, O procul est, profani.*' 'Away, away, you uninitiated,' cries the Sybil, guide to the Underworld, as she points the hero towards the open cavern to see 'things buried in the dark and deep of earth'.

> love you better than I supposed I did, when I promised to love you for ever – and I will add what will gratify your benevolence, if not your heart, that on the whole I may be termed happy. You are a tender, affectionate creature; and I feel it thrilling through my frame giving and promising pleasure.
>
> Fanny wanted to know 'what you are gone for,' and endeavours to pronounce Etruria. Poor papa is her word of kindness . . .
>
> I find you can write the kind of letter a friend ought to write, and give an account of your movements. I hailed the sunshine, and moonlight and travelled with you scenting the fragrant gale – Enable me still to be your company, and I will allow you to peer over my shoulder, and see me under the shade of my green blind, thinking of you, and all I am to hear, and feel when you return – you may read my heart – if you will . . . Yours truly and tenderly,
>
> Mary . . .

Godwin, in turn, was overcome. 'No creature expresses, because no creature feels, the tender affections, so perfectly as you do,' he replied from Etruria on 10 June. After all his philosophy it had to be confessed that her love was 'extremely gratifying'. He went on: 'One of the pleasures I had promised myself in my excursion, was to increase my value in your estimation, & I am not disappointed. What we possess without intermission, we inevitably hold light; it is a refinement in voluptuousness, to submit to voluntary privations. Separation is the image of death; but it is Death stripped of all that is most tremendous, & his dart purged of its deadly venom. I always thought that St Paul's rule, that we should die daily, an exquisite Epicurean maxim. The practice of it would give to life a double relish.'

It's odd that, even as Godwin spells out this deepening attachment, he remains opposed to marriage for anyone else. He did not approve Montagu's engagement (it ended), and observes to Mary that he regards friends about to be married as if 'they were sentenced for life to hard labour in the Spielberg' (a Moravian prison). The only hope for 'the unfortunate captive' is that 'the despot may die' – rather tactless talk to a pregnant wife nearing her term in an age when six to seven per cent died in childbirth.

At that very time, Mary was penning a second letter, this one more teasing. Fanny had made their friend Mrs Reveley laugh by telling her, when she could not find her green monkey, 'that it was gone into the country'. Mary's belly, in the seventh month, was growing ever larger, 'and my appearance no longer doubtful – you, I dare say, will perceive the difference. What a fine thing it is to be a man!'

Trouble raised its head as Godwin began a fourth letter, on 12 June. Mary's second letter had not yet reached him, and a sliver of hoarfrost returned. When Godwin was put out he tended to become righteous. He would continue to inform her regularly of his movements, he told her: 'I am the less capable of altering my method, if it ought to be altered, as you have not dealt fairly by me this post.' From now on he confined himself to travel letters, assured as he had been of her imaginative company. There is a comic portrait of the none-too-obliging Mrs Darwin who, having no notion of Godwin's celebrity, suspects him of lionising her important husband (with an earning power of £1000 a year); there is a prize booby with whom the lovely, accomplished daughter of Godwin's friend Dr Parr has eloped in obedience to her mother's advice to marry a fool; and there are bungling actors reducing the *School for Scandal* and *The Taming of the Shrew* to the counterpart of a puppet show at a country fair. Nothing of this pleased Mary. His later letters might have been addressed to anybody, she thought crossly, ignoring the fact that he had sent no less than six long letters, to her two shorter ones.

'Another evening, & no letter,' he complains. 'This is scarcely kind . . . What am I to think?'

Common sense told him that he would have heard of any serious accident. His anxiety was that she had succumbed to 'sickness of heart, a general loathing of life & of me. Do not give place to this worst of diseases! The least I can think is, that you recollect me with less tenderness & impatience than I reflect on you. There is a general sadness in the sky: the clouds are shutting round me, & seem depressed with moisture: every thing tunes the soul to melancholy. Guess what my feelings are, when the most soothing & consolatory thought that occurs, is a temporary remission & oblivion in your affections.' Scarcely had he set this down when Mary's second letter arrived.

Godwin's lingering on the return road upset Mary further, prolonged as it was by his wish to view a Lady Godiva (the legendary benefactress of the poor who had to pay by riding naked through Coventry). This was to be re-enacted at a Birmingham fair. The deferred date of return was Monday 19 June. Though he had warned Mary not to fix on a particular hour, she was still waiting at nearly midnight when the scab that had formed over Imlay's defection broke open afresh:

> Previous to your departure I requested you not to torment me by leaving the day of your return undecided. But whatever tenderness you took away with you seems to have evaporated in the journey, and new objects – and the homage of vulgar minds, restored you to your icy Philosophy.
>
> . . . I . . . approved of your visit to Mr. Bage [who had praised Mary Wollstonecraft in his novel *Hermsprong*] – But a *shew* [Lady Godiva] which you waited to see & did not see, appears to have been equally attractive . . . In short – your being so late to night, and the chance of your not coming, shews so little consideration, that unless you suppose me a stick or a stone, you must have forgot to think – as well as to feel . . . I am afraid to add what I feel – Good-night. –

This was not the last of their tiffs. But Godwin was not Gilbert Imlay, and Mary had yet to see the full extent of his devotion.

To be a wife and mother was Mary Wollstonecraft's last role, again unorthodox. Her marriage was, in Virginia Woolf's words, 'an experiment, as Mary's life had been an experiment from the start, an attempt to make human conventions conform more closely to human needs'. At times it seemed to Mary, as she mused in the shade of her green blind, that her experiment in marriage was not the new narrative she had meant to forge. 'I am . . . thrown out of my track,' she thought, 'and have not traced another.' How easy to feel sidelined when Godwin dined with Johnson or dropped in on Fuseli and others who had formed part of her single social life, from whom she was now cut off by ostracism or by the agreement that she and Godwin should not go out together. Mary had to explain to him that she wanted company.

In testing out a new form of marriage, both made mistakes. If Mary neglected to write to Godwin as often as he to her and if she was excessively provoked by the Godiva delay, he reproached her for 'savage resentment', calling it 'the worst of vices'. Mary reproached him in return as a lover: 'There is certainly an original defect in my mind – for the cruelest experience will not eradicate the foolish tendency I have to cherish, and expect to meet with, romantic tenderness.' Imlay's counter-drama had been a careless sensuality. Godwin's counter-drama was a measured reserve – a reserve coexisting with his kindness and loyalty. Mary responded either by abasing herself in ironically feminine terms, or pre-empting Godwin with expectation. Either way, romantic drama was not easy to put into action – going back to the time when they had bungled their first attempts at making love.

Then, too, the heat of her jealousy took him by surprise: her dislike of Mrs Perfection, and her resistance in June–July 1797 when Godwin felt free to visit and correspond with a certain Miss Pinkerton who intended more, Mary warned, 'than comes out of her mouth'.

On the day of Godwin's return from the Midlands, with tensions running high, Miss Pinkerton arrives for tea. There she sits, cup lifted, smiling sweetly while Mary fumes. Godwin called women 'Fairs', and Pinkerton was, in Mary's words, 'a Fair in intellectual distress'. Such women continued to pursue Godwin, and Mary, who herself had been a 'Fair' in distress, detected in her husband a vanity that encouraged pursuit. She went so far as to call him an intellectual 'cocotte' with soft words for the pretty ones (like Miss Pinkerton), and nothing less than 'insanity' for Hays, who was plain. At last, on 9 August, with the baby due in three weeks, Mary persuaded Godwin to back a rebuff to Miss Pinkerton, forbidding her the house unless she consented to 'behave with propriety' towards a married man. With Godwin against her, pretty Pinkerton was forced to apologise – bathed in tears, so she claimed. All the same, she was a predator, beckoning Godwin while his wife's 'portly shadow' met her eye as she walked. Her walk had slowed. In her eighth to ninth month, she was experiencing discomfort, even some pain, as the full-grown baby pushed against the diaphragm.

During this second pregnancy, Wollstonecraft was planning a book to be called 'Letters on the Management of Infants'. The first of the seven Letters was on pregnancy and the management of childbirth. In defiance of the custom for a mother to stay in bed for a month after a birth, she intended to come down for dinner the day after, much as she'd done after Fanny was born.

Wollstonecraft ridiculed the pathologising of women's bodies, the enfeebling invalidism of the middle and upper classes that was fashionable and remained so until the advent of 'natural' childbirth in the mid-twentieth century. Later Letters would offer advice on the infant's diet and clothing, and its changing needs to the end of its second year. Her aim was to make infancy 'more healthy and happy' at a time when a third of infants died. 'I must suppose,' she wrote, 'that there is some error in the modes adopted by mothers and nurses.' She had long thought 'that the cause which renders children as hard to rear as the most fragile plant, is our deviation from simplicity'. By simplicity, Wollstonecraft meant cleanliness and the importance of bathing – not generally recommended in the eighteenth century. Dr Darwin, hearing her views, sent a message via Thomas Wedgwood to caution Mrs Godwin against cold bathing for children. He considered it 'a very dangerous practise'.

The date of this message, 31 July, shows that Wollstonecraft was airing her ideas at the end of her eighth month of pregnancy. She discussed her 'advice' with another friend of Godwin's, a young surgeon of Soho Square called Anthony Carlisle, whose hair stood up in a high crest and whose large ears projected attentively on either side of a gentle, enquiring face. He promised to vet her call to mothers to treat childcare as a profession to be learnt.

She also prepared ten 'Lessons' for Fanny, filled with familiar domestic scenes (including times with 'Papa') designed to hold a child's attention. The 'Lessons' create an imaginary Papa, an omnipresent father figure who takes his domestic character from Godwin. Richard Holmes had an appealing idea that 'Papa' could be Imlay, but Godwin was the man Fanny called

'Papa', and Fanny is constantly invited to compare her able self with helpless baby 'William'. The scenes are projected in the future when Fanny will be four and the baby about six months old. The presence of little William proves that the Lessons were devised while Mary was pregnant with William Godwin's child. This means that Godwin was wrong to suggest they were written as a legacy to Fanny when her mother was about to end her life in October 1795 – a time when he and Mary were not yet friends. The Lessons were certainly a legacy for Fanny but conceived later, when Mary faced the dangers of childbirth. On the back of the manuscript is written: 'The first book of a series which I intended to have written for my unfortunate girl'. Amanda Vickery's research on women's lives in Georgian England reveals that 'it was still not uncommon for pregnant women to prepare themselves for death to the extent of drawing up conduct letters' for children too young to read. Wollstonecraft's Lessons are said to be 'one of the most graceful expressions in English prose of the physical tenderness of a mother's love'.

If 'Papa' had been Imlay, the Lessons would have opened up a trove of untapped memories. We should not be disappointed by the necessity to shift the date. A trove there is, though its scenes derive not from the Imlays at home in Le Havre but from the Godwins at home in London, the domestic unit that formed in April 1797 when Fanny's mother delighted in the affection growing between the child and her new father. This rosy child, 'Fannikin' or 'Lambkin', can draw baby-talk from the sober Godwin as she goes 'plungity-plunge' early in the morning when she jumps out of bed. 'Kiss Fanny for me,' Godwin writes from the Wedgwoods. 'Tell Fanny, I am safely arrived in the land of mugs.' The mug he chose for her had an F in a garland of flowers. Meanwhile, he sends bantering messages about her lost green monkey – the one who, Fanny said, 'went into the country' – who would be back where he belonged when Godwin returns. He admires Fanny as the product of her mother's ideas, a quick, sensitive child who is learning the primary lessons of Wollstonecraft's educational system as set out in *Thoughts on the Education of Daughters* and tested in Ireland on Margaret, Caroline and Mary King. Godwin thought these Lessons 'struck out a path of her own'. The voice is all-important: it has the breath of life in it, the breath of private attachment – unlike school readers. It reminds Fanny that she used to cry when her face was washed and demonstrates the appeal of cleanliness: 'Wipe your nose. Wash your hands. Dirty hands. Why do you cry? A clean mouth. Shake hands. I love you. Kiss me now. Good girl.'

Here, in another education book, Wollstonecraft turns her imagination to the burgeoning consciousness of the very young child, teaching her how to be, as well as how to read. Her relation with Fanny was robust, physical, and anxious to 'prepare her body and mind to encounter the ills which await her sex'. Godwin preferred to lay off the pressure while he backed the praise. 'The first thing that gives spring & expansion to the infant learner,' he said, 'is praise; not so much perhaps because it gratifies the appetite of vanity, as from a . . . satisfaction in communicated & reciprocal pleasure. To give pleasure to another produces . . . the most animated & unequivocal consciousness of existence.'

Fanny's second lesson is to sense the needs and feelings of others. 'You are wiser than the dog, you must help him.' Theirs is a home where parents help each other when either is ill. 'Papa opened the door very softly, because he loves me . . . When I brought him a cup of camomile tea, he drank it without . . . making an ugly face. He knows that I love him.' When Papa falls asleep on the sofa (Godwin, in fact, was prone to sleep in company), Fanny goes on tiptoe. 'Whisper – whisper', as she asks for her ball. 'Away you went. – Creep – Creep.'

The Lessons don't mimic the child's voice – we are spared the bleating of school readers, cheery or sentimental or righteous. We hear instead the voice of an attentive adult who knows a child for an eager, responsive creature. So Wollstonecraft went on exploding genre after genre: *Thoughts on the Education of Daughters* is not platitude for obedient girls; her *Travels* are not a travelogue; and the naturalness of her children's reader makes nonsense of damped-down girls on the one hand and, on the other, a hothouse exhibit like Queenie Thrale whose head was stuffed with unfelt knowledge. Her mother, Mrs Thrale, had boasted that two-year-old Queenie could repeat the Pater Noster, the three Christian virtues, the signs of the Zodiac, and all the Heathen Deities according to their attributes. No wonder

Queenie grew up to be a chill character who hated her mother. Wollstonecraft's passion for domestic education as well as more attentive childcare was carried on, as we shall see, by her pupil, Margaret King, but not borne out by subsequent feminists who, as Christina Hardyment says, 'have not distinguished themselves in the ranks of practical advisers on child-raising'. Germaine Greer has suggested brightly that mothers could dump brats kibbutz-style in a farmhouse, to be visited when time allows. 'Running through the women's liberation movement has been a thread of hostility to mothers and babies,' comments Ann Daly in *Inventing Motherhood*. 'Few had children themselves when they tapped out their militant demands for equality.'

Mary chose, again, a home delivery. She still thought a midwife more appropriate to women's 'delicacy', but a stronger reason was fear. The establishment of lying-in hospitals in the seventeenth century, and many more in eighteenth-century London, had led to epidemics of childbed fever. Everyone knew that many more women died in hospital – now and then as many as eighty per cent. By the 1790s deaths from home deliveries were, by contrast, five or six per thousand. Mary's own dark view is reflected in Jemima's story: 'I cannot give you an adequate idea of the wretchedness of an hospital . . . The attendants seem to have lost all feeling of compassion in the bustling discharge of their offices; death is so familiar to them, that they are not anxious to ward it off. Every thing appeared to be conducted for the accommodation of the medical men and their pupils, who came to make experiments on the poor . . .'

In 1797, during an epidemic in an Aberdeen hospital, Dr Alexander Gordon wondered if the fever was introduced into the womb by a woman's attendants. The medical establishment took no notice. Perish the thought that doctors themselves might be the cause of death. It was not until the work of Semelweiss, Lister and Pasteur later in the nineteenth century that bacterial infections began to be understood. At the Vienna teaching hospital, Semelweiss noticed that mortality from fever in one birth ward was astronomic, while there were fewer fevers in a ward closed to students. He begged fellow-obstetricians and students to wash their hands – in vain. Wollstonecraft's insistence on cleanliness, later the cornerstone of

Florence Nightingale's revolution in hospital management, was common sense. For what made hospitals so lethal for women was that doctors and particularly students would come straight from dissection of cadavers or from infected patients in other parts of the hospital, and insert an unclean hand deep into the woman's birth canal. The prouder of anatomical know-how doctors became, the more given to internal examination, the more at risk were women in labour. They had actually been safer from sepsis when midwives did little more than catch the baby – though ignorance of course had other dangers. In *The Midwives Book* (1671), a manual still in use in the eighteenth century, the experienced Jane Sharp gave some sound advice, disregarded by the more manipulative doctors who took over the field. Sharp notes that women are in as great if not more danger *after* the birth, and warns midwives to 'be very gentle, for the woman is now grown weak and her womb is quick with feeling'. If the afterbirth does not come away, she goes on, herbal remedies are preferred to manual tearing ('the mid-wife's nails'). She is alert to the sensitivity of the patient and the need for comfort at this moment: 'put the woman to as little trouble as you can, for she hath endured pain enough already'.

Mary's labour began at five on a Wednesday morning, 30 August.

'I have no doubt of seeing the animal today,' she informed Godwin after he went to his rooms. Matter-of-factly she asked him to send a newspaper and a novel to while the time away.

At nine Mrs Blenkinsop, matron and midwife from the Westminster New Lying-In Hospital, arrived. She had seen Mary several times during her pregnancy. Nothing further is known of Mrs Blenkinsop, but midwives, who had low status in England, did not have the training they received in France. A hospital midwife performing a home delivery was a compromise between an untrained midwife and a doctor's manipulations. The hospital where Mrs Blenkinsop was matron had been founded on the southern periphery of London to serve the needs of those who could not afford a home delivery, mainly the wives of tradesmen, soldiers and sailors reduced to want. Unmarried mothers, refused by the fashionable hospitals, were admitted at the Westminster. Though the fashionable maternity hospitals were closer to the Godwins' northern suburb – there was one in

Store Street where Mary used to live – her identification with the poor may explain her choice. Certainly, she and Godwin had no money to spare.

Mrs Blenkinsop confirmed that the child was on its way, but so far contractions were slow. 'I am very well,' Mary reassured Godwin. 'Call before dinner time, unless you receive another message from me – .'

At two in the afternoon, she went upstairs to her bedroom on the top floor. At three she sent a third reassuring note to Godwin: 'Mrs Blenkinsop tells me that I am in a most natural state, and can promise me a safe delivery – But that I must have a little patience.' Calm as she sounds, the last words echo at difficult times in her life, going back to her mother's deathbed.

Unable to work, Godwin took *Mary* off his shelf and reread two-thirds of it. He dined at the Reveleys, and supped with a radical editor, John Fenwick, and his wife Eliza. At seven in the evening Godwin returned to Evesham Buildings. At ten, he walked back to the Polygon, accompanied by Mrs Fenwick, to find Mary in the final stages of what had turned out a long labour – different from the ease of Fanny's birth. Mary had asked Godwin not to come to her room until all was over. While he waited in the parlour on the ground floor, Mrs Fenwick went upstairs to help. Until the end of the eighteenth century women usually gave birth in a crouching position that had the advantage of straightening the unfortunate curve in the human birth canal; from about this time it became customary for women to lie supine for the convenience of their attendants. At last, at 11.20, Godwin notes in his diary, a girl – Mary – was born. And still he waited. An hour passed, and then another, and no invitation came to go upstairs.

By unlucky chance, the placenta had not come away. Normally this organ that supports the baby in the womb is expelled soon after its birth. At two in the morning a worried Mrs Blenkinsop sent Godwin for Dr Louis Poignand, a French physician at her hospital. Godwin had no coach, and may not have found one at that hour of the night. If he did not, it was a long walk from Somers Town to the doctor's house in Parliament Street, Westminster, and an even longer walk to the hospital on the far side of Westminster Bridge, but Godwin did find the doctor.

For the next four hours Dr Poignand tore out the placenta. He pulled it out in bits, torturing and resuscitating his patient, and almost certainly introducing an infection. The placenta, thick with blood vessels, is prone to haemorrhage. Godwin went into the room at four in the morning to find Mary fainting repeatedly from blood loss and the trauma of an operation without anaesthetic. She told him afterwards that she had not known before what pain was.

'I would have died,' she said afterwards, 'but I was determined not to leave you.' For a moment she smiled.

Dr Poignand finished at about eight, believing that he had extracted every piece. Later that morning, Mary expressed her distrust of this rough treatment. She therefore asked Godwin to call in Dr Fordyce. Poignand bristled at what he regarded as interference, when Fordyce arrived at three that afternoon. There was some truth in Poignand's contention that Fordyce was more a lecturer than a practitioner. So angry did Poignand become that three days later, on Sunday morning, he walked out on his patient. That same morning she had a spell of shivering, the first sign of sepsis. Godwin, who was out for a lot of the day, came back to anxious faces and Mary's asking him why he had left her so long. He 'felt a pang' of self-reproach. That evening he witnessed a second shivering fit. 'Every muscle of the body trembled, the teeth chattered, and the bed shook under her.' This went on for about five minutes. When it was over, she said it had felt like a struggle between life and death. Even so, Godwin was not too alarmed: 'her cheerings were so delightful, that I hugged her obstinately to my heart'.

An unanswered question is the delay in calling in the leading authority on puerperal fever, Dr John Clarke, on the staff of the General Lying-In Hospital in Store Street and author of *Practical Essays on . . . the Inflammatory and Febrile Diseases of Lying-In Women*. He had published this book with Joseph Johnson in 1793. One reason for the delay could be that Mary herself refused to see Clarke, known for his attack on midwives' continued presence in a field whose status he wished to raise. At this stage Mary may still have controlled the situation, for Godwin too was in the anti-midwife camp.

Following her three calm messages to Godwin during labour, there are echoes of her voice during the days that followed: her call for Fordyce; her plea for her Berkshire friend, Mrs Cotton, to come and nurse her; her gratitude to Eliza Fenwick for untiring attentions; and her agreement with Dr Fordyce that if her milk were contaminated, the baby must stop feeding at the breast. The abrupt removal of the baby on the Monday – she was taken to join Fanny at Mrs Reveley's – would have left Mary's breasts rock-hard within hours. A bizarre solution was to put puppies to her breasts to draw off the milk. This would have granted some relief in a situation where gradual weaning was not deemed in the baby's interest. While the puppies did their work, Mary managed 'some pleasantry' with Godwin and others in the room.

'Nothing,' Godwin said, 'could exceed the equanimity, the patience and affectionateness of the poor sufferer. I intreated her to recover; I dwelt with trembling fondness on every favourable circumstance; and, as far as it was possible in so dreadful a situation, she, by her smiles and kind speeches, rewarded my affection.'

Joseph Johnson, who had seen Mary on the Friday when she had appeared better, returned that Monday to see her growing worse. At a guess, it was Johnson – her trusted adviser and seasoned publisher of medical texts – who persuaded her that Clarke should take a hand.

Clarke was brought in next day, nearly a week after the birth. It's likely that he blamed the midwife, for Mrs Blenkinsop, who had been called on Sunday and Monday, now disappeared from the scene. The agreed diagnosis was that Dr Poignand had been mistaken in thinking he had removed all the pieces of the placenta. Dr Clarke arrived with some idea of surgery, but judged it too late. Sepsis had spread through the body. In fact, no surgery could have counteracted its fatal course, and Mary's resistance to the surgeon would have been sound in that it spared her the unspeakable agony of another operation without anaesthetic. In desperation, the following day Godwin called in yet another doctor, their friend Carlisle, who prescribed the routine palliative, a wine diet with an idea of keeping up the patient's strength. Godwin was opposed to wine. Spirits would demean the spirit his wife continued to show, teasing Godwin for his refusal of divine consolations.

'I feel in Heaven,' she told him.

'I suppose, my dear, that that is a form of saying you are less in pain.'

It wasn't unfeeling rationalism, as it seemed later to Victorians; it was their old repartee of mind and feeling. Even in the shadow of death, they still joked. They were still themselves. Reluctantly, Godwin resigned their spark to the doctors' distorting regime and, starting on Wednesday afternoon at about four, fed Mary as much wine as she would take over several hours. Of course, it was futile. By ten o'clock on Thursday night, Carlisle, who never left the house, had to tell his friend that his wife was dying.

Whatever the effects of the wine on Wednesday and Thursday, on Friday Mary's head cleared. For Godwin's diary records a 'solemn communication' as she faced the 'Idea of Death'. Mary always had a robust constitution, and lingered longer than doctors predicted. On Saturday morning, Godwin spoke to her of Fanny and Mary. What did she wish for her daughters? He would try to fulfil that wish, even though she – the author of *Thoughts on the Education of Daughters* – had been supremely fitted to care for them, while he, alas, was not. His reasonable voice, addressing a prone and silent form, asked her guidance. No answer. Then, after he repeated the question a few times, she managed to reply, 'I know what you are thinking of.' Her last conscious words were that Godwin was 'the kindest, best man in the world'.

That night he stayed by her side till one in the morning, when the faithful Carlisle persuaded him to sleep, promising to call in case of a change. The doctor woke him at six, and he sat with Mary until she died at twenty minutes to eight. His diary for 10 September 1797 sets down the precise time, underscored by three wordless lines across the page.

Fanny was returned home for dinner later that day, and then sent back to Mrs Reveley. It would have been customary to take the child upstairs to kiss her mother goodbye. Godwin sat with a copy of *Thoughts on the Education of Daughters* to hand – though he could not bring himself to read. The 'poor animals' his wife had left behind were 'hers'; he could not feel as a father, and spoke of the newborn as 'it'. During the following week 'it' became alarmingly ill at Mrs Reveley's; doctors were called; then little Mary took a turn for the better when she came home to Mrs Fenwick's skilled arms.

Mrs Fenwick stayed on for two nights to ensure her recovery. At Godwin's request, she wrote to Everina, who was to tell Bess and assure her of Mary's fondness. In this letter Mrs Fenwick reports Godwin's 'unremitting and devoted attentions' to his wife, and her gratitude for 'the kindest, best man in the world'. The baby, she adds, is the finest she ever saw.

The Wollstonecraft sisters did not melt. Godwin received no word from them.

'Be sure I feel it,' he declared to a friend. 'Be sure I am not the fool to look for that happiness in any future vicissitude of life, that I was beginning to enjoy, when I was thus dreadfully deprived of it. My understanding was enlarged, my heart was improved, as well as the most invaluable sensations of admiration & delight produced in me by her society.' Privately, he thought Mary had been too quick in conceiving resentments, yet 'they left no hateful and humiliating remembrances behind them'. His passion had always been to associate with the intellectually great. Godwin never thought to exclude women from this category, and when it came to his wife, he 'honoured her intellectual powers and the . . . generosity of her propensities'. It was characteristic of him to slight 'mere' tenderness. 'Mere tenderness,' he ruminated, 'would not have been adequate to produce the happiness we experienced.' Mary had been glad, she once told him, 'to discover great powers of mind in you', but remained convinced 'that the strongest affection is the most involuntary'. She wished he would find himself, rather, 'enamoured of some fugitive charm that seeking somewhere, you find every where: yes,' her voice went on, teasing, transforming the austerity of his singleness, 'I would fain live in your heart and employ your imagination – Am I not very reasonable?'

15

SLANDERS

A contest for Mary Wollstonecraft's memory began with her death. On the day she died, Joseph Johnson wrote to Godwin: 'I know her too well not to admire and love her. Your loss is irreparable. May you know the same strength of mind which . . . she exerted for your support.' Two days later Johnson's voice changed. He laid a claim, an undeniable claim of prior knowledge, when, on 12 September, he discovered that Godwin meant to exclude Henry Fuseli from the circle of legitimate mourners – a circle including Johnson himself, her portraitist Opie and her early adviser John Hewlett – who were to be invited to her funeral. Here is Johnson's protest:

> Dear Sir
> In the list you shew'd this morning I did not observe the name of Fuseli. It is true that of late he was not intimate with Mrs. Godwin, but from circumstances that I am acquainted with I think he was not to be blamed for it; before this they were so intimate and spent so many happy hours in my house that I think I may say he was the first of her friends, indeed next to ourselves I believe no one had a juster sense of her worth or more laments her loss.

The first of her friends: this is about more than an invitation to a burial; it's about constructions of the past. Fuseli's spitfire speech lurks behind

Johnson's persuasiveness when he asks Godwin to grant Fuseli his part in Mary Wollstonecraft's life. Who owns the great? Who has the right to mourn and remember? Already, a struggle for possession had begun, starting with the question: what version of this woman's life will be transmitted to posterity? In his quietly firm way, Johnson reminds Godwin there had been others who had loved Mary Wollstonecraft and helped make her what she was.

On 15 September she was buried in the churchyard of Old St Pancras, where she had married five months before. Godwin did not attend. Instead, he sat brooding in Marshall's lodgings, a return to his bachelor companion. 'I firmly believe there does not exist her equal in the world,' he wrote to Holcroft. He picked up *Mary* where he had left off the day she had been in labour, and read the unfinished manuscript of *The Wrongs of Woman*. Then he plunged into her letters. Those to Imlay he thought the most touching love-letters in the language. It came to him that his wife had been a female Werther, suicidal, doomed. They had read Goethe's novel *The Sorrows of Young Werther* aloud on 29 August, the night before she had gone into labour, and its arch-Romantic image of surrender to destructive emotion was presented to the public in Godwin's *Memoirs of the Author of a Vindication of the Rights of Woman*, published in January 1798. The resilience so marked in the course of her life was somewhat obscured by Godwin's own sorrow. In the weeks following her death – while he reeled off the memoir – it was he who gave way to melancholy, finding in it a Werther sort of relish foreign to his wife's temperament. A white-faced portrait shows him in mourning at the age of forty-two, less angular than as a bachelor, and neater; his thinning hair, cut short, is parted in the centre. He looks out above his jutting chin with sombre eyes. Godwin's unaccustomed surrender to emotion was all the more dangerous for his illusion of control.

'I love to cherish melancholy,' he confided to Mary's friend Mrs Cotton. 'I love to tread the edge of intellectual danger, and just to keep within the line which every moral and intellectual consideration forbids me to overstep, and in the indulgence and this vigilance I place my present luxury.' His vigilance did not prevent his offence to Fuseli, nor rows with Mary Hays and Mrs Inchbald, nor his frostiness towards the Wollstonecraft sisters which spared no thought for their status as Fanny's aunts, the only blood relatives who might one day interest themselves in the girl's welfare. Godwin's grief in the last quarter of 1797 took him out of his measured habits in ways that were to have consequences for the reputation of Mary Wollstonecraft and for the fate of her beloved Fanny.

When Mary Hays had visited Mary Wollstonecraft five days before her death, Godwin had not encouraged her to stay. From his point of view he had enough attendants, but all were habitués of his own circle. As Wollstonecraft's loyal friend and the one who had brought the Godwins together, Hays protested that she was 'not altogether insignificant'.

Godwin jumped on what he mistook for vanity. 'To speak frankly, I think you have forgotten a little of that simplicity & unpresuming mildness, which so well becomes a woman.'

To someone of spirit this smacked of 'tyranny'. So Hays rebuked a thinker who had made tyrants his target.

Godwin said she 'poisoned' the roots of friendship and deserved to lose him as her mentor. 'We are at present, twin stars that cannot shine in the same hemisphere.' Despite the awful solemnity of this pronouncement, they did visit now and then. Mrs Inchbald was cast off more completely. Godwin wrote to four people on the day of Mary Wollstonecraft's death, and Mrs Inchbald was one of them. 'My wife died at eight this morning,' he wrote. 'I always thought you used her ill, but I forgive you.' Mrs Inchbald offered polite condolences without the intensity of regret Godwin required from a one-time friend who had snubbed Mary in public. Suddenly, control snapped.

'I must endeavour to be understood as to the unworthy behaviour with which I charge you towards my wife,' he began. 'I think your conversation with her that night at the play base, cruel, and insulting . . . I think . . . that you have an understanding capable of doing some small degree of justice to her merits. I think you should have had the magnanimity and self-respect to have shewed this. I think that while the Twisses and others were sacrificing to what they were silly enough to think a proper etiquette, a person so out of all comparison their superior, as you are, should have placed her pride in acting upon better principles, and in courting and distinguishing insulted greatness and worth; I think that you chose a mean and pitiful conduct' – and so on.

Mrs Perfection was unrepentant. She needled Godwin's wound by reminding him of her reluctance to know Mary Wollstonecraft. Perfection could not, as she put it, 'sufficiently applaud my own penetration in apprehending, and my own firmness in resisting, a longer and more familiar acquaintance'. She proceeded to put 'an end to our acquaintance *for ever*'. This was accompanied by an offer of 'the most perfect forgiveness of all you have said to me'. She added, 'I respect *your prejudices*, but I also respect *my own*.'

By the time Godwin received this on 26 October, he had already determined on a public vindication of his wife. In the past he had attacked public wrong through his pen, most effectively with the collapse of the Treason Trials in 1794. At that time he had been prudent enough to act anonymously; this time he meant to use his wife's name and fame in defiance of gendered morals. He burned to express his outrage that 'the firmest champion, and, as I strongly suspect, the greatest ornament her sex ever had to boast' had been snubbed by those who condoned 'the dull and insolent dictators, the gamblers, and demireps of polished society'.

So a fortnight after his wife died, Godwin began to set down his version of a life he had known only in its final phase – less than two of Mary's thirty-eight and a half years. Needing basic information, especially on her early years, he turned to a man who had known Mary Wollstonecraft only at second hand, apart from her one month in Lisbon in November–December 1785. In his most formal manner, Godwin wrote to Fanny Blood's husband Hugh Skeys in Dublin. This letter appoints Skeys to question the Wollstonecraft sisters, and Skeys may have felt flattered by Godwin's dependence, but it's hardly an approach to appeal to the sisters. Only two weeks after Mary was buried, Godwin, who has not so far addressed a word to her sisters himself, fails to imagine *their* loss. Instead he makes a case for biography and for himself as biographer, a case where logic gets lost in insistence. For even as he claims to 'know a good deal respecting every period of her life', he is forced to own, 'I am not well prepared on the subject', and has to limit his surprisingly uninformed queries as to the factual shell:

[early] Oct 1797

. . . I should be glad to be informed respecting the schools she was sent to, & any other anecdotes of her girlish years. I wish to obtain the maiden name of her mother, & any circumstances respecting her father's or her mother's families. Her sisters probably could tell some things that would be useful to me respecting the period when they lived together at Newington Green. I am doubtful, for instance, whether she did or did not frequent Dr. Price's meeting house. You must know many things respecting her . . . I think the world is entitled to some information respecting persons that have enlightened & improved it. I believe it is a tribute due to the memory of such persons, as I [am] strongly of the opinion that the more intimately we are acquainted with their hearts, the more we shall be brought to respect and love them.

Remember me in a very kind manner to [the second] Mrs Skeys & my wife's sisters.

I am, sir,

With very great respect,

W. Godwin

Mrs Fenwick . . . wrote rather a long letter to Miss Wollstonecraft, about a fortnight ago. You do not mention whether she has received it.

Why does Godwin not communicate directly? In part he may be sore that the sisters have not seen fit to reply. Neither they nor Godwin seemed able to heal the estrangement, and were compelled to communicate through intermediaries. The coldness during Everina's last visit to Mary in February 1797 had reappeared in Godwin's report on Everina to Mary when he visited the Wedgwoods in Staffordshire: 'Your sister would not come down to see me last night at supper, but we met at breakfast this morning. I have nothing to say about her.' Mary had been rueful: 'I supposed that Everina would assume some airs at seeing you – she has very mistaken notions of dignity of character.' Neither makes allowance for the inferior position of a governess meeting her employers' guests, and Mary seems to forget her own trials of that kind. Eventually, it was Everina who crossed the barrier and wrote directly to Godwin about his 'hurry'. By 17 October, the posthumous manuscripts were already with the press. A month after the funeral, he was not only well into the *Memoirs*, he had also performed the massive editing job required by the half-finished *Wrongs of Woman*. By 15 November, two months after the funeral, Godwin had a complete draft of the memoir. He proceeded to give it a scant four days for revision. His rationale was 'that the public curiosity was most excited relative to an eminent person by publications that appeared in no long time after their decease'.

The Wollstonecrafts were taken aback to discover that Godwin intended a more candid and detailed life than they had been led to believe. Everina's first and just concern was biographic accuracy:

> When Eliza and I first learnt your intention of publishing immediately my sister Mary's life, we concluded, that you only meant a sketch to prevent your design concerning her memoirs from being anticipated. We thought your application to us rather premature, and had no intention of satisfying your demand till we found that Skeys had proffered our assistance without our knowledge – he then requested us to answer his questions, and give him dates, which we complyed with, though reluctantly. At a future day we would willingly have given whatever information was necessary; and even now we would not have shrunk from the task, however anxious we may be to avoid reviving the recollections it would raise, or loath to fall into the pain of thoughts it must lead to, did we suppose it possible to accomplish the work you have undertaken in the time you specify. The questions you have addressed to me confirm this opinion; and I am sorry to perceive you are inclined to be minute, when I think it is impossible for you to be even tolerably accurate.

My sister Mary's life: Everina reminds Godwin – as Johnson and Hays had tried to remind him – that he alone did not own Mary's memory. Of these protests, Everina's is the most pointed. Godwin took no notice, and excluded his wife's sisters from the memoir, except as peripheral recipients of her largesse. He used friction between them and Mary to press his own 'maxim' against 'cohabitation'; and he saw fit to publish a judgement on the sisters' jealous inferiority. Another biographer might have tried to draw out their invaluable knowledge, as well as the trove of letters in their possession. He might have noted, for a start, their choice of independence over marriage, and they could have alerted him to his error in blaming the caustic note in Mary's voice on the influence of Fuseli. As her sisters would have known, that voice went back to her schooldays: 'I have a heart that scorns disguise, and a countenance which will not dissemble,' she had said at fifteen. Godwin possessed a different Mary from what he called the 'harsh', 'rigid' author of the *Rights of Woman.* For Godwin had fallen in love with a woman wounded almost to death – a suicidal Dido abandoned by her sea-going lover, a 'female Werther', a creature of 'exquisite sensibility' in a state of intractable depression, who had come to him for help. This familiar plot – the rescue of a Fair in distress – Godwin laid like a grid over an uncategorised creature with a plot of her own in the making. That grid has fixed Mary Wollstonecraft's image in the public mind for more than two centuries.

There were two bases to Godwin's power over posterity. In positioning himself as her intimate and vindicator, he gained one form of authority; in positioning himself as a man who must tell the truth, even if it told against the wife he meant to vindicate, he gained another. And these two forms of authority reinforced each other with a conviction that continued to distract readers from the fact that some of his views are skewed. The second *Vindication,* he thought, contained sentiments 'of a rather masculine description . . . There are also, it must be confessed, occasional passages of a stern and rugged feature', offset, he hastens to assure us, by passages of 'trembling' feminine delicacy. 'The *Vindication of the Rights of Woman*', he concludes, 'is undoubtedly a very unequal performance, and eminently deficient in method and arrangement. When tried by the . . . long-established laws of literary composition, it can scarcely maintain its claim to be placed in the first class of human productions.' Such attitudes were rooted in Godwin's first clash with Mary Wollstonecraft at Johnson's dinner table in November 1791. Her rebuke of Burke still seemed to him 'too contemptuous and intemperate'. How dared a woman attack a 'great man', whatever he had done? He was willing to grant that *men* of her persuasion were justifiably 'inflamed' by Burke's desertion of human rights. Godwin is plainly speaking for himself when he reports how readers expected to find 'a sturdy, muscular, raw-boned virago' (hardly softened in the second edition to 'a rude, pedantic, dictatorial virago'), but found instead 'a woman, lovely in her person, and . . . feminine in her manners'.

Godwin did point to what he calls 'genius'. 'She felt herself alone . . . in the great mass of her species', he recalls, and elsewhere touches on Mary's 'firmness of mind, an unconquerable greatness of soul, by which, after a short internal struggle, she was accustomed to rise above difficulties and suffering'. But these rays are lost in fogged versions of Fanny Blood and Fuseli. Godwin grants too little to what didn't fit a female Werther. Armed with ready-made fictions, he offers doomed narratives from Virgil and Goethe in place of the open-ended narrative of a woman who refused to walk 'the beaten track'. Everina, we recall, was the person to whom Mary had confided her venture ten years before. She had trusted Everina to understand something of the character who would take such a course. When Godwin shut the valves of his attention to Everina's warning, he was engaged in bringing Mary's exploratory path into line with recognised fictions and with his own doctrines of free love and atheism. When he reports her attendance at Dr Price's sermons, he makes a gratuitous point that she did not succumb to 'superstition' – not a word Mary herself would have used, and certainly not in relation to a minister she revered. Godwin's claim that Mary manifested no faith on her deathbed proved damaging. Not only was the claim, again, gratuitous – as we know, she was barely able to speak during the last twenty-four hours of her illness – it was also emotionally untrue, for as he himself explains earlier, she held to a faith of her own:

> When she walked amidst the wonders of nature, she was accustomed to converse with her God. To her mind he was pictured as not less amiable, generous and kind, than great, wise and exalted. In fact, she had received few lessons of religion in her youth, and her religion was almost entirely of her own creation. But she was not on that account less attached to it,

> or the less scrupulous in discharging what she considered as its duties. She could not recollect the time when she had believed the doctrine of future punishments. The tenets of her system were the growth of her own moral taste, and her religion therefore had always been a gratification, never a terror to her. She expected a future state; but she would not allow her ideas of that future state to be modified by the notions of judgment and retribution.

Godwin's ability, here, to see into character matched a subject who required it. At times, though, another manner undercut this gift. Godwin believed that biographic honesty honoured the Enlightenment ideals he and Mary shared, but when he compels himself to state a private fact, that they became lovers, he is blunt – almost defiant: 'We did not marry.' Then, too, he ignores Imlay's more complex character as an author of the American frontier and, as such, bound to appeal to a woman who had turned from the beaten track. Godwin was not likely to forget Mary's confession of the 'rapture' she had found in Imlay's arms; he was certainly gripped by the love-letters that came into his possession after her death. In narrowing Imlay's appeal to sexual charm, Godwin did further unintended damage to his wife's reputation. For he assumed, in his bluntest manner, that a virgin of thirty-four must have 'panted in secret' for a man, before he goes on to explain that, for Mary, love was sacred, imaginative, intensely loyal. Readers, fixed on the former, disregarded the latter. They saw a wanton, wafting from Fuseli, to Imlay, to Godwin himself. In 1798 a posthumous image appeared of a brash, unintelligent face under a masculine hat, said to have been engraved after a vanished portrait of Mary Wollstonecraft. Its authenticity has gone unquestioned, but it's nothing like the intellectual seriousness of the authenticated Opie portraits or the austere Liverpool one. Her case for women's rights was dismissed as 'scripture . . . for propagating w[hore]s', and Thomas J. Mathias, a leading antifeminist, reduced her range to shallow caprice:

> Fierce passion's slave, she veer'd with every gust,
> Love, Rights, and Wrongs, Philosophy, and Lust.

The even more venomous 'Vision of Liberty', published anonymously in 1801 in the *Anti-Jacobin* (a counter-revolutionary journal put out by Pitt's Treasury), taunts Godwin for his 'simple' candour, and even for misplaced grief:

> For Mary verily would wear the breeches –
> God help poor silly men from such usurping b——s.

Her friends were dismayed. Southey thought Godwin had 'stripped her naked'. Roscoe, her supporter in Liverpool, spoke of his 'heart of stone'.

Godwin shocked readers in a different way with Mary's obstetric complications. His witness to the agony of an operation without anaesthetic and their exchanges during her last days draws us into that tragedy. In his restrained way he is wonderfully true to his own tie with Mary Wollstonecraft. But he remained ill informed about her contacts in France. Joel Barlow or Tom Paine could have told him that she was distanced from the prime actors in the Revolution, and mixed almost entirely with other expatriates. As for Mary's summer in Scandinavia, Godwin was forced to sidestep the business of that journey, and censor her letters to Imlay. During the very month the *Memoirs* came out Godwin was in touch with Imlay's English associate, Mr Cowie, who had advanced funds to Mary Wollstonecraft in view of what was due to her for her efforts in Scandinavia. Either Godwin chose not to question Cowie, or suppressed what information he had. Nor did he test the evidence of Mary's real letters against the fictional letters of the *Travels* – a pity, for he destroyed the real letters when the censored copies were published. Nor do the *Memoirs* explore her entry into Hamburg's heart of darkness. This is because Godwin narrowed her motives to emotional fits and starts: in his version, Mary goes to France not because she wished to see the Revolution, but to escape her crush on Fuseli; and she abandons Hamburg not because she was appalled by fraud (as the *Travels* tell us), but solely in response to Imlay's neglect.

Again, this was entirely well meant. Godwin had lent himself to the emotional expressiveness of her role in their private domestic drama, but it led him to play down her rationality and the amount of time she devoted to the life of the mind. The claim of his Preface to transmit word-of-mouth fact, on the basis of notes made during Mary's lifetime, remained unquestioned until 1876 when Godwin's first biographer, Charles Kegan Paul, cast doubt on Godwin's version of Fuseli's story. Richard Holmes revived this doubt in *Footsteps* in 1985. The time has come to go back to Everina's warning of wider inaccuracy.

Godwin lacked the letters that would have filled in Mary's schooldays in Yorkshire; her friend Jane Arden, now Jane Gardiner – for she too had married in 1797 – produced them a year late, in January 1799. Afterwards she met Godwin, but those were only social occasions; he didn't come to know her. Fanny Blood's letters to the Wollstonecraft sisters would have revealed a more humorous and professional character than Godwin's notion (derived most likely from what became a common misreading of *Mary: A Fiction*) that Fanny was a feeble figure. Godwin's portrait was designed to offset Mary's energy of purpose: 'Fanny, on the contrary, was a woman of a timid and irresolute nature, accustomed to yield to difficulties, and probably priding herself in this morbid softness of her temper.' The real Fanny was almost the opposite: steady, sensible, guiding her family and screening her pains out of consideration for others (in the manner of Elinor Dashwood in *Sense and Sensibility*). Mary's letters to her sisters would have lit up the scene of Fanny's death in Lisbon, and, even more vividly, scenes in Ireland: Mary's midnight reveries at Mitchelstown Castle, her dislike of Lady Kingsborough, her satire on Dublin society from behind her mask, the Viceroy's ball, and the radicalisation of Margaret King. The letters in the Wollstonecrafts' possession would have led Godwin to the Revd Dr Gabell. The young Oxford graduate who had debated theology with Mary Wollstonecraft on the packet to Ireland in the autumn of 1786 was another who kept her letters. In time, he became headmaster of Winchester College. Jane Austen, who spent her last weeks in Winchester in 1817, could view 'Dr Gabell's garden' from her window.

For all this, Godwin had two advantages. There were indeed things Mary had told him. She had talked of her formative years at Newington Green: the goodness of Mrs Burgh, and Dr Price as prime mover of her political life. Of her time in Ireland, she had singled out George Ogle as the most perfect gentleman she had known. The persistence of these memories reminds us how much she loved benevolence – how reassuring it was to find it in the world to which an unsupported woman in the eighteenth century was exposed. Godwin's other advantage was access to Mary's fourth benefactor, Joseph Johnson. He read her letters to Johnson, and gave *Mary: A Fiction* its due.

Unfortunately, Godwin errs in his judgement of everything else Mary Wollstonecraft produced for Johnson during 1787–92, nor does he commend her subsequent history of the French Revolution, though this book was astute enough for President Adams to read it twice with minute attention. When Wollstonecraft politicises the importance of women and domestic relations, Adams concedes the amoral basis of standard political dramas of greed and commerce demanding the increase of armies and navies in order to protect property. Is this not 'the true secret' of French and English policies? Adams asks himself in the margin. 'And is not this the secret of the inscrutable conduct of our American Congress?' Adams fears he might be 'visionary' if he accepts the idea. He might find himself tugged outside his usual habits of mind by Wollstonecraft's portrait of an alternative civilisation governed by domestic affections – a superior domesticity of a kind he has known with Abigail, another champion of women's rights. When Wollstonecraft looks to a glorious era when 'fool' and 'tyrant' will be synonymous, Adams echoes her. 'Amen and Amen! Glorious era come quickly!'

Godwin's training at the Hoxton Academy had stressed a slow, measured logic. He was astonished at Wollstonecraft's speed and insistence, and did not see that she proceeds less as philosopher and more as an orator for whom repetition is part of the impact. Godwin's influential reservation about a want of order in the *Rights of Woman* became almost obligatory for future critics. Godwin also found the book 'amazonian'. From about 1795 this word came into use, as the counter-revolution turned against boldness in women. Erasmus Darwin said, 'great eminence in almost anything is sometimes injurious to a young lady; whose temper and disposition should appear pliant rather than robust'. Women themselves preached a retreat from independence, experiment and politics. Anti-heroines based on Mary Wollstonecraft abound. Harriot Freke in Maria Edgeworth's *Belinda* (1801), Adeline Mowbray in a novel of that name (1804), and Elinor Joddrel, the rash spokeswoman for 'the Rights of human nature' in Fanny Burney's *The Wanderer; or Female Difficulties* (begun before 1800, published in 1814) are amongst numerous warnings against independent girls who wreck their lives. Mrs Opie, the author of *Adeline Mowbray* was none other than Mary Wollstonecraft's one-time friend Amelia Alderson. When Mary had married in the spring of 1797, Opie had turned his attention to Amelia Alderson, and they had married. Social pressure led Amelia Opie to show up a mistaken heroine who refuses marriage and lives with the man she loves. But the most influential opposition came from Hannah More's best-selling *Coelebs in Search of a Wife* (1808). This novel more than any other voice of the counter-revolution turned minds back to the separate spheres of the sexes. During the nineteenth century women's sphere was modelled in some ways on the dependent role of wives in the propertied upper classes, and unmarried mothers lost the right to sue the father for maintenance.

The scandal of Godwin's revelations happened to coincide with a greater scandal. The youngest of Mary Wollstonecraft's charges at Mitchelstown Castle was Mary King, who had been six years old in 1786–7. Her mother Lady Kingsborough, won over by Wollstonecraft's methods of reason and affection, had granted her a free hand. Her ladyship even took to showing off her daughters' governess – author of *Thoughts on the Education of Daughters* – in Dublin society. In September 1797, at the time of Wollstonecraft's death, Mary King, aged sixteen, was living at Windsor with her mother. Viscount and Lady Kingsborough, ill-matched (as the governess had observed), had separated in 1789, and Kingsborough had introduced a mistress, Elinor Halleran, into his wife's ancestral home. She had borne him two children.

In Wollstonecraft's time with the Kings, an extra boy had been attached to the family. His name was Henry Gerald Fitzgerald, and he was probably a first cousin of Lady Kingsborough, the son of her father's brother, who was declared by the family to have been illegitimate and orphaned at about fourteen in 1781. The Kings brought him up with their own children –

though they didn't send him with their sons to Eton, followed by Exeter College, Oxford. So he was set apart educationally, despite superior abilities, nor was he to share their prospective wealth. Kingsborough must have bought him a commission in the army, where he rose rapidly through the ranks. Still, it may have rankled that he lacked the advantage of the King son destined for the army – another Henry – who was sent to Oxford in 1794. In September 1797, as Henry Fitzgerald enters our story, he was about thirty, a lieutenant-colonel in the Guards, married three years, with two children, and living further up the Thames from Windsor, in Bishopsgate. Mary King – parted from her errant father – fell in love with this relation about twice her age.

She had a graceful figure and a profusion of long hair like her mother's. Fitzgerald, who was a frequent visitor in Windsor, began to pay Mary attentions – the servants noticed; no one else. Suddenly, Mary disappeared. A hue and cry followed. Advertisements were placarded about London offering a massive reward for information. One day, a servant appeared at the King's door with the story that, at the time of Lady Mary's disappearance, a gentleman had brought a girl to the lodgings where she worked at Clayton Street, Kennington. He had slept with her that Sunday night, departed at six next morning, and returned often. The maid had noted the luxuriant hair (mentioned in the advertisement), and spied the girl cutting hers off. Then, at the very moment the servant was relaying her suspicion to Lady Kingsborough, Fitzgerald turned up.

'There is the gentleman who visits the young lady!' the servant cried.

Fitzgerald dashed from the house, pursued by Mary's brother Colonel Robert King, the second of the Kingsborough sons, who then challenged Fitzgerald to a duel. (It was the customary way to settle differences, especially in Ireland where a young man who had not 'blazed away' was rather looked down on in Dublin society.) The duel, with pistols, took place on Sunday 1 October near the Magazine in Hyde Park, even though Fitzgerald was unable to find a second. We need to pause at this surprising fact. Why would no fellow-officer support him? A sexual escapade was hardly unusual. Kingsborough, after all, was a copulating rabbit. His eldest, Big George, had whisked a Miss Johnstone to the West Indies and produced three illegitimate children before he came back to marry Lady Helena Mount Cashell. So, we might ask, is Fitzgerald cast out because of the superior status of Lady Mary, or because he betrays the family who took him in? Is it the power game of the outsider – the adopted son – to inseminate a girl of the true line? If Henry Fitzgerald had been loved and looked up to as an older brother, elements of emotional incest and exploitation would explain the level of public shock. What is certain is that this was no gentlemanly duel: it was rage of brother against brother.

The duel ended with no injury to either side. Fitzgerald refused to grant he was in the wrong. Meanwhile, Mary's maid was dismissed and the girl restored to her mother, who carried her off to Mitchelstown Castle. Her father followed after warning Fitzgerald: 'If you ever presume to appear where I or any part of my family may happen to be, depend upon it the Consequences will be Fatal.'

Undeterred, Fitzgerald, disguised, followed the girl in December. He put up at the King's Arms, a hotel near the Castle's gate. Barry, the innkeeper, became suspicious of a man who left his room only at night. Scenting danger, Fitzgerald then moved to the Kilworth Hotel, seven miles from the Castle and close to where Mary's eldest sister Margaret now lived, at Moore Park. It's hard to be invisible as a stranger in a village. A hint of his presence brought Kingsborough and Robert hotfoot to the hideout. On the night of 11 December they forced Henry's door and burst in. As he reached for his pistol and Robert grappled for it, Kingsborough shot Henry dead.

'God!' he exclaimed to Margaret. 'I don't know how I did it; but I most sincerely wish it had been by some other hand than mine.' Margaret later spoke of the family's ' horror'. Kingsborough was duly arrested. Just then, his father died, and he succeeded to the title as the 2nd Earl of Kingston. As a peer he could claim a trial by fellow-peers in the Irish House of Lords. This celebrated trial, decked out with pageantry, took place on 18 May 1798. The prisoner was brought in from the Castle, with the axe carried before him. Dressed in black for the relative he had killed, he knelt to hear the charge. There were no witnesses for the prosecution, and the assembled peers gave a unanimous verdict of not guilty.

One of those present, Bishop Percy, passed on Dublin gossip to his wife together with a copy of Godwin's *Memoirs*, 'a curious Life of a woman who was governess to Lady Kingsborough's . . . Daughters[,] who professed to discharge the Marriage Duties, without submitting to the Marriage Ceremonies. This was exactly what her Ladyship's unfortunate Daur did with her seducer. – They say she lives concealed in this Town, where she has been brought to bed of a Daughter. – Lady K. is said to have discharged the governess after one year's trial, because she wanted to discharge the Marriage Duties, with that lady's husband – such is the report.* – However Lady M[oun]t C[a]sh[e]ll the eldest Daur [*née* Margaret King] glories in having had so clever an Instructress, who had freed her mind from all superstitions . . .' He adds that Margaret Mount Cashell 'made violent complaints', as well as a caricature, of a fellow-bishop who warned against Mary Wollstonecraft in a sermon.

Mary King's baby was disposed of. She was sent to Wales under an assumed name, and placed in the care of a clergyman. Her conversational powers made her a favourite. In 1805 she married George Galbraith Meares of Meares Court, Clifton, and had two daughters. Nothing further is known of her life beyond the fact that she died young, in 1819. It was pure chance that Godwin's memoir was reviewed in the period between the murder in December 1797 and the trial the following May. Sooner or later, the wider public was going to connect Mary Wollstonecraft with the Earl's scandalous daughter.

'A female unrestrained by the obligations of religion, is soon ripe for licentious indecorums,' growls the *European Magazine* in April 1798. Mary Wollstonecraft, it reports,

> accepted the office of Governess to the daughters of Lord Viscount Kingsborough . . . and wonders are told of the salutary effects of her system of education; but when we reflect on what Mr. Godwin is silent about, the misconduct of one of her pupils, who has lately brought disgrace on herself, death on her paramour, risk to the lives of her brother

*This is gossip. Mary Wollstonecraft had a poor opinion of Lord Kingsborough and his womanising.

> and father, and misery to all her relatives; when we consider also Mrs. Godwin's subsequent conduct; we hesitate in giving implicit credit to the eulogium.

It followed that her conduct 'must consign her name to posterity . . . as one whose example . . . would be attended by the most pernicious consequences to society; a philosophical wanton, breaking down bars intended to restrain licentiousness'.

How easy to find a scapegoat for the Earl's murder of his adoptive son in a one-time governess.

Scandal necessitated a revised edition of the *Memoirs* in June 1798. Godwin had to omit the names of Wollstonecraft's contacts who wished to dissociate themselves: the Gascoynes, who had been the Wollstonecrafts' neighbours in Barking, Essex; the Easts in Berkshire; and even Mrs Cotton. Amazingly, Mr Wollstonecraft, alive at the time, is still the 'despot'; and Imlay is still named, on the grounds that Mary's identity as 'Mrs Imlay' could not be concealed. Godwin handles 'Mr. Imlay' with velvet politeness. The effect is rather distantly flat. Wollstonecraft's journal, the *Analytical*, regretted 'the paucity of [Godwin's] means of information', as well as his emphasis on Mary's emotions at the expense of her mind. There was 'no correct history of the formation of Mrs. G's mind. We are neither informed of her favourite books, her hours of study, nor her attainments in languages and philosophy.' The *Analytical* foresaw the scandal, and blamed Godwin. There was too little use of her letters as a corrective to the seeming 'versatility of her attachments'; those who read her letters would 'stand astonished at the fervour, strength and duration of her affection for Imlay'.

Johnson continued to act for Fanny after her mother's death. In April 1798, he shamed Imlay with a 'remonstrance', and began to investigate whether Imlay had an 'attachment' to a mistress unsuited to act as Fanny's mother. The uncertainty of Fanny's future would explain why Johnson was looking into her father's private life. Johnson pursued his enquiries through Imlay's London associate, Mr Cowie, and made him an offer.

Godwin, who wished to keep Fanny, called on Cowie on 16 April. He thought of Fanny as a child his wife 'left behind her who has no friend upon whose heart she had so many claims as upon mine'. Imlay agreed on condition Godwin took over financial responsibility. In February Imlay had lost a court case in Gloucestershire in which he had sued for a residue of £1750 owing to him as dividends from a West Country coal mine he had leased. (The defendant had turned on Imlay with a counter-claim for unpaid promissory notes.) After Imlay and Mary had parted when Fanny was four months old, Mary had tried to stir a father's feeling – in vain, it seems, though Imlay always did acknowledge his duty to support the child. That duty had to do with shame – Mary's last resort when she had placed Fanny in front of him at the Christies. For, as we have seen, Imlay prided himself as an 'upright' citizen.

'I beg you will let the bearer have the trust deed* of Imlay's,' Johnson asked Godwin. 'No satisfactory information from Mr. Cowie respecting an attachment.'

Imlay made a plea to Johnson not to shame him further, and Johnson sent this on to Godwin, on 22 April, with an acid comment that Imlay's letter was 'the most intelligible one that I have seen from that quarter':

> I like to be where I can be most useful – I believe the opinion of the world is always sufficiently secured by an upright & unequivocal conduct, though one should refuse to conform to some of its narrow & censorious maxims – I believe this observation will apply in my case – if so, I beseech you to reflect how brotherly & considerate a conduct it is, to begin the senseless cry of scandal, & incite the world to consider that conduct as faulty, which, I believe, it is inclined to consider as innocent – observe also, that what is done is irremediable, & that, if I were to change my situation upon such a remonstrance as yours, I should encourage, not silence, the tongue of calumny[.]

*Presumably the bond Imlay had offered to take out when he and Mary had parted in 1796, the interest of which was to support Fanny.

Here, for a moment, Imlay stands out in the light, as he fights to return to the shadows. Most telling is his claim to 'upright conduct' without a thought for his child – except to save himself the cost of a bond. We see the defensive wall, the false logic ('what is done is irremediable'), the irrepressible buoyancy he shared with Barlow. It may soothe us to isolate Imlay as scoundrel, but increasing information about the spins and swindles of commerce shows that Imlays, alas, are always with us.

On 14 April 1798 Dorothy Wordsworth notes in her journal that 'Mary Wollstonecraft's life, &c., came'. The weather was stormy, so that evening sister and brother 'staid indoors', perhaps reading Godwin's *Memoirs*. In 1799 Wordsworth composed a poem, 'Ruth', about the contest for the soul of an American in England. If this bears on Imlay, it's more open than Godwin to the ambiguity of the American entrepreneur turning on his moral axis.

In part he is linked with frontier scramblers; in part he retains a 'high intent' that wins Ruth's love. But her beautiful man 'to whom was given/ So much of earth – so much of heaven,/ And such impetuous blood' is corrupted by his associates, whom he corrupts in turn. In Ruth's company he is restored; then his 'better mind' would vanish: 'low desires' and 'new objects did new pleasure give,/ As lawless as before'.

Ambiguity is more disturbing than a confirmed scoundrel. So, too, the real Imlay. Additions to a third edition of his *Topographical Description of the Frontier* in 1797 show the influence of Mary Wollstonecraft was not forgotten. Smuggling corrupts the morals, Imlay concedes, and commerce has taken too deep a root in America 'as in some instances to militate to the injury of philosophy, and the happiness of mankind'. He bears out Dr Price's warning to the Founding Fathers of the United States. Mary Wollstonecraft's protest against fraud when she acted as Imlay's agent makes her the front-runner of present-day protesters in the name of the millions over the globe who are defrauded by corporate greed.

The Imlay–Barlow–Brissot plot to wrest Louisiana from Spain, initiated by Imlay in December 1792, did eventually transpire, though not in an undercover Imlay way. France did regain the territory, and then, in 1803, Napoleon allowed President Jefferson to make the Louisiana Purchase.

Imlay's old associate General Wilkinson, who had once dreamed of a separatist empire in the West, now took possession of these lands – home to several nations of native Americans – in his capacity as commander of the Western Army. He reigned supreme as governor in St Louis, the new capital of the vast territory of Louisiana.

Imlay continued to buy up frontier land. On 19 November 1810 he was granted a deed for 3400 acres in Kentucky. Unlike others in the old records, he gives no place of residence. The burial of one Gilbert Imlay aged seventy-four in 1828 is recorded in the parish register at St Brelade's on the isle of Jersey. As he grew old, did he recall Mary Wollstonecraft's warnings that commerce would wither his heart? 'In the solitude of declining life, I shall be remembered with regret,' she had said. After he gave up his child to Godwin, Imlay disappears from sight. An edge fading from sight is where he had his habitations: the Kentucky frontier; the borders of neutral territory during the European war; and last, the border world of an island lying between France and England, a smugglers' haven. It was a dodging, risky life in which home, wife and child had no place. Rarely seen to act, he worked through agents – including Mary Wollstonecraft.

Imlay's Finnish agent, Elias Backman, found himself rewarded. On 27 February 1797, in the year 'of the Independence of the United States of America the twenty-first', Washington (in his last months as President) and Timothy Pickering as Secretary of State, appointed Backman Consul for the port of Gothenburg and adjacent areas – the first American Consul in Sweden. His ship, the *Maria and Margrethe*, once Imlay's, continued to sail until 1798. That year it was destroyed by a French privateer in a Spanish harbour. Again, Backman had access to high-level diplomacy, but though he lobbied for years, compensation was denied – Talleyrand, acting on behalf of France, dismissed him curtly. A new American Consul was appointed in 1804, when Backman was declared bankrupt. He even had to resign his status as burgess. Some comfort came much later when Sweden crowned a French general, Bernadotte (or Carl XIV Johan, as he was called): Backman's three sons (who had once opened their nursery to little Fanny Imlay) became court officials – the third son secretary to Queen Desirée and eventually to Oscar I.

Another of Imlay's agents, Barlow, who had prospered from his dealings in Hamburg, returned to Paris in October 1797. A further $6000 was due to him for his time in Algiers, where he had successfully represented American (and his own) interests. In 1798 he bought the Hôtel de la Trémoïlle for 430,000 francs. It had a frontage of a hundred and forty feet opposite the Luxembourg Gardens, at 50 rue de Vaugirard. There were eleven bedrooms, cellars that could hold five hundred barrels of wine, and stables for twelve horses. Its courtyard could accommodate five carriages. The front gate was the most magnificent he had seen in Paris and is now a historical monument. Another fortune-hunter, Richard Codman from Boston, who had visited Imlay and Mary Wollstonecraft in Le Havre in the spring of 1794, bought the historic Château de Ternes in 1800. Every so often the Barlows talked of returning to America, but President Adams thought Barlow a 'wretch' and talked of his 'blackness of heart'. So the Barlows stayed on in France, as Napoleon became First Consul and crowned himself Emperor. Barlow's return to America and his later return to France as American Ambassador to Napoleon is another story, marking his triumph over what he termed 'malicious calumnies' in American papers, and marking too an increasing distance from the mirror of Imlay he had been in his Louisiana and silver-ship days. Ruth Barlow, that intelligent and warm-hearted friend who used to breakfast *à deux* with Mary Wollstonecraft and was her confidante while she lived with Imlay, would have known – had Godwin approached her – facts about Imlay and his business with Joel Barlow at the time Mary Wollstonecraft handled the silver in 1794.

Knowing little of Mary's years in France, Godwin makes no mention of the Barlows, and only a passing reference (in a footnote) to her 'particular gratification' from her friendship with the Irish fugitive, Archibald Rowan. Reviewers took 'gratification' to indicate Rowan as yet another of Mary's lovers. Rowan wrote from his log cabin in Wilmington, Delaware, to congratulate Mary on her marriage – by chance on the day of her death. Revolutionary action, he confessed, for him a matter of principle, his wife considered 'wild ambition or foolish vanity'. Mrs Rowan's prudence made it possible for her husband to recant, and after ten years to return to his estate, which she had maintained in his absence.

Meanwhile, Mary's other Irish friend, George Ogle, had retired from the Irish House of Commons to his estate at Belle Vue in County Wexford in 1796. With the Irish Uprising in 1798, he re-entered Parliament as member for Dublin. He voted against legislative union with England in 1800. Despite this, he was returned to the united Parliament of 1801, again as member for Dublin, though he soon resigned his seat, and retired once more to Belle Vue. Godwin refers to him also only in passing, for Mary's anecdotes of Ogle and Irish society would lie buried, for decades to come, in letters owned by the Wollstonecraft sisters.

When Godwin delivered his final draft of the *Memoirs* to Mary's publisher, Johnson again protested on Fuseli's behalf. Godwin's portrait of him seemed untrue. Godwin's reply shows how stubborn he could be, bent on blaming Fuseli for certain elements in Mary's character.

> Dear Sir,
>
> . . . With respect to Mr Fuseli, I am sincerely sorry not to have pleased you . . . As to his cynical cast, his impatience of contradiction, & his propensity to satire, I have carefully observed them; & I protest in the sincerity of my judgment, that the resemblance between Mary's [taste?] of this kind to his, was so great, as clearly to demonstrate that the one was copied from the other . . . You see no faults in your friend. I do not blame you for this; it raised my idea of your temper & character; but (as you justly intimate) I am to state the report, not of your eyes, but of mine . . .
>
> Yours truly,
> W Godwin
>
> Sat., Jan. 11. 1798

Later that year, at the height of the counter-revolutionary crack-down, Johnson was sentenced to ten months in the King's Bench Prison. The charge was publishing a so-called seditious pamphlet by the political activist Gilbert Wakefield against the conservative Bishop of Llandaff. In fact, the government had trumped up the charge to punish Johnson for providing a forum for radical ideas. He went to prison with his usual equanimity, but his health was undermined, and from this time he withdrew from the full stress of business. He died (of an asthma attack) in 1809, and was buried in Fulham churchyard. Those who knew him best 'never heard him say a weak or foolish thing', Godwin recalled in the *Morning Chronicle* on 29 December 1809. 'Accordingly, his table was frequented . . . by a succession of persons of the greatest talents, learning and genius.' Maria Edgeworth – his best-selling author since *Castle Rackrent* came out in 1800 – celebrated his rarity as publisher:

> Wretches there are, their lucky stars who bless
> Whene'er they find a genius in distress;
> Who starve the bard, and stunt his growing Fame
> Lest they should pay the value for his name.
> But JOHNSON rais'd the drooping bard from Earth
> And fostered rising Genius from his birth
> His lib'ral spirit a *Profession* made,
> Of what with vulgar souls is vulgar Trade.

Godwin's *Memoirs* do less than justice to the generosity with which Johnson had fostered Mary Wollstonecraft since 1787. By the time Godwin married her ten years later, she was famous; when Johnson took her up she had been an obscure governess longing, like her sisters, to leave that track.

A month before his marriage, Godwin had published an essay, 'Of Posthumous Fame'. It tells a story of Sir Walter Raleigh writing his *History of the World* while he was a prisoner in the Tower of London. One morning Raleigh heard a commotion under his window, but could neither see the combatants nor hear what they said. He asked one person after another what had happened, and each told a different version. It struck Raleigh that if he could not make sense of an incident that had happened an hour before under his window, how could he expect to understand the history of Hannibal and Caesar? 'History is in reality a tissue of fables,' Godwin concludes, a year before he devised one of his own.

His *Memoirs* coincided with two unfortunate events. First, he was writing them at the height of anti-Jacobin ferment. As Napoleon swept into Italy,

Switzerland and the Low Countries in 1797, alarmed Englishmen feared a conspiracy of liberal intellectuals would bring revolution to all of Europe. After 1796 no Jacobin novel was published – apart from *The Wrongs of Woman*. Its author seemed to exemplify what Britain was at war against. The villain of the counter-revolution was the person who questioned the established hierarchies of law and gender. Then, too, by unlucky chance, the *Memoirs* coincided with the Mary King scandal, casting doubt on Wollstonecraft's pre-eminence as educator of daughters.

The image of reckless intemperance persists to this day, varied by accusations of prudery. Both judgements perpetuate the prude/whore caricature of womanhood, oddly unwilling to engage with a woman who accepts our nuanced sexual nature – its romance, its modesty, its warmth, its capacity for pleasure – as nature's endowment. A new imputation at the outset of the twenty-first century is that Mary misbehaved on her deathbed – manifesting the 'self-centredness of the dying'. This contradicts the fondness and gratitude witnessed by those who were there. 'Self-centredness' over the course of her life can't be denied – all art, all endeavour, is selfish to some extent – but deathbeds vary, and Mary's was gracious, as Godwin testified in the *Memoirs*, as did Mrs Fenwick to Mary's sisters and Hays in her obituary. So it happens that an untrue 'self-centredness' slips in to confirm the failure narratives imposed on Mary Wollstonecraft's life: the doom of the fallen woman; the comedy of the dizzy enthusiast who presumes to pick up a pen – ephemeral fame, bound to come to grief; and the Victorian melodrama of Mrs Fuseli shutting her door on a sexual intruder.

Biography, as we practise it now, will endorse Godwin's principle that truth must be revealed. Accordingly, he steeled himself to expose his wife's relationship with Fuseli. Godwin's diary shows that he continued to see Fuseli often while he composed the *Memoirs*, so Fuseli had ample opportunity to insinuate his self-flattering version: Mary as sick with love for him. It's a curious fact that this was the sole occasion when Johnson could offer Mary no comfort. Critics have concluded that her conduct put her beyond sympathy, but Mary's response to Johnson's rebuff was not the shame we might expect. It's more like embarrassment – she jokes about wearing a fool's cap. So might there have been another scenario? Is it possible that Mary stumbled on love at the heart of the lifelong intimacy of Johnson and Fuseli? Same-sex love could explain Mary's silence on the subject of Fuseli after that talk with Johnson. It remained a capital offence in the eighteenth century. More often, convicted men were punished by exposure on the pillory where they were liable to be stoned and vilified, pelted with mud and excrement, and sometimes killed by a homophobic mob. Johnson's author William Beckford was one of the homosexuals who had to flee England rather than face the threat of a criminal charge.

It did not take long for a man like Fuseli to feel out Godwin's susceptibility as husband and biographer to an untold story. Under normal circumstances Godwin would have been too imperturbable to give way to jealousy. It was not an emotion that troubled him at any other time. But in the disturbed aftermath of Mary's death he was, briefly, the plaything of Fuseli, subject to a warped image of the reasonable person he knew his wife to be. Woman as sexual predator usurps attention: Odile, as it were, blots out Odette. Fuseli knew his man: a philosopher too truthful to conceal what he was made to believe. In the dark, melancholy, loathing Fuseli but attentive to his story, Godwin lent it undue credence in his *Memoirs*. There, the poison infiltrates his measured prose with a slander in Fuseli's favour, and so seeps into the consciousness of future generations. Eighty years later, when Browning announced a forthcoming poem on 'Mary Wollstonecraft and Fuseli', Kegan Paul, one of the last people to see their correspondence (destroyed about two years later), asked Browning to reconsider the evidence.

Kegan Paul, Trench & Co. Publishers
1 Paternoster Square,
London.
January 12. 1883

Dear Mr Browning
I see among the Literary Notices of this week, that you are about to publish a Poem on Mary Wollstonecraft and Fuseli. Of course I cannot

> even guess how you are treating it, and have no doubt that you have considered all the evidence in the matter . . . May I ask if you will very kindly . . . accept my assurance that having read as I believe every letter of Mary's or Godwin's which now exist, and they are legion, I utterly disbelieve that there was anything whatever in the relations of Mary and Fuseli, than those of a young woman to an elderly *fatherly* married friend, with whose wife she was on most affectionate terms . . . Godwin in part adopted [the slander], but he really had known next to nothing of his wife's early life. He is even demonstrably wrong in much that he says which he might have known . . .
>
> The letters [between Mary and Fuseli] which exist are of the most common place character, and I have read them all . . .
>
> With kind regards to Miss Browning
>
> I am yours most sincerely
>
> C Kegan Paul.

Browning was not deterred. His dramatic monologue updates the old slanders as if it were from Mary's own lips. She is made to confess to Fuseli that she is womanish in a helpless, Victorian way ('Mine are the nerves to quake at a mouse'), futile as a writer, and a scary femme fatale. There's no vestige of Wollstonecraft's actual voice:

Much amiss in the head, Dear,
I toil at a language, tax my brain
Attempting to draw – the scratches here!
I play, play, practise and all in vain . . .

Strong and fierce in the heart, Dear,
With – more than a will – what seems a power
To pounce on my prey . . .

A 'devouring' love labours to win a lover, but has to concede defeat: 'I have not quickened his pulse one beat.'

This came out at a time when women in America and England were campaigning for the vote, contriving higher education and starting to enter the professions. Elizabeth Blackwell in the United States, followed by Elizabeth Garrett in England and Sophia Jex-Blake in the Edinburgh of 1869, had pioneered women's medical training against stiff opposition, and these doctors were now in practice. A scandalous link between prostitution and women's advance – disseminated in the late 1790s and renewed by Victorians – led a new generation of feminists to distance themselves from Wollstonecraft, lest her supposedly dissolute life should damage the Cause. Instead of questioning the transmission of fact, many (like political economist Harriet Martineau) chose to side with detractors. In 1885, when Karl Pearson founded the high-minded Men and Women's Club to discuss the nature and relations of the sexes, he wished to name the club after Mary Wollstonecraft. Her *Vindication* was discussed and circulated, but women members refused. Wollstonecraft was not respectable.

Volleys of slanders lined up to finish off her new genus. 'She fancied that she who was merely a woman of lively, but neither strong nor profound genius, was a phenomenon of nature, born to give new direction to human opinion and conduct,' was the close shot from Pitt's propaganda machine in July 1798 (in response to the second edition of Godwin's *Memoirs*). The most effective damage is when detraction creeps close to truth, killing it with elegant economy. 'Fancied' and 'merely' are all it takes. Wollstonecraft's benevolence to her family and the poor, 'hurried on by her feelings' and uncontrolled by reason, was the second target. The third shot followed automatically: her constitution 'was very amorous'. So the Woman Question was born anew – unconnected with Wollstonecraft – in the mid-nineteenth century, after a hiatus of fifty years. This myth ignores the transmission of the new genus through Mary's daughters and heirs in the next generation, an independent set who seeded new species and kept them going underground.

16

CONVERTS

Wollstonecraft's was an interrupted life. The eager vehemence of a voice breaking off in mid-sentence and the originality of unfinished books continued to shape lives bound up with her own. Nearly three years after her death, in the summer of 1800, Godwin travelled to Dublin to pick up some ends. There, on 5 July, he met for the first time Mary's sister Bess whose fine brown eyes and civilities put him at ease. As much as he likes her, he reports to Marshall, he 'hates' Everina – though 'hate' is not a word to repeat to her nieces, Mary aged two and a half and Fanny aged six. More curious was the prospect of an encounter with his wife's favourite pupil whom she had known as Margaret King, now Lady Mount Cashell.

She was in Dublin for the summer, Godwin heard. Her friend John Philpot Curran, the lawyer who had defended Rowan and other United Irishmen, reported to Godwin that she 'speaks of you with peculiar regard, mixed with a tender and regretful retrospect to past times and to past events with which you have yourself been connected'.

They met on 9 July when they dined with the politically independent Lady Moira, her daughter Selina, and her son-in-law the Earl of Granard, in a house overlooking the Liffey. At twenty-nine, Margaret had recently begun to bring her life into line with Mary Wollstonecraft's. She was keen to discuss questions of politics and education, and welcomed Godwin to Mount Cashell House on St Stephen's Green. He dined there on 13, 21 and 28 July. On the last of these days, Margaret took him together with two of her children to the Devil's Glen, surrounded by 'stupendous' rocks and mountains. Beside her in her cabriole, Godwin eyed her, half-amused, half-wary. She was handsome, but her folded, 'brawny' arms were bared to the shoulder and her dress lacked the 'linen' to hide her cleavage – up to now, Godwin had associated its absence with the grey and grinding poverty of Mrs Fenwick. As his new friend's breasts obtruded on his notice, he may not have realised she was six months pregnant. He observed her irregular teeth and 'white eyes' while she declared herself a democrat and republican. Her intelligence was undeniable, but instinctively he recoiled from a woman with much to say. Here was a 'singular' being who could be 'gibbeted' in a play. The manner of the great lady consorted oddly with her radical opinions, yet Margaret was genuinely keen to learn from Godwin.

'In what you say concerning the propriety of treating children with mildness, kindness, and respect, you express exactly the opinions I have long entertained,' she said with a veiled reference to the blend of severity and emotional neglect she had known at the hands of Lady Kingsborough.

She confided to Godwin how much Mary Wollstonecraft had changed her.

'I am convinced that had it not been my peculiar good fortune to have met with the extraordinary woman to whose superior penetration and affectionate mildness of manner I trace the development of whatever virtues I possess, I should have been, in consequence of the distortion of my best qualities, a most ferocious animal.'

After her governess had left, Margaret was soon of an age to 'go to market'. Mary Wollstonecraft had taught that early marriage is a bar to improvement: 'many women, I am persuaded, marry a man before they are twenty, whom they would have rejected some years after'. The mistake of Margaret's life was to marry Stephen Moore, 2nd Earl of Mount Cashell, in September 1791 when she was twenty. He, at twenty-one, was a good-looking young man with gentle manners. Apart from his seat at Moore Park near Kilworth, in flat agricultural land near Mitchelstown, he also owned a grand Dublin house where he stayed while the Irish House of Lords was in session. In truth, he had little interest in politics, and preferred the country. His horizons were bounded by turnips, rams, stallions, cows and bulls. Marriage locked Margaret to a husband whose education, she said, 'had been of the meanest sort: his understanding was uncultivated and his mind contracted'. Wealth and titles were all he cared for. She had been aware that his character was 'perfectly opposite' to hers, and had thought to change it, as she herself had been changed by Mary Wollstonecraft. In the freshness of youthful confidence, Margaret had expected to govern him – 'the silliest project that ever entered a woman's mind', she owned later.

Like her mother, Lady Kingsborough, Margaret was fertile. There were six births in the eight years between 1792 and 1800, and five children survived: Stephen (Lord Kilworth), Robert (whose amiable nature made him her favourite), Helena, Edward and Jane. As Margaret read her way through her twenties in the quiet of Moore Park, she drew apart from her husband. They could agree neither on politics nor on the education of their children. The Earl was disinclined to waste funds on something so worthless as education, since he meant his sons to follow him as country gentlemen – running the estate and making agricultural improvements. Determinedly, Margaret engaged United Irishman John Egan as a tutor. Politicised by Mary Wollstonecraft, who had pointed to the poor as victims, Margaret questioned the inequities and incompetence of landowner government.

In May 1794, at the time Rowan was tried for treason and fled to Paris, United Irishmen were suppressed, and the society went underground as a revolutionary network bound by secret oaths. When a popular uprising came in the spring of 1798, Margaret, aged twenty-seven, was said to have declared herself a United Irishwoman and republican – but this is hearsay. Whether she actually took the oath (a capital offence) can't be verified, nor a rumour that the cellars of Moore Park were used as a hideout. Safe to say, her sympathy – perhaps no more – with radical change fitted Mary Wollstonecraft's teaching. It may not be entirely fortuitous that Wollstonecraft's death coincided with Margaret's first break with a family allied to the pro-Crown politics of Dublin Castle. The dead only die if we let them. Some, like Margaret Mount Cashell, 'are born with the dead' whose communication 'is tongued with fire beyond the language of the living'.

Margaret's turn against the Anglo-Irish Ascendancy has a different shading from the glamour of Lord Edward Fitzgerald, a rebel-leader who knew Margaret and stayed occasionally at Moore Park. R. F. Foster has described his francophilia as 'the current form of radical chic', associated with a small intelligentsia 'who preferred not to recognise the sectarian underpinning of all political activity in Ireland'. Fitzgerald, who had been in Paris in 1792, had a francophile associate in Arthur O'Connor, born in Mitchelstown eight years before Margaret. She acquired his 1798 pamphlet *The State of Ireland*, a protest against the 'plunder' of Irish resources, including England's refusal to allow the colony to develop manufactures from its raw materials, which might compete with those of the imperial power. Meanwhile, the 'land-sharks', like the 'tithe-sharks', prowled after their 'prey' – *not* the sort of thing a landowner like Mount Cashell would welcome in his library. The mind of Europe had changed, O'Connor claims: 'See the armies of despotism advance against France and be annihilated by the armies of liberty.'

On 17 May 1798, at the start of the rising, an idea was floated to kill all members of the Irish House of Lords when they collected for the trial of Margaret's father. The plan was rejected by one vote, for fear of massive casualties as the United force broke through the soldiers on College Green. Fitzgerald was caught and died of wounds. It's not known what Margaret thought of this, but we can gauge how far she diverged from her eldest brother, Big George, as commander of the notorious North Corks.

His havoc outdid Mr Wollstonecraft, whose violence was confined to home. George promoted a form of torture known as 'capping', calculated to humiliate – to wound self-respect as well as the body. Men who cut their hair short in imitation of French republicans would have 'caps' of molten tar jammed on their heads by George and his men. One witness described their hideous acts:

> As a gentleman of respectability was passing near the old Custom House [Dublin], in the afternoon of Whit-Sunday, 1798, two spectacles of horror

> covered with pitch and gore, running as if they were blind through the streets arrested his attention. They were closely followed out of the old Custom House by Lord Kingsborough [Big George, who had succeeded by this time to the title], Mr John Beresford and an officer in uniform. They were pointing and laughing immoderately at these tortured fugitives, one of them John Flemming, a ferry-boatman, and the other Francis Gough, a coachman. They had been mercilessly flogged to extract confession, but, having none to make, melted pitch was poured over their heads and then feathered. Flemming's right ear was cut off, both were sent off without clothes. Lord Kingsborough superintended the flogging, and almost at every lash asked them how they liked it.

At the height of the Rising, when women came to plead for the lives of their kin, George raped some as fair exchange. He joked about his disappointment at having 'only had two Maidenheads'. After the Rising was crushed in June, the British government resolved on Union, dooming Ireland to impotence and further poverty. Margaret defended her country's independence in three anonymous pamphlets, a woman taking political action as Mary Wollstonecraft had dared nine years earlier with her pamphlet exposing the self-interest of Burke. Margaret was thinking along similar lines when she published this warning:

> I cannot perceive what advantage it would be to Ireland to have a servile, artful and ambitious native of that country pursuing his own interest in the British Cabinet, nor how it could benefit our island to have him reproached with being an Irishman. Would this produce any commercial advantages to our cities? Would this occasion any civilization in our provinces?

Political union could not be deflected, and by the time Margaret met Godwin in July 1800, reaction had 'set hard'. Yet defeated as she was, alongside more famous defenders of Irish independence, she did assent to a different sort of union.

This was the only kind open to women: a private bond between the Irish

Margaret and the English Mary Wollstonecraft, beginning in 1786 and carried through to the end of Margaret's life. Together they present an alternative to the routine narratives of 'dominant history'. A counter-narrative to dominance is this women's story of successive generations, as Margaret, nearing thirty, starts to command her own voice. The voice comes first; next, the challenge to leave a beaten track. Between 1800 and 1804 she lingered in an old track, much as Wollstonecraft had done during her period as Margaret's governess.

After Godwin returned to London, Margaret opened a correspondence on the subject of children. Both were committed to the methods of Mary Wollstonecraft: children were not to be forced into obedience like horses to be broken; they were not to be beaten. They were to be loved and understood and encouraged to judge and question, even though judgements and questions were not encouraged by Europe's threatened monarchies in 1800.

Margaret, like many, approached Godwin as a mentor. 'I should be very happy to have my errors pointed out by you,' she said.

Godwin responded with alacrity. He had observed that, in practice, she did not treat her children with enough 'tenderness'.

Such candour struck Margaret, in turn, as 'singular' – and the singular pleased her. She was keen to clarify her position: children must not have 'that frivolous exhibition of tenderness which makes them appear to themselves and others more like the playthings of capricious fancy than the objects of rational attachment'. Her insistence on being 'rational' was not necessarily as cool as it sounds; it signalled participation in Wollstonecraft's mission to rescue women from silliness and restore them to the dignity of rational creatures. Margaret also worried whether she should continue with a tutor for her two elder boys or whether they should be sent to school. 'I should wish to teach children as early as possible to think for themselves.'

Godwin's similar position is set out in his preface to his first children's book, called (in his diary for 1801–2) 'Jewish Histories' and published in 1803 as *Bible Stories* under the pseudonym of William Scolfield. Godwin was proud of this statement on education, and asked the executors of his will to include it in his collected works. It shows him tongued with the fire of Mary Wollstonecraft: her idea of learning through the affections, as a pulse of imaginative sympathy called out from within the child:

> Everything is studied and attended to, except those things which open the heart, which insensibly initiate the learner in the relations and generous offices of society, and enable him to put himself in imagination into the place of his neighbour, to feel his feelings, and to wish his wishes.
>
> Imagination is the ground-plot upon which the edifice of a sound morality must be erected. Without imagination we may have a certain cold and arid circle of principles, but we cannot have sentiments: we may learn by rote a catalogue of rules . . . but we can neither ourselves love, nor be fitted to excite the love of others.
>
> Imagination is the characteristic of man. The dexterities of logic or of mathematical deduction belong rather to a well-regulated machine; they do not contain in them the living principle of our nature. It is the heart which most deserves to be cultivated . . . the pulses which beat with sympathy . . .

Margaret wished to view the workings of a republic. France beckoned, as it had beckoned to Wollstonecraft nine years before. For these nine years England had been at war with France and travel was blocked. With the Peace of Amiens in 1802, there was a stampede to Paris. The great coroneted coach of Lord Mount Cashell led the way, rolling along the roads even before peace was signed. Inside were the Earl and Countess, their two daughters Helena and Jane, and Catherine Wilmot, a lively, well-connected young woman from Cork who was to be Margaret's companion. Behind the travellers, in another carriage, came four servants; all (said Miss Wilmot) 'driving at full speed, nine Irish Adventurers, to the French dominions'.

Their first stop, from late September to late November 1801, was London. Margaret wrote to Godwin in advance. 'It would give me great pleasure to see your two little girls, a gratification I shall hope for.' She made half a dozen visits to the Polygon, where she would have met Fanny aged seven and a half and Mary just four. It's not known to what extent she was aware of a change in their life.

Four months back, a woman had leaned from the neighbouring balcony at the Polygon, and said: 'Do I behold the immortal Godwin?'

He turned and saw a dark-haired woman in green glasses, calling herself Mrs Clairmont. She was supposedly a widow of French extraction – her family name was Vial – and she had a son, Charles, born of a Swiss father called Charles Gaulis (who had died in 1796), and a daughter, Clara Mary Jane ('Jane' at home, later 'Claire'), of indeterminate paternity. Jane, born in 1798, was eight months younger than Mary Godwin. Mrs Clairmont's history has remained shady. Might she have taken her charming name from the title of a gothic novel, *Clermont* (1799), popular enough to be amongst those devoured for their far-fetched terrors in *Northanger Abbey*?

From early June 1801, Godwin's diary begins to record weekly and sometimes daily dinners and teas with his neighbour. Mrs Clairmont worked for the well-established children's publisher Benjamin Tabart as editor of his series of nursery stories, and she also earned her living as a translator; but the main attraction for Godwin was that her expressiveness reminded him of Mary Wollstonecraft. On 11 July he reread his wife's *Travels* before supping with Mrs Clairmont. Later, he spelt out to her his sense of the similarity of her letters. '. . . The same sensibility irradiated them, the same warmth of feeling, the same strength of affection, the same agonising alarm, the same ardent hope, as I trace with such unspeakable delight in the letter before me.' Godwin and Mary Jane Clairmont married in December 1801, soon after Margaret's visits to the Polygon. Deep currents, welling from the life of Mary Wollstonecraft, ran between these lives with far-off repercussions for Margaret Mount Cashell, Fanny Imlay, Mary Godwin and their new stepsister, Jane Clairmont.

In Paris, Margaret looked up Mary Wollstonecraft's old associates, Joel Barlow, Tom Paine, Thomas Holcroft and Helen Maria Williams. Helen, still living with Stone, conducted a bi-weekly salon in the library of her *hôtel*. The guests were literary lions, senators and members of the National Institute in blue, embroidered coats. Helen combined an eager welcome with sophisticated languor, 'like an invalid' to an Irish ear. She appeared in perpetual mourning for a sister, in black with a gauzy black veil thrown over her head and reaching to her feet. Margaret looked 'like a frosty moon' amidst the rouged French ladies. She and Miss Wilmot felt like trussed fowls in their stays and straight pelisses, compared with the floating drapery of French fashion – unseen for a decade. The Parisians were wearing diaphanous fabrics to give the effect of Greek statues, with circlets across their brows. A portrait of Margaret in Paris shows off her profile with an aquiline nose, and hair circled and caught up in the Greek manner. Joel Barlow visited her frequently in May–June 1802, when she was eight months pregnant. '. . . Every other night now, [we are] very thick together,' he told Ruth, 'that is, she is very thick, & [your] hub is not thin.' He entertained her with American stories and thought her 'certainly a most excellent woman, manages her family with the greatest dignity & least affectation'.

At the end of June, Barlow slipped into England, incognito, on a secret mission. He was not too occupied to peek at the absurd shape of stays in Bond Street: 'long, labored, stiff, & armed with ribs of whale. It is a frightful to think of. They don't walk so handsome as our ladies do.' Public walks were filled with white chip hats (made out of thin strips of wood) – 'why you might as well look into a bleach-field – quelle différence!' One day, he ran into Joseph Johnson who saw through Barlow's disguise and seized his hand, swinging affably from side to side.

'Well, you could not get me hanged,' Johnson laughed. 'You tried all you could.'

Barlow, nervous to be identified with such ease, declined Johnson's invitation to dine.

He returned to Paris to find that Margaret had produced her sixth child, Richard – 'another little republican Lord', as Barlow put it to Ruth on 4 July. On his first visit to Margaret after the birth, he sized up her mismatch with Mount Cashell.

'He is an aristocrat, & has not a great deal of sense, and what is worse for him, he seems to know it – & what is worse still *she* seems to know it – & what is worse than all, she seems to try to make her friends know it.'

In the Paris of the day women liked to dress as men. Dressed as a Turk, Margaret tried this out at a masquerade at the Opéra. Her six-foot height, extended with a turban, made her manliness so convincing that a gallant commiserated with Miss Wilmot's 'subjection' as the Turk's wife, and proposed to snap her chains.

Through Miss Wilmot's admiring eyes, we see Margaret's performance of her society role, so at odds with the burgeoning successor to Mary Wollstonecraft. Short, brusque General Berthier, the Minister of War, has tall Lady Mount Cashell on his arm at the American Embassy. Beautiful in black crepe and diamonds, Lady Mount Cashell is entertained by the First Consul, Napoleon, at a banquet in the Tuileries. He puts himself out to charm, smiling, in plain clothes, while Josephine blazes in purple under a canopy. The British Ambassador hands her ladyship in to dinner, while Miss Wilmot has to make do with Talleyrand whose paunch, she comments, led the way, and whose cunning expression repelled her almost as much as his greed. His mouth did not close for two hours, she observes. 'Oh! such a cormorant.'

In Ireland, the Mount Cashells would have been perceived as Protestant English; in France they were Irish. Wherever they went, crowds gathered about their coach with 'a kind of republican homage', for the Irish aristocracy was thought friendly to the French republic. Lady Mount Cashell gave a ball for her new friends, and a *thé à la française* for sixty people. Another scene at the American Embassy: Lady Mount Cashell as guest of Robert Livingston, Jefferson's newly arrived Ambassador who was to bring off the Louisiana Purchase the following year – a diplomatic alternative to the coup planned by Imlay, Barlow and Leavenworth a decade earlier.

As the English filled Paris, they packed Helen Maria Williams's salon, now held nightly for thirty, forty or fifty visitors – soirées so breathy with mouth-to-mouth banalities that Barlow began to tire. He was popular with Helen and Margaret and often chosen to sit with them at supper. At Helen's he met the Swiss banker Schweizer and his wife Madeleine (who had sat out the Terror along with Mary Wollstonecraft). Every second evening Barlow spent in more select company at the Mount Cashells. There, in August, he met Opie and his wife Amelia – 'a woman of *beaucoup d'esprit*, and Opie is very proud of her'. Mrs Opie inscribed a copy of her *Poems* (1802) for Margaret. Come September, and Fuseli arrived in Paris. Each month seemed to bring more of the Wollstonecraft past, as though her milieu reconstituted itself during that brilliant Parisian summer while Margaret moved in her shadow.

The Mount Cashells lingered for nine months, unlike many of the English who, according to Barlow, were restless: 'The glare of the arts that they find here soon satiates.' Not so for Margaret and Miss Wilmot, who were agog at the ballets of Vestris, and startled by the speed and brilliance of the pirouette, in contrast with the eighteenth-century 'moderation of grace', as they understood the ways of the body. Used to the chatter of English theatres, they were struck by the rapt silence of the French audience.

They wintered in the Italian states, travelling between Florence, Rome and Naples. From Florence, in April 1803, they made an excursion to the old trading port of Livorno (Leghorn, as it was called by the English), where they visited a synagogue and the small English cemetery. The latter was full of women who had died in childbirth far from home, and children who had died of exposure to unaccustomed ills in the course of their parents' adventures. On their way back, the Mount Cashells stopped for three or four hours in Pisa, a town renowned for its medical school. That spring, baby Richard had measles, and Margaret a period of debility that went on for a long time. Although she appeared 'brawny', her health was poor. Mary Wollstonecraft, who had nursed her through a fever at fifteen, had feared Margaret could be tubercular, and had tried to protect the girl from her mother's exactions. Margaret herself ascribed her weakness to extended breast-feeding. She followed Wollstonecraft's teachings on the value of mother's milk to set up infants' constitutions and initiate their crucial education in tenderness. Yet frequent births had meant that she had been feeding almost continuously for eleven years.

On 1 June, war broke out again. France was again out of bounds, and Miss Wilmot returned to Ireland via Germany, while the Mount Cashells and their children turned south for another stay in Rome. Their British circle never mixed with Italians, and seldom left the area of the Piazza di Spagna. Margaret found them as closed-off as Henry James would find his compatriots eighty years later, when he called Rome 'the American village'. But in 1804 the dullness of expatriate society was broken by the arrival of a reading and thinking man.

Margaret was thirty-one and pregnant with her seventh child when she met George William Tighe, aged twenty-eight. He was a friend of her husband, and came from the same Ascendancy circle. His family traced its descent from Edward III, but the titles had been in the female line and so not passed on. Tighe had been at Eton with Margaret's brother Henry King (both had entered the school in 1785, which means that Tighe had been there – though not in Prior's house – when Mary Wollstonecraft had stayed at Eton in 1786). When he had appeared in Dublin society in the mid-1790s, a captain in the 7th Dragoon Guards, he was thought a 'very handsome man & a great Beau'. Margaret fell in love. Her passion was overwhelming in a woman who had never expected to feel desire. Tighe called her 'Laura', after Petrarch's beloved. She called him 'Vesuvius'. 'Laura' and 'Vesuvius' wrote ardent poems to each other. Laura writes of a man who has 'taught my heart unknown delights'. 'High beats that heart at thy renewed embrace:/ Whilst present joys the past once more retrace.'

They belonged to a class (extending through the European aristocracies) where, once an heir was born, discreet adultery was almost *de rigueur*. Lady Oxford, for instance, produced six children of miscellaneous parentage, known to the wits of London as 'the Harleian Miscellany' (after a collection of manuscripts inherited by her husband). This is where Margaret broke with her class. She took sex seriously. Tighe was 'Beloved beyond existence, health and fame', and though at first he echoed her willingness 'to abandon friends & fame' for 'vagrant ways', he was stirred even more by a classicist's passion for the land of 'Tully's wisdom' and 'Maro's strains'.* The love affair, for him, was a consummation of the Grand Tour. Margaret

*Tully (Marcus Tullius Cicero) (106-43 BC); Virgil (Publius Vergilius Maro) (70-19 BC).

was uneasily aware that Tighe's passion fell short of her own, and she longed for convincing words.

In the summer of 1805 she remained with four of her children in the German principalities, while the Earl departed for England, having agreed to place their three eldest sons in school. To have gone part of the route with her husband means that Margaret was, at this point, still attached – if not to her husband, certainly to her children. She made excuses about her health and, for about nine months, hesitated. Should she return to Moore Park, or do something else with her life? If she didn't act now, at thirty-four, she would go back to more pregnancies and increasing debility. Should she return to the health-giving climate of the Italian states – and Tighe? But could she part with her children? These were her 'days of adversity'. As she continued to waver in the spring of 1806 a 'fit of illness' decided her to go south, and to send Helena (aged eleven), Jane (nine) and 'precious' Richard (four) to join their father in London. She parted from them in Dresden.

Mount Cashell allowed it now to be known that he was furious. He ordered his wife to return in order to sign papers of separation. Divorce was not discussed. Expense would not have been an issue for aristocrats – the only class who could afford it – but divorce still required an Act of Parliament, with public airing of intimacies and two witnesses to adultery. Mount Cashell then demanded that his wife give up their last child, Elizabeth, who was almost two. If she did not, he would stop the money and take the child by force. The law was on his side. Margaret asked her Irish lawyer Denys Scully if a mother did not have the right to keep a child until the age of seven. The answer had to be no. (Such a right would not become law until 1839.) An alternative was to return to Italy, under Napoleon's rule, where Margaret might hold on to her 'dear little Elizabeth' beyond her husband's reach.

'I am resolved', she said, 'that nothing shall force me to relinquish the performance of this duty but death.'

Margaret's mother Caroline, the Dowager Countess of Kingston, urged her daughter to return, making light of her ill-health, while the philandering and torturing George, having inherited the title of 3rd Earl, rebuked his sister for her folly. During the second half of 1806 Margaret remained in doubt about access to her children. She saw her husband as 'a very weak man' whose friends had convinced him 'that his character would rise on the ruins of mine. Whether he really intends to shut my own doors against me & separate me from my children I know not, but nothing shall prevent me from endeavouring to be of use to them for whose sake I have endured more than anyone has notion of, in the way of petty tyranny and trifling opposition.'

Tighe joined Margaret as she wandered with her remaining child from town to town: Dresden, Carlsbad, Eger, Ratisbon. His poem 'Jena (20 September 1806)' gives a cool answer to a lover's reproaches and tears. 'Yes, Laura', he says, love's summer is o'er. Her emotional storms have withered his 'desolate heart'. Friendship, at least, remains. Another 'Yes, Laura' poem shows the compunction of a decent man. 'Let us cease to reproach & complain . . ./ oh come let me wipe off thy tears,/ And press thee once more to my Heart.' It meant joining Margaret as outcast from home and country, 'a stranger at each door'. The date of his Jena poem indicates that by September 1806 they were in a battle area, where Napoleon would defeat the Prussians the following month. Tighe's poem 'The Fall of Jena' records 'the burning torrents' of gunfire, the 'mangled' bodies and 'the spoiler's' abuse of women – on all of which 'Glory drops her meretricious wreaths'. Jena appears a bizarre destination for an Irish pair travelling with a two-year-old in 1806, but Margaret had a secret purpose.

Mary Wollstonecraft had taught that women might 'study the art of healing, and be physicians as well as nurses'. Margaret's plan to study medicine was an ambition of this kind. With seven healthy children (all of whom would live full lives), she already had practical experience. She had not been one of those privileged women, like her mother, who left her children to hirelings. 'In one thing alone have I succeeded for their advantage,' she said, 'and that is in giving them good constitutions, which I am convinced they owe to my good care during the first four or five years of their lives.' It was an agony she never permitted herself to express to have surrendered six of the children, yet she was determined to build on the observation and expertise she had gained from motherhood in order to be 'of use' to other children. Jena was known for its medical school and liberal student body. Disguised as a man, Margaret Mount Cashell attended medical lectures at the university. Again, her height helped her disguise, and her German must have been good enough by this time; it was certainly good enough to translate, as she did, some medical texts.

In years to come, Margaret revealed only the bare facts of her cross-dressing to two or three intimates. But a quasi-fictional manuscript (discovered amongst her papers by Cristina Dazzi, wife to an Italian descendant) allows a glimpse of the persona Margaret may have developed as a male. It purports to be the travel notes of a Frenchman whose chatter is as mundane as his history: he's a harmless, rather sickly and plaintive traveller who goes to Jena soon after the battle and proudly inspects the positions of the French cannon on the battleground between Jena and Weimar. Jena's surroundings, he finds, 'are very pretty . . . and the town itself comes across as rather romantic with its old remaining walls and gothic towers'. He is marked wherever he goes for his French nationality. Why did Margaret choose to develop this character?

Part of the answer may be her disguise as a man. French would have allowed her to pass as a stranger without revealing who she was. Her Italian at this point would have been negligible, and her German would have been recognisably not that of a native. It's just conceivable that she wrote this travel piece to work up a man's voice and account for her foreignness with a concocted history – a cover for an Irishwoman of thirty-five, seated amongst the tiers of medical students in a lecture theatre. She was not one to lower her head; in her breeches she would have sat as tall as at her tea-table in Dublin, arms characteristically folded across her breasts, and ready with the practised phrases of her story.

Margaret's husband stopped her funds. This is fictionalised in the piece which ends with a brave joke about encroaching poverty – far from the cosseted life of the Countess, Lady Mount Cashell: '. . . I have not received the money here[,] the money which I was expecting . . . I am beginning to experience what it is to be poor. I was advised by doctors to eat good food . . . and not to tire myself out. I am obliged to dine in a small way . . . and cannot get to my room without having to go up 89 steps. I am rather poor and close enough to the heavens to become a good poet but if that is not enough, I do not wish to become any poorer and climb any higher.'

Poet she was not, but Margaret, pressed for funds, did begin to write for children. *Stories of Old Daniel* and *Continuation of the Stories of Old Daniel* were published, the first in 1807 and the sequel in 1820. As in Wollstonecraft's stories, children were to learn from 'real life', but nothing in the Daniel stories is as sad as the fate of little Eliza whose mother was forced to hand her over to a father she scarcely knew. In 1807 Margaret made the journey to Ireland to give up this last child. Margaret never spoke of this except to say that she had been justly punished for breaking the laws of society. There is much pain that is soundless. Only the smallest hints may reach us: the fact that the two youngest Mount Cashell children, deprived at four and three of their mother, never married; or the fact that when Margaret and Tighe had a daughter called Laura ('Laurette') in 1809, she could not let this child out of sight or hearing. For many years Laurette slept in her mother's room or adjoining it, and was never allowed to spend a night away.

At length, in November 1812, a legal separation between Lord and Lady Mount Cashell was signed in London. She was to get £800 a year, and the settlement of her accumulated debts. Some advised her to stay in her husband's London house – fortified of course by her title. But Margaret preferred independence as 'Mrs Mason' (the good governess in Wollstonecraft's *Real Life*). Wollstonecraft had judged accurately the words she'd put into the mouth of Margaret's counterpart in those stories: 'I wish to be a woman and to be like Mrs Mason.' Twenty-three years later, fiction became fact: Margaret as 'Mrs Mason' was ready to enter 'that middle rank of life for which', she said, 'I always sighed'. It marked the end of her aristocratic identity.

The defeat of Napoleon in 1814 opened the way through France to Tuscany. Margaret's choice was a town she had seen briefly in 1803. Her lungs had continued to trouble her, and one of Pisa's attractions was its reputation as a health resort. Another advantage was the decline, there, of English society. She wished to avoid Florence for that reason, and her determination to close her door to English connections suggests the exclusion she suffered during her stays in England between 1807 and 1814. Her choice of Pisa, though, had still more to do with medical ambitions. She was drawn not only by the renown of its medical school, she also wished to learn from Andrea Vaccà Berlinghieri (1772–1826), the famous professor of surgery at the University of Pisa known as '*Dio della Medicina*', whose reputation attracted patients from England, Lisbon, the German principalities, and as far away as St Petersburg and Egypt. Vaccà (as she calls him) had studied under leading surgeons, Hunter and Bell in London and Pierre Joseph Dessault in Paris. As a student in Paris he had witnessed the fall of the Bastille. His radicalism had the same humane character as her own, so humane indeed that his republicanism did not prevent his being called in when a daughter of George III gave birth. He is one of the distinguished Italians in Byron's dedication for canto IV of *Childe Harold*.

In one way Margaret did not put her past behind her. Her attachment to Ireland speaks through the books she took to Italy: Spenser's *View of the Present State of Ireland* (1596); *A Narrative of what Passed at Killalla* [County Mayo] *during the French Invasion in the summer of 1798* (Dublin, 1800); Arthur O'Connor's *The State of Ireland* (London, 1798); James Gordon's *History of the Rebellion in Ireland in the year 1798* (Dublin, 1801); and *Historical Memoirs of the Irish Bards* by Joseph Walker (Dublin, 1786). Later, Margaret acquired *Irish Melodies, National Airs* by Thomas Moore (1823) and *Fairy Legends and Traditions of the South of Ireland* (1825).

On 7 August, Margaret paid her final call on the Godwin family. At this time she also made a will, leaving all she had to Tighe. They arrived at Pisa on 18 October. It was a quiet city, with only eighteen thousand inhabitants in 1814. The family settled on the south side of the River Arno, in Casa Silva, a house with a garden shaded by walnut trees in the Via Mala Gonella, now 2 Via Sancasciani near the Church of S. Maria del Carmine. The garden remains, topped by high pines, on a corner facing the Via Pietro Gori as it runs south from the river. There, Margaret began to practise her 'cures' in consultation with Vaccà. He encouraged her prime idea that medical treatment of non-emergencies should be gentler, less interventionist, and doctors should avoid unnecessary drugs. In this she was ahead of much medical practice today. Like Mary Wollstonecraft she knew that sound care at home could prevent numerous deaths in childhood.

Both women avoided the common extremes of neglect on the one hand and damaging treatments on the other – we recall how Mary shunned doctors, braving gossip in Le Havre, when Fanny had smallpox.

Margaret also deplored the lucrative links between doctors and druggists, an issue George Eliot would bring to public notice through the scientific integrity of Dr Lydgate, bitterly challenged by the medical establishment in *Middlemarch*. Far in advance of her age, Margaret promoted the body's own healing powers, and proposed preventative medicine.

At the time she embraced Pisa at the age of forty-three, another child – her last – was conceived, the girl who was to be the ancestor of a line of Italian descendants. On 20 June 1815 Catherine Elizabeth Ranieri ('Nerina') was born in Pisa. The girl's middle name speaks more loudly than words of her mother's pain at being forced by law to give up Elizabeth Mount Cashell eight years before – never to see her again. Margaret fancied that Nerina looked like her youngest son, Richard Mount Cashell, the boy born in Paris. Nerina was also thought to bear some resemblance to her mother's favourite son, the amiable Robert, the only one to write and see his mother over the last ten years. He had joined the British Army, and the month after Nerina's birth was badly wounded at Waterloo– it's not known when, or how soon, or even *if* his mother was told. So effectively did her caste close ranks that slander had little to say to those outside the privileged circle – unlike the lasting public slander of Mary Wollstonecraft. When Miss Wilmot's travels were published in 1920, the worst the editor could discover about her patron, Lady Mount Cashell, was that she was 'imbued with what were then the most extravagant political notions'. The editor transmits a cover-up tale: in time, he says, she became 'more prosaic. She was content to share her husband's retirement at Moore Park, where she lived till his death.' In other words, Lady Mount Cashell, as she had been, was no more. An entirely different and purposeful woman called Mason practised medicine – unofficially – in another country.

It was her '*religion* to make the best of every thing' – this was her strength together with work. 'I have always my medical studies & my patients, which do not allow me to be very idle,' she said. Her consolations were her dispensary for the poor, her writing, and the two children who remained to her. She produced reading-books for children, dedicated to Nerina: *Simple Stories for Little Boys and Girls* and *Stories for Little Boys and Girls, in Words of One Syllable*. These are modelled more on the correctional aims of Mary Wollstonecraft's *Original Stories* than on the maternal intimacy of her 'Lessons' for Fanny. Little Bob who, at four and a half, will not learn his ABC, is shamed at a fair by a pig who can spell; while Sal, who is 'not a bad child' but silly enough to parade her lace frock and plumed hat, wrecks them in a farmyard. Cross Ruth is shut off from friends, and orphaned Nell who slaps her cousin is put in 'a dark room, at the top of the house' until she learns to offer her hand with a smile. Compared with Wollstonecraft whose voice is warm with humorous affection – the familiar tone of parents today – Margaret's reading-book is dated. Godwin, who had witnessed Wollstonecraft with Fanny, was probably right to think Margaret deficient in tenderness. A letter from Nerina is respectfully formal, addressing an exacting mother – not unlike Margaret's own mother, Lady Kingsborough.

Yet the loss of her seven Mount Cashell children did make Margaret anxious for Laurette and Nerina. Since her own health remained shaky, she wrote a set of 'Memorandums respecting the children', drawing on her medical experience. If fevered, children should be given plenty of liquids, she recommends. For a sore throat, the child must gargle with vinegar and honey in warm water. Vomiting should be encouraged at first with warm water, as it may be nature's effort to liberate the body from excessive or infected food. Of course, should vomiting or any other symptom continue, it's advisable to send for a good physician. She was not, then, 'alternative' in a dogmatic way; just sensible, as Mary Wollstonecraft had been in her insistence on cleanliness when Fanny was a baby. After Mary's death, Godwin testified to her skill in childcare when he looked at Fanny: 'She has . . . left a specimen of her skill in this respect in her eldest daughter . . . who is a singular example of vigorous constitution and florid health.' When Margaret and the other King children had fevers, Mary had observed how they languished beneath the bedclothes in the presence of their mother's indifference. As a mother herself, Margaret was attentive to what patients said, especially inarticulate children, and took quick action where necessary instead of prescribing palliatives that could postpone treatment until it was too late.

Her cures were simple, homey, unthreatening to children, and refreshingly harmless. They compare favourably with eighteenth-century medical books filled with witches' brews (viper was a common ingredient in tonics, and easily obtained from the apothecary); and they present a huge advance on misguided medical tortures inflicted on millions throughout the ages. 'Madame Mason' continued to defend midwives against the incursions of obstetricians: 'the Midwives are very skilful & it is a mistake to suppose the aid of men necessary,' she wrote. In childbirth, nature should be allowed to take its course. The placenta, she believed, should not be forced out – she would have known of the fatal consequences for Mary Wollstonecraft; it should take its own time to emerge, whether it be the usual ten minutes, or five hours, or even five days. Contrary as this is to present-day practice, her advice makes sense in an age when infection (not then understood) was the chief danger.

In the early autumn of 1819, a set of visitors from London broke into this self-contained, professional life. The group consisted of Mary Wollstonecraft Godwin, aged twenty-two, and her stepsister Claire (Clara Jane) Clairmont, aged twenty-one, together with Mary's husband Percy Shelley. On 30 September they came to the door of Casa Silva: Mary grave, milky-pale, and swollen in late pregnancy; Claire dark with a high colour, a cluster of black curls and wide black eyes; and the eccentric figure of the poet, showing the full length of his white neck – tinged with rose – from an open shirt, set off by a grey coat, like a dressing-gown, flapping at his heels. They were passing through Pisa on their way from Livorno to Florence. Mary presented a letter of introduction from her father.

'Mrs Shelley, my daughter,' Godwin had written, 'thought it possible that in the course of her travels she might accidentally arrive at some town where one of her mother's dearest friends had taken up residence.' He went on hesitantly, unsure how Mary's vagabond group would be received. If she were unwelcome, he adds, 'you have only to put this billet in the fire; she will consider that is an answer sufficient'.

The fears were groundless. Mrs Mason (as they called her) declared herself too 'a vagabond on the face of the earth'. She kissed Claire twice, and rejoiced with Shelley in 'the frankness of your or my character . . . You cannot abhor *cant* more than I do.' Shelley's complaints about his kidneys she judged to be harmless pains brought on by stress over the recent Peterloo Massacre, when, that August, workers in Manchester had gathered to petition for the reform of Parliament.

'I have a sad opinion of the British Parliament,' Margaret agreed. 'Since my country sank never to rise again, I have been a cool politician, but I cannot forget how I once felt, & can still sympathize with those capable of similar feelings.'

Letters pursued the Shelleys in Florence, urging their return, with offers of company, help, and medical advice for Mary's newborn Percy Florence. Her advice again takes its sage tone from the voice Wollstonecraft had cultivated in Mrs Mason. This vein was not far off from the opinions of the great lady – and opinions came easily to a woman who had been Lady Mount Cashell. She thought Shelley should consult Dr Vaccà, happy though she was with her own diagnosis. 'I am not sure that I should not myself be as good a physician for Mr S as any one, were not the first requisite wanting – I mean the confidence of the patient.' So it happened that, late in January 1820, Mary, Claire and Shelley settled in Pisa. Their home was Casa Frassi on the Lung'Arno. Over the next three years, a daughter, a pupil and two followers of Mary Wollstonecraft formed an outcast society of their own.

17

Daughters

A portrait of Mary Wollstonecraft had looked down on three girls as they grew up: on Godwin's adopted daughter, Fanny Imlay; on his own daughter, Mary Wollstonecraft Godwin; and on his stepdaughter Clara Jane Clairmont. Opie's grave portrait of 1797 was to become the lasting public image of Mary Wollstonecraft, on permanent exhibition in the National Portrait Gallery. In the opening years of the nineteenth century it still hung in what had been her study at the Polygon. After her death this room became Godwin's study, and here the continued presence of the dead broke through the limits of the lifespan. For girls growing up with Godwin, this thoughtful image of Mary Wollstonecraft was the model of what they must strive to be. No slander reached them; no questioning of marriage; no atheism. Faith was not absent from their home. The second Mrs Godwin was Catholic, and her own two children had attended the Chapel of St Aloysius in Somers Town. She took the view that one form of Christianity did not differ in essentials from another, so after her marriage to Godwin all the children attended Sunday service at St Paul's. Godwin tested them on the sermon when they returned. In this home, religious conformity was endorsed, while the respectability of the second wife was ensured through two marriage ceremonies, in different places, to cover and legalise the alternate names she had used.

Mrs Godwin was animated and hardworking, and these qualities seem to have sufficed in the main for so constant a man. His friends disliked her: a clever, second-rate woman without the finer sensibilities, Marshall judged. This new wife was given to 'baby-sullenness', Godwin discovered. Equable as he was, he did have to remonstrate over 'the worst of tempers'. Twice, she threatened to walk out on the family. When she came back the second time Jane capered, but Fanny stood still in dismay. Her stepmother was moody, outspoken, and given to dramas. Similar, it might be thought, to the flaws of Mary Wollstonecraft, were it possible to separate those from the eager sympathy and truth. Once, while Godwin was away, he asked his wife to kiss her stepdaughters – but only if she could do so wholeheartedly. Fanny and Mary called her 'Mamma', though Mary loathed her – Mamma at once divined Mary's romantic attachment to her father – while Fanny tried to see Mamma's good points. Godwin was attracted by what he called her 'unsinking courage under calamities that would have laid any other person level with the earth'. She told him a dramatic tale of going abroad alone to seek her fortune at the age of eleven. He believed her – but can we? She was liable to twist facts, and determined to fix her husband's mind on money.

One reason why Godwin had meant to be a bachelor was that he could not support a family through writing. When he remarried he had four children to maintain, and then there were five, for in March 1803 little William was born. It became necessary to find new means. Back in 1800–1 Godwin had been amongst the contributors to a high-flying publication for children called *The Juvenile Library*, a cross between an encyclopaedia and a magazine, specialising in moral philosophy, botany, geography and the manners of nations, ancient and modern languages, mathematics, scientific experiment, and biography. Included were lives of intellectual women like Margaret Roper (daughter of Sir Thomas More); Katherine Parr (who entered into the theological debates of her age and published devotional books in the 1540s); and Mme Dacier, a translator of Greek and Latin classics in the time of Louis XIV. Peacock, De Quincey and Holman Hunt, then schoolboys, were amongst the monthly prizewinners; also, a number of girls, in nearly every subject. Godwin and his wife set up their own Juvenile Library, a publishing firm for children's books, in 1805. In

1807 the family moved from the salubrious air of the Polygon, on the outskirts of London, to a corner site in the centre of the City, occupying four floors above their bookshop at 41 Skinner Street. It was the right home for Margaret Mount Cashell's *Stories*, published anonymously in December of that year.

There is a drawing in Godwin's *Fables Ancient and Modern*, where Aesop holds up an explanatory forefinger to two little girls. When Godwin tested his storytelling powers on Fanny (eleven), Charles (ten), Mary (eight), and Jane (seven), he found that Aesop's moral came too abruptly. The children would ask, 'What happened then?' So Godwin (under the pseudonym of Edward Baldwin) expands the fable. He also adds fables of his own. In December 1805 the *Anti-Jacobin* recommends this book – if it had only known – by the enemy. A bust of Godwin's Aesop went up above the door of the shop in Skinner Street, where beasts went lowing and snorting to their end at the Smithfield Market, and crowds rushed by to view public hangings at Newgate Prison – unlikely readers for the Godwins' fine list which included Godwin's *Life of Lady Jane Grey* (1806), the model of a learned girl who at the age of twelve knew eight languages; Charles and Mary Lamb's *Tales from Shakespeare* (1807); and an English grammar by William Hazlitt (1810). Mrs Fenwick, who had assisted with the delivery of little Mary and nursed her dying mother, contributed *Lessons for Children*, which taught the rudiments of humanity in manners and morals. Mary Jane Godwin did the first English translation of *The* [*Swiss*] *Family Robinson* (1814), wrote (anonymously) *Dramas for Children*; and there was a performance version of *Beauty and the Beast* in verse, accompanied by Beauty's Song set to music. Margaret Mount Cashell's reading-books were popular.

Amongst the most successful publications were several schoolbooks by Godwin: histories of England, Greece and Rome. He believed that history should raise questions, not provide answers, and that the true aim of education was not rules and imitation but to stimulate a pupil to reach beyond the limits of his lessons. The respected schoolmasterish author called Baldwin was able to get away with more emphasis on republican virtues than was customary at the time. In his *English Dictionary* the word 'revolution' is defined as 'things returning to their just state'. The bookshop also carried stationery, maps, games and gifts, including a series of shilling booklets.

One of these was by Mary Godwin, a prodigy at the age of eight in 1805. She took up a current song about the adventures of an absurdly mistaken John Bull in France. Addressing the French in English, he takes their '*Je n'entends pas*' to be information about a grandee called Nongtongpaw. So pervasive does this grandee appear that John Bull begins to doubt the French Revolution. The hilarious quatrains of *Mounseer Nongtongpaw* dramatise, in effect, why the word 'insular' entered the English language at this time.

When Fanny was eleven, the Wollstonecraft sisters proposed to take her over. Godwin refused. Fanny was his – a charge Mary Wollstonecraft had left him. She was known as Fanny Godwin, though the French registration of her birth had recorded her surname as Imlay. It might have been a prudent move to send Fanny for a visit at this time, when the Wollstonecrafts reached out to her, but the coolness between the sisters and Godwin remained. The effect of Godwin's *Memoirs* had been to taint the name of Wollstonecraft and this, the sisters feared, might undermine their Dublin schools. To start schools had been risky – Johnson had shaken his head – but they did survive: Everina had a boarding-school for girls and Bess had a day-school for boys. Whatever lessons were to be learnt from the failure at Newington Green, they had learnt them. They had to make their way with no help beyond the sisterly support they always gave each other.

Another attempt to benefit Fanny came when Joseph Johnson died in 1809. His will left fifteen-year-old Fanny £200, which Godwin was to give her to pay his debt to Johnson. Fanny was not given this bequest. Though the Godwins worked hard, their press was never solvent. There was no capital; the whole venture was based on loans, and under pressure from his wife to keep it going Godwin turned into an inveterate borrower. The need to scrounge what he could on the strength of a past reputation constricted somewhat the character of a thinker who had once opened himself to the 'heart'. A need for £3000 was in the air when, in 1812, an apparent solution appeared in the shape of a young poet who was heir to a great fortune.

Percy Bysshe Shelley was the grandson of an American named Bysshe who presents another prototype of Gatsby. Bysshe had been no one in New Jersey. Then he migrated to England, grew rich, sent his son to Oxford and his grandson to Eton. Percy Shelley was at Eton in the heyday of Dr Keate, known as 'Flogger'. The boys were worse, and from 1804 to 1810 Percy Shelley was bullied for his oddity and gentleness. During this period, Bysshe became Sir Bysshe. Part of his gains had been to elope with two heiresses (while he kept a mistress with four children, one of whom was also called Bysshe). His grandson, Percy, expelled from Oxford for atheism, eloped at a very young age with a Jewish schoolgirl, Harriet Westbrook, and married her. Too late, he caught up with Mary Wollstonecraft's rationale against marriage. Like her, he detested political oppression and was a non-violent revolutionary who believed change must be gradual to be secure. He was a convert to *Political Justice*, and at the age of twenty approached Godwin with the awe of a disciple. He also had an eye to Fanny as Wollstonecraft's daughter.

In July 1812, before they met, Shelley invited Fanny to join him and Harriet in a rented cottage at Lynmouth in Devon. Godwin forbade it. He said he did not know a man until he had seen his face.

Shelley was puzzled: surely Godwin knew him through his ideas and letters? But Godwin would not budge.

When, at length, Shelley and Harriet arrived to dine at Skinner Street on 4 October, the two younger girls were away, and it was Fanny, a dark young woman of eighteen with long brown hair, who found herself drawn out by a poet's regard. Harriet set down their first impressions:

> There is one of the daughters of that dear Mary Wolstoncroft living with [Godwin]. She is . . . very plain, but very sensible. The beauty of her mind fully overbalances the plainness of her countenance. There is another daughter . . . who is now in Scotland. She is very like her mother, whose picture hangs up in his study. She must have been a most lovely woman. Her countenance speaks her a woman who would dare to think and act for herself.

For the next six weeks there was daily contact with Shelley – the Godwins saw almost no one else. Their new friend was tall and stooped with long legs, narrow chest and shoulders, a prominent blue-veined brow and dishevelled light locks. His head was all front – straight behind – and his face and features small. Even his brow, though striking in a 'marble' way, wasn't broad like Godwin's. He spoke in a high tenor that seemed to come from the back of his head, like a child's. This strange creature had grown up surrounded by five sisters, and would often try to reconstitute a female court. He was sensitive to women – as well as susceptible. 'Then it was Fanny Imlay he loved', her stepsister reported later. She thought he loved 'as women love', and she believed that 'Fanny loved Shelley'.

Fanny was taken aback when Shelley, always restless, took off for Wales on 13 November. It was ungrateful not to say goodbye, she said in her direct way; she still had some questions to put to him, though she was not sure if this was proper.

Shelley replied on 10 December 1812: 'So you do not know whether it is *proper* to write to me. Now, one of the most conspicuous considerations that arise from such a topic is – who & what am I? I am one of those formidable & long clawed animals called a *man*.'

He could assure Fanny that he was a tame representative of the species, lived on vegetables, '& never bit since I was born'. As such, he adds, 'I venture to intrude myself on your attention.'

His attentiveness reflects a different Fanny from the plain girl. Others saw her as upright and generous yet 'odd in her manners and opinions'– a girl with 'nothing' of her mother in her. Though Fanny may have continued to look like Imlay, she was, in truth, much like her mother in the honesty and independence of her opinions, her domestic affections and acute feelings. Jane joked that the hero of *The Man of Feeling* (leading the cult of sensibility forty years before) would have made a perfect husband for Fanny. Her tender-hearted and hopeless attempts at keeping the peace provided some amusement for the younger children.

At a guess – it can only be a guess – Mary and Jane were jealous of her closeness to the woman of the portrait, the icon of the household. After all, it was Fanny whom Mary Wollstonecraft had adored unreservedly, who was praised for her bloom and intelligence, who had slept in her mother's arms, and whose pulses of imaginative sympathy had been cultivated in accord with *Thoughts on the Education of Daughters*. Wollstonecraft had worried that it might unfit this sensitised child for her future life, but it had reassured her that Fanny – 'gay as a lark' as they set sail for unknown lands far north – took so naturally to humane training. It was Fanny to whom Mary Wollstonecraft had clung for comfort when Imlay left, and to whom she had hastened on her return journey from Norway. 'At Gothenburg I shall embrace my *Fannikin*,' she had written to the father of her 'babe'. 'I never saw a calf bounding in a meadow, that did not remind me of my little frolicker. A calf, you say. Yes; but a *capital* one, I own.' All this was in print in the prized relics of her *Travels* and *Letters to Imlay*.

'Les Goddesses', the three girls were called by Aaron Burr, who came to visit, the American who had read the *Rights of Woman* through the night and wished his daughter, Theodosia, to benefit. When a stranger asked Godwin if he was bringing up Wollstonecraft's daughters according to her teachings, Godwin was too truthful not to admit that he had ceded control to his second wife, who was 'not exclusively a follower'. He and Mrs Godwin lacked 'leisure enough for reducing novel theories of education to practice'. Although Godwin retained a special bond with Fanny, he was caught up in the slog of a failing business, and allowed Mrs Godwin to sideline the willing girl as her helper. Here, Godwin appears to forget that Fanny had already taken the imprint of Wollstonecraft's teachings – the compassion, the directness – an inward shape none could change.

Formally, the girls were set apart from dutiful Charles Clairmont, who was sent for six years to Charterhouse, a London school for boys. Les Goddesses were taught largely at home – not necessarily a disadvantage with its publications, its political debate, and visitors like the Lambs and Coleridge. Godwin took them to Coleridge's lectures at the Royal Institution. Mary Godwin wrote later: 'There is a peculiarity in the education of a daughter brought up by a father only, which tends to develop early a thousand of those portions of mind, which are folded up, and often destroyed, under mere feminine tuition.' Jane and Mary had alternate spells at boarding-schools, and then in 1812–13 Mary joined the household of Godwin's friend Baxter in Dundee.

Fanny stayed at home, an unrescued Cinderella at the mercy of Mrs Godwin's temper – 'the bad baby', the Lambs called her. Someone – it sounds like Mrs Godwin – told Fanny that others wore themselves out working to keep her. Godwin's catalepsy came back; Fanny watched him anxiously. Henry Reveley, a playmate in childhood, recalled that this 'amiable and loving little girl' was pitted with smallpox. It may not have been much, since to children any blemish looms large, but, if true, it would have been impossible to shield a girl. At some point Fanny began to suffer from depression. She apologised to her sister Mary for her 'torpor', with resolutions to transform her 'faults'. 'So young in life & so melancholy' – Jane shook her head over Fanny. Without looks, dowry or feminine wiles, Fanny had small chance of marriage. The brutal economics of Jane Austen demonstrate the alternatives for young women of that generation. They are visible in *Emma* in the misery of the polished pianist Jane Fairfax who is due to become a governess, and in *The Watsons* a girl unable to contain her tears when her brother reproaches her as 'a weight upon your family'. Fanny's charity status, compounded by her 'torpor', licensed the similar reproach.

Godwin's answer to the enquiry about Wollstonecraft's daughters reflects the partisan atmosphere that had settled on a household where every child, bar Fanny, had a blood tie to one parent. Godwin pictures Mary – 'my own daughter' – as 'bold', 'imperious', 'invincible' in her perseverance, and Fanny's superior in her appetite for knowledge. Fanny is 'somewhat given to indolence' – hardly recognisable here as the lark-like child whose intelligence had intensified Mary Wollstonecraft's love for her. Godwin sees a subdued girl who is quiet, unshowy, observant, and disposed to follow her own thoughts. Though he wasn't asked about looks, he again puts forward his own daughter. She is 'very pretty', compared with Fanny who, he allows, is 'in general prepossessing'. Godwin boasted to Shelley that Mary was like her mother. Actually, Mary's pallor, high forehead, and long, elegant nose resembled her father's; and this resemblance would become more pronounced in her forties in the moon-pale Rothwell portrait with the smoothed and parted hair of the Victorian period. But a miniature in her teens, angled to mirror her mother in the Opie portrait, does show the likeness at that time: the same heavily dented upper lip, curled at the left corner, the marked brows, the whisps of hair clustering on the forehead. Her grey eyes are dreamier than her mother's, her lips thinner, and she's more delicate, with fine, nut-brown hair that spun about and tangled as she turned her head.

In the spring of 1814, les Goddesses were all three at home when Shelley returned to Skinner Street. He came this time without his wife, who was in Bath. On 23 May, Fanny – just twenty – was sent off for the summer to Pentredevy, near Swansea. This fact gets passed over, but could be important. Until then, Fanny almost never left home – partly reluctance to leave what home she had; partly the legacy of her mother's protectiveness towards 'Papa' – in his fits of sleep – as spelt out in Wollstonecraft's 'Lessons'; and partly acquiescence in the lack of plans for her. So why Wales just now? One purpose was to meet her Wollstonecraft aunts who were to come over from Ireland, but that would hardly have occupied the whole summer. Another possible reason lies in a set of notebooks in the Peabody Museum in Salem, Massachusetts. The notebooks belonged to a Shelley enthusiast, Captain Edward Augustus Silsbee, who elicited Jane's remarkably acute and detailed recollections in old age. She recalled that when her mother, Mrs Godwin, thought that 'Shelley was in love with [Fanny] or might be', she sent Fanny away. Jane was not always reliable, but this sounds plausible, given Shelley's worship of Wollstonecraft and willingness to discern her in Fanny. Whatever the truth, Fanny's absence that June and July had life-changing repercussions for all three girls. Jane also recalled Shelley's telling Mary that his wife Harriet no longer loved him and that the child she was carrying was not his.

The more Mary loathed 'Mamma', the more she took 'pride & delight' in her real mother. At sixteen, it became Mary's habit to commune with her in St Pancras churchyard. On the plain table-tombstone, beneath the willows Godwin had planted, the girl saw the same name as her own:

MARY WOLLSTONECRAFT GODWIN
Author of
A VINDICATION
OF THE RIGHTS OF WOMAN:
Born 27 April, 1759:
Died 10 September, 1797.

Shelley, meeting her there on Sunday 26 June, embraced this imaginative inheritance. 'I would unite/ With thy beloved name, thou Child of love and light,' he said in his secret self. Mary would recall 'that churchyard, with its sacred tomb, was the spot where first love shone in your dear eyes'. She was to be his 'spirit's mate', a channel of inspiration from a mother whose life and fame seemed to clothe this daughter in a 'radiance' that shone also from the shelter of her father's own 'immortal name'. This girl with her great white tablet of a forehead was beautiful and free as, calmly, she confessed her love, ready to break conventions for his sake. He thought she had 'the subtlest & most exquisitely fashioned intelligence', and told her that 'among women there is no equal mind to yours'. Poor Harriet could not compete.

'I am in want of stockings, hanks, and Mrs W[ollstonecraft]'s posthumous works,' Shelley let Harriet know soon after he left her. Stockings and handkerchiefs were sent. Wollstonecraft's works Harriet retained.

His passion was for Mary's aura as Wollstonecraft's daughter, Harriet protested.

'It is no reproach to me that you have never filled my heart with an all-sufficing passion,' he told her. Shelley expected his pregnant wife to condone his abandonment as the action of a higher being. 'I murmur not if you feel incapable of compassion & love for the object & sharer of my passion.'

He had 'no doubts of the evils of marriage, – Mrs Wollstonecraft reasons too well for that'. It would have been from Shelley that Mary, five years younger, learnt her mother's theory. Godwin was aghast at Shelley's blithe proposal to make off with his teenage daughter. Godwin argued for marital fidelity, he thought convincingly, ever ready to believe that reason must carry the day. But the lovers, accompanied by Jane – Fanny was still in Wales – stole away at 5 a.m. on 28 July, and made for France. The girls wore black silk dresses, hoping to look grown-up, Mary in fact already pregnant and rather wan.

After a stormy Channel crossing she saw 'the sun rise broad, red, and cloudless over the pier' at Calais, and stared at women in high bonnets with their hair pulled back. As they travelled on through France, they were appalled by the pillage of war: 'a plague,' Mary wrote, 'which in his pride, man inflicts on his fellow'. Like her mother twenty years earlier, Mary Godwin observed brutalised faces as well as roofless houses, gardens covered with the white dust of torn-down cottages, and sullen, filthy lodgings. All this she set down in a *History of a Six Weeks Tour*, modelled on Mary Wollstonecraft's *Travels*. Her party carried Wollstonecraft's writings and Godwin's *Memoirs* in their bags. As they travelled by boat along the Rhine from Basle to Baden, Shelley read the *Travels* aloud.

'This is one of my very favorite Books,' Jane enthused in her new journal on 30 August 1814. 'The language is so very flowing & Eloquent & it is altogether a beautiful Poem . . . The Rhine was extremely rapid – the Waves borrowed the divine colours of the sky.'

Mrs Godwin did not blame Jane for running away. She blamed Mary Godwin, in a moan to one of her authors, Lady Mount Cashell (as she continued to call her). From now on, Mrs Godwin set up Mary as the cause of everything that went wrong in her life and prime target for her rage. Jane, won over to the cause of free love, refused to go home, and to everyone outside that home became from this time 'Clara', 'Clare', 'Clary' and eventually 'Claire'. Mrs Godwin was no match for Shelley's high-flown eloquence backed by the starry shade of Mary Wollstonecraft. In truth, his self-righteousness over his susceptibility to a succession of women shows how far he mistook Wollstonecraft who never practised free love. Claire, aged seventeen, was a convert to this misapprehension when she approached Byron in the spring of 1816: she told him she believed in nature and detested marriage, reducing Wollstonecraft's reseeding of women's nature to an offer of guilt-free sex. 'I shall ever remember the gentleness of your manners & the wild originality of your countenance,' she wrote to her lover afterwards.

In her sexual freedom did Claire model herself, not on the real Wollstonecraft, but on a dead celebrity distorted by report? What some see in Claire Clairmont is a perpetuation of a contemporary view of Wollstonecraft: a caricature of rash passion. Fanny, in contrast, had a deep-souled sense of her mother's 'superior being'. In 1816 Fanny Blood's brother George returned to London after an absence of twenty-six years. 'Everything he has told me of my mother has encreased my love and admiration of her memory,' Fanny wrote to Mary and Shelley. 'I have determined never to live to be a disgrace to *such a mother*.' Only an inward 'revolution' could overcome her 'faults' – her 'torpor' – and so find beings 'to love and esteem me'. Fanny's depression is usually explained as an inheritance from her mother, though her status in the home is cause enough. This inferiority was reinforced by her exclusion from the Shelley party, underlined by her stepmother's report that Fanny was their 'laughing stock' – intended to detach Fanny further from her sister.

Fanny did defend herself to Mary and Shelley, who had yielded to Claire's wish to follow Byron to Geneva.

'Mary gave a great deal of pain the day I parted from you,' Fanny wrote on 29 May 1816, 'believe my dear friend's that my attatchment to you has grown out of your individual worth, and talents, & perhaps also because I found the world deserted you I loved you the more. What ever faults I may have I am not *sordid* or vulgar. I love you for *yourselves alone*[.] I endeavour to be as frank with you as possible that you may understand my real character.'

Mary brushed off Fanny's pain with another reproach. A silence preceded Fanny's apology, on 29 July: 'I plead guilty to the charge of having written in some degree in an ill humour.' She offered a summary of current events – 'mixed up with as little spleen as possible'. Spleen. Indolence. Torpor. Ill-humour. These were the words for what Fanny called 'the dreadful state of mind I generally labour under & which I in vain endeavour to get rid of'.

Quickly, she shuts this off, and moves to the case for the unemployed in the riots following the final defeat of Napoleon at Waterloo. She sees that talk of a change of ministers 'can effect no good; it is a change of the whole system of things that is wanted'. She relays a conversation with Robert Owen, the Lanarkshire cotton manufacturer, factory reformer and pioneer trade unionist. 'He told me the other day that he wished our mother were living[,] as he had never before met with a person who thought so exactly as he did – or who would have so warmly and zealously entered into his plans.' Fanny was sceptical of Owen's optimism. His expectation 'to make the rich give up their possessions and live in a state of equality is too romantic to be believed'. Fanny takes a dim view of Owen's rhetoric, and interjects with stuttering conviction: 'I hate, and am am sick, at heart at the misery I see my fellow beings suffering – but I own I should not like to live to see the extinction of all genius, talent, and elevated generous feeling in great Britain, which I conceive to be the natural consequence of Mr Owens plan . . .' Here, Fanny demonstrates her own more sophisticated grasp of Mary Wollstonecraft's politics.

Fanny's letter slips in a covert plea to join her sister. 'I had rather live all my life with the Genevese as you and Jane describe than live in London with the most bril[l]iant beings that exist.' Contrite, Mary bought Fanny a Swiss watch. It was a gesture – no more; Mary did not offer Fanny a home despite the comfort Fanny could offer. (A year and a half before, in London, Mary's first baby had died. While Claire and Shelley had skipped off as usual to town, Mary notes in her *Journals*, 'Fanny comes, wet through; she dines, and stays the evening; talk about many things . . .') So it happened that a caring sister was ousted by a careless stepsister who confessed to Byron: 'I cannot say I had so great an affection for [Fanny] as might be expected.' Though Claire and Mary quarrelled, Fanny was never preferred, for Shelley had come to need Claire, her vivacity and sense of adventure a complement to Mary's kindred spirit. This jostling but (for Shelley) fertile unit of Mary and Claire was therefore closed to Fanny, who had been the focus of Shelley's attention before and during his first visits to Skinner Street. Yet Fanny too continued to love poetry, and asked Shelley to send his work, saying in 1816, 'It is only the poets who are eternal benefactors of their fellow-creatures' – a source for Shelley's famous words in 1821: 'Poets are the unacknowledged legislators of the world.'

Fanny continued to worry about Godwin's poverty, and saw herself as an extra burden. Imlay never showed any wish to know her, and she remained as unaware as her mother of a flourishing clan of Imlays across the

Atlantic. If this were a romance, the modest English girl would have arrived on the doorstep of the Imlay Mansion in Allentown, New Jersey, and lived happily ever after. In actuality a harder fate awaited. In the spring of 1816, when she was almost twenty-two, Aunt Everina proposed that Fanny join her and 'Aunt Bishop', by now in their fifties. Everina detailed her 'sufferings' and Aunt Bishop's poor health. Hardly an enticing prospect, but Fanny expressed a suitable gratitude, and hoped, she added humbly, to deserve her aunt's affection as long as she lived. Fanny could now contemplate, as she put it to Mary, 'my unhappy life' as her aunts' drudge.

The Wollstonecraft sisters arrived in London later in the summer, and it's possible that by the time they departed on 24 September, it was settled that Fanny would leave Godwin's household and join their schools in Dublin. Fanny may have met Shelley on a visit to London – if so, she would have relayed her aunts' plans and her own reluctance. The Shelley party had its own troubles. During the second half of 1816, they were living in Bath to keep Claire's pregnancy a secret from her London circle. In Bath she styled herself 'Mrs Clairmont'. Unfortunately, Shelley's withdrawal of financial help from Godwin in late September moved Fanny to write to Mary, on 3 October, to defend Godwin's philosophy that those with means are obligated to support worthy ones who are in want.

'Forgive me if I have expressed myself unkindly,' she wrote. 'My heart is warm in your cause – and I am *anxious most anxious* that papa should feel for you as I do both for your own, and his sake – I have written in a great hurry and have not had time to consider and round my sentences – But I am so direct in all my thoughts and opinions that I cannot but believe every one must like frankness as much as myself.' The following day Mary received this 'stupid letter from F'.

Fanny was trapped between warring Godwins, their love conditional on her serving their wills. Four days later, on 8 October 1816, Fanny passed through Bath, on a westward journey. She had dressed carefully in a blue striped skirt and white bodice, with a brown pelisse and brown hat. She took the Swiss watch Mary had given her, and put a meagre eight shillings – not enough for a passage to Dublin – in her reticule. Next to her skin, she wore her mother's stays.

There's a gap in what's known when Fanny stops in Bath. What happens during that hour or two when she arrives by the morning mail in the fashionable spa where her mother had served long ago as a companion and where her sister is now living? She has wished to see Mary, for that very day Mary receives word from Fanny, presumably about her arrival. Godwin, who banned all contact, is too far away to know. But Mary can't invite her sister to their lodgings because she can't reveal that Claire is six months pregnant. And Mary may still be annoyed with 'stupid' Fanny whose compassion for Godwin can't fail to touch a daughter who has also worshipped him – and who, in his eyes, is disgraced and banished. Fanny has taken her place as prime daughter. Not Mary, then, but Shelley – possibly – comes to see Fanny at the inn. There's no invitation to join him, no alternative on offer. This would have been her last hope of rescue from the 'sufferings' Aunt Everina has invited her to share.

From her next stop in Bristol, she sends warning notes to Shelley (to come and see her buried) and to Godwin ('I depart immediately to the spot from which I hope never to remove'), but she doesn't say where she is. Both, alarmed, set out to trace her. Fanny, in the meantime, has taken a coach to Swansea. There, at the Mackworth Arms, she sips some tea. By now it is night, and she asks for her candle. By its dim light she writes:

> I have long determined that the best thing I could do was to put an end to the existence of a being whose birth was unfortunate, and whose life has only been a series of pain to those persons who have hurt their health in endeavouring to promote her welfare. Perhaps to hear of my death will give you pain, but you will soon have the blessing of forgetting that such a creature ever existed as

Before she swallows the overdose of laudanum or before oblivion overtakes her, she thinks of those she loves, and to save further trouble, she tears off her name and lets the candle consume it.

Next morning, Fanny's body was discovered. It was not identified, not the stockings marked 'G', nor the stays marked 'MW'. Godwin gave out that Fanny had gone to join her aunts in Dublin. Privately, he told Mrs

Gisborne (who had looked after Fanny and baby Mary) 'that the three girls were all equally in love with [Shelley] and that the eldest put an end to her existence owing to the preference given to her younger sister'. This has to be a simplification, distancing the suicide from any responsibility on Godwin's part, though it may explain why he 'half-expected it'.

Shelley recalled their last meeting in 'On Fanny Godwin' (1816):

Friend, had I known thy secret grief
Her voice did quiver as we parted
Yet knew I not that heart was broken
From which it came – and I departed
Heeding not the words then spoken –
    Misery, oh Misery,
This world is all too wide for thee!

On the back of the fragment is a sketch, maybe a design for a grave under a tree and surrounded by urns with flowers. On this Shelley has scrawled: 'It is not my fault – it is not to be alluded to.'

This was not the only casualty (in whatever sense) of Shelley's and Mary's union. Two months later Harriet Shelley, who thought her husband a 'monster', drowned herself and her unborn child in the Serpentine in Hyde Park. Shelley was more distressed by Fanny's death, he confided to Byron. Another poetic fragment pictures Fanny as the child reunited with Mary Wollstonecraft on her return to Gothenburg from Norway in the summer of 1795:

Thy little footsteps on the sands
    Of a remote and lonely shore – . . .
The laugh of mingled love & glee
    When *one* returned to gaze on thee

These footsteps on the sands are fled
    Thine eyes are dark – thy hands are cold
And she is dead – and thou art dead –

Mary, her sister, dressed in mourning, reread their mother's *Rights of Woman*, and continued to work on her novel *Frankenstein*. Did Fanny's 'unfortunate' birth strike Mary and Claire, who were bearing children outside marriage? Mary, with a year-old son, William ('Will-Mouse') – called after her father, despite his refusal to see her – now wished to be married, and Shelley consented out of consideration for her, much as Godwin had consented, long ago, to marry Mary's mother.

Byron, who was married already (though separated from his wife), chose to believe that only a bad girl from a Godless background would have offered him her virginity.

Claire protested, 'I cannot pardon those who . . . believe because I have unloosed myself from the trammels of customs & opinion that I do not possess within[,] a severer monitor than either of these.'

Byron dismissed Claire's novel 'The Ideot', about a Robinson Crusoe sort of girl, isolated from social conditioning. Acting on impulses that arise from herself alone, Claire's heroine flouts custom, yet is filled with affections and sympathies. The narrator, an old clergyman, attributes the girl's errors to a neglected education, but sophisticated readers were expected to grasp what looks like a gender experiment in authenticity. Byron's derision led Claire to kill her work as 'that hateful novel thing I wrote'.

Most of Byron's women were of his own class, protected by money and privilege, and accustomed to discreet adultery. Claire, by contrast, had been a virgin in love with a jaded Don Juan. Ten years later she looked back on herself as the 'victim' of ten minutes' happy passion that had withered any other possibility.

If passion withered, other kinds of love flourished. She had two near-perfect attachments, the first to Shelley who opened her mind to a freedom that breathed through his looks and manners with a beauty, she said, no picture could express. Claire lived out an unconventional scenario Wollstonecraft had conceived but could not put into practice: her plan to join the Fuselis. Claire's triumph was her sustained closeness to a protector-genius without the drawbacks of being his wife. Mary was depleted by sick and dying children, and also had to bear Shelley's romantic enthusiasms for other women, his irresponsibility, and the self-centredness that left Mary to her griefs, then blamed her for coldness. Mary saw herself rather as a person whose eager sympathies entangled her, and her love for women friends became intense in later years. Her letters have a beating pulse like her mother's, a similar playfulness and candour, with a learned edge from her Godwinian education. Mary Shelley was always political, an opponent of tyranny and monarchy and a supporter of individual liberty, but she was also open to rarer influences: her mother's 'greatness of soul' and Shelley's search for knowledge.

Shelley had encouraged her to expand her idea for *Frankenstein*, conceived on Lake Geneva when she was nineteen. Dr Frankenstein, a solitary scientist, expects to invent a man, and finds that he's made a monster. This killer is the unnatural son of an unnatural father, for Frankenstein has detached himself from domestic ties, relying on ingenuity alone. In dramatising this point, the young author confirms her mother's case for parental nurture, together with Wollstonecraft's attack on 'the cold workings of the brain'. Frankenstein is an irresponsible creator who abandons his creature. The monster's testimony is similar to that of the criminal Jemima in *The Wrongs of Woman*: deprived of domestic affections, Jemima perceives herself 'an outcast from society'. 'I hated mankind,' she says. 'Whoever acknowledged me to be a fellow-creature?' It's now thought that the primary pattern underlying feminist writing 'is that of *Frankenstein*', a world in which cerebral man and monster are one. The oneness of uncaring creator and killer-creature aims at 'refashioning an entire sex'. The uncaring might of democracy corrupting its hideous progeny and the monster-Moslem who is careless of life, reflect each other's fantasy of power, bearing out the continuing relevance of Mary Shelley's fable where the natural alternative of maternal nurture has no public status. Its very absence from the political arena, the horrors its absence inflicts, call for an alternative order that can collapse the boundaries between the domestic and the cerebral – in a sense, the lost but retrievable possibilities of the Godwin–Wollstonecraft union.

Claire owned to envy of Mary's achievement, but found that envy 'yields when I consider that she is a woman and will prove in time . . . an argument in our favour. How I delight in a lovely woman of strong & cultivated intellect.'

Shelley praised the recessed stories and framing plot of exploration. They seemed to emanate from the buried nature of the author. 'What art thou?' he asks at the time she completes *Frankenstein* in 1817. 'I know but dare not speak.' It is something for the future to define. Her distinction burns 'internally', seen in eyes 'deep and intricate from the workings of the mind'; in 'thy gentle speech, a prophecy/ Is whispered'. This alertness, his wish to know women worth knowing, made Shelley irresistible – irreplaceable – to les Goddesses.

During the restless years when Claire and Mary moved with Shelley from place to place – France, the Rhine, London, Geneva, Bath, Naples, Rome, Pisa – they learnt Greek and Latin, and read history and literature day by day. Both undertook a programme of reading, copying, letter-writing and journal-keeping. While Mary Shelley wrote fiction, Claire Clairmont produced two lost works: a satire on Byron called 'Hints for Don Juan' (conceived in February 1820) and 'Letters from Italy' (between April and August 1820). But her public triumph lay in music. Her voice inspired one of Shelley's greatest poems, 'To Constantia*, Singing':

> Thy voice slow rising like a Spirit lingers
> O'ershadowing it with soft and lulling wings,
> The blood and life within thy snowy fingers
> Teach witchcraft to the instrumental strings
> – My brain is wild – my breath comes quick –
> The blood is listening in my frame,
> And thronging shadows, fast and thick,
> Fall on my overflowing eyes;
> My heart is quivering like a flame;
> As morning dew, that in the sunbeam dies,
> I am dissolved in these consuming extacies.

*'Constantia' was Claire's nickname, from the novel *Ormond* by Charles Brockden Brown.

Where Byron jeered at the 'prancing' advances of 'the odd-headed girl' with her black curls, Shelley recognised yet another creature 'Upon the verge of Nature's utmost sphere'.

'Constantia turn!' he commands. 'Yes! In thine eyes a power like light doth lie.' Even as her voice is laid to sleep between her lips, that power lingers in her breath and springing hair. In Pisa, Shelley hired a music master, Zannetti, who trained Claire to performer's standard. She tries out the overture to Rossini's new opera, *La Cenerentola*; she sings arias from Mozart; and her voice is like 'the breath of summer's night' on starry waters. Her lingering breath suspends a poet's soul 'in its voluptuous flight'.

Claire was too close to Shelley for Mary's comfort. Mary could never forget the ease with which he had left his first wife. It's not known if Claire and Shelley were lovers, but undoubtedly there was some understanding. Claire claimed that she knew Shelley in some ways more than his wife did. When Mary's grief for her dead children displaced Shelley from the centre of her attention, he thought of going off with Claire to the Middle East.

Claire's second perfect bond was with her daughter, Allegra, a fair-haired child with a cleft chin like her father's. Claire's tragedy was not Byron but parting from Allegra, who was to belong to her father and to have the advantages of a peer's daughter. In Venice, in Byron's dissipated Palazzo Mocenigo, the once merry Allegra grew quiet and started wetting her bed. Byron, fancying she was not his child after all, proposed to make her his mistress when she grew up. A solution could have been for Claire to marry someone who would provide a home and adequate support, but she turned down an offer from the novelist Thomas Love Peacock; another from Maria Gisborne's son, Henry Reveley, now a steamship engineer; and yet another from Shelley's and Byron's friend, the Cornish adventurer Trelawny.

When the heirs of Wollstonecraft came together in 'Carissima Pisa', they visited almost daily. Shelley used to arrive at Casa Silva at nine in the evening (when the children were in bed). Wollstonecraft's Margaret – dressed in chintz as 'Mrs Mason'– read Aeschylus and Sophocles with him until eleven. Shelley's poem 'The Sensitive-Plant' (composed in March 1820) praised her at the age of forty-nine:

> A Lady – the wonder of her kind,
> Whose form was upborne by a lovely mind
> Which dilating, had moulded her mien and motion,
> Like a sea-flower unfolded beneath the Ocean –

The Lady's way with plants recalls Mrs Mason's use of the gentlest herbals for her cures. She sprinkles water on plants which are faint, lifts their heads 'with her tender hands', and props them up. She presides over the chrysalis of transformed lives: 'many an antenatal tomb/ Where butterflies dream of the life to come'. The Shelley party came to Pisa bereft of five children: the Shelleys' premature baby girl had died soon after birth early in 1815; another girl, Clara, who had been much like Shelley her mother thought, had died in the summer of 1818; Allegra had been taken over by Byron, who refused Claire's pleas to see her; another girl of uncertain parents, registered as Elena Shelley, had been left in Naples where she was born; and then, the previous June, 'Will-Mouse' – another of Shelley's 'blue-eyed darlings' – had succumbed to Roman fever.

Mary Shelley had spent the summer of 1819 at Livorno with her mother's friend Mrs Gisborne, while Godwin turned his face away and reproached his daughter for lack of fortitude. 'Alas!' Mary Shelley wrote in a new novel about a daughter's yearning for her father: 'he, my beloved father, shunned me, and either treated me with harshness or a more heart-breaking coldness.' Something 'malignant' seemed to have blinded him to her – in real life this would have been Mrs Godwin, who disliked her step-daughter so much that she disliked Mrs Gisborne for her sympathy. Mary Shelley's novel *Matilda*, welling out of a dark pit during the second half of 1819, is a deathbed confession by the daughter of a disconsolate man who had lost his wife, the love of his life, in childbirth, shortly after their marriage. 'Oh, my beloved father! Indeed you made me miserable beyond all words, but how truly did I even then forgive you, and how entirely did you possess my whole heart while I endeavoured . . . to soften thy tremendous sorrows.' This was Mary Shelley's message to her father, who thought the incestuous love at the centre of the story too scandalous for publication. She had the discernment not to blame him: 'often did quiescence of manner & tardiness in understanding & entering into the feelings of others cause him to chill & stifle those overflowings of mind from those he loved which he would have received with ardor had he been previously prepared'.

The loss of so many ties – mother, sister, children, as well as the chill of her father – left Mary shut off, even to Shelley, while the proximity of Claire, always a source of friction, was becoming intolerable. Mrs Mason, as 'Minerva' (goddess of wisdom), saw the need to detach Claire from dependence on Shelley, and at the same time relieve one cause of Mary's depression while she nursed her sole remaining child, Percy Florence.

'I made the most terrible mistakes so long as I was with the S[helley]s,' Claire admitted later. 'As soon as I got into Lady Mountcashell's [Mrs Mason's] hands . . . I succeeded in all I undertook – but then I had confidence in her and obeyed her implicitly . . . She set me on the right path, which I have followed ever since.'

Mrs Mason arranged for Claire to enter the household of a physician, Dr Antonio Boiti, who lived near the Pitti Palace in Florence. There, Claire gained experience teaching English, while learning German from Boiti's German mother-in-law. The idea was to prepare herself as an English teacher for a future with her brother who was teaching in Vienna.

Relieved of Claire, Mary still had to contend with the difficulties of living with a poet who reinvented himself repeatedly. He might rescue Emilia Viviani from her convent; better, he might send Mrs Mason dressed as a man to marry Emilia and spirit her away; or he might vibrate to the guitar he presented to a languid newcomer from British India, Jane Williams. On the top floor of the Palazzo Scotto overlooking the Arno, with the poet and essayist Leigh Hunt and his family below, Shelley complained to his companions that marriage was hell.

Mary turned to writing: it was her form of survival. A semi-satiric story, 'The Bride of Modern Italy', recreates Emilia as an empty-head who draws a bemused young Englishman (a boyish Shelley) into the overblown romantic banalities of her dramas. Her convent is squalid; the nuns venal; and the girl too silly to invest herself with more interest than an Italian situation – a convent bride for sale – which Mary Shelley exposes with Godwinian acumen. She stood ready to tear the veil 'from this strange world & pierce with eagle eyes beyond the sun – when every idea strange & changeful is another step in the ladder by which I would climb'. So she writes in her *Journals* on 8 February 1822. On Tuesday the 19th she reads her mother's *Travels* yet again, and the following Monday affirms her venturesome inheritance: 'Let me fearlessly descend into the remotest caverns of my own mind – carry the torch of self knowledge into its dimmest recesses . . . Read Wrongs of Woman – . . . Claire departs – . . . spend the evening at Mrs Masons.'

Claire's departure from Pisa coincided with Byron's arrival. Claire hoped to convince him that she was a responsible person, independent of the Shelleys (whose capacity for childcare Byron questioned), and thus worthy to receive visits from Allegra. Byron, unconvinced or (in Claire's view) vindictive, placed Allegra beyond her mother's reach in a convent at Bagnacavallo, twelve miles from Ravenna. The four-year-old was labelled 'vain' and 'stubborn', faults the convent must correct 'so far as nature allows'. Its rules required Allegra to remain to the age of sixteen. Byron ignored Claire's reminder of their agreement that their child would never be left without a parent till she was seven. He ignored, too, her pleas to be allowed to see Allegra ('my life', Claire said), and the child's own pleas to see him.

With Mrs Mason's help, Claire wrote a letter to Byron to suggest placing Allegra in a different establishment in Pisa or at school in England, offering to pay all the expenses. No answer.

When Claire had premonitions of Allegra's death in the early months of 1822, Mrs Mason offered to fund a daring scheme: Claire was to enter the convent, steal the child and go to Australia. Sadly, Claire listened to the Shelleys, who counselled patience.

The Pisa coterie came to an end in a series of disasters in 1822. In April, Byron's banker in Ravenna informed him that Allegra was ill. He did not go to her, nor tell Claire. She received the news of Allegra's death with the calm of one to whom nothing worse could happen. Claire would hate Byron for the rest of her life – beyond life, she said. The loss of her child, followed in July by Shelley's drowning in the Bay of Spezia (north of Pisa), stripped Claire of all that mattered. Vaccà diagnosed tubercular glands, but she could not afford to nurse her health. For Shelley's death put an end to support, and the equally sudden death of Mount Cashell in October 1822 put a stop to his wife's stipend. That autumn, Claire boarded the coach for Vienna, expecting to teach English alongside her brother. Charles Clairmont received her kindly and found her employment, yet no sooner did she begin to establish herself than she was tracked by Metternich's spies. They proclaimed her a dangerous radical for her ties with Shelley and Godwin, refused her a teaching licence, and gave her notice to leave Austria. That winter Claire grew 'skeleton thin'. Mrs Mason, who feared 'there was little to hope' of this 'treacherous malady', and uneasy about the medication (sixteen pills a day), wrote twice to Byron to ask for aid. She was rudely refused.

'That he should hate her as he does, I cannot understand,' Mrs Mason confided to Mary Shelley, ' – I could not hate the Devil so, if he had torn out my heart with his claws & trampled on it with his cloven foot.'

Those looking for the influence of Mary Wollstonecraft on Claire Clairmont might be drawn to her unconventional relations with Shelley and Byron, and the illegitimate child she had adored. More significant, though, for women of the future is Claire's silence in 1823 when she went to Russia to become a governess. It seemed insane for a tubercular young woman to travel north into the Russian winter, 'my ice cave' she called it. Still, she had to earn her living, and earn it in a far-off place where no one knew her 'dark history'. At this point, when her life was almost destroyed, there was this silence. Friends thought they would never see her again, but silence, in so powerful a character, has a different quality, as Mary Shelley, back in London, guessed. Mary did believe in Claire's endurance, and she was right. For Mary too had a character concealed, of necessity, underground. 'My thoughts,' she told herself, 'are a sealed treasure which I can confide to none.'

The surface silence, the non-events as lives seem to unravel, can be their most momentous events – however unrecorded. When Claire Clairmont lived on with her secret in the alien ice cave, cut off and alone, distrusted at first as a stranger, she shed tried and familiar narratives of sexual surrender, abandonment and unrequited love, together with the motherhood that had been taken from her. During her five years in Russia, she stripped herself of feminine hopes and the marriage story. She knocked M. de Villeneuve's verses to her 'out of his hand and would not read them,' she said, 'for I am tired of learning that I am charming'. A fellow-tutor in Moscow, a German intellectual called Herman Gambs, fell in love with her. He reminded her of Shelley, he was delightful, but she had to refuse him. For she had become something else, a fiercely independent creature who could endure. 'My soul,' she jotted, 'seems to have been regenerated in the fountains of adversity into which it fell; there is a vigour and elasticity in my spirit which it never knew even in the spring of life.'

It's this strength to remake herself that's of the same ilk as Mary Wollstonecraft's, as well as her educational theories that Claire now put into practice. Mary Wollstonecraft had mocked a showcase education. She had pictured gratified Mammas who listen with astonishment to unintelligible words, 'the parrot-like prattle, uttered in solemn cadences, with all the pomp of ignorance and folly. Such exhibitions only serve to strike the spreading fibres of vanity through the whole mind.' In the same way, Claire mocked Russian education. 'They educate a child by making the external work upon the internal, which is, in fact, nothing but an education fit for monkies, and is a mere system of imitation – I want the internal to work upon the external; that is to say, that my pupil should be left at liberty as much as possible, and that her own reason should be the prompter of her actions.'

As Claire taught day by day in Moscow and St Petersburg, managing the demands of employers and mourning in silence for Allegra, she learnt a professional discipline, and came to regard herself as a champion of her sex. This silence in her life recalls the midnight reflections of Mary Wollstonecraft when she was a governess and a stranger in Ireland, and recalls too Margaret Mount Cashell's 'days of adversity' when she had sloughed off her aristocratic life and wandered through German towns till she came to the medical school at Jena.

Claire Clairmont's letters have not yet had their due. 'You write the most amusing & clever letters in the world,' Mary Shelley acknowledged. 'If your letters are ever published, all others that were published before, will fall in the shade, & you will be looked on as the best letter writer that ever charmed their friends – Is this glory?' Here Claire recounts her domestic trials in Russia:

> One's intimate friend here is sure to live nine versts* off, and such as have many acquaintances, or go to many parties, pass whole days and nights in their equipages – Neatness in a Russian Woman's dress can never be expected; for the paving is so bad, the ruts, holes and mud so numerous and excessive, that before one arrives at the end of one's journey, one's whole dress is in disorder; every pin in it has jumped out; every curl has been jolted out of its place, and to finish the list of grievances, of which I have spoken of the hundredth part, tho' last not least, come the troops of black-beetles [cockroaches], bugs and ear-wigs, which swarm in every Russian house – I have foresworn sleep in Russia – never any where did I sleep so little. I see nothing but these horrid animals crawling about all day, and my imagination is so affected, that it seems to me always as if my bed were covered by a troop of black insects . . . my letter blackens to my eyes, even as I write –

The apparent spontaneity of her letters modulates into a Queen of the Night attitudinising, the comic edge of coloratura extravagance, recalling the stylised plaint of the Mozart and Rossini arias that Claire Clairmont continued to perform in the drawing-rooms of St Petersburg and Moscow.

Now and then, a confession breaks, volcanic, through the orchestration of her control. 'From Morning till Night, month after month and year after year never to see a person one cares a pin to see . . . to have one's soul ever full of rage and despair . . . Has Hell any thing worse to offer?' As one

---

*A verst and a half is roughly a mile.

who had spent her youth in the company of Godwin and the Shelleys, it was nothing less than 'misery' to endure talk of cards and servants. She hoped to cultivate 'philosophic indifference to all external circumstances'.

Her relief was writing. Claire's private writings – her journals and letters to the friends of her youth – are earthquakes of unspoken words. When Mrs Godwin blurted out her daughter's past to a fellow-teacher in Russia who visited London, Claire's reputation as a teacher collapsed. Again, scandal; again, Claire moves on; again, she learns the prudence of keeping her past to herself. In 1827, the fifth year of her exile, she writes in her *Journals*: 'The world is closed in silence to me.'

What caught the public eye, of course, were sexual freedoms. By flouting discretion, Wollstonecraft, her daughter Mary, and followers Margaret Mason and Claire all appeared to confirm licentious womanhood. In the privacy of her journal, Mary Shelley defends her attempt to act out women's desires despite the prospect of disappointment: 'the most contemptible of all lives is where you live in the world & none of your passions or affections are called into action'.

Passion has tended to obscure what else Wollstonecraft passed on to the next generation: intellectual ambition and independence through work. Public events like the Terror had their enormous impact, but in another sense never distracted Wollstonecraft from the cumulative strength of continuing education. Self-teaching and the tests of expatriation were central to her life, and to those of the next generation who consciously re-enact and re-explore the course of existence she had pioneered. 'I believe that we are sent here to educate ourselves,' Mary Shelley says in her *Journals*. After Shelley died, she earned her living by her pen, encouraged by Mrs Mason: 'Composing is certainly the best antidote to melancholy.' Mary preferred to support Percy Florence herself, rather than surrender him to the Shelley family. When Byron advised the opposite course, she was shaken.

'I can, I do, live in solitude, I can act independently of the opinion of others; but the expression of that opinion if it be in opposition to mine shakes my nature to its foundations. I differed from L[ord] B[yron] entirely,' she confided to Jane Williams, 'but I literally writhed under the idea that one so near me should advise me to a mode of conduct which appeared little short of madness & nothing short of death.' Byron had not taken in Mary Shelley's point in *Frankenstein*: to create and not to nurture one's creature is nothing short of monstrous. It was hard to go alone yet retain Mary's degree of sensitivity. Godwin indeed reproached her, saying 'you are a Wollstonecraft'; but this reproach of mother and daughter could take a different colour. These resourceful natures who don't lose their sensitivities pose the possibility of an improved breed over hardened people who seek power.

Generosity was another trait. Mary sent Claire £12 when the two went separate ways after Shelley's death, even though Mary's own situation at the end of 1822 was equally threatened by the scandals of the past eight years. In contrast, Jane Williams wheedled sums of money from Claire, while shunning contact. As an Englishwoman who had left her husband for a lover, Jane Williams had lived amongst the outcasts in Pisa, but did not take on their character. She betrayed Mary's friendship, perpetuating the myth of a cold wife, despite Mary's confidence to her a year after Shelley's death: 'I cannot live without loving and being loved, without sympathy; if this is denied to me, I must die. Would that the hour were come.' Though Mary Shelley had more reserve than her mother, there is the same depth of feeling – the voice is the same. The myth of coldness fused with slander of Mary Shelley as social climber, betrayer of her radical inheritance. 'I have ever defended women when oppressed,' she protested. It was not so in every instance: she had been blind to Shelley's first wife, Harriet; she had not encouraged him to visit the children of that marriage; and she had been unfeeling to Fanny, left to bear the domestic tyrannies of their stepmother. Yet during her lonely years back in England, Mary Shelley did indeed befriend women like Caroline Norton, who fought the law that deprived separated wives of the right to their children. She also proposed to her publisher a history of women or a book on the lives of celebrated women.

Margaret Mason too confirmed the formative sway of Mary Wollstonecraft, whose 'mind appeared more noble & her understanding more cultivated than any others I had known'. Her *Advice to Young Mothers* 'by a Grandmother' (1823) might be said to complete the work Wollstonecraft had outlined at the time she died. Mrs Mason invented a gentle form of paediatric practice long before paediatrics was established as a branch of medicine. Much illness, she thought, was the result of neglect in childhood. No sign of trouble in a child should be thought trifling; it was dangerous for children to exercise a stiff upper lip. Clearly, *Advice* did not reach British boarding-schools like the Clergy Daughters' School at Cowan Bridge whose repressive ethos killed off two coughing Brontë girls in 1825. *Advice* predicts a future when preventative medicine should be the pre-eminent treatment.

The far-sightedness of its cures is most apparent in what it has to say about mental health. Mrs Mason recognised that a child's state of mind affects physical health. No punishment, she says, should make a child feel contemptible: 'Teach a being to despise himself, and prepare the mind for the reception of every vice.' She forbids adults to lock children in the dark. 'Terror is a sensation against which they should be protected with the greatest care: the injuries done by fear, to the physical and moral health, are incalculable.' At the same time, courage should be the first quality to be cultivated, in girls as much as boys. *Advice* rejects the notion 'that all women have a right to be cowards. This opinion is extremely injurious to the health of young girls, who would often try to conquer their fears, if they were not taught to believe it a thing impossible, and that they even appear more amiable as helpless than as independent beings.' Older girls must be relieved of 'the tortures of fashion'. Mrs Mason deplores constricted or down-sized constructions of womanhood. Attempts to produce excessive smallness were more likely to produce the 'ugliness' of bad health. Many handsome women, she insists, have broad hips.

Mrs Mason gave her royalties to Claire. What she wanted was recognition, as when Vaccà and other doctors praised her medical stance. 'A woman of good sense, who studies that book will want no physician for her children,' Vaccà said, and added, 'there is a great deal in this book of which many physicians are ignorant'.

'Yes,' Mrs Mason agreed, ' – for it is impossible for any man to know all those little things which a mother who watches her children carefully can observe.'

'I don't mean that,' Vaccà said. 'I mean the medical part of the book. None but a medical man could be perfectly aware of the merit of that part of the book which treats of diseases.'

In 1825–6 she corresponded about *Advice* with George Parkman of Boston who sent doctor-to-doctor comments: 'here's my variation on your remedy . . . I'd try . . .' and so on. As a young Harvard graduate, Dr Parkman had accompanied Joel Barlow when he had sailed for France in 1811 as US Ambassador. There, Parkman had studied mental illness in women, under the aged Pinel at the Salpêtrière. He had returned to Boston determined to introduce optimism about recovery and a similar humane mildness in the mental hospital he established (now merged with Massachusetts General Hospital). His colleagues, he tells Mrs Mason, shunned him for opposing their lucrative links with drugstores. (And later, a fellow-physician went so far as to murder him). He questioned the doses doctors, in their ignorance, administered, and his letters concur with Mrs Mason that doctors' intervention too often denied the body's capacity to heal itself. In this way, *Advice* opened up a dialogue with a world that had cast her out.

'Vesuvius' subsided early in Tighe's relations with Margaret Mason. He regarded Wollstonecraft's ideas of equality as suitable only for a small percentage of the upper class. Sex should be moderate: an 'emission' once a week sufficed. It was an urge to be curtailed as long as possible, and when it had to happen, offered no more than physical relief. He didn't share Margaret's exalted views of Italy. To Tighe, it was as small-minded, as given to social pretension, as any other country. No radical, he believed with Margaret they had been at fault in cutting loose from the laws of their society.

Tighe remained an exile who tried to put down roots in his adopted country. He composed travel pieces suitably embellished with Latin tags, and embarked on a year-by-year history of Pisa. But his favourite occupation was growing potatoes, reminiscent of Ireland. Though he was scientific and wished his variety of potato to advance agriculture in Italy, he blinded himself to the Italian habit of feeding *patate* to animals. His family nicknamed him 'Tatty'. Increasingly reclusive, Tatty lived on his own floor of their rented house, and his scaled-down needs made him think his retired way of living must suit them all.

Though Tatty was fond of his daughters, their mother became uneasy about her will leaving all she had to him. She feared his absorption with potatoes might lead him to discount what was due to their daughters if she should die – as her continued ill health seemed to threaten. She therefore took two steps.

In April 1818 she revealed her real identity to Laurette and Nerina, and related how Mary Wollstonecraft had changed her. It was a sombre declaration: she blamed herself for her marriage, and said that she had deserved to suffer as she did. Tight-lipped about the loss of her Mount Cashell children, she warned the girls, then aged nine and three, they must expect nothing from their grand relations. After the death of Mount Cashell in 1822 she was free to marry Tighe, but though she did need to legitimise their daughters, it was more urgent to secure them financially. In 1823, she inherited £6000 on the death of her mother. There are further indications that she was not as cut off as she gave out. She retained ties with her second son, Robert Mount Cashell; with her sister Diana who lived in Florence; and with her cosmopolite brother, John King. Then, in 1824, she contacted her eldest daughter, Helena, who had married Richard Robinson of Rokeby Hall, County Louth, in 1813.

Before Margaret married Tighe, granting him legal authority over their children and all she owned, she took her second decisive step. On 15 June 1824 she made a new will in her real name as the Dowager Countess of Mount Cashell, leaving everything she owned to 'my excellent and beloved daughter Helena' in trust for 'two young friends' of the Countess who were residing with her at Pisa in Tuscany. The Countess made this bequest, 'relying on my beloved daughter the said Helena Eleanor Robinson to act in this matter as she knows I should wish her to do'. The will was witnessed by the Honourable John Harcourt King, who resided in Paris. The 'young friends' were, of course, her illegitimate daughters and Helena's half-sisters, Laurette and Nerina. The will proves that Margaret had complete trust in her thirty-year-old-daughter. A stray scrap amongst the Tighe papers records the expenses of a twenty-day journey that Helena, her husband and two little girls, together with governess and servants, undertook from Paris to Pisa. *Advice*, the previous year, had called its author 'a Grandmother' but this grandmother had never seen her grandchildren. Now she met them. Behind the bare statements of £120 for two carriages with stops in Turin and Genoa, is this dramatic meeting with a daughter who had been eleven years old when she parted from her mother in Dresden nearly twenty years before. When mother and daughter evade the bias of the law, which kept money as far as possible in male hands and punished the illegitimate, they move united in the underworld of *The Wrongs of Woman*.

In Margaret Mason's anonymous novel, *The Sisters of Nansfield* (1824), an outcast mother, Mrs Maynard, has two daughters called Harriet and Fanny who grow up aware they have noble relations who will not recognise them. Harriet is an indolent beauty who elopes – disastrously – with a nobleman, while Fanny is a brown girl whose animated and intelligent expression makes her more attractive as a person. Sensitive Fanny is a comfort to her mother and attached to home. Why did Margaret Mason call the sisters 'Harriet' and 'Fanny', the names of the two young women lost to the post-Wollstonecraft milieu? In her Lady Mount Cashell days, Margaret had met Fanny Imlay on her visits to the Godwins, and she could well have met Harriet Shelley there in 1812. Odder, even, is that the characters and to some extent the fates of the fictional sisters forecast those of Laurette and Nerina who were at the time fifteen and nine.

By the age of eighteen, Laurette was bored. She was a beauty. The elaborate coils of her high-piled hair recall the abundant tresses of her grandmother, Lady Kingsborough, and her runaway aunt, Lady Mary. She was a fashionplate with sloping shoulders and balloon sleeves tapering to a tiny, belted waist in the style of the magazines that, at this time, captivated the Brontë children. A portrait of Laurette bears an ominous similarity to the languid ladies of Charlotte Brontë's imagination, who wait around for Byronic poseurs to spoil their lives.

In 1827 the Tighe family was living at Villa Archinto in the suburb of San Michele degli Scalzi. It was an unsuitable place for girls reaching marriageable age. Tighe and his wife, on their separate floors, were quarrelling over the future of Laurette and Nerina. At this point Mrs Mason (as she continued to be called – though now married to Tighe) made the most crucial of her decisions relating to her daughters. She must take them back to Pisa, leaving Tatty in the suburb. Since Tatty opposed this move, she would use all her own income to dress and present them well. They had to mix in society if they were to marry.

Accordingly, back in Pisa, Mrs Mason founded a society of her own. Her new home, an apartment in the large and handsome Casa Lupi (known affectionately as the Cave of Wolves), became the venue for an Accademia di Lunatici consisting of forty-six members, including the Italian poet Giacomo Leopardi who stayed for six months a few streets away in the Via della Faggiola. Other members were his friend Giovanni Rosini, Giuseppe Giusti, and a Frenchwoman, Sophie Vaccà (Dr Vaccà's wife). The *Lunatici* met every fortnight on Mondays (Moon days, appropriate to loonies). The name may have revived a theatrical Lunatica Accademia that had flourished in Pisa during the first half of the seventeenth century. Mrs Mason's *Lunatici* had an altogether more secret character at a time when Tuscany was ruled by the Austrian Archduke. 'Madness', as with Hamlet, was a cover for deviance. Each member adopted the name of a constellation, and received a certificate of membership with the motto: '*se non son matti, non ce ne volemo*' (if you aren't mad we don't want you). Ostensibly, it was a literary society, consisting mainly of students from the University of Pisa, but the secrecy may have had a political vein. Several of the *Lunatici* – poet Angelica Palli, Francesco Guerrazzi, Antonio Mordini and Ferdinando Zannetti – went on to become political activists for the unification of Italy after 1848: links in a revolutionary chain from Mary Wollstonecraft and the French Revolution, to Margaret Mount Cashell and the Irish Rising, to the Risorgimento.

In 1832 when Claire returned to Pisa, she was invited to live with Mrs Mason, who had moved into the Via della Faggiola in the centre of town.

'Nothing can equal Mrs. Mason's kindness to me,' she told Mary Shelley. 'Hers is the only house except my mother's, in which all my life I have ever felt at home. With her I am as her child.' Claire was delighted to find a like character in seventeen-year-old Nerina, who 'will walk by herself and think for herself'. Together, Claire and Nerina were 'as wild as March hares', scoffing at men and love. Laurette's fashionable style had attracted a princely rogue, while Nerina was more suited to the subversive *lunatici*. She was darker and slighter than her sister, a playful talker, who pretended to moan over her education in languages and literature, since it would prevent her marrying 'some blackbeard of an Italian' in search of a sewing wife.

Claire continued to teach long hours, serving 'the tyrants', the English Bennets, from nine in the morning. At ten each night she returned to the Tighes, 'a singular family – of the females of which it may truly be said, they form among their species, an oasis in the grand desert of society'. After many depleting years, Claire at last grew 'fat and ruddy'. Mrs Mason, she felt, 'understands me so completely – I have no need to disguise my sentiments, to barricade myself up in silence as I do almost with every body for fear they should see what passes in my mind and hate me for it because it does not resemble what passes in theirs'.

In 1834 Nerina, aged eighteen, married a member of the *Lunatici*, Bartolomeo Cini, a law student of literary, political and scientific tastes from a papermaking family in San Marcello Pistoiese. 'Meo' was a good man whom everyone, including Claire, loved and trusted. He had a thin, sensitive face, with a large nose and delicate but firm lips. Nerina's aunt Diana came to help at the birth of her first child Margharita. Nerina's descendants lived in San Marcello, generation after generation. In the summer of 2001, my husband drove us through a blinding rainstorm up the perilously winding road of the Apennines, to meet Nerina's present-day descendant Andrea Dazzi and his wife Cristina in their book-filled rooms. A woman in her nineties entered with quiet dignity, Signor Dazzi's mother, Nerina's great-granddaughter, born Cini in 1909. Her eyes were so pale a blue they were almost white. Lady Mount Cashell's eyes, as Godwin described them to his friend Marshall in 1800. Was it fancy or did Giovanna

Dazzi look like the earliest portrait of Margaret King in her mid-teens? The rambling old house has been divided, but tucked away, doors open on a nineteenth-century library – untouched, intact – for Cristina Dazzi has refused to sell it off. There, still, are the Irish books of Lady Mount Cashell, her manuscripts and the love letters of 'Laura' and 'Vesuvius'.

In 1834, Tighe, distancing himself further from Lady Mount Cashell, required her to resume that name. She died at the age of sixty-four the following year. A portrait painted by her sister Diana, not long before her death, shows her level eyes and fresh complexion under the bonnet of an old woman – her persona in *Advice*. On her lap she holds the manuscript of 'The Chieftains of Erin', one of her unpublished novels: a rather tedious tale of Ireland in the age of the Elizabethan invaders. Her Cini son-in-law remembered her as Irish, with her *amor della patria* and her talk of *oppressi compatriotti*.

Her practical benevolence recalls George Eliot's tribute to the unrecorded work of nineteenth-century women. Half the good in the world, George Eliot says, comes from those who lie in unvisited graves. This pupil and disciple of Mary Wollstonecraft lies in the English graveyard at Livorno, which she'd visited in 1803 with Lord Mount Cashell. The graveyard, off the Via Verdi, has been closed for a century and a half: the key turns in the rusty lock and the old iron gates open into an expatriate past. The ground is covered by the leaves of many winters, and the steady blue sky looks down through the pines on those who died far from home. At the back, two pediments have fallen from a large table-tombstone, with weeds growing between exposed stones. Much of the marble is stripped, but the inscription is still intact:

HERE LIE THE REMAINS
OF MARGARET JANE
COUNTESS OF MOUNT CASHELL
BORN A.D. 1773
DIED 29 JANUARY 1835

By her side is Tighe's matching tombstone, their estrangement engraved in the fact that she was not buried as 'Mrs Tighe' but in her former identity as the wife of Mount Cashell. Though she lived the second half of her life in obscurity – her name absent from the many editions and translations of her *Advice* – her influence on childcare was widely disseminated. One of her readers was an Italian expatriate in London, Gabriele Rossetti, who bought a copy to guide him with his children, Dante, William, Michael and Christina.

When William Godwin died, aged eighty, in 1836, he asked to be buried with Mary Wollstonecraft. So Mary Shelley stood once more at her mother's tomb as the gravediggers dug twelve feet down. 'Her coffin was found uninjured,' she wrote to her mother's old friend Hays, 'the cloth still over it – and the plate tarnished but legible.' By the time she herself came to die in 1851, the railway had broken through St Pancras churchyard. Her son Percy Florence moved the bodies of Mary Wollstonecraft and Godwin to Bournemouth where he lived. Mary Shelley was buried between her parents.

Independent, existing by her own skills and wits, Claire Clairmont lived on and on into the age of Henry James. He heard her story, and recreated her as the secret survivor – the beloved of a long-gone Romantic poet – in *The Aspern Papers* (1887). The artist John Singer Sargent remembered her in her seventies playing the piano for his dancing class in Florence. She ended there a Catholic, as she had begun, and at the age of eighty-one was buried with Shelley's shawl. She would have preferred to lie with Allegra in her unmarked grave, in unconsecrated ground near the entrance to the chapel at Harrow, Byron's old school.

Buried with these lives are stories of promise: counter-narratives to the cut-off plot of Wollstonecraft's death. Biography is ceasing to make death more final than it is. Continuities turn the focus from deaths, disasters and slanders to heirs in the next generation who manifest the staying power of her self-making. Three managed to survive through mutual support. Ever since reading Mary Wollstonecraft in 1814, Claire had been gripped by an idea of 'the subterraneous community of women'. 'The party of free women is augmenting considerably,' she remarked to Wollstonecraft's daughter in 1834. 'Why do not they form a club and make a society of their own.'

18

GENERATIONS

There's no end to the reverberations of far-reaching lives. Mary Wollstonecraft can't be dissociated from her daughters – her biological daughters and those who come under her influence, her political descendants over subsequent centuries. A past experience revived for its meaning is 'not the experience of one life only/ But of many generations'. Present-day generations, in the choices and opportunities open to us, are Wollstonecraft's heirs.

Women and men of the first generation after her death – mostly workers during the first half of the nineteenth century – took up the socialism of Robert Owen, a follower of Wollstonecraft, as he told Fanny Imlay in 1816. One of the earliest claims for women's enfranchisement was made by an Irish supporter of Owen, William Thompson, in his *Appeal of One Half of the Human Race*. Thompson's aim was to 'raise from the dust that neglected banner which [Wollstonecraft's] hand nearly thirty years ago unfolded boldly, in the face of prejudices of thousands of years'. A practical outcome was that women could join a trade union. Owenite tracts and newspapers, followed by the *Chartist Circular*, regularly reprinted passages from *A Vindication of the Rights of Woman*. A new translation also came out in France in 1826 (as part of the run-up to the 1830 revolution).

Whilst this was going on in England and France, a daughter of a New England clergyman published her thoughts on 'The Natural Rights of

Woman' in the *Boston Monthly Magazine* in 1825. The Great Creator, she argues, crowned his labours by giving being to the most intelligent of his creatures:

> Male and female created he them; but declared them of 'one bone – one flesh'- one *mind.* To *them* he directed his divine commands – and gave *them* rule over all he had made . . .
>
> But it seems that *man* soon became wiser than his Maker, and discovered that the Almighty was mistaken, or had *made* a mistake, and that all the mind, or the greatest part of it, had been bestowed on *himself*, and that *woman* had received only . . . the mere leavings, and scrapings that could be gathered after his own wise brain was furnished.

This sermon examines small-mindedness from an American perspective. Ten years before Mount Holyoke, the first women's college, the author is hopeful of the schoolhouse with its custom of equal education, and fewer inducements to phoniness. To be sure, girls still leave with nothing more than 'a smattering of *terms*', but 'we feel the influence of the female character' in some shift from modish sensibility towards 'sympathy for real distress'.

Curiously, this author's name was Mary Wollstonecraft. It was not an invention or coincidence. This American Wollstonecraft, a botanist, was in fact the widow of Mary's youngest brother, Charles. A dictionary of *Distinguished Women*, published by the owner of the *Boston Monthly*, finds the second Mary Wollstonecraft as eloquent and bent on domestic values as the first.

Her impression of the better-educated American girl was borne out by Alexis de Tocqueville. During his stay in 1832, he was 'surprised and almost frightened' by the distance of the American girl from her European counterpart: an extraordinary absence of shyness and ignorance. Instead of training a girl to distrust herself, education was drawing forth her voice. Tocqueville finds a 'singular skill' and 'happy audacity' (happier than a philosopher stumbling along a narrow path) as this girl steers her thoughts and language through the traps of conversation.

While these voices carried Wollstonecraft's ideas further into the public arena, certain writers tried out the character she had brought into being. The future poet Elizabeth Barrett was only twelve in 1818 when she read the *Vindication.* At fourteen she declares her 'natural' independence of mind and 'spurns' the triviality of women's lives. Her outsider heroine Aurora Leigh, displaced from Italy to England, sees that the fatal lesson is imitation. She holds to 'the inner life with all its ample room', aware that 'feebler souls/ Go out', fading in the glare of obligatory artifice. This narrative poem of 1857 revives Wollstonecraft's unwavering focus on education, on the soft soul of a girl as she takes the imprint of what she and her speech are to be – be it the gush of affected sensibility in Wollstonecraft's time, or the docile 'so be it' of Victorian womanhood.

Copies of the *Vindication* were scarce in Victorian England, George Eliot observed. She read it with surprise to find Wollstonecraft 'eminently serious, severely moral' in her impatience with silly women. In *Daniel Deronda*, when the despairing singer Mirah dips her cloak in the river, George Eliot looks back to Mary Wollstonecraft soaking her clothes that rainy October night on the brink of the Thames. Mirah, like Mary, is restored to life; each lives to know a good man and unfold what she has to offer. George Eliot wrote an admiring leader for the *Westminster Review* on the common ground between Wollstonecraft and the New England reformer Margaret Fuller, with her 'calm plea' in *Woman in the Nineteenth Century* 'that the possibilities of her nature may have room for full development'.

The nineteenth century pressed the issue of women's education; the twentieth century, that of professional advance; but the subtler question of our nature is still to be resolved. In 1869 John Stuart Mill says, 'what is now called the nature of women is an eminently artificial thing'. In 1915 Virginia Woolf predicts it will take six generations for women to come into their own – if so, we're not there yet. 'The great problem is the true nature of woman,' she alerts students at Cambridge. If this century is to solve it, the 'new genus' of Mary Wollstonecraft offers a start. What is this genus? What will it contribute to civilisation?

A new-found creature – 'almost unclassified' – has been crawling out from under the stone of history. 'I am a rising character' is the reply of

Charlotte Brontë's Lucy Snowe when people ask 'Who are you?' '"Nature" is what we know – / Yet have no art to say,' says Emily Dickinson as poems pour from her pen during her *annus mirabilis*, 1863. In the late 1870s Olive Schreiner, a young governess in a lean-to room on a rocky stretch of the veld, wrote a novel about a New Woman. Though she appears an oddity on a backward colonial farm, she does not yield her conviction of who she is in order to pursue the mediocre plots open to her sex. An authentic self seems to speak out of a stark and timeless landscape. *The Story of an African Farm* won over the intelligentsia when Schreiner brought it to London. Her eloquence broke in on the earnest deliberations of the Men and Women's Club as they sifted the nature of the sexes. Her manner was visionary, her gestures emphatic, her dark eyes glowed as she spoke. Mary Wollstonecraft, she says, is 'one of ourselves'. So she puts it in an unfinished introduction for a centenary edition of the *Vindication*. Wollstonecraft 'knew'. She had foreseen a transformation of gender, 'the mighty sexual change that is coming upon us'. In the 1880s the two sexes seemed to Schreiner still a mystery: 'what in their inmost nature they are . . . Future ages will have to solve it.'

Henry James approached this mystery through his vibrant cousin, Minny Temple. He looked on her as 'an experiment of nature', surrounded by a circle of gifted Harvard men who were drawn to her honesty and hunger for life. 'Let us fearlessly trust our whole nature,' Minny urged before her early death. She became the model for *The Portrait of a Lady*, where James tests the '*grande nature*' of an American girl who turns down marriage to an English lord. As she does this, she senses some undefined destiny: 'Who was she, what was she, that she should hold herself superior? . . . If she would not do this, then she must do great things, she must do something greater.' A century before, when Wollstonecraft's 'Mary' rejects the practice of 'giving' a bride in marriage, the author herself was germinating the new character.

As we trace her experiments, above all 'that most fruitful experiment', her relation with Godwin, we see everywhere a single purpose: to centre the domestic affections as a counter to violence. Miseducation undoes those ties: the iron jargon of sacrifice fed to a female terrorist, sealing her off from the beating heart of the baby she holds; invaders of other countries who 'are the real savages', proving 'how unfit half-civilized men are to be entrusted with unlimited power'; and bombarding images of bodies decked with prostitute-chic dispatched down the catwalks of the present. The traits Wollstonecraft outlaws threaten us as never before: the terror of violence too close to home, and the more insidious callousness of unregulated greed. She herself had to pierce through mask after mask, not just to observe but to experience in the most jarringly intimate way her father's violence, the blood on the cobblestones of Paris, and habits of fraud in the free port of Hamburg.

Her protest, when she heard the lash resound on slaves' naked sides, came fifty years ahead of the World Anti-Slavery Convention in 1840 where 'the woman question' – a question of their wish to speak in so public a forum – took up the opening day. A group portrait of that London event includes Wollstonecraft's friend Amelia Opie in a towering black bonnet, seated beside Lady Byron, with Lucretia Mott from America, a leader of the silenced women delegates, just visible at the back.

Wollstonecraft is nearby, in the next room of the National Portrait Gallery. She 'saved her soul alive', mused pioneering anthropologist Ruth Benedict in 1913, as she came face to face with her.

'She is alive and active,' Virginia Woolf agreed, 'she argues and experiments, we hear her voice and trace her influence even now amongst the living.' That was 1929, the year Woolf wrote her essay 'Mary Wollstonecraft'. She republished it in 1932, the year she planned to speak out for the untried possibilities of women – 'a great season of liberation', her diary says. Her own boldest move was to walk away from the edifice of power by calling on an 'Outsider Society', heirs of that 'party of free women' that Claire Clairmont proposed to Mary Shelley. This line of inheritance leads less to the limited suffrage movement of the nineteenth and early twentieth centuries – the joiners of the dominant parties – than to the Virginia Woolf who refused heroes, honours, medals and war, and strode more freely, 'enfranchised till death, & quit of all humbug'.

For all that, Mary Wollstonecraft does not terminate in disciples. Her 'new plan of life' is growth, not fixity of form. Though she speaks an eighteenth-century language of 'rights', she looks beyond rights as an end in themselves, beyond the feminist goals of the last two centuries, towards an evolving intelligence – listening, sensitivity, tenderness – still untapped in public life.

Women who imitate men lack ambition, goes the old phrase. Though Mary Wollstonecraft came into contact with an array of able men, not least her husband, she held to a course of her own. Tugged off-course by an Adam of the American frontier, she came back from the brink of extinction, once, twice, with inventive renewals. 'I am not born to tread in the beaten track,' she said, 'the peculiar bent of my nature pushes me on.' Active, articulate, she takes the lead, takes it still, as she cuts her way to the quick of life.

ABBREVIATIONS

Used in the 'Sources, Contents, Questions' section
and in the Bibliography

Abinger	Wollstonecraft, Godwin, Mount Cashell and Shelley manuscripts, on loan from Lord Abinger to the Bodleian Library, Oxford
AR	*Analytical Review* (MW was a reviewer and editor)
Beinecke	The Beinecke Library, Yale University, New Haven, Conn.
Bodleian	The Bodleian Library, Oxford
BW	Elizabeth (Eliza/Bess) Wollstonecraft (sister of MW)
CC	Clara Mary Jane ('Claire') Clairmont (stepsister of MW's two daughters, FI and MWS)
CCJ	*The Journals of Claire Clairmont 1814–1827*, ed. Marion Kingston Stocking (Cambridge, Mass.: Harvard University Press, 1968)
Cini–Dazzi	The Cini–Dazzi archive in the private library of Andrea and Cristina Dazzi in San Marcello Pistoiese, Italy
ClCor	*The Clairmont Correspondence*, i–ii, ed. Marion Kingston Stocking (Baltimore: Johns Hopkins University Press, 1995)
EB	Elias Backman
Education	MW, *Thoughts on the Education of Daughters: with Reflections on Female Conduct, in the more important Duties of Life*, in *MWCW*, iv
EW	Everina Wollstonecraft (sometimes Averina, sister of MW)
FI	Fanny Imlay
FR	MW, *An Historical and Moral View of the Origin and Progress of the French Revolution; and the Effect It Has Produced in Europe*, in *MWCW*, vi
GI	Gilbert Imlay (MW's American partner)

HF	Henry Fuseli (artist)
HMW	Helen Maria Williams (English friend of MW in France)
Houghton	The Houghton Library, Harvard University: the main repository for the Barlow Papers and those of Joel Barlow's wife, Ruth Baldwin Barlow
JB	Joel Barlow (American poet and diplomat)
JJ	Joseph Johnson (publisher)
KP	Charles Kegan Paul, *William Godwin: His Friends and Contemporaries*, 2 vols (1876). Includes a revisionist biography of MW, the first since WG's *Memoirs* and the subsequent scandal.
Mary	MW, *Mary; A Fiction*
Memoirs	WG, *Memoirs of the Author of 'The Rights of Woman'*, in *Mary Wollstonecraft and William Godwin*, ed. Richard Holmes (Penguin, 1987)
MM	Margaret King, later Lady Mount Cashell ('Mrs Mason')
MW	Mary Wollstonecraft
MWCW	*The Collected Writings of Mary Wollstonecraft*, i–viii, ed. Marilyn Butler and Janet Todd (London: Pickering & Chatto, 1989)
MWL	*The Collected Letters of Mary Wollstonecraft*, ed. Ralph M. Wardle (Ithaca: Cornell University Press, 1979)
MWletters	*The Collected Letters of Mary Wollstonecraft*, ed. Janet Todd (London: Allen Lane, 2003; repr. Penguin, 2004)
MWS	Mary Wollstonecraft Godwin, younger daughter of MW, who became Mary Shelley
MWSJ	*The Journals of Mary Shelley*, ed. Paula R. Feldman and Diana Scott-Kilvert (repr. Baltimore: Johns Hopkins, Softshell Books, 1995)
MWSL	*The Letters of Mary Shelley*, i–iii, ed. Betty T. Bennett (Baltimore: Johns Hopkins University Press, 1980–8)
PBS	Percy Bysshe Shelley (husband of MWS)
Pf	The Carl H. Pforzheimer Collection, papers of Shelley and his circle; letters of Mary Hays and Jane Williams; Cini Papers. New York Public Library
RB	Ruth Barlow (American friend of MW's, wife of JB)
Real Life	MW, *Original Stories from Real Life; with conversations, calculated to regulate the affections, and form the mind to truth and goodness*, in *MWCW*, iv
RM	MW, *A Vindication of the Rights of Men*
RW	MW, *A Vindication of the Rights of Woman*

SC	*Shelley and his Circle, 1773–1822*, 10 vols, ed. Kenneth Neill Cameron, Donald H. Reiman and Doucet Devin Fischer (Harvard University Press, 1961–2003)
SPP	*Shelley's Poetry and Prose*, ed. Donald H. Reiman and Neil Fraistat (New York: 2nd Norton Critical Edition, 2002)
Travels	MW, *A Short Residence in Sweden, Norway and Denmark*, in *Mary Wollstonecraft and William Godwin*, ed. Richard Holmes (Penguin, 1987)
WG	William Godwin (MW's husband)
WW	MW, *Maria; or The Wrongs of Woman*

SOURCES, CONTEXTS, QUESTIONS

1 VIOLENCE AT HOME

1 *'I want to see . . .'*: *MWL*, 227; *MWletters*, 217.
2 *'the lash'*: *RM*. See below, ch. 7.
2 *'I am . . . pushes me on'*: *MWL*, 164; *MWletters*, 139.
2 *wildness*: *English Woman's Journal*, cited by Caine, 'Victorian Feminism', 268. Harriet Martineau called her a 'poor victim of passion' in 1855.
2 *'romantic sentiments'*: *RW*, ch. 13, sect. ii.
2 *'firmest champion' and 'the greatest ornament'*: *Memoirs*, ch. 9.
2 *'Europe was rejoiced . . .'*: *The Prelude*, Book VI.
3 *'Every day . . .'*: Woolf, 'Mary Wollstonecraft'.
3 *'a foolish consistency . . .'*: Ralph Waldo Emerson, 'Self-Reliance'.
4 *'an experiment from the start'*: Woolf, 'Mary Wollstonecraft'.
4 *'mad' . . . 'licentious'*: John Adams, marginalia in his copy of *FR*.
4 *'little short of monstrous'*: 'A Heart that Scorned Disguise', *Times Literary Supplement* (21 Apr. 2000), 36.
6 *Ned Wollstonecraft*: Named after a relative, Edward Bland, who had been first officer on Grandfather Wollstonecraft's part-owned ship, the *Cruttendon*.
6 *'apparent partiality'*: *Mary*, ch. 2.
6 *'harsh'*: MW to BW (17 Aug. [1781?]), *MWL*, 76; *MWletters*, 31.
6 *'women . . .'*: Kames (1696–1782) participated in the Scottish Enlightenment as a writer on law, history and farming. His *Elements of Criticism* was published when MW was three years old.
7 *divide . . . between worker and gentleman*: A telling instance of this divide is what happened to John Ruskin when he arrived as an undergraduate at Christ Church, Oxford, in 1837: though he had the gifts to be a scholar, his social-climbing parents had insisted on entering him instead for the higher class of gentlemen-commoners, who proceeded to ridicule Ruskin when his first essay was chosen to be read aloud. Gentlemen scorned work, and it was customary to bribe a scout (a college servant) to write the required weekly essay. (John Batchelor, *John Ruskin: No Wealth but Life*, London: Chatto, 2000, 36–7.)
8 *'a gentleman's daughter'*: Jane Austen, *Pride and Prejudice*, ch. 56.

8 *'an old mansion . . .'*: Elizabeth Osborne, *The History of Essex* (1817), cited in *SC*, i, 40. Her information came from WG as well as hearsay.

8 *New Farm and neighbours*: Alternatives could be Reeves Gate Farm or Hay Hill Farm. Todd, *Wollstonecraft*, assumes plausibly that Grandfather Wollstonecraft provided for this farm, and notes Mr Gascoyne's origins in London trade.

9 *'reveries'*: MW, *Travels*, letter 8.

9 *MW's attempts to protect her mother*: *Memoirs*, ch. 1.

9 *'Mary was continually in dread'*: *Mary*, ch. 2.

10 *father beat her*: WG to JJ (11 Jan. 1798). Abinger: Dep. b.227/8.

10 *status of women in Anglo-Saxon England*: Whitelock, *Beginnings*, 45, 87, 94, 150–1. Legal historian Anthony Honore tells me that slaves in the Roman Empire, women as well as men, had a mass of legal rights which we might think of as human rights.

10 *The Lawes Resolutions*: Fraser, *Weaker Vessel*, 527.

10 *Burney on marriage*: *Early Journals*, cited by Harmon, *Fanny Burney*, 72.

11 *law and domestic violence*: The Hardwicke matrimonial law was only slowly unpicked in the latter half of the nineteenth century. As late as the end of the twentieth, legal redress and protection from domestic violence were still at issue.

11 *farm at Walkington*: Todd, *Wollstonecraft*, 459, notes a local tradition that the farm was the present mixed farm of Broadgate just off the road from Beverley to York, but there is no firm evidence of this.

11 *'a very handsome town . . .'*: *Memoirs*, ch. 1.

12 *isolation of child victims of violence*: Herman, *Trauma*, 99–100, 105. I am grateful to Margaret Bluman for recommending this innovative study.

12 *letters to Jane Arden while they were at school*: *MWL*, 51–64; *MWletters*, 1–18.

13 *Yorkshire idiom*: Used to GI in 1794, *MWL*, 247; *MWletters*, 244.

14 *on John Arden*: His granddaughter Everilda Gardiner, *Recollections*.

14 *'The good folks . . .'*: *MWL*, 66; *MWletters*, 23.

14 *leaving Yorkshire*: Durant's Supplement, gives the date as Sept. 1774 though the date of Henry's apprenticing suggests 1775 as more likely.

15 *Henry's fate*: Insanity or criminality suggested by Sunstein, *Wollstonecraft*, 36ff.

15 *WG avoiding Henry's name*: WG does note that one of MW's brothers predeceased her; i.e. he tells us indirectly that Henry was dead when his memoir of 1798 cites the brothers and sisters of MW who 'are still living'. How Henry died remains unknown.

15 *Henry and Hoxton's asylums*: Todd, *Wollstonecraft*, 21.

15 *'amiable Couple . . . took some pains'*: MW to Jane Arden, *MWL*, 66; *MWletters*, 24.

16 *meeting Fanny*: *Memoirs*, ch. 2.

16 *Blood family background*: Todd, *Wollstonecraft*, 23.

16 Flora Londinensis: Bodleian.

17 *the Revd Mr Bishop*: EW to WG (Nov. 1797) distinguishes him from the Mr Bishop who was to marry BW in 1782.

17 *MW's education*: Wagner and Fischer, *Pforzheimer Collection*, 25, suggest plausibly that education was more important for MW than feminism.

18 *'the prejudices'*: *Memoirs*, Appendix, 277. WG's revision of ch. 10 for the second edition.

2 'SCHOOL OF ADVERSITY'

19 *'school of adversity'*: *Education*, *MWCW*, iv, 36.
19 *'keen blast of adversity'*: *MWL*, 69; *MWletters*, 27.
19 *'right of directing'*; *'forwardness'*: *WW*, ch. 7.
20 *bequest from their Dickson grandfather*: MW sent 'several letters' to a Mr Dickson in the summer of 1780 which, Wardle's notes to her letters suggested, had to do with money.
20 *Balthazar Regis*: Educated at Trinity College, Dublin. Chaplain to the King 1727–57 and Canon of Windsor 1751–7. Died in 1757.
20 *entering Bath*: There is a profusion of contemporary detail in Jane Austen's *Persuasion*, ch. 14.
20 *'flattery . . . wear a cheerful face, or be dismissed'*: *Education*, 25.
20 *pay for companions*: Hufton, *History of Women*, 79.
21 *MW's management of Mrs Dawson*: *Memoirs*, ch. 2.
21 *'very good understanding'*: To Jane Arden, *MWL*, 69; *MWletters*, 27.
21 *'A mind accustomed . . .'*: *Education*, 48–9.
22 *'Women are . . .'*: Lord Chesterfield, *Letters* i, 330. A play on Dryden's 'Men are but children of another growth' in *All for Love*.
22 *Waterhouse*: W. H. S. Jones, *St Catherine's College*, 129. In 1798, when he rigged votes so as to elect himself Master, Waterhouse instigated the worst quarrel in the history of the college. Three of the five Fellows who voted made the following declaration: 'We whose names are underwritten declare this to be no Election, no one of us having voted for Mr Waterhouse.' The Lord Chancellor declared against Waterhouse. In 1801 the Lord Chancellor was called in again over the diversion of a share in the dividends of a vacant fellowship into Waterhouse pockets. Ousted eventually from the college, he lost his glamour, and became mean and slovenly. As Rector of Little Stukely in Huntingdonshire, he was murdered on 3 July 1827 by Joshua Slade, a dismissed servant.
23 *'I knew a woman . . .'*: *Education*, 29; *SC*, i, 46–7; Nitchie, 'Early Suitor', 163–9.
23 *MW and the Ardens*: Quotations not noted separately below are from the second batch of letters to Jane Arden in 1779–c.1783; see *MWL*, 64–80; *MWletters*, 19–39.
23 *Jane Arden as governess*: Everilda Gardiner, *Recollections*.
23 *'I should be glad . . .'*: *MWL*, 67; *MWletters*, 26.
24 *visit to Southampton*: It's probable that she stayed with a Wollstonecraft relation. *MWL*, 70, notes the existence of an Edward Bland Wollstonecraft of Gloucester Square, Southampton, who died in 1795 – likely to be (as *MWletters*, 26, adds) the first officer of Grandfather Wollstonecraft's ship the *Cruttendon*, a share of which was inherited by Edward Bland's namesake, Ned Wollstonecraft.
24 *'truth . . .'*: *Education*, 16.
25 *Gunning sisters*: Murray, *High Society*, 253.
25 *stays*: It was considered indecent not to wear them. CC's mother and stepfather, WG, were relieved when she took to wearing stays in 1817 after she ran away (see below, ch. 17). The sense of indecency remained as late as the early years of the twentieth century. When the feminist Olive Schreiner walked without stays down Adderley Street in Cape Town, after the Boer War, she heard hoots and insults.

26 *sombre grandeur*; *'measured pace of thought'*; *'Life . . .?'*: *Travels*, letter 7.
28 *MW's efforts for mother*: *Memoirs*, ch. 2.
28 *father's view of mother's illness*; *'I shall not dwell . . . disagreeable'*: Draws on *WW*, ch. 8.
28 *date of Mrs Wollstonecraft's death*: Discovered by Tomalin, *Mary Wollstonecraft*.
29 *James . . . went to sea*: Aboard HMS *Carysfoot*: Durant's Supplement, 155.
29 *'the dear County of Clare'*: To Jane Arden, *MWL*, 78; *MWletters*, 37 .
29 *'Few men . . .'*: *Education*, 26.
30 *Her favourite song*: MW to WG [3 July 1797], *MWL*, 402; *MWletters*, 426, notes: 'Allan Ramsay's poem *The Kind Reception*, sung to the tune of *Auld Lang Syne*. It opens with "Should auld Acquaintance be forgot/ Though they return with Scars?"' .
30 *'canker-worm'*: *Mary*, ch. 5. Some details of the character Ann are based on Fanny's history.
31 *marriage versus MW's wish to be free*: In English law the unattached woman was termed the *feme-sole* as distinct from the woman who had no legal existence apart from her husband, the *feme-covert*. To declare aversion to marriage was seen as resistance to performing the only function – to propagate the species – for which woman existed. To earn a living in trade, to be a mantua-maker for instance, was sometimes looked on as a form of prostitution. As late as the 1840s, Charlotte Brontë's friend Mary Taylor emigrated to New Zealand in order to start a business. It was still unthinkable in England for a woman to do so on her own, and though Mary Taylor did amass a modest sum in the new colony, she had to put up with jeers.
31 *baby Mary*: The child's full name was Elizabeth Mary Frances, but she was called Mary according to a letter from MW to EW.
31 *held Bess in her arms*: MW to BW (23 Sept. [1786]), *MWL*, 113; *MWletters*, 78: 'I could have clasped you to my breast as I did . . . when I was your nurse.'
31 *Her ideas are all disjointed . . .'*: *MWL*, 80; *MWletters*, 39 (replacing Wardle's (*MWL*) reading of 'unconnected' thoughts with 'uncorrected'). Wardle's makes better sense (in the context of senselessness) and fits the repetition of 'unconnected' in the similar situation of *WW*, ch. 2.
31 *'Poor Eliza's situation'*: *MWL*, 81; *MWletters*, 40.
34 *'One of the most terrible . . .'*: Stone, *Marriage in England*, 168.
35 *not the only loving mother*: One of Jane Austen's brothers would be given up to better fortune; while Byron gained custody of his daughter Allegra from her extremely loving mother (ch. 17 below).
37 *'monstrous intervention'*: *Times Literary Supplement* (21 Apr. 2000).
37 *MW unsure*: This continues to be overlooked in a persistent demonisation of MW (see ch. 15 below), lasting from her lifetime to a renewed round of attacks in the year 2000. An exception was Kate Chisholm in the *Sunday Telegraph* (30 Apr. 2000): 'What else could she have done?'
38 *Fanny Blood's letter*: Abinger: Dep. b. 210 (9).

3 NEW LIFE AT NEWINGTON

40 'humane': *MWL*, 93; *MWletters*, 55.

40 *'exertions . . . twenty scholars'*: EW to WG (24 Nov. 1797). Abinger: Dep. c.523; extract of letter in *SC*, i, 45.

41 *Disney family*: Included the liberal Unitarian theologian, John Disney (1746–1816).

41 *Miss Mason*: *SC*, i, 87, identifies Mason as MW's servant. It is true that to use a surname alone, as MW occasionally does in speaking of 'Mason', was usual with servants, but could there be other reasons? Men used surnames, and MW imitates this when she publishes extracts from her works, later, in an anthology (see below, ch. 7). An interesting suggestion appears in Todd's notes to *MWletters*, 4: a minister at Driffield, about ten miles north of Beverley in Yorkshire, was William Mason (1724–97). He was one of the 'Driffield Bards' mentioned in a poem MW copied out for Jane Arden in 1773. Was 'Mason' a relative? The clergy background fits MW's later linking of her with Mrs Gabell, a clergyman's wife who has 'clearness of judgement'. (See below, ch. 9.)

41 *Fanny confided*: *Memoirs*, ch. 3.

42 *Sowerby*: He eventually contributed two and a half thousand illustrations to Sir Edward Smith's *English Botany* (1790–1814), and brought out his own *Coloured Figures of English Fungi* (1797–1815).

42 *past and present inhabitants of Newington Green*: Defoe studied at Morton's Academy in the 1670s, and farmed civet cats (for perfume) in 1692. D'Israeli was the father of the Victorian Prime Minister, Disraeli. Anna Laetitia Aikin, afterwards Mrs Barbauld, who published *Lessons* for children (1781), grew up in Newington Green. Anne Stent and James Stephen were the great-grandparents of Virginia Woolf.

42 *Poe*: Recalls Newington Green and the Revd Dr Bransby's school (where he studied 1817–20) in his tale 'William Wilson'.

42 *MW attended services*: It was not uncommon for Latitudinarian Anglicans to mix with Dissenters.

43 *Burgh's tomes*: Collected as *The Dignity of Human Nature* (1754–67) and culminating in his three-volume *Political Disquisitions* (1774–5), which took a reformer's view of taxation without representation – he, too, published a pamphlet in support of the American revolt against the Crown.

43 *'I am sick . . .'*; *'I wish . . .'*: *Education*, 11, 21.

44 *'words of learned length'*: From Oliver Goldsmith, *The Deserted Village* (1770), lines 211–14: '. . . words of learned length, and thund'ring sound/ Amazed the gazing rustics rang'd around'.

44 *'A florid style'*: *Education*, 21.

44 *'Each child'*: Preface to *Real Life*, 360.

45 *Jane Austen*: Tomalin, *Jane Austen*, 34–7.

46 *'mass of flesh'*; *'superior dignity'*: Burgh, *Human Nature*, 76–7, 276.

46 *'as if I had been her daughter'*: MW to BW (23 Sept. [1786]), *MWL*, 113; *MWletters*, 78.

47 *'I love most people best . . .'*: (20 July [1785]), *MWL*, 92; *MWletters*, 54.

47 *'When I think . . .'*: Gardiner, *Recollections*, 4, quotes this letter (2 July 1785).

47 *'With children . . .'*: *Memoirs*, ch. 3.

48 *religious utopianism*: During MW's stay in Newington Green, Dr Price was preparing a new edition of his most famous work, *A Review of the Principal Questions in Morals* (1758).

48 *'next to the introduction of Christianity . . .'*: *Observations on the Importance of the American Revolution* (1784).

49 *couldn't have cared less*: Priestley, 'Death of Dr Price'.

50 *Mirabeau co-opts Dr Price*: Tyson, *Joseph Johnson*, 85–6.

50 *'when the Dissenters . . .'*: Price, *Correspondence*, ii, 236, to an American friend, Ezra Styles, from Newington Green (15 Oct. 1784).

50 *the unprecedented setting of New World republicanism*: Robert A. Ferguson, 'The American Enlightenment 1750–1820', in *Cambridge History of American Literature*, ed. Sacvan Bercovitch, i, 380–5.

50 *'Blue-stocking Club'*: A vivid portrait in Harmon, *Fanny Burney*, 180–1. See Le Doeuff, *Sex of Knowing*, for the history of the term. Blue worsted stockings, knitted in thick, warm wool, were originally worn in England by men at home. In the seventeenth century the term evolved to refer to the Parliament of 1653 – suggesting that Cromwell's supporters were unconcerned with matters of dress. A century later the term referred to a group of men who preferred literary debate to cards, linking intellect with socially aberrant appearance. Both in English and French (bluestocking: *bas-bleu*) there was then a transfer of meaning across the sexes to a woman interested in intellectual matters – retaining a connotation of ignorance of social niceties. Le Doeuff places it with 'intuition' in a category of terms, once used for men, that get transferred as cast-offs to women and function as a deterrent to women's intellectual aspirations.

51 *'never thrive'*: *MWL*, 98; *MWletters*, 61.

52 *the American model of 'rights'*: Exercised in a series of declarations by the first Continental Congress of American states in October 1774, which had led to their nine-year battle to free themselves from British rule.

53 *'to the United States . . .'*: Cited in article on Price in the old *Dictionary of National Biography*.

53 *the London Friends of the People*: Marilyn Butler, *Romantics, Rebels & Reactionaries*, 42.

53 *'the whole scope of my life'*: McCullough, *Adams*, 417.

53 *'This is the 3d Sunday . . .'*: Abigail Adams, *Adams Family Correspondence*, vi, 196.

54 *'a new order . . .'*: Quoted in McCullough, *Adams*, 348.

54 *'snearing'*: Abigail Adams to Cotton Tufts (18 Aug. 1785), *Adams Family Correspondence*, 283.

54 *'titled Gamesters'*: Abigail Adams to sister (15 Sept. 1785), ibid. 361.

54 *'If I live...'*: 'Abigail Adams, 'Return Voyage', 215.

55 *'I thank you Miss W.'*: Adams read *FR* twice (in 1796 and 1812), adding about twelve thousand words of combative marginalia (more than in any other of his books). Here was a conservative versus a radical, a man versus a woman who dares to enter the masculine arena. He did, though, respect her enough to concede her status as an honorary male. See below, ch. 10 and internet site for this book.

55 *Jefferson to Price*: Price, *Correspondence*, iii, 261–2. See also Washington's response to Price's advice, 324–5.

56 *'Circular to the States'*: This was known at the time (1783) as 'Washington's Farewell to the Army', and Ferguson, 'American Enlightenment', calls it a text Americans 'no longer know how to read', though arguably the most important document in the America of the 1780s.

56 *contrast of the Declaration of Independence and the Constitution*: I owe this point to Sacvan Bercovitch, in conversation in Newton, Mass. (June 2000).

56 *'the abominable traffick'*: *RW*, ch. 9. On this, the economy of the Southern states depended. Because of their pressure, the clause condemning slavery (acceptable to Washington and Jefferson) had been deleted from the Declaration, and would remain unresolved until Lincoln ruled against slavery in 1863 in the course of the American Civil War. The Thirteenth Amendment to the Constitution, abolishing slavery, was ratified in 1865 at the close of the Civil War.

56 *Judge Mansfield's ruling*: The conservatism and limited intentions of Judge Mansfield have been proved beyond doubt by Gerzina, *Black England*, ch. 4, 'Sharp and Mansfield: Slavery in the Courts', 106–23.

57 *Price to Jay*: Price, *Correspondence*, ii, 292–93.

57 *MW and Price sharing ideas on education*: She read Rousseau only later (see below, ch. 5 and 6), when she came to question his attitudes to women, which is given powerful expression in *RW*. During her career she both absorbed and knocked against Rousseau's ideas.

58 *'Poor tender friendly soul'*: MW to BW (23 Sept. 1786), *MWL*, 115; *MWletters*, 79.

58 *sizar*: See footnote on Waterhouse at Cambridge (above, ch. 2). Since Hewlett came from a landed background, this poor status suggests a problem about money. Either his father had become impoverished or had cut him off for some reason, perhaps his marriage.

58 *Dr Johnson's eccentricities*: See James Boswell, *Life of Johnson* (1791), Mrs Thrale's letters, and for a caricature, Virginia Woolf, *Orlando*, ch. 4.

59 *Dr Johnson on beggars*: Boswell, *Life*. Johnson was sixty-seven in 1776.

59 *Dr Johnson and MW*: Only record is in *Memoirs*, ch. 3.

59 *Dr Johnson as conservative who opened up brave new world*: Dr Paddy Bullard, lecture, 'Dr Johnson versus Lord Chesterfield', Oxford University, 7 May 2003.

59 *Dr Johnson's romantic tenderness*: Harmon, *Fanny Burney*, 126–7.

59 *Dr Johnson as rationalist*: Prof. Roger Lonsdale, lecture on Johnson's *Lives of the Poets*, Oxford University, c. 1998.

59 *Dr Johnson on Pope*: In the last of his *Lives of the Poets*.

59 *melancholy admired*: Todd, *Wollstonecraft*, 75.

59 *Dr Johnson on melancholy*: *Rambler*, nos 85, 103.

60 *'constant nature'*: 'The Natural Beauty', publ. in the *Gentleman's Magazine*, Feb. 1784; repr. by MW in *The Female Reader*, *MWCW*, iv, 142.

60 *Smallweeds*: Dickens, *Bleak House*.

4 A COMMUNITY OF WOMEN

62 *Fanny's letter to the 'dear lasses'*: Abinger: Dep. b. 210(9).

64 *'He is such a man . . .'*: *MWL* (20 July [1785]), 91; *MWletters*, 53.

64 *'She is still very ill'*: *MWL* (14 Aug. [1785]), 97; *MWletters*, 59.

64 *'difficulties'*; *'grown indefatigable'*: *MWL* (4 Sept. [1795]), 98–9; *MWletters*, 61.

65 *letter from Lisbon*: *MWL*, 100–1; *MWletters*, 63–4.

66 *'looks I have felt . . .'*: *Travels*, letter 6; cited in *Memoirs*, ch. 3.

66 *'by stealth'*: *Memoirs*, ch. 4.

67 *Palmer's fraud*: KP, i, 175.

67 *'the most uncivilised nation . . .'*: *Mary*, ch. 14. MW quotes Dr Johnson's saying 'they have the least mind'. She used her experience of Portugal for her review of Arthur Costigan's *Sketches of Society and Manners in Portugal*, *AR* (Aug. 1788); *MWCW*, vii, 29–32.

68 *wreck and rescue*: *Mary*, ch. 20.

69 *'very disagreeable'*: This and most subsequent details of the following months come from MW's letters to George Blood. *MWL*, 101–12; *MWletters*, 66–77.

70 *Dr Johnson's elegy*: 'On the Death of Dr. Robert Levet' (1783), line 4: 'Our social comforts drop away.' *MWL*, 93; *MWletters*, 55.

70 'cordial *of life*': *MWL*, 110; *MWletters*, 74.

70 *'tenderness'*: *Education*, 37.

70 *'whole train . . .'*: MW to George Blood (27 Feb. [1786]), *MWL*, 103; *MWletters*, 67.

71 *'replete'*: John Hewlett, *Sermons* (London: Rivington & J. Johnson, 1786). Published by subscription (subscribers included various people at Newington Green, Shacklewell and Islington: Mr Church and a Mr Price in Islington; Mrs Burgh, Thomas Rogers, Mr and Mrs Cockburn at the Green; and two teachers at Christ's Hospital, Mr Benjamin Green, a drawing-master, and Mr William Wilcox).

71 *'the issues of life'*: Proverbs 4: 23. 'Keep thy heart with all diligence; for out of it are the issues of life.' Quoted in Hewlett's sermon, and echoed in MW's chapter 'On the Misfortune of Fluctuating Principles', *Education*, 42.

71 *'sublime harmony'*: Ibid., 18.

72 *'Lothario'*: BW to EW (17 Aug. 1786). Abinger: Dep. b. 210(7).

72 *'plunge'*: MW to George Blood (1 May [1786]), *MWL*, 105; *MWletters*, 68.

72 *provincial narrowness*: BW to EW (17 Aug. 1786). Abinger: Dep. b. 210(7).

73 *'very affectionate' letter'*: *MWL*, 113; *MWletters*, 78.

74 Education *as the fruit of long thought*: It is often said that she wrote too fast. This is the age-old putdown (satirised by George Eliot in the voice of a provincial schoolmaster in *The Mill on the Floss*) that if a woman be quick, she must be shallow.

74 *'so different from nature' ; 'those who imitate it'*: *Education*, 20.

74 *'The passion . . .'*: Ibid., 28.

74 *editorial note*: Cited by Todd, *Wollstonecraft*, 77.

75 *daughters' education less constricted in the past*: Deft summaries in Hufton, *History of Women*, 421–37; and in Stone, *Marriage in England*, 142–4, on 'The Education of Women' in the sixteenth and seventeenth centuries.

75 *Katherine Parr's intellectual attainments*: Weir, *Six Wives*, 514–23, 549–50.

75 *strong women emerging during the Civil War*: Stevie Davies, *Unbridled Spirits*: *Women of the English Revolution 1640–1660* (London: Women's Press, 1998).

75 *Basua Makin*: Fascinating details in Fraser, *Weaker Vessel*.

75 *Catchat*: Wright, *Female Vertuosos*, III, i, cited by Fraser, 378.

75 *'affected'*; *'designed to hunt . . .'*: *RW*, ch. 5, sect. ii. MW refers here specifically to Fordyce.

76 *kinds of advice*; *Halifax and Fordyce*: One of these advice books was *Advice to a Daughter* by George Savile, Marquess of Halifax; MW would publish an extract on the dangers of passion in her anthology, *The Female Reader* (1789). A popular preacher of the Scottish Enlightenment, Dr Fordyce's *Sermons to Young Women* was the most widely read in this genre in the England of the later eighteenth century. There were twenty reprints by 1800.

The main women to precede MW with innovative advice are Mary Astell, in *An Essay in Defence of the Female Sex* (1696), which declares that as souls are equal, women's minds should be developed to assist their salvation (the 'Cause . . . of the defects we labour under is, if not wholly, yet at least in the first place, to be ascribed to the mistakes of our Education, which like an Error in the first Concoction, spreads its ill Influence through all our Lives'); the anonymous Sophia, a Person of Quality, whose *Woman Not Inferior to Man* (1739) argued that gender difference was a construct of education and custom, and that all professions should be open to women; and the respected educator Hester Chapone, who presents learning as a dignified consolation for women's hard lot in *Letters on the Improvement of the Mind, Addressed to a Young Lady* (1773). Mrs Chapone also insisted that women aren't naturally inferior to men. MW used an extract from the last in *The Female Reader*.

76 *mocks one of Mrs Barbauld's poems*: 'To a Lady, with some painted flowers', *RW*, ch. 4, and *MWCW*, v, 122–3, 125.

76 *'bold . . . spirit'*: More, *Essays*, 145.

76 *'I must be independant'*: MW to George Blood (18 June [1786]), *MWL*, 107; *MWletters*, 71.

77 *MW realistic about the status of governess*: Not for her the fantasy of high society, which would drop Charlotte Brontë into deep pits of disillusion.

77 *''tis an unweeded garden'*: *Hamlet*, I, ii, 135.

78 *A lawsuit*: *SC*, i, 85–6. *MWL*, 111, 112n.; *MWletters*, 76.

78 *a creditor*: MW to George Blood (6 July [1786]), *MWL*, 109; *MWletters*, 73.

78 *'uncommonly friendly'; 'Mrs Burgh's kindness'*: *MWL*, 113–15; *MWletters*, 78–9.

79 *Clarissa dying*: Richardson, *Clarissa*. Her protracted dying takes up the last third of the longest novel in the English language.

5 A GOVERNESS IN IRELAND

80 *Mr Prior*: I am grateful to Michael Meredith, Librarian at Eton College, for these facts.

80 *'actually on the road'*; *'witlings'*: (9 Oct. [1786]), *MWL*, 117; *MWletters*, 79–80 .

81 *a warped specimen*: The higher a child's prenatal testosterone, the less eye contact that child will make, unless inheritance be transformed – as up to a point it might – by training at home (Simon Baron-Cohen, *Extreme Male Brain*). See debate with Lynne Segal in 'Sex on the Brain', *Guardian* (3 May 2003).

81 *numbness*: Joan Smith, *Moralities*, 97–8.

82 *'hotbeds'*; *'libertinism'*; *'tyranny and abject slavery'*: *RW*, ch. 12, 'On National Education'.

82 *'new ideas'*: To EW (9 Oct. [1786]), *MWL*, 119; *MWletters*, 83.

83 *the 'castle'*: Mitchelstown Castle is no more. In Sept. 2000 I saw milk trucks trundle up this rise to the Dairygold butter factory which has replaced the last castle (rebuilt by Lord Kingsborough's heir George, the 3rd Earl, after his mother's death in the 1820s). It was burnt down during the Troubles in 1922. Its stones were then carted off to build the Cistercian monastery of Mount Melleray. All that remains on the site is a pile of rubble on the edge of the hill overlooking what is now Mitchelstown's sewage plant. Cattle graze across the valley on the next rise. It's a

tamer landscape than I had expected from Arthur Young's books (see Bibliography) and MW's letters – 'mountains' topped by clouds had suggested more formidable peaks.

83 *the Ascendancy*: Irish historian Roy Foster calls these the golden years, referring to the period from 1785 to the end of the century – terminating presumably with the Uprising of 1798 and the Act of Union in 1801 when the Ascendancy lost its separate Parliament.

84 *'the Bastile'*: MW to BW (30 Oct. [1786]), *MWL*, 120; *MWletters*, 84. The fear of live burial, and the wall as barrier between being and nonexistence, is wonderfully imagined by Dickens, *A Tale of Two Cities*, and repeated by Schama, *Citizens*, 394.

84 *violence in North Cork*: Roy Foster in conversation, 20 Sept. 2000.

84 *a 'solemn kind of stupidity'*: *MWL*, 120; *MWletters*, 84.

85 *'decide with stupid gravity . . .'*: *Mary*, ch. 11.

85 *servants dancing*: *MWL*, 122; *MWletters*, 85. Young, *Tour*, i, 446, describes dancing as almost universal amongst the poor. Dancing masters travelled across Ireland, at 6d a quarter, from cabin to cabin with a piper or blind fiddler. There were few who would not, after a hard day, walk seven miles for a dance.

86 *'the prettiest French expressions . . .'*: Similar critiques of Lady K reappear in several of MW's works. Here, from *Mary*, ch. 1.

86 *'Every part' and quotations below*: Except where indicated, all quotations about MW's experience in Ireland are from her letters to her sisters and to George Blood, *MWL*, 120–54; *MWletters*, 84–127.

86 *'factitious' femininity*: See another critique of Lady K in the portrait of the dog-loving lady of fashion in *RW*, ch. 12.

87 *'unreadable' . . . bardic*: Foster, *Modern Ireland*, 4–6.

88 *George's official birth date 'wrong'*: King-Harmon, *King House*, 30, gives George King's birth date as 1770. Many useful quotations from the King-Harmon Papers including some details from the 1st earl's letters to his son.

90 *'the face of desolation'*: Young, *Tour*, i, 460–2. Maria Edgeworth claimed that this book contained the most faithful picture of Irish peasantry ever to have appeared.

90 *Young and Lady Kingsborough*: Young, *Autobiography*.

91 *Kingsborough's Volunteers*: Power, *White Knights*, 19.

91 *'gamy'*: Foster, *Modern Ireland*, 169.

91 *Clonmell*: Ibid., 176–7.

91 *'stupidity . . . victory'*: Quoted by King-Harmon, *The Kings*, 31.

91 *Romney's portrait of George*: George Romney was a celebrated English portraitist of the day, along with Joshua Reynolds and Gainsborough. It was common at the time to present Eton College with a leaving portrait. George looks surprisingly innocent at the age MW knew him in this portrait. Thanks to the Provost and English master Roland Martin.

92 *Mrs FitzGerald*: The Colonel's second wife, Mary Mercer, daughter and heir of Fairfax Mercer, of Fair Hill, Louth. As Caroline's stepmother, she exercised a good influence, to judge by MW's observations in 1786–7. The second Mrs FitzGerald lived till 1830, outliving her husband by fifty-four years.

92 *'just off to market'*: As it turned out, only the youngest, Margaret, married.

92 *FitzGerald girls in Mary's first novel*: *Mary*, ch. 11.

93 *'baneful effects'*: MM to WG (8 Sept. 1800), *SC*, i, 84.
93 *'a little* fun *not refined'*: *MWL*, 123; *MWletters*, 87. Power, *White Knights*, 38, speaks of his exercise of the *droit de seigneur*. This is folklore, but could be borne out by this observation on the part of MW.
94 *overturned the regimentation*: *Memoirs*, ch. 5: WG presumably heard this from MW herself.
94 *MW's stories for children*: The tradition of collections of moral tales for children originated with Sarah Fielding's popular *The Governess* (1749). Each of nine girls tells her life story, illustrating a moral strength or weakness. The short lives use different genres, ranging from the moral fable to the fairy-story. Fielding's strength lies in individualising the girls, whereas MW has more of an intellectual edge, and a wider social conscience, with a more radical view of diseased power. Both are withering about the airs of aristocratic girls. In Fielding's book, a perfect model of wise womanhood is found in the girls' governess Mrs Teachum, head of the Little Female Academy. There is no evidence that MW read this. Mme de Genlis published her miscellany of tales told to a family of children in 1784.
94 *Mme de Genlis on 'luxury' and 'magnificence'*: MW later inserts this extract in *The Female Reader*.
95 *'explain the nature of vice'*: MW's Preface to *Real Life*, *MWCW*, iv, 359.
95 *'profligate lord';'state'*: Ibid., iv, 385, 387.
95 *'friendship and devotion'*: Ibid., iv, 372.
95 *'the society of my father's house'*: From MM's revelations about her history for her youngest daughters. (See ch. 17 below.)
96 *'I wish to be . . . like Mrs Mason'*: *MWCW*, iv, 389.
97 *alerted her pupils to falsehood*: Ibid., iv, 384. This extract from *Real Life* is repr. in *The Female Reader*, *MWCW*, iv, 276–7.
97 *'narrow souls'*: *Mary*, ch. 11.
98 *Ogle's songs*: The song was still popular at the start of the twentieth century. Burns described 'Banna's Banks' as 'heavenly' and 'certainly Irish', but it was included in Wood's *Songs of Scotland* in 1851. Some of Ogle's songs appeared in Crofton Croker's *Popular Songs of Ireland*.
98 *the Ogle family*: Ogle, *Ogle and Bothal*, 213; also 133, 135, 215, 220. Further details in Burke's *Peerage*.
99 *conversations of Ogle and MW*: I'm drawing here on *Mary*, chs 12 and 18 . In the novel, 'Mary' falls in love with 'Henry', but I'm assuming that's fiction. All the other evidence suggests that MW found Ogle appealing as an intellectual who encouraged her, but that she was shocked by what appears in her letters to be his adultery with an unnamed woman.
99 'of subtiler essence . . .': MW quotes from Edward Young, *The Complaint: or, Night-Thoughts*, 'Night the First', line 100. Cited in Gary Kelly's note to *Mary*, 213.
99 *'polished manners'*: Ogle's fictionalisation in *Mary*, ch. 24.
100 *distant connection*: Assumed to be distant because it's not traceable from the family records in Ogle, *Ogle and Bothall*. Dame Isabell Ogle had married, first, her cousin, an English admiral called Sir Challoner Ogle. The history of the Ogle line shows many different branches of the family.
101 *visit to the Baillies*: *MWL*, 135; *MWletters*, 101.

102 *MW to JJ*: *MWL*, 129–30; *MWletters*, 94–6.

102 *'into a consumption'*: *MWL*, 139; *MWletters*, 108. This was in March 1787, three months after the fever.

102 *advocates the study of 'physic'*: *Education*, 34–5.

102 *'a GREAT favourite'*: To BW, *MWL*, 131; *MWletters*, 97.

6 THE TRIALS OF HIGH LIFE

103 *court case*: *SC*, i, 85–6, has details of the suit.

103 *'disappointment'*: To EW (15 May ([1787]), *MWL*, 153; *MWletters*, 126 . Quotations about MW's experience in Dublin, Feb.–May 1787, are from letters to her sisters, *MWL*, 134–54; *MWletters*, 101–27, unless otherwise indicated.

103 *without liberty, she would die*: *MWL*, 130; *MWletters*, 97.

104 *'neglected in the education of women'*: de Montolière, *Caroline de Lichtfield*, iii, 5.

104 *the house in Merrion Square*: I have been unable to verify where MW stayed in Dublin. No address is given in her letters. In accepting Merrion Square I have followed Claire Tomalin. Biographers Janet Todd and Diane Jacobs follow *SC*, i, in assuming that MW and her employers stayed with the Earl of Kingston in Henrietta Street. Todd notes (*MWletters*, 82) that Henrietta Street is listed as the residence of Lady Kingsborough in *A List of the Proprietors of Licences on Private Sedan Chairs, 25th March, 1787*. It was the more prestigious of two possible addresses. Yet given Lady Kingsborough's hatred of her mother-in-law and refusal to continue to live with her some years back, I think it likely that she would have insisted on living, for the most part, in a Dublin home of her own, i.e. the alternative address in Merrion Square. What seems to weigh the balance towards Merrion Square is the fact that the Earl comes to dine with his daughter-in-law. 'Her father-in-law had dined with her, and she repeatedly requested me to come down to the drawing-room to see him,' MW wrote to EW on 25 Mar.: there would be no point in saying this unless the Earl came from another house.

105 *The Rotunda*: Its profits helped Mosse's Lying-In Hospital next door. The maternity hospital is still there, itself now called the Rotunda. The old Rotunda became a movie-house which is now closed and looks rather derelict.

105 *the most hospitable city*: *Travels*, letter 2 .

106 *Robert Home*: Worked in Dublin in the 1780s, and went later to India where he made his fortune as painter to the Maharajah of Oude.

106 *a gift from George Blood*: In *MWL*, 139, Wardle notes that the publisher, Bell, was bringing out two editions of Shakespeare in 1787: one in twelve volumes, ed. by Samuel Johnson and George Steevens, and another in sixteen volumes with Johnson's and Steevens's additional notes. I don't agree with Wardle's view that Blood's gift would have been 'more modest', a single-volume edition of 1784 (a view repeated in Todd's notes). It was in Blood's nature to be extravagant; MW had long supported him and his parents in innumerable ways, and he'd finally settled into a position where he had a steady income; and finally, MW says clearly that it was a 'new' edition.

106 *Blair on genius and his influence on MW*: *Lectures*, i, 52–4; ii, 36–7, 40–3, 47; iii, 19–21.

107 'l'exercice . . .': Epigraph to *Mary* – 'The exercise of the most sublime virtues raises and nourishes genius'.

107 *'a genius will educate itself'*: *Émile*, book ii, cited in a letter to EW (24 Mar. [1787]), *MWL*, 145; *MWletters*, 114–15.

107 *eighteenth-century Henrietta Street*: Information from the 'Conservation Recommendations for Individual Building Elements for Henrietta Street, Researched for Dublin Corporation Historic Area Rejuvenation Project (Harp) by Dublin Civic Trust', kindly photocopied in Sept. 2000 by Liam McNulty of the Society of Pipers at no. 15. Henrietta Street flourished 1720–1820. It began to decline when Leinster House (the Irish Parliament) was built, and fashion moved from north to south of the Liffey. What remains – now inhabited by the Society of Pipers – is half of the double-fronted house which the Kings inhabited. No. 16 was knocked down about fifty years ago.

107 *only 'Miss King's governess'*: *Memoirs*, ch. 4.

107 *MW's manners and conversation*: *Gentleman's Magazine* (Oct. 1797).

107 *received registers of . . . pronunciation*: Roy Foster tells me that a Cockney accent would have jarred in Dublin society, but that it would not have recognised much difference between middle- and upper-class accents. Aristocratic accents in England were often regional until about the 1920s, as evidenced by Victoria Sackville-West's *The Edwardians*. (In conversation, Oxford, 28 Sept. 2000.)

109 *He urged her*: This letter has not survived, but its content can be deduced from MW's reply, quoted in full here. *MWL*, 148; *MWletters*, 118–19.

111 *Mary's ingratitude*: To suggest too strongly that MW was behaving badly implies that she should have known her place. As for the bleakness of a governess's position, we have only to read Jane Austen on the just fears of Jane Fairfax (in *Emma*) or Charlotte Brontë on Jane Eyre. Ch. 17 below relates the wretched experiences of CC as a gifted and spirited governess in the Wollstonecraft mode.

111 *St Werburgh's Church*: I'm grateful to the Revd Canon David Pierpont, Vicar of the Christ Church group of parishes, for opening the church on Saturday 9 Sept. 2000, and providing helpful information. That day, there was another visitor from County Wexford, to see the tomb (beneath the church) of his hero Lord Edward Fitzgerald, one of the leaders of the 1798 Rising.

111 *Handel in St Werburgh's*: Handel may have played on the original organ, as he lived in the area for eight years. He brought the half-finished *Messiah* with him to Dublin, but his programme at the church did not include that work, which had its first public performance in 1742 in a hall only fifty yards away in Fishamble Street. (It could not take place in a church because Protestants thought it improper to put the Word to music.)

111 *Caroline Stuart Dawson*: Known as Lady Portarlington. She was an artist as well as a singer.

112 *In London she had ridiculed*: *Education*, 46–7.

112 *Calista*: In Nicholas Rowe's *The Fair Penitent*.

112 *'really great merit'; events of 23–25 Mar. 1787*: MW to EW (25 Mar. [1787]), *MWL*, 146–8; *MWletters*, 116–9.

113 *'Is it not . . . Alas!!!!!!!!'*: The final PS of a letter to EW, written late at night after her evening with the Ogles (24 Mar. [1787]), *MWL*, 145; *MWletters*, 116. She alludes to Gray's *Elegy Written in a Country Churchyard*: 'Full many a flower is born to blush

unseen/ And waste its sweetness on the desert air'; and to *Hamlet*, II, ii, 97f.: 'That he is mad, 'tis true 'tis pity/ And pity 'tis 'tis true.'

114 *Solitary reading for women*: Todd, *Wollstonecraft*, 416, interestingly, compares rabbis' disapproval of reading the Torah alone, canto 5 of Dante's *Inferno*, and Dutch paintings of women reading love-letters which can be associated with the libidinal dangers of novel-reading.

114 Émile *and the new status of the child*: I'm indebted to an illuminating lecture on this subject by Angelica Goodden (Fellow in French at St Hilda's College, Oxford), given at the university on 3 Nov. 2000.

116 *'I long to go to sleep'*: MW to George Blood (c. Jan 1787), *MWL*, 134; *MWletters*, 100.

116 *'to make any great advance . . .'*: To EW, *MWL*, 140; *MWletters*, 109.

117 *'adventitious' rights*: Paley, *Principles*, in a chapter on 'The Division of Rights', 75–6.

117 *'Nature may have made'*: Ibid., 279.

118 *biographic basis of the novel* Mary: The form may provide a missing link between Dr Johnson's *Lives of the Poets* (completed in 1781) and the autobiographical schema of Wordsworth's *Prelude: or, Growth of a Poet's Mind* (1805). Earlier fictions like *Tom Jones* or *Tristram Shandy* do include scenes from childhood, but these are more snippets of anecdote than measured biography.

118 *'carefully attended'*: *Mary*, ch. 1.

118 *the fictional Ann*: Ibid., ch. 11. It's suggested that MW demeaned Fanny by having a fashionable English family in Lisbon call Ann a 'beggar'. It hardly needs to be said that, far from endorsing this, MW is making a point about the worldly.

119 *'neither marrying'*: Matthew 22: 30: 'For in the resurrection, they neither marry, nor are given in marriage, but are as the angels in heaven'. Repeated *RW*, ch. 2.

119 *restless – with no institutional habitation*: As such, 'Mary' is a precursor to Catherine Earnshaw in *Wuthering Heights*.

119 *'darted into futurity'*: *Mary*, ch. 4.

119 *answer to Sophie*: Taylor, 'Wild Wish', 216.

119 *flash of contempt*: *Mary*, ch. 3.

119 *'knowledge of physic'*; *'medicine of life'*: Ibid., ch. 23. 'Mary' also studies 'physic' in chs 6 and 17.

119 *'men past the meridian'*: Ibid., ch. 8.

119 *the nature of 'Mary's' mind*: Ibid., chs 10 and 23.

119 *the novel* Mary *indebted to Dr Price*: Gary Kelly, Notes to *Mary*, 211. MW asks after 'Le Sage' in a letter to EW (24 Mar. 1787).

119 *'witching time'*: *Hamlet*, III, ii, 406.

119 *'I think and think'*: To EW (24 Mar. [1787]), *MWL*, 144; *MWletters*, 113.

120 *'Still does my panting soul . . .'*: *Mary*, ch. 20.

120 *'lost in stupidity'*: *MWL*, 156; *MWletters*, 130.

121 *'the only Rt Honourable . . .'*: MW to BW (27 June [1787]), *MWL*, 155; *MWletters*, 129.

121 *governess was away*: There is no evidence of where MW went, but it seems likely she would have paid a visit to her father at Laugharne on the south coast of Wales, not far from Bristol but not close enough for a day-trip.

7 VINDICATION

123 *eighteenth-century drawing of the north side of St Paul's Churchyard*: Etching by T. Horner in the British Library, reproduced in *MWL*, 180.

123 *best friend*: I agree with Tomalin, Woof and Hebron, *Hyenas*, 2–4.

125 Vathek *as a private cult*: Noel Annan's recollections of *The Dons*, largely at Trinity and King's in Cambridge during the early years of the twentieth century, tell us that the Fellows sometimes chose students and colleagues on the basis of their responses to questions about their taste for *Vathek*. Before homosexuality was legalised in the 1960s the novel served as a private code.

125 *early Wordsworth and Coleridge*: Wordsworth's *Descriptive Sketches* and *An Evening Walk*, and Coleridge's *Frost at Midnight*.

125 *JJ's* 'tenderness': BW to EW (7 July 1794). Abinger: Dep. b. 210.

125 *JJ's sympathies*: Tomalin, Woof and Hebron, *Hyenas*, 4.

125 *compassionate conduct*: Godwin's obituary for JJ (1809).

125 *new 'plan of life'*: To JJ (13 Sept. [1787]), *MWL*, 159; *MWletters*, 134 .

125 *'The Cave of Fancy'*: *MWCW*, i, 191-206

126 *'determined'*: (13 Sept. [1787]), *MWL*, 159; *MWletters*, 134. WG heard that MW was depressed in this transition period, and ascribes it to renewed grief for Fanny in the wake of writing *Mary*. This must have been hearsay. I don't see evidence of depression in the letters of this period.

126 *'new systems'*: To JJ (13 Sept. [1787]), *MWL*, 159; *MWletters*, 133.

126 *writing* Real Life *in the autumn of 1787*: What appears a speedy production depended on material gathered slowly and tested in practice over a substantial preceding period. Her *Education*, written over the summer of 1786, drew on two years' experience in her own school at Newington Green; *Mary*, written in the summer of 1787, drew on the long years of Fanny's decline, the voyage to Lisbon in 1785, and Mary's growing conviction of her own 'genius' in the course of 1787; and then *Real Life* drew on her outdoor curriculum at Mitchelstown Castle, designed to awaken the conscience of privileged adolescent girls like Margaret and Caroline King before it was too late.

127 *'painful emotions'*: MW to JJ (20 Sept. [1787]), *MWL*, 162; *MWletters*, 137.

127 *a situation nearer her*: MW reporting to EW, *MWL*, 165–6; *MWletters*, 140–1.

127 *the organ and her resolves*: Ibid.

127 *'Without your humane and* delicate *assistance'*: To JJ, *MWL*, 186; *MWletters*, 159. Todd dates this undated letter to early 1789; Wardle dates it to late 1789 or early 1790 because MW asks JJ for a German grammar at a time; Wardle assumes, she would have been working on her German translation. But there are three reasons for dating this letter as early as the autumn of 1787: the letter makes it clear that she's only beginning to 'attempt to learn that language' as a preparatory act – there's no translation as yet at hand. Second, she signs herself 'Mary W.', indicating an earlier stage in the relationship, for by late 1787 or early 1788 she was using her Christian name alone. Third, the assessment of JJ as 'delicate' echoes the wording of a letter to Everina in mid-Nov. 1787.

127 *'how warmly and delicately'*: To EW [c. mid-Nov. 1787], *MWL*, 166; *MWletters*, 159.

128 *Blackfriars Bridge*: Begun in 1760, the bridge was initially called after the Prime

Minister William Pitt the Elder. In 1791 the mill burnt down. The spot is now Rennie Garden.

128 *panoramic view*: The word 'panorama' entered the language in 1796 soon after a panorama of London was drawn from this vantage point.

128 *'whim'*: *MWL*, 166; *MWletters*, 141.

128 *'vehement'*: *Memoirs*, ch. 5.

128 *'You can conceive . . .'*: *MWL*, 164–5; *MWletters*, 139–40.

129 *often the sole woman*: Mrs Barbauld (Anna Laetitia Aikin) and later the novelist Mary Hays sometimes attended JJ's dinners.

129 *'I often visit . . .'*: To George Blood, *MWL*, 171; *MWletters*, 149.

129 *better health*: *MWL*, 170; *MWletters*, 146.

129 *'past tumultuous scenes of woe'*: To George Blood, *MWL*, 176; *MWletters*, 156.

129 *Hewlett's sermon for MW*: *MWL*, 171; *MWletters*, 149.

129 *'Mrs S is sunk . . .'*: *MWL*, 172; *MWletters*, 149–50.

130 *Cowper*: *The Task* (1785).

130 *Bonnycastle*: Leigh Hunt, *Lord Byron and Some of His Contemporaries* (London: 1828), ii, 34.

130 *George Fordyce*: Son of the author of an advice book for girls (see above, ch. 4).

131 *MW's movements after dinner*: Apt question from Elizabeth Crawford (author of *The Women's Suffrage Movement: A Reference Guide* and biographer of the pioneering women in the Garrett family) on a rainy Sunday, 30 April 2000, as we followed MW's footsteps from St Paul's to her home across the river.

131 *'lank'*; *'a philosophical sloven'*: Knowles, *Life of Fuseli*, i, 164. Pope similarly slandered Lady Mary Wortley Montagu as dirty, and the mud stuck till Isobel Grundy's scrupulous biography in 1995. Fuseli's slander has stuck too, repeated into the twenty-first century.

131 *Henry Adams to his wife*: Patsy Vigderman of Cambridge, Mass., supplied this anecdote.

131 *'an excellent preservative of health'*: MW used this quote from the *Spectator*, no. 15, for her *Female Reader*.

131 *Hays on MW*: 'Memoirs of Mary Wollstonecraft', 460.

132 *Fuseli helped*: Weinglass, 'Letter of Enquiry', 144–6.

132 *spent . . . £200 on her family*: JJ, 'A few facts'.

132 *gain £200*: *MWL* 174; *MWletters*, 154.

133 *the obscure, the rude . . .* : MW is in line with the tendency of the age, represented by the rude forefathers of the hamlet who sleep in the churchyard of Gray's *Elegy*. Marilyn Butler, *Romantics, Rebels, and Reactionaries*, 17, shows parallels in music: in Gluck's 'noble simplicity' and condemnation of superfluous ornament in his dedication of his opera *Alceste*, and in Haydn's attraction to the folk-dances of Eastern Europe.

134 *'I cannot bear . . .'*: *MWL*, 167; *MWletters*, 143.

134 *'deeply immersed'* . . . *MWL*, 173; *MWletters*, 152.

134 *Necker*: MW remained unconvinced by his combination of finance with spirituality.

134 *'what does this mean?'*: *MWL*, 177; *MWletters*, 158.

135 *'almost rewrote'* Young Grandison: JJ, 'A few facts'.

135 The Female Reader: Her model was William Enfield's *The Speaker; or, Miscellaneous Pieces*, the popular anthology published by JJ for use in schools.

135 *'Negro woman'*: From hymn VIII, *Hymns*, repr. in *The Female Reader*, *MWCW*, iv, 189.

135 *as yet no schoolbooks for girls*: Gillian Avery, author of *The Best Type of Girl*, in conversation (Feb. 2001).

136 *Most girls . . . protected . . . from serious books*: Carol Shields, *Jane Austen* (Weidenfeld & Nicolson, 2001).

136 *Mr Cresswick*: He gave public readings, and died in 1792, the year he published an imitation volume: *The lady's preceptor; or, a series of instructive and pleasing exercises in reading; for the particular use of females; consisting of a selection of moral essays, . . .* It's hard to see why a jingle that appeared in it entitled 'On Breaking a China Quart Mug Belonging to the Society of Lincoln College, Oxford' could be appropriate. Mr Cresswick opts for undemanding compositions close to 'easy and elegant conversation, upon topics interesting the Fair Sex' which would fill their minds with harmless and pleasing thoughts. He does, though, include 'Observations on Reading by Miss Wollstonecraft and other writers'.

136 *the first anthology for and about women, and in part, too, by women*: I owe this information to Elizabeth Crawford.

136 *contents of* The Female Reader: In 2001 Elizabeth Crawford discovered a Dublin edition of *The Female Reader*, published the same year as the first English edition (1789) but with an additional 198 pages consisting of 'A Complete System of Geography Not in the London Edition'. It remains to be verified whether the addition was by MW or from some other source (such as her schoolteacher friends, John Hewlett, the Revd Mr Gabell, Jane Arden, or a Dublin teacher). There is no indication that it is by another hand. The British Library has a copy of the Dublin version, republished in 1791. It is not listed in the general catalogue.

136 *'Dying Friends'*: *The Complaint*, iii, as published in *The Female Reader*, *MWCW*, iv, 183.

137 *'During her stay . . .'*: JJ, 'A few facts'.

137 'intimate'; *'crimes'*: *MWL*, 178; *MWletters*, 166.

137 *'how often I teazed you . . .'*: *MWL*, 221; *MWletters*, 206.

137 *'trash'*; *'I seemed . . .'*: *MWL*, 178–9; *MWletters*, 156–7.

137 *short-notice professional reviewing*: Waters, 'Literary Critic', 415–34.

137 *reviewed her own translation*: *AR* (Jan. 1789).

138 *'Address to the Bastille'*: Cowper, *The Task*, book v.

139 *Enclosure Acts*: Porter, *English Society*, 208–13.

139 *implications of the Enclosure Acts*: Joan Smith, *Moralities*, 87.

139 *the Revolution debate initiated by Price and Burke*: I am indebted to Prof. David Wormesley's Oxford lecture on this subject (1 May 2003), stressing contested historical narratives.

139 *'Behold . . .'*: Price, *Political Writings*, 195–6.

139 *'an addition of nondescripts . . .'*: *The Writings and Speeches of Edmund Burke*, viii, 'The French Revolution' (1790–4), ed. Paul Langford (Oxford: Clarendon, 1981), 63. Dr Paddy Bullard drew attention to this passage in his Oxford lecture on Burke (15 May 2003).

140 *'Upon that . . . stock of inheritance'*: Burke, *Revolution in France*, 117.

141 *'Man will not be brought up . . .'*: Paine, *Rights of Man*, 230.

141 *MW stopped writing*: WG mistook this as a fit of 'indolence', *Memoirs*, 230.

141 *perfectly judged*: *Memoirs*, 300, Holmes's notes.

141 *'coming warm from the heart . . .'*: MW writing of Dr Price to George Blood (17 Jan. [1788]), *MWL*, 170; *MWletters*, 147.

143 *publication of* RM: A second edition expanded by nine pages appeared on 14 Dec. with Mary Wollstonecraft's name. (Crawford, 'Mary Wollstonecraft', 14–19.

143 *MW's exchange with Catharine Macaulay*: MW's letter to Macaulay was first published in 1995, repr. in *SC*, ix, 1–2.

143 *Price to MW*: Abinger: Dep. c. 514.

144 *the title of RW*: MW's switch from the plural (*Men*) to the singular (*Woman*) in her title of the sequel may be influenced by the success of Paine who used the singular form to effect in his *Rights of Man*, published in March 1791 between MW's two *Vindications*.

144 *to meet Paine*: In March, WG had helped to bring out the first part of Paine's more famous *Rights of Man*, against a threat of banning and prosecution.

144 *'I had little curiosity . . .'*: *Memoirs*, ch. 6.

144 *Woolf spoofed*: In *Three Guineas*.

145 *the woman as preacher*: Dinah Morris in *Adam Bede*, ch. 2 ('The Preaching'). I'm grateful to Dr Isobel Rivers of St Hugh's College, Oxford, for alluding to this after Taylor's lecture, cited next.

145 *MW and the Scottish Enlightenment*: Deftly summarised by Taylor in 'Mary Wollstonecraft and the Enlightenment'.

146 *'my book . . .'*: To EW (23 Feb. [1792]), *MWL*, 210; *MWletters*, 198.

146 *Joan Smith*: *Moralities*, ch. 5.

146 *the new marriage laws as property laws*: Ibid., 86–7.

146 *the law's sanction of beating and rape in marriage*: Ibid., 93.

146 *'From the respect paid to property . . .'*: *RW*, ch. 9.

147 The Ladies Dispensatory: Vivien Jones, 'Sex Education', cites the extract.

147 *Mrs Mason linked with model of womanhood in* RW: Moore, *Mary Wollstonecraft*, 42.

147 *debate on surrendering sexual pleasure for friendship in marriage*: Ibid. Well-judged summary by Jane Moore, who reads *RW* 'as an early attempt to bridge the sexual chasm' between men and women.

147 *sex-based subordination based on educational disabilities*: In the twenty-first century, there remain vast numbers of women who may be equal citizens yet remain victims. One-third of girls in South Africa – despite an advanced constitution – are raped in school, in the main by teachers and headmasters, and though in 2002 the Minister for Education Kadar Asmal commissioned a book from social scientist Anne-Marie Wolpe, addressing teachers in very tactful terms (the book takes account of a culture of entitlement amongst school authorities and compliance traditionally expected from girls), the Department of Education neglected to distribute it. The press exposed an undiminished level of school-hours rape in 2004.

147 *domestic affections cut across distinctions of gender*: Moore, *Mary Wollstonecraft*, 39.

148 *'ridiculous falsities'*: *RW*, ch. 7 (on modesty).

148 *Preface to* Elements of Morality: 'Address to Parents'. A 'new and improved' edition in 1821 eliminated sex education together with the opening 'Advertisement', which had carried MW's name in the 2nd edn, 1792.

148 *no precedent for naming the sex organs*: Vivien Jones, 'Sex Education'. Dr Jones led the way in shedding an accretion of prejudice about MW's supposed prudery, in order to demonstrate what was innovative.

148 Aristotle's Complete Master-Piece: Ibid.

148 *'Children very early see . . .'*: *RW*, ch. 7.

148 *'not only to enable [women] to take proper care . . .'*: Ibid., ch. 12, on national education.

149 *hard fact in copious footnotes*: Marilyn Butler, comments after Dr Jones's lecture, 'Sex Education'.

150 *change institutions . . . change human nature itself*: Schama, *Citizens*, denies any advantage resulting from the French Revolution, despite this psychological effect (described memorably in the American context by Henry Adams in his chapter on 'American Ideals' in his *History of the Administrations of Jefferson and Madison*). In France the psychological consequence is overlaid by violence, but had its impact on the mind of MW, which in turn is having its impact now in the transformation of women's agency.

150 *Talleyrand on girls' education*; *'the will of nature'*: Girls were to be allowed to attend primary schools up to the age of eight, where they would learn handiwork 'suitable to their sex'. (Tallyrand, *Rapport sur l'instruction publique*.)

150 *'the natural emotions . . .'*; *'pretty superlatives'*: *RW*, Introduction to the 1st edn.

150 *'a premature unnatural manner'*: Ibid.

151 *'a road open . . .'*: Ibid., ch. 9.

151 *Catharine Macaulay*: Comments on women's education in letter 4, *Letters*, 46–50. Reviewed by MW in *AR* (1790). She expressed her admiration in *RW*, ch. 5, when Macaulay (Mrs Graham) had recently died.

151 *'Brutal force . . .'*: *RW*, ch. 2.

151 *'Man accustomed to bow down to power . . .'*: Ibid., ch. 3.

152 *'It would puzzle a keen casuist'*: Ibid., ch. 9.

152 *'moral agents'*: Ibid., end ch. 12.

152 *'It is time to effect a revolution . . .'*: Ibid., ch. 3.

152 *mainstream of British and American politics*: I've taken this wording from Barbara Taylor's innovative study of the impact of MW on Owenite feminism in the first half of the nineteenth century, a period in which MW is still often taken to have been in eclipse (*Eve and the New Jerusalem*, 1). Owen's meeting with MW's elder daughter in 1816 will figure in ch. 17.

152 *debated by women in the British provinces*: Hufton, *History of Women*, i, 450.

152 *Mrs Grant's response to* RW: Noted by Elizabeth Crawford, 'Mary Wollstonecraft: "the first of a new genus"'.

152 *Catherine, or the Bower: The Juvenilia of Jane Austen and Charlotte Brontë*, ed. Frances Beer (Penguin Classics, 1986), 136–77. The connection between Aunt Philadelphia and the fictional orphan Miss Wynne, forced to sell herself to an unattractive man twice her age, was made originally by Chapman, and is sensitively developed in Tomalin, *Jane Austen*, 80. *Catherine* was written a few months after Aunt Phila's death. Later, in *Mansfield Park*, a slave-owner, Sir Thomas Bertram, tries to force his niece Fanny into a lucrative marriage to a faithless flirt – again a demand that a dependant should prostitute her person for 'a maintenance'. Kathryn Sutherland's introduction to the Penguin Classics edition of *Mansfield Park* suggests that 'in the intensity and even violence of her feelings, Fanny can seem the heir of a Romantic revolutionary feminine tradition, of heroines like Wollstonecraft's Maria [in *WW*]'.

153 *'Upon my word . . .'*: *Pride and Prejudice*, ch. 29.

153 *'Her manners . . .'*: Ibid., ch. 8.

153 *Fanny Price as legatee of* RW: Hufton, i, 450.

153 *Anne Elliot*: *Persuasion*, ch. 23.

153 *'remember the Laidies'*: *Adams Family Correspondence*, i, 329, 370: 'That your sex are naturally tyrannical is a truth so thoroughly established as to admit of no dispute . . . Why then not put it out of the power of the vicious and lawless to use us with cruelty and indignity. Men of sense in all ages abhor those customs which treat us only as vassals of your sex.' 'I desire you would remember the Laidies, and be more favorable to them than your ancestors. Do not put such unlimited power into the hands of husbands.'

Adams replied (382): ' I cannot but laugh', noting complaints that American freedoms had fomented disobedience amongst 'Indians' and 'Negroes'. 'But your letter was the first intimation that another tribe more numerous and powerful than all the rest were grown discontented . . . Depend on it, we know better than to repeal our masculine systems.'

153 *Abigail Adams and* RW: Adams MS correspondence: John Adams in Philadelphia to Abigail Adams in Quincy, Massachusetts (22 Jan. 1794); Abigail Adams to John Adams (2 February 1794); and on women rulers (26 Feb. 1794). Elements of this exchange cited in Akers, *Abigail Adams*, and Paul C. Nagel, *The Adams Women* (New York: Oxford University Press, 1987), 57.

154 *Aaron Burr's response to* RW: To Mrs Burr (16 Feb. 1793), *Memoirs of Aaron Burr*, i (New York: Harper, 1855), 363; extract in *SC*, i, 328.

154 *Hannah More to Walpole*: 18 Aug. 1792, Walpole, *Correspondence*, xxxi, 370.

154 *the gentry . . . 'shocked'; effigy of Paine; 'immortalizing Miss Wollstonecraft'*: BW to EW (20 Jan., 10–20 June 1793). Abinger: Dep. b. 210. BW was a governess in Pembroke at this time.

154 *William Roscoe applauded MW*: 'Life, Death and Wonderful Achievements of Edmund Burke' (quoted Chandler, *Roscoe*, 390). Roscoe (1753–1831), elected MP for Liverpool in 1806, backed the bill to abolish the slave trade which became law in 1807.

155 *Roscoe commissioned a portrait*: Walker Art Gallery, Liverpool. Painted in Oct. 1791. The artist is unknown, but a possible clue may be in an unpublished letter (20 Aug. 1834) from WG to his second wife (Abinger: Dep. c. 523). He is pestered for money, he says, by William Perry, who did a portrait of JJ '& another of first mamma' (as MW was called after WG remarried).

155 *' a book'*: *MWL*, 202–3; *MWletters*, 190.

8 RIVAL LIVES

156 MW's letters to George Blood and her sisters between Nov. 1787 and early Dec. 1792 are in *MWL*, 163–223; *MWletters*, 138–213.

156 *'a* dreadful *situation'*: 1 Jan. [1788].

157 *'I wish when I transact . . .'*: 15 Sept. [1789].

157 *Mrs Bregantz's 'snarling' . . .*: BW to EW (Apr. 1791). She writes to George Street, where EW was staying with MW after her return from Paris.

158 *Tasker and Rees*: I'm dependent here on Todd, *Wollstonecraft*, 170–1, having myself not come to satisfactory conclusions. BW's rage is so pervasive that she doesn't make the exact nature of the relationships sufficiently clear.

159 *'It is happy . . .'*: BW to EW (29 Mar. 1792).

160 'Mrs Wollstonecraft *is grown quite handsome*': BW to EW (3 July 1792).

160 *'while the sad hours of life . . .'*: BW to EW (10 Feb. 1793).

161 *'threw some money away'*: MW to Roscoe, *MWL*, 203; *MWletters*, 191.

161 *James Wollstonecraft's future*: According to JJ, 'A few facts', James afterwards served 'on board Lord Hood's fleet as a midshipman, where he was presently made a lieutenant'. This doesn't take account of the shadier aspect of his history, which includes a spell in a French prison; money borrowed from his sister's friend, John Barlow, and not returned, to JB's outspoken fury; and four subsequent letters to WG with excuses for not repaying a debt.

161 *MW and Ann*: Unclear whether this was always to be a temporary arrangement. It seems that Ann was later returned to Mrs Skeys.

161 *Mrs Skeys none too kind*: EW to BW (n.d. but post-1798, because the letter refers to the publication of *WW*). Abinger: Dep. b. 210/5.

163 *JB's revisionist view of 'nature'*: Beinecke: Za Barlow folder 5: MS ch. 6, headed 'Means of Subsistence', of an unidentified work.

164 *MW meets Mrs Leavenworth*: Stiles, *Diary*, iii, 502–3.

164 *'the easy . . . behaviour' of American women*: MW's review of J. Brissot, *Nouveau Voyage dans les États-Unis de L'Amérique*, *AR* (Sept. 1791). *MWCW*, vii, 391. This issue came out one month after RB's arrival in London in Aug. 1791, which makes it just possible for MW to have met her through JJ.

164 *'What I feel, I say'*: RB to JB (Jan. 1796). Houghton.

164 *RB's concern with private integrity*: J.B. Cutting to JB (19 May 1810). Beinecke: Pequot M 969.

164 *'could never contrive to make any boys . . .'*: MW to EW (23 Feb. [1792]).

164 *'heavy expence'*: MW to EW (20 June [17]92).

165 *'snap'*: *MWL*, 208; *MWletters*, 196.

165 *'Mrs. B. . . .'*: 20 June [17]92.

165 *resemblance of GI and JB*: Seelye, *Beautiful Machine*, 117.

166 *JB's family*: The son of Samuel and Esther Hull Barlow, JB was a descendant of farmers who had settled in Connecticut in the 1650s.

166 *JB's war record*: He fought in the battle of Long Island and was chaplain to the 4th Massachusetts Brigade.

166 *'the tenderest . . . of* Lovers': From Hartford (13 Aug. 1781). Houghton, bMS Am 1448 (537).

166 *'the Hartford Wits'*: Timothy Dwight, John Trumbull, Lemuel Hopkins, and David Humphreys who reappears later in JB's correspondence as one of Washington's spies in Europe. See below, ch. 14. During these early years in Connecticut, Barlow tried law – with little success – and then was employed to revise Isaac Watts's version of the Psalms for the Congregational Church.

166 *JB's epic*: In its foretelling the triumph of liberalism, a forerunner of Constantin Volney's *Ruins* (1791) and Shelley's *Queen Mab* (1813).

167 *Scioto Land Company*: On 1 Mar. 1784 Jefferson proposed a plan for the government of the entire West which led to the Northwest Ordinance of 13 July 1787, providing for government of the territory north-west of the Ohio River. Under the direction of the Revd Manasseh Cutler, a group of Revolutionary War veterans from New England organised the Ohio Company. The group signed a contract

on 27 Oct. 1787 for 1, 500, 000 acres on the Ohio and Muskingum Rivers.

167 *JB's first visit to London*: Diary: Houghton, bMS Am 1448 (9). See also JB in London to Hazard (27 Aug. 1788), Beinecke: MS Vault Pequot M886–M937.

167 *JB to Sargent*: Winthrop Sargent Papers, Massachusetts Historical Society, Boston.

168 *'I thot you were dead . . .'*: JB to RB (9 Mar. 1790). Houghton, bMS Am 1448 (181).

168 *RB's voyage*: RB to Mary Dwight (3 Oct. 1790) from Paris. Houghton.

168 *JB enamoured of Mrs Blackden*: Morris, *Diary* (May 1789).

168 *'I have not slept with any body but God . . .'*: JB to Mrs Blackden (11 July 1790). Houghton.

168 *Scioto scheme*: Trans. of the Paris agreement in is Beinecke: Za Barlow, folder 8.

169 *JB and a Virginia merchant*: JB to Mr Fitzgerald of Alexandria, Va, from 162, rue Neuve des Petits Champs, Paris (17 Jan. 1790). (Scioto Land Co. Papers, MSS Division, New York Public Library.) This letter exhibits the high-toned graces Barlow could command. On the Internet site accompanying this book.

169 *agreement with Hallet*: Trans. in JB's hand (14 Aug. 1790). (MSS Division, New York Public Library: Ohio box, Scioto Land folder.) JB's address is the Hôtel de la Grande Bretagne, rue Jacob, Paris. Agreement dated 28 June 1790, between JB and M. Hallet, by which Hallet entered the employ of the Scioto Co. for two years at an annual salary of 1500 livres.

169 *'taken infinite pains'*: He goes on: 'If the first 100 people find themselves happy, the stream of emigration will be irresistible, they may be followed by a million of European settlers into the western country. This will greatly increase the value of all those lands, & enable Congress to sink the national debt by the sale of lands.'

169 *'the period of our deepest difficulties'*: JB's reflective will-letter to RB from Algiers during the plague of 1796. See below, ch. 14. Draft of letter in Beinecke, Za Barlow folder 13.

169 *JB to Baldwin*: On 3 May 1791 he promised Baldwin that if the Scioto schemers started up again, 'I shall have nothing to do with them.' Houghton, bMS Am 1448 (65).

169 *JB's new role*: To Baldwin, again. Ibid.

170 *JB invited MW to tea*: WG's diary: 'Tea at Barlow's with Jardine, Stuart, Wollstencraft [*sic*], and Holcroft.' Abinger: Dep. e. 201.

170 *JB's* Advice *burnt*: Reports contradict.

170 *Jefferson to JB*: 20 June 1792. Houghton *56M–52.

170 *'The visit to the king . . .'*: JB to RB from Paris (25 June 1792). Houghton, 1448 (650).

170 *'fondness for tracing . . .'*: *MWL*, 162; *MWletters*, 137.

171 *'pent up'*: RB to Mary Dwight (3 Oct. 1790). Houghton, bMS Am 1448 (650).

171 *coveted by Lady Hamilton*: JB described the house in detail to Lady Hamilton. Letterbooks, Houghton.

171 *'will be handed you by Mr Wollstoncraft'*: 1 Oct. 1792. Beinecke: Za Barlow folder 29.

172 *'The exertions . . .'*: *Memoirs*, 228.

172 *correspondence with MM*: *MWL*, 167, 172, 176; *MWletters*, 143, 150, 156.

172 *'From the time she left me . . .'*: *SC*, viii, 909–11.

172 *Mrs Mason takes leave*: *Real Life*, ch. 25, *MWCW*, 449–50.

172 *'do not suppose'*: c. mid-1788. *MWL*, 177.

174 *planned to visit Paris*: In June, JB sent word to RB that he had found lodgings for MW in Paris. Houghton.

174 *'the world . . . married me . . .'*: To Roscoe (12 Nov. [17]92), *MWL*, 218; *MWletters*, 208.

174 *Adam and Eve*: *Paradise Lost*, iv, 411–504.

175 *Lavater*: MW was translating Lavater. HF's translation ousted hers. See ch. 7.
175 *'grandeur of soul' and '. . . comprehension'*: MW's letters to HF, quoted by Knowles, *Life of Fuseli*, i, 163.
175 *'brimful'*: BW to EW (4 Oct. [1791]). Abinger: Dep. b. 210.
175 *'palsied'*: JJ, 'A few facts'. MW managed only a few reviews.
176 *'grotesque mixture'*: Haydon, *Autobiography* (1853).
176 *WG on HF*: Letter to Knowles (28 Sept. 1826), cited in Knowles, *Life of Fuseli*.
176 *Opie's portrait of HF*: National Portrait Gallery, London. At one time it hung opposite Opie's 1797 portrait of MW.
176 *exchange at JJ's dinner*: 'Memoir of HF' in *18 Pamphlets on British Art 1797–1934*, Bodleian.
176 *HF and JJ*: In 1979 JJ's biographer, Tyson, contested Claire Tomalin's plausible 1974 suggestion that JJ could have been homosexual. Tyson, *Joseph Johnson*, xvii, argues (rather unconvincingly) that a lot of London shopkeepers never married because of economic insecurity, and that HF had a lot of single friends who are assumed to be heterosexual. Since it can't be proved either way I have left the matter open.
177 *HF as pornographer*: Drawings at the Victoria and Albert Museum, London.
177 *'women have no character . . .'*: Pope, *Moral Essays*, Epistle II. HF agrees in his *Remarks on Rousseau* (1767), which MW almost certainly would have read.
177 *HF on women*: Fuseli, *Aphorisms* (1788–1818), 225–7. Selections in Fuseli, *Mind of Henry Fuseli*.
177 *HF's address*: Information from Elizabeth Crawford. HF was still there in 1794.
177 *home visits of HF and MW*: *Memoirs*, ch. 6.
177 *'loved the man'*: To George Blood [c. 1791].
178 *'Like Milton . . .'*: MW to Roscoe, *MWL*, 206; *MWletters*, 194.
178 *'fugitive' and 'intangible'*: Fuseli, *Aphorisms*.
178 *'I hate . . .'*: Knowles, *Life of Fuseli*, i, 363, quotes this snippet from MW's letters to HF. Knowles put it about that MW's letters were too ardent for their own good. This one seems decidedly unardent.
178 *not to be trusted*: MW to JJ [c. spring 1790], *MWL*, 189; *MWletters*, 170.
178 *Sophia as model*: Knowles, *Life of Fuseli*, presents a more respectable image: Sophia, he says, was visiting an aunt in London when she met HF. In fact, HF often drew and painted her as something of a dressy courtesan – probably a fantasy.
178 *needy of paternal protection*: Did MW's fixation on HF have to do with her father? Looking at HF's *Nightmare*, as MW did every week at JJ's dinners, the viewer is put in the position of voyeur of a woman in danger, with no protector. This repeats MW's position as a child watching her father abuse her mother. Could MW have been drawn to HF's awareness of a woman's vulnerability? Judith Herman, *Trauma*, 111, has done a convincing study of the after-effects of domestic violence (aligning it with military trauma). For all that, we must take care not to impose on this distinctive woman a pattern of neurosis in place of the caricature of lust.
179 *snips*: *MWletters*, 205, a speculative assemblage of a few separate and questionable snippets selected to fit the prevailing slander and quoted out of context by Knowles, *Life of Fuseli*, i, 162. Richard Holmes, in notes to *Memoirs*, 301, calls MW's involvement with HF 'something of a puzzle' and suggests, I think rightly, the innocence of her proposal as another experiment in living. JJ recalls her 'love' in a note to HF on the day of MW's death (see ch. 15 below), but I suspect that JJ's view was coloured by what HF reported to him. Kegan Paul, in his prefatory memoir,

Letters to Imlay, rebuts slander. 'The slander stuck sufficiently to make even Godwin surmise that had Fuseli been free, Mary might have been in love with him. But in fact Godwin knew extremely little of his wife's earlier life.' Kegan Paul finds the strongest indication against Knowles in the fact that MW remained friends and corresponded with Mrs Fuseli for the rest of her life. No one questions the plan of the young stage-struck Hannah More to live as a permanent guest with Garrick and his wife.

179 *acquired elegant furniture and better clothes*: Knowles, *Life of Fuseli*, i, 166.

179 *apologised to HF*: Ibid., i, 168.

179 *winterly smile; 'fool's cap'*: *MWL*, 221; *MWletters*, 205–6.

180 *'the temerity . . .'*: Knowles, *Life of Fuseli*, i, 167–8. More myth of MW, 162–3.

180 *MW–HF letters have vanished*: Knowles was Fuseli's executor, and MW's letters to HF passed into his hands. Fearing unkind use, Roscoe requested them but Knowles refused declaring them 'chiefly but not entirely amatory . . . for the sake of all parties [they] had better be consigned to oblivion'. All the same, he quoted from them (against Godwin's plea) in *Life of Fuseli*. In 1870 his heir, a nephew called the Revd E. H. Knowles, announced that the letters were in his possession. In 1884 they were bought by MW's grandson Sir Percy Florence Shelley, who refused Elizabeth Robins Pennell permission to use them in her biography of MW in 1885. It's thought that the Shelleys destroyed them. See Richard Garnett, *Letters about Shelley* (London: Hodder, 1917). Transmission and the slanderous consequences are discussed further in ch. 15.

180 *ran away to Paris*: An anonymous 'friend' who published a supposed 'Defence of the Character and Conduct of the late Mary Wollstonecraft Godwin' in 1803 propagates this gossip: she made 'a sacrifice' of her private desires and 'prudently resolved to retire to another country, far remote from the object who had unintentionally excited the tender passion in her breast'. KP, i, 207, who had seen MW's harmless letters to HF before they vanished, tried in vain to dispel this myth. He said that the story elaborated by Knowles 'is supposed to be confirmed by extracts from her letters which are given. But . . . Mr Knowles is so extremely inaccurate in regard to all else that he says of her, that his testimony may be wholly set aside.'

180 *'I intend . . .'*: To Roscoe (12 Nov. [17]92), *MWL*, 218; *MWletters*, 206–7. Both editors of MW's letters are amongst those who misread 'desire' to conform with the myth.

180 *women's rights in France*: Excellent detail in Tomalin, *Mary Wollstonecraft*, ch. 13.

181 *'a just opinion . . .'*: MW to EW (Mar. 1794).

181 *'neck or nothing'*: *MWL*, 218; *MWletters*, 207. According to a letter from HF to Roscoe, she set out on 8 Dec.

9 INTO 'THE TERROR'

182 Uncited quotations are from MW's letters to EW, JJ, Roscoe, GI and RB between Dec. 1792 and Sept. 1793 in *MWL*, 225–35, and *MWletters*, 214–29. The letters are poorly dated; the order uncertain. I follow the order in *MWL*, which allows for a more flexible interpretation of her conduct.

182 *rue Meslée*: Now the rue Meslay in a drab, lifeless immigrant area.

183 *'edged tools'*: She took the phrase from a poem by one of the Connecticut Wits, David Humphreys, a friend of JB: 'The Monkey Who Shaved Himself and His Friends'. Another source is Dryden's 'Men are but children of another growth', *All for Love*, IV, i, 43.

183 *MW witnessed*: When Richard Holmes looked at the narrow rue Meslay (as it's now spelt), he was at first puzzled how MW could have watched the procession; then he realised (*Footsteps*, 99) that a back window would have given a grandstand view.

183 *renewal*: The trial had begun on 11 Dec. 1793, when MW would have been en route from Calais to Paris.

183 *the Temple*: A medieval keep.

183 *British reaction to the King's guillotining*: RB to JB (1 Feb. 1793). Houghton.

184 *'place of fear . . .'*: Wordsworth, *The Prelude*, Book X.

184 *Mirabeau*: Honoré-Gabriel Riqueti, comte de Mirabeau (1749–91).

184 *JB's political impressions*: Diary (3 Oct. 1788), Houghton, bMS Am 1448 (10).

184 *women's march on Versailles*: According to Schama, *Citizens*, 633, it was led by Stanislaus Maillard who liked to swagger around as the captain of a paramilitary troop of strong-armed men at the service of the most militant sans-culottes. He was commissioned to undertake summary 'trials' during the September Massacre in 1792.

185 *massacre as turning-point for English opinion*: Analogous to English revulsion over the Massacre of Saint Bartholomew in the sixteenth century, and again in the seventeenth century with the revocation of the Edict of Nantes by Louis XIV when Protestants were forced to flee France.

185 *no. 7 passage des Petits Pères*: Tiny, leading into the Place des Petits-Pères dominated by the Basilique de Notre Dame des Victoires, at the convergence of several streets. The passage looked too small to admit of a number 7, until I realised that it must have once extended across a present road into a development that took place in 1820, the Galerie Vivienne, today an elegant shopping complex.

185 *French citizenship for JB*: Procès-Verbaux de la Convention (7 Nov. 1792, year one of the republic). (Archives Nationales, Paris.) JB had offered a critique of the 1791 constitution: *'Un membre fait hommage à l'Assemblée, au nom de Joel Barlow, Citoyen Anglais, d'un ouvrage sur les vices de la Constitution française de 1791, & sur les bases à donner à la Constitution nouvelle.'*

186 *MW's February article*: The first of her intended series for JJ was this 'Letter on the Present Character of the French Nation'. If she sent this to JJ, he didn't publish it, but since her response to France was unsure or ambivalent at this moment, it's likely she held it in reserve. Eventually published in Wollstonecraft, *Posthumous Works*.

186 *'a plan of education'*: Nothing of her contribution in the records of the Committee.

186 *Condorcet's support for women*: In *Lettres d'un bourgeois de Newhaven* (1787) and *Sur l'admission des femmes au droit de Cité* (1790).

186 *HMW . . . support for the Revolution*: *Letters from France*; Todd, *Wollstonecraft*, 212.

186 *'Authorship'*: MW to EW (24 Dec. [1792]), *MWL*, 226; *MWletters*, 215.

187 *HMW's salon*: In the rue Helvétius (now the rue Sainte-Anne).

187 *Brissot de Warville*: Author of *New Travels in America* (1788, trans. JB, 1792).

187 *MW and Mme Roland*: There is conflicting evidence as to whether MW met Mme Roland. I. B. Johnson (who saw MW in Paris from Apr. to Sept. 1793) says that she

did know her. Hays, in her 'Memoirs of Mary Wollstonecraft', says that MW regretted not meeting Mme Roland.

187 *MW's impressions of Frenchwomen*: *FR*, book iii. *MWCW*, vi, 148.

188 *I.B. Johnson's recollections of MW in Paris*: WG, composing *Memoirs*, approached I.B. Johnson. His reply is in Abinger. An extract is quoted in *SC*, i, 125–6.

188 *MW dining with Paine and militant female*: Ibid. No. 63 Faubourg Saint-Denis is now 142, near the Gare de l'Est.

189 *Théroigne de Méricourt*: Hufton, *History of Women*, 478–9; Schama, *Citizens*, 462–4, 530, 611, 873–4; Linda Kelly, *Women*, 49.

189 *'the lowest refuse . . .'*: *FR*, *MWCW*, vi, 196–7 .

190 *'– here you cannot return'*: RB to JB (1 Jan. 1793). Houghton.

190 *RB's wish to return to America*: RB to JB (7 Jan. 1793) Houghton.

190 *JB's letters to RB*: Houghton, bMS Am 1448 (64–330).

191 *JB in* Savoy: A new department of France on its border with Italy. While JB had been away, the Convention had ratified a proposal of 7 Nov. to grant him French citizenship and admit him to its rights (*'on a proposé . . . d'inscrire Joel Barlow sur la liste des Étrangers à qui on doit accorder le titre & les droits de Citoyen Français'*), a reward for his advice on transforming a constitutional monarchy into a republic on the American model (in his *Letters to the National Convention of France*, 1792). JB's letter to RB from Savoy is dated 13 Feb.

191 *a tall, handsome American*: EW to BW, reporting James Wollstonecraft's meeting GI in London.

191 *first mention of GI to RB*: Houghton. Cited by Flexner, *Mary Wollstonecraft*, 181.

192 *GI's family background*: I've been unable to trace the exact relation of the most successful member of the family, James Henderson Imlay, born in Imlaystown in 1764 – Gilbert's junior by ten years – who taught classics at Princeton and was admitted to the bar in 1791; Speaker in the New York Assembly, 1793–6; then elected to Congress.

192 *St Thomas*: Norwegian historian Gunnar Molden tells me that this island was Danish-Norwegian, and that many merchants in Copenhagen and in Arendal, Norway, had connections there. Could be relevant to GI's later connections in Scandinavia.

192 *Imlay mansion*: It remains in Allentown, next to the library, now with downstairs rooms converted into shops. There was a public sale of its contents on 2–3 June 1936. I am grateful to Joan Ruddiman, member of the Historical Association of Allentown, for her ingenuity in finding details of the original contents of the house as set out in the announcement of this sale.

193 *GI's war service, and that of other Imlays:* Imlay, *Index*, ii, 1416; Imlay, Records, Van Kirk Collection.

193 *'omitted'*: Casualty book of Forman's Regiment, MS 3777, 4.

193 *Washington's spies*: Information is hard to find. There are disappointingly few details in Jeffreys-Jones, *Cloak and Dollar*, 17, and in Christopher Andrew, *For the President's Eyes Only: Secret Intelligence and the American Presidency from Washington to Bush* (London: HarperCollins, 1995, repr. 1996), 8. Not much luck, either, with Washington's papers or Internet investigation through the Library of Congress. What's always repeated is the Culper spy ring, set up in New York in July–August 1778, but one would like to know more of Washington's other set-ups.

193 *GI as fashionable beau*: *Pennsylvania Magazine of History and Biography*, xxiv, 417–18, cited by Rusk, 'Gilbert Imlay'.

194 *dealings of Boone and GI*: Faragher, *Daniel Boone*, 246. Boone, trustingly, had endorsed the deal in August 1785: 'I do hereby assign my right and title of the within survey to Gilbert Imlay and his heirs and assigns.' When GI couldn't pay his debt he assigned the tract to James Wilkinson, leaving Boone, it seems, the loser. Wilkinson used Boone's fame to advertise the land as 'located and surveyed by Col. Daniel Boone'. Faragher thinks Imlay gained 'an undisclosed sum' from Wilkinson for the Boone land.

195 *GI's note to Lee*: Bullet is described by Wilkinson as a man of fortune but very changeable. Beinecke.

196 *William* Cooper: *A Guide in the Wilderness* (1810). New York Historical Association, Cooperstown.

197 *GI's speculations in Kentucky*: See the Internet site accompanying this book.

197 *'massacres'*; *'safety'*: Imlay, *Topographical Description*, 16, 19.

198 *code names of Wilkinson's agents*: Wilkinson, spy-letters to Miró. Col. Bullet is also listed amongst potential backers of separatism who are due for Spanish bribes ('pensions'). The possibility that GI may have been a secret agent was raised by St Clair, *Godwins and Shelleys*, 159.

198 *GI's deed of 1789*: May Papers, Filson Historical Society. This is one of three grants of deeds for Kentucky lands by GI: on 1 Aug. 1786 (2148 acres), on 27 Dec. 1789, and on 19 Nov. 1810 (3400 acres), listed in *Old Kentucky Entries and Deeds*. Only the 1786 entry gives a place of residence: Virginia.

198 *Wilkinson's message to 'Gilberto'*: From Louisville, Rapids of Ohio (19 Feb. 1789).

198 *'constitute a Barrier'*: Ibid. (17 Sept. 1789).

199 *Colonel Conelly*: Mentioned in Wilkinson, spy-letters to Miró.

199 *Wilkinson's advance on Spanish Florida*: Hammond to Grenville, PRO.

199 *GI emerged in London*: Wilkinson too switches careers, appointed by Washington in Nov. 1791 to his army position. He led the frontier wars of the 1790s.

199 A Topographical Description: The London edition of 1792 was enlarged in 1795. Published in two volumes in New York in 1793, with a supplement by John Filson, a portrait of Boone. Another and further enlarged edition appeared in London in 1797. Seelye, *Beautiful Machine*, 89, notes the 'anthology effect', especially in the 3rd English edn, which adds some fifteen documents. In this way, Seelye says, GI gave ephemeral publications additional circulation.

200 *GI's secret plans for the Louisiana scheme*: 'Observations du Cap Imlay', trans. in *Annual Report of the American Historical Association*, i (1896), 953–4; and the longer 'Mémoire sur la Louisiane', trans. in *American Historical Review* (Apr. 1898). Includes JB's separate proposal. It is clear that GI intended to take an active part in this expedition, and JB's plan to travel to America with expenses paid suggests the same. See also Frederick Jackson Turner, 'Policy of the French'. Was GI acting in concert with Wilkinson? See Emerson, 'Notes on Imlay'. Faragher too, *Daniel Boone*, 247, suggests that Wilkinson 'was connected with Imlay in a number of intrigues'.

200 *Crèvecoeur*: Michel-Guillaume Jean de (1735–1813). In his youth had explored the Ohio River and Scioto frontier, and celebrated the pioneer farm in his *Letters from an*

American Farmer, calling himself Hector St John de Crèvecoeur. This early American classic was published in London in 1782, the year that its author's Loyalist alignment in the war had forced him to flee his adopted country. He and Otto returned to France. MW later alludes to GI's acquaintance with Crèvecoeur in her *Travels*.

200 *Genêt's instructions* : Archives des Affaires Étrangères, États-Unis, xxxix, f. 144. Archives Nationales, Paris.

200 *Cooper's defence of MW*: *A Reply to Mr Burke's Invective* (London: 1972), 98–9, cited in Miriam Brody Kramnick, Introduction to *RW*. Cooper's dates (1759–1840) make him an exact contemporary of MW. Connected as he was with JJ's circle, he almost certainly knew her. In 1794 he published, with JJ, *Some Information Respecting America*. Later, migrated to America; in 1816 was appointed Professor of Mineralogy and Chemistry at the University of Pennsylvania; eventually became president of the University of South Carolina. President Adams referred to him as 'a learned, ingenious, scientific, and talented madcap'.

200 *JB chosen as thinker*: Archives des Affaires Étrangères, Espagne, vol. 635, doc. 313.

202 *'Miss W. is massacred . . .'*: BW to EW (24 Apr. 1793). Abinger.

202 *GI's belief in America*: Imlay, *Topographical Description*, 20, 28, 29, 168, 179. GI makes one of the first contrasts of American simplicity with corruption, suggests Seelye, *Beautiful Machine*, 91.

203 *GI averse to standing armies*: *Topographical Description*, 16.

203 *GI on slavery*: Ibid., letter viii. Imlay, *Emigrants*, 61.

204 The Emigrants: There are repeated suggestions that MW was the author of this book (Robert R. Hare's edition of 1963; Faragher, *Daniel Boone*; and John Cole, 'Imlay's Ghost: Wollstonecraft's authorship of *The Emigrants*', in *Eighteenth-Century Women: Studies in Their Lives, Work, and Culture*, ed. Linda V. Troost (New York: AMS Press, 2001), 263–98). Janet Todd considers the argument in her notes to *MWletters*, 222–3, yet though she is not finally persuaded, she does re-order letters to allow MW to have met GI at an earlier date, and influenced him. This takes away his own initiative in writing polemical passages on the victimisation of women and the possibly more complex basis of his appeal for MW. The strongest reason for MW's not being the writer of GI's novel is that if she were, it would not be so tedious.

204 *Jefferson's response to Louisiana schemes*: Jefferson–Genêt letters are enclosed in Jefferson's instructions to Morris (especially letter of 16 Aug. 1793). Morris, Papers.

205 *asked for his recall*: M. Fauchet had replaced Genêt by March 1794.

205 *rhyme quoted by Adams*: Abigail and John Adams, MS correspondence (6 Jan. 1794): microfilm reel 377.

205 *Hichborn's proposal*: Hichborn intended to explore possibilities with a ship and cargo broker called Henry Bromfield of 1 Size Lane, London – a dangerous proposal given the Traitorous Correspondence Bill (1793) which named anyone supplying France with goods a traitor and punishable by death. Hichborn travelled incessantly between Paris and London, carrying diplomatic letters between Pinckney, the American Minister in London, and Morris, his counterpart in Paris.

205 *British navy's retaliation*: Pinckney to Grenville (1793). PRO: F05/3 and 7.

205 *MW's four letters to RB*: Lost, but mentioned in JB to RB (7 May 1793). The boat service between Dover and Calais had been suspended. It is likely that these letters, like others MW wrote at the time, did not reach their destination. She was aware of the

problem, and sometimes followed up one letter with another in quick succession, in the hope one of the two would make it. When possible, letters were carried by Americans. Another method was to send letters via EW in Ireland.

207 *'great book'*: MW to BW (13 June [1793]), *MWL*, 231; *MWletters*, 226.

207 *'to trace the hidden springs . . .'*: Book 1, ch. 4. Relevant ideas of history may be found in historians of the 1770s and 1780s which MW excerpted for *The Female Reader*: Robertson and Jardine, whom she had reviewed. Rendall, 'History and Revolution'.

207 *'How silent is now* Versailles!': *MWCW*, vi, 84–5, 328.

207 *the history of the present*: Taken from the title of essays by Timothy Garton Ash.

207 *JB and the* Hannah: Correspondence between Captain Parrot and JB, showing JB's involvement from 15 May 1793. Beinecke: MS Vault Shelves: Pequot (Barlow) M992.

208 *the* Cumberland: Letter from W. Harrison in London, an agent of Bromfield, to JB in New York (18 Sept. 1805), after Leavenworth in Paris asked Bromfield's firm to seek compensation for losses in 1793. Beinecke: MS Vault Shelves: Pequot (Barlow) M982.

208 *a hundred foreign ships*: Stephen Cathalan, US consul at Marseilles, to Morris (22 Feb. and 15 Apr. 1794). Morris, Papers. The embargo remained until Apr. 1794.

208 *£20 from JJ*: MW cashed the bill through Christie. On 2 May, she had received a draft of £30 from JJ. This could have been money mentioned in her sisters' correspondence, which they planned to send her with a view to an exchange favourable to the pound. BW contributed £10 to the sum in the hope that MW would find her a post in Paris or Geneva, which, as we saw, MW tried to do – it was BW who decided against Geneva. (*SC*, i, 121–3, 127–30.) One common aspersion of MW is that she helped herself to her poor sisters' money (it has even been suggested that her sisters funded her month by month), but an unpublished letter from BW to EW (29 Feb. 1793) seems to make it clear that the money was for Bess's profit: MW, she says, 'wishes to take advantage of the exchange that would be greatly in *my favor* (meaning me) at present' (Abinger: Dep b. 210/7). Ch. 8 above gives instances of MW's taking responsibility for the money troubles of all members of her family.

208 *'almost impossible'*: *AR*, iv (1792), *MWCW*, vii, 424–30. Cited Jacobs, *Her Own Woman*, 111.

208 *MW on Paris as a prison seen from the barriers*: *FR*, 215–16.

209 *'the french had undertaken . . .'*: Ibid., 219–20.

210 *'a Lady of . . . Understanding'*: Adams, marginalia, *FR*. Quoted in Durant's Supplement, 267.

210 *thousands . . . lost their heads*: Different historians give different estimates, ranging from 2650 (Steel, *Vive la Révolution*) to Schama, *Citizens*, 791–2, who challenges a conservative estimate of loss of life during the Terror, and points to the near quarter million massacred in the Vendée.

210 *Revolution . . . active in her own blood*: Woolf, 'Mary Wollstonecraft', 195.

211 *MW's relationship with GI*: Tomalin, 'Fallen Woman', comments on this kind of semi-formal relationship as the most difficult to conduct.

211 *'We are soon to meet . . .'*: *MWL*, 233–4, follows WG in giving Aug. 1793 as a conjectural date. MW was at Neuilly when she wrote this letter (she talks of the barrier).

211 *'Why cannot we meet . . .'*: MW to RB (Friday afternoon [summer 1793]). Internal evidence shows she was still at Neuilly: 'I do not wish to spend a whole day in Paris.'

211 *Chinese Baths*: Morton, *Americans in Paris*, 103–4.

211 *MW's foot slipped on blood*: This story circulated at the time, and found its way into Amelia Opie's novel, *A Wife's Duty*. According to Holmes, *Footsteps*, 112–13, it was seen as a symbol of MW's courage during the Terror.

212 *the Maison de Bretagne*: Now an antique shop. The York Hotel is still at no. 56, towards the rue de l'Université and the rue du Bac.

212 *'barbarous' marital laws*: Imlay, *Emigrants*, 46.

213 *'I have so many books . . .'*: I have kept to the sequence of letters set out by WG in 1st edn of *Letters to Imlay* in Wollstonecraft, *Posthumous Works*, and followed by Wardle in *MWL*. Todd, *MWletters*, shifts this letter to an earlier, pre-Neuilly date, April–May, even though the letter refers to returning to Saint-Germain and needing a carriage for all the books, which would fit the August move. I'm not convinced that there is sufficient evidence to justify the shift of this letter. Its effect is interpretative: its intimacy, its sense of MW's life bound up domestically with that of GI, bolsters a long-held idea of MW's quick plunge into sexual intimacy. This is an idea I question, especially as it leads to accusations that MW was inconsistent, throwing off the chastity expressed in *RW*. The myth of MW as reckless wanton (discussed below, ch. 15) was reinforced by a parallel history of misreading her attachment to HF.

213 *soon was pregnant*: It is assumed too literally that sex took place at the barrier since the child, Fanny, was later called 'the barrier child'. MW would have used the phrase metaphorically – it was where desire stirred. Conceivably, it was at one of their meetings at the barrier that she and G I first spoke of having children.

213 'Tant pis pour vous . . .': Recorded in Crabb Robinson, *Diary* (2 Sept. 1817), after he visited HMW, who repeated MW's friend von Schlabrendorf's report of the exchange. Crabb Robinson, *Books and their Writers*, i, 209.

10 RISKS IN LOVE

214 Uncited quotations from MW's letters to GI, RB and EW from Nov. 1793 to Sept. 1794 are in *MWL*, 237–62; *MWletters*, 232–63.

214 *The prisoners*: The four Williamses were arrested on 12 Oct. The Frenchman who intervened was Athénèse Coquerel, who later married Cecilia Williams.

214 *GI's news and MW's faint*: MW described this scene to Amelia Alderson, recorded in Brightwell, *Memorials*, 49. I change 'suppose' to the American 'guess' in order to rectify the transmission through an English vocabulary.

215 *closed women's clubs*: Proposed in the Convention of 30 Oct. Hufton, *History of Women*, 479; Schama, *Citizens*, 802. Their precise dates do not agree, but this happened between the end of Oct. and early Nov. 1793.

215 *executed Mme Roland*: JJ subsequently published her memoirs, which MW may have helped to edit in 1795. The work on Mme Roland's *Appeal to Impartial Posterity* (1795) was suggested by Tomalin, *Mary Wollstonecraft*, 179, followed by Todd, *Wollstonecraft*, 482, who notes similarities in opinion and tone with MW's *Travels*.

215 *rising cost of soap*: Schama, *Citizens*, 708.

215 *GI to Le Havre*: JB's friend Nathaniel Cutting was appointed US consul there in Mar. 1793. Morris, Papers.

215 *Wheatcroft's safety*: He travelled on a passport issued by the Committee of Public

Safety in 1795. His destination was Scandinavia and the purpose was to give evidence in a case to do with one of GI's ventures, the silver ship. See below, ch. 12.

216 *alum*: A whitish transparent mineral salt used for printing and tanning, and an ingredient in potash, soda, ammonia and iron.

216 *'money-getting face'*; *'honest countenance'*: c. Dec 1793.

217 *GI on commerce*: Imlay, *Topographical Description*, 75.

217 *Wentworth taking prizes*: This was the most accepted way for those without a fortune to make one.

218 *'Speculation'*: Jane Austen, *Mansfield Park*, ch. 25. Appropriately, it is the player (in every sense) Henry Crawford who inducts the uninitiated – the incorruptible Fanny Price and stupid Lady Bertram – into the game. Incidentally, WG attended a play called *Speculation* on 7 Nov. 1795.

218 *'We know not . . .'*: Appropriated from Hardyment's fascinating chapter (*Perfect Parents*, 29) on the medical and domestic applications of 'Nature and Reason 1750–1820'.

218 *'The way to my senses . . .'*: Soon after they began living together, c. Sept. 1793.

219 *truth and facelessness*: Truth versus face exerts a similar moral power in the facelessness of Fanny Price in *Mansfield Park*, but only if the reader can resist the amoral allure of the siren Miss Crawford.

219 *GI on the soul's sympathy*: See above, ch. 9. He promised a society in which 'sympathy was regarded as the essence of the human soul'.

220 *'I do not want . . .'*: 2 Jan. 1794.

220 *glassware*: A letter to GI and Leavenworth in Paris (22 Mar. 1795) from William Jackson (1759–1828), once a major in the Continental army, now Secretary to the Federal Convention and President Washington, asks GI why the glassware he has ordered and paid for in advance has not arrived. He argues that there must be numerous vessels sailing from France to Philadelphia. (*Collector*, lxiii (Feb. 1950), item D358 (dealer's catalog: 4207).)

220 *public sales*: Another took place at Marly between 6 Oct. and 25 Nov.; another at Saint-Cloud beginning on 29 Mar. 1794; and yet another at Fontainebleau beginning in June of that year, with similar sales in Paris.

221 *Swan unprincipled*: Monroe, then American Minister in France, to James Madison (30 June 1795). Swan, born in Scotland, fought at Bunker Hill, and then married a wealthy Bostonian who kept him in some comfort in prison. He refused to pay the debt against his release.

221 *Swan and JB*: Swan advised Washington on 21 Dec. 1793 that the present American Minister, Gouverneur Morris, was unpopular with the French. Swan proposed JB as the person to replace him. One of Swan's agendas had to do with the failure of Morris to effect an end to the embargo of American ships at Bordeaux (see above, ch. 9). Swan wrote similarly to General Henry Knox, US Secretary for War (21 Dec. 1793): 'Should there be virtue enough left, to respect merit & talents in the election of diplomatick men, altho' they are not rich, Mr Barlow . . . possesses every quality, that could render an agent usefull to the United States & this Republique, and who in the highest degree has the esteem of all[,] but in daring to give you this hint, I can assure you that it did not originate in me, nor come from him; for I believe the last thing he thinks of is that.' (Knox, Papers, reel 35/2.)

221 *GI and Copenhagen*: Another contact was Christer Skaarup Blacks Enke & Co.

221 *A member of the Paris conspiracy*: His name was Lyonnet.

221 *Genêt denounced*: On 11 Oct. 1793 the Committee of Public Safety had recalled Genêt (who had the sense to save his neck by remaining in America).

221 *renewed plan, November 1793*: Archives des Affaires Étrangères, Espagne, vol. 636, f. 391.

222 *the generals*: Another general, Clark, was co-opted for the abortive Genêt Affair.

222 *'expatiating'*: Imlay, *Emigrants*, 69.

222 *'the hard-hearted savage romans'*: *FR*, 160.

222 *American Minister as spy*: A letter from JB to Abraham Baldwin in Congress (4 Mar. 1798) offers his opinion that the highly conservative Morris 'acted as a secret agent & spy' for the British and Austrian Cabinets after their ambassadors left Paris. (JB's Letterbook (1797–1803), Houghton: bMS Am 1448 (4), 86.) After Morris left office, the French intercepted a 1795 letter from Washington to Morris as secret agent to the Cabinet in London.

223 *Paine's arrest and petition*: Paine, Dossier. The agents of the Terror included Doilé, another Commissioner called Gillet, and a policeman; an admirer, Achille Audibert, who had invited Paine to represent Calais in the French National Convention; and citizens Jean-Baptiste Martin and Lamy from the Committee of Public Safety.

224 *Paine languished in prison*: See his plea to the American Minister. Morris, Papers.

224 *Americans' petitions to Morris*: On 18 Nov. 1793 Angelica Church in London appealed to Morris on behalf of her imprisoned friend, Miss Catharine Herring of Albany, who had gone to France to learn the language in order to improve her credentials as governess. Morris was repeatedly asked to verify such prisoners' claims to having been born in America.

224 *von Schlabrendorf's escape of the guillotine*: Crabb Robinson, Diary, *Books and their Writers*, i, 300.

224 *von Schlabrendorf recalled MW*: Notes in his copy of *Memoirs*. Relayed by his friend Carl Gustav Jochmann and trans. in Durant's edition of *Memoirs*, xxvii, 251–2.

224 *'every brute'*; *'haunted'*: BW to EW (5 Nov. and 4 Dec.1793). Abinger.

225 *'I am grieved . . .'*: *FR*, *MWCW*, vi, 444.

225 *'Alas!' . . .* : *FR*, book I, ch. 3.

226 *Havre-Marat*: JB's description of Le Havre a few years earlier in his diary. He had a sharp eye. Houghton: bMS Am 1448(9).

226 *lodgings*: Address discovered by Tomalin, *Mary Wollstonecraft*, ch. 14.

227 *'I could not sleep'*: GI was in Paris briefly in March.

227 *'View' vs institutional history*: Gary Kelly, *Revolutionary Feminism*, 154–5; Moore, *Mary Wollstonecraft*, 52–3.

227 *Jane Austen's attack on institutional history*: In the innocent voice of Catherine Morland in *Northanger Abbey* (completed in 1799; published posthumously in 1817). Virginia Woolf, a century on, also questions history's focus on war, and caricatures the staginess of kings with golden teapots on their heads (in her feminist treatise *A Room of One's Own* (1929)). See also her early fable, unpublished in her lifetime, 'The Journal of Mistress Joan Martyn' (1906), which questions a male bias in historical record.

228 *Clarissa vs lusty, conniving woman*: Hufton, *History of Women*, 445.

228 *MW on the French character*: *FR*, 213.

228 *'sober matron graces'*; *'maternal wing'*: Ibid., 115, 22. MW overlooks the violence of the

American Revolutionary War in favour of the welcome to immigrants and America's more benign political system. This is close to what Henry Adams would write in his chapter on 'American Ideals' and their transformative effect on immigrants in his *History of the Administrations of Jefferson and Madison*. The second President John Adams (Henry Adams's great-grandfather) didn't find the mother bird over the top. See Adams, marginalia, *FR*: 'I thank you Miss W.'

228 *MW on servility and the retaliation of slaves*: Ibid., 126, 234.

228 Grande Terreur: Began in June, and lasted until Robespierre's fall on 27 July 1794.

229 *Paine condemned*: On 24 July, three days before Robespierre met his end. The fall of Robespierre may not be unconnected with a French naval defeat in June (as the fall of the Girondists in the first six months of 1793 had been connected with earlier defeats).

229 *'My God . . .'*: 8 July [17]94.

229 *registration of Fanny Imlay's birth*: Discovered by Tomalin, *Mary Wollstonecraft*, ch. 14. The witnesses were Wheatcroft and his wife, Marie Michelle.

230 *MW's letter to RB about childbirth and breast-feeding*: Carried by hand, as so often in this time of war, by an unnamed Imlay contact who presumably was to intercept the Barlows. The latter are hard to place at this time, shifting between Amsterdam and Hamburg. Ruth may have returned briefly to London, but it's unlikely that Joel could have joined her, given his reputed sedition.

230 *'The suckling of a child'*: 'The Nursery', *Education*, 7. I infer that she watched Fanny before MW hired a wet-nurse for Fanny's baby, since she is unlikely to have watched anyone else. Possibly, she recalls her mother, though her mother was not a figure of tenderness.

231 *statistic on breast-fed babies in the 1780s*: Schama, *Citizens*, 145–8.

231 *Galen and wet-nursing*: Hardyment, *Perfect Parents*, 4–5, 16–17.

231 *'raven mother'*: MW told this to von Schlabrendorf on her return to Paris. Relayed by his friend, Jochmann. Repr. in Durant's Supplement, 251–2.

231 *Dr Haygarth on smallpox*: *An Enquiry How to Prevent Small-Pox* (London: Johnson, 1784). *MWletters*, 263, notes his sequel, *A Sketch of a Plan to Exterminate the Casual Small-Pox from Great Britain* (1794), but it's unlikely to have reached MW.

11 THE SILVER SHIP

232 Uncited quotations are from MW's letters to GI, EW, BW and Archibald Hamilton Rowan from Sept. 1794 to May 1795. *MWL*, 262–89; *MWletters*, 263–94 .

232 *newly discovered letter*: To Danish Prime Minister, Bernstorff (5 Sept. 1795), written in Copenhagen. Discovered by Gunnar Molden in 2003 in the Danish National Archives. I am grateful to him for sending a photocopy for verification. Transcribed in ch. 12.

232 *JB the only American in Europe . . . honorary French citizen*: I exclude Paine, a naturalised American, because the French treated him as English.

232 *JB's dealings*: Woodress, *Yankee Odyssey*, 145.

232 *export of luxury goods from France under the Terror*: Pierre Verlet, *French Royal Furniture* (New York, 1963), 56–7, and Gerald Reitlinger, *The Economics of Taste*, ii (1963), 130. Cited by Fraser, *Marie Antoinette*, 393.

233 *Elias Backman*: Came from Lovisa in Finland, which had been under Swedish dominion before it was taken by Russia. Having spent much time in France, his brother Pehr Backman was suspected of being a French spy when he settled in Gothenburg. Nyström, *Scandinavian Journey*, 29.

233 *Backman's petition to the Crown*: To the Swedish Regent, Carl. (Riksarkivet, Stockholm: Biographica.) Dated 21 Mar. 1794. Enclosed is a letter to Baron Carl Bonde (15 Mar. 1794), who is to present the petition. On 16 June, Backman took the oath as a burgess which gave him a licence to trade. Landsarkivet, Gothenburg.

233 *English cutter . . .*Rambler: Obtained secretly by a French consul in Sweden.

233 *Backman as owner*: EB was the formal owner. As Molden puts it in an email (July 2004), 'who was the real owner is of course another question'. Is that GI in the shadows?

234 *secrecy and GI's approach to EB*: Their association at this point is speculative, a matter of circumstantial evidence. Because of the secrecy, proof is difficult. Scandinavian sources for the *Rambler* are confirmed by British letters (originally in cipher) which can't as yet be cited.

234 *France's rejection of the coup*: The French were attempting to heal relations with the US. Early in Feb. 1794 it was decided that Genêt's successor as French Minister to America, Fauchet, must issue a proclamation ending the Louisiana expedition (see chs 9 and 10 above) against Spain. He was instructed to announce: 'Every Frenchman is forbid to violate the neutrality of the US.'

234 *dating the Barlows' departure for Hamburg*: American citizens had to be authenticated as genuine by their Minister if they were to obtain a pass to leave Paris. Morris, Papers, show that passports were supplied to RB and JB on 10 Mar. (Letter from Henry W. Livingstone to Gouverneur Morris (10 Mar.).) Colonel and Mrs Blackden, JB's associates, were supplied with passports at the same time.

234 *MW to RB, backing the joint enterprise*: 27 Apr. [1794], *MWL*, 253; *MWletters*, 251–2. Their friendship may have also mattered to RB at a time when a brother and sister suddenly died of fever in Connecticut.

235 *Bourbon platters*: MW notes in *FR*, 173, that in 1789 'the king sent his rich service to the mint' as a donation 'to relieve the wants of the country'. Several others made similar donations of jewels and plate. It's not known if GI's thirty-six plates were saved from the mint, or whether they came from another silver service.

235 *£3500*: GI to EB (24 Oct. 1794), reported by EB (18 Nov. 1794) in a letter transcribed by Gunnar Molden. (According to Judge Wulfsberg – who presided over the subsequent criminal case – the value was 17, 000 to 18, 000 *riksdaler*.) The sum was first cited in Nyström, *Scandinavian Journey*, a study that marks a turning-point in our knowledge of this phase of MW's career. The author initiated research on the fate of the treasure ship and the subsequent trial. He surmised that the silver was to be exchanged for grain, one of the commodities Backman exported (and some of the money was to pay for repairing or re-rigging the ship).

236 *Ellefsen . . . pointed out a ship*: Bought by GI from the Laïent brothers of Le Havre.

236 *oak*: The accounts of the repairs later carried out in the Swedish port of Strömstad mention that oak was needed. Enclosed with the ship's papers in the Riksarkivet, Stockholm.

236 *GI's disguising of the ship*: Buus, 'Promethean Journey', 228, makes clear that Scandinavian neutrality allowed resourceful types from all European countries to carry on business with France despite the British blockade and England's Traitorous Correspondence Bill of 1793. It was common for French ships to be re-registered as neutral Scandinavian ones.

236 *Coleman*: MW's spelling of the name. Sometimes spelt 'Colman'.

236 *draped the tricolour about his waist*: When Ellefsen was interrogated in Arendal, Norway, on 28 Apr. 1795, he recalled this charming detail.

236 *Algerian pirates*: They had recently captured twelve ships from the rich Hanseatic (north German) towns. Britain permitted these pirates to cruise the Atlantic to prevent France getting supplies from America and to punish the latter for recognising the French republic. (Before the US broke away, they had enjoyed Britain's protection). Britain, France and other countries paid tribute to the pirates to ensure their ships' safety.) Swan told the American Secretary of War that these pirates were expected as far as Elsinore, Denmark. (Swan to Knox (21 Dec. 1793): Knox, Papers.) In mid-July Ellefsen registered the ship with the Danish consul, Mr Pickman, in Rouen, the nearest inland town from Le Havre. Nyström, *Scandinavian Journey*, says that Ellefsen told the Danish consulate the ship was bound for Copenhagen. Elsewhere it is said be Elsinore. But since Ellefsen actually sailed to Norway, this was probably a sop to the Danes in order to be accredited.

236 *naming of the ship*: There is no basis for the idea that GI had the ship named after MW and the Frenchwoman Marguerite who looked after Fanny. The latter was employed only later when MW went to Paris. The spelling is unstable: 'Margrethe' appears as 'Margarethe' or 'Margareta' . Possibly, Peder Ellefsen was also recalling his baby sister Margrethe, who had died aged eight in 1790.

236 *Ellefsen and the mate . . . loaded the silver*: Details from Judge Wulfsberg's report (18 Aug. 1795) delivered to Danish Prime Minister Bernstorff and copied to the *Stiftamtmann* in Oslo, outlining Ellefsen's actions as part of a case against him . The report emphasises that the silver was loaded 'without the knowledge of the rest of the crew'. Report discovered by Gunnar Molden in the Oslo Regional Archives.

236 *receipt and other vital papers*: GI's instructions to Ellefsen, Kristiansand Town Magistrate, *Notarialprotokoll* 8 (1794–1804). The instructions (dated 13 Aug. 1794) don't specify Norway. The orders are to procure Danish papers for the ship with 'the utmost dispatch & economy'; EB will reimburse Ellefsen for expenses. This might, in fact, imply that Ellefsen was to go to Norway for properly legal papers, but he did already have papers from the Danish consulate before leaving France. Ellefsen's agreement is witnessed by 'Wheatcroft jun'. If produced in court this letter would prove that Ellefsen was Imlay's subordinate and not the owner of the ship. Later, Wheatcroft would be called on to testify in court.

236 *letter to EB*: The US Vice-Consul for Le Havre, Francis Delamotte, appears to introduce GI to EB. GI is said to be a native of the US who is worthy of his confidence in a 'joint commission'. The tone of the letter, co-signed by Delamotte and Imlay, besides being deadpan, is also too deliberately vague to ring quite true. It would be unlikely that GI would be dispatching treasure to a stranger – with his ship due to sail in the next day or two. Could this letter have been a kind of safety net or blind that, if opened, would seem to prove no scheme was afoot, and the relation with Backman as yet nonexistent?

This is another brilliant discovery by Gunnar Molden, deepening the mystery of GI's dealings. (Molden does not agree with my idea that this letter was a blind.) Kristiansand State Archive: Town Magistrate: Notary Protocol 8 (1794–1804), pub. by the Chief of Police, Sorensen (Dec. 1794). Presumably trans. from the English or French of the original into Norwegian.

Delamotte was a big businessman and conceivably a player in GI's game, though as yet not enough is known. His correspondence (in French) with the American Minister (Morris, Papers) reveals that Delamotte was arrested on 16 Feb. 1795 for a reason he never spells out. He had an English wife, and may have been suspected of spying. He had had an English business partner who had returned to England two years earlier. His fluency in English and the American ships at his disposal led to suspicions. His English correspondence was scrutinised. His nights were spent in prison; during the day he was allowed to work from 8 a.m. to 6 p.m. He remained in prison for four months – that is, till June – fearful for the safety of his wife. Married to a Frenchman, she was automatically protected, but not if that Frenchman was deemed a traitor.

237 *date of ship's departure and nine days at sea*: A later interrogation of Ellefsen at Arendal, Norway, on 28 April 1795, refers to his statement that a letter from GI on 13 Aug. had instructed him to sail to Gothenburg. Ellefsen testifies that the voyage took nine or ten days, and by 25 Aug. he was signing the ship over to his stepfather in Norway. This accords with new evidence in the letter from EB on 18 Nov. 1794 in which he says that the ship sailed from Le Havre on 14 or 15 Aug. It was therefore at sea c. 14–24 Aug.

237 *guillotining of leaders of the Terror*: Eyewitness report in Rowan's *Autobiography*, 238.

238 *dates of JB's visit to Paris*: Letters dated from Paris, 17 and 20 Aug. Houghton.

238 *MW's 'indignation' with the knave*: 19 and 20 Aug. *MWL*, 260; *MWletters*, 258–9. The dash in place of the name was inserted either by MW herself, or by WG in 1798 when he edited these letters for publication, destroying the originals.

238 *'fully acquainted . . .'*: MW's newly discovered letter to Bernstorff.

239 *Paris after the Terror*: Linda Kelly, *Women of the French Revolution*, 153.

239 *Paris fashions in 1795*: Laver, *Taste and Fashion*, 18; Murray, *High Society*, 245, 247, 253–7.

239 *women's protest suppressed*: In May 1795 radical women laid siege to the Convention so as to break the new order. The military was called in, and the Convention banned all unaccompanied women from its meetings and forbade more than five females to walk together in the street. (Jacobs, *Her Own Woman*, 177.) Women repeatedly attempt to enter politically into the Revolution, only to be controlled by the forces in power, whether Robespierre or his adversary Tallien .

240 *the Williams women had fled*: HMW joined her lover John Hurford Stone in Switzerland, and from then they lived as a couple.

240 *Paine's letter to the Convention*: Paine, Dossier. In English with French translation.

240 *James Monroe*: Replacing Morris as Minister, he had arrived at Le Havre at the end of July 1794, when the Terror ended. Conceivably, GI and MW, plus GI's associate Delamotte the Vice-Consul, would have participated in a welcome for Monroe.

241 *von Schlabrendorf on MW in Paris*: His recollections were in the form of notes in German in his copy of the *Memoirs*; relayed by his friend Carl Gustav Jochmann and reprinted in Durant's Supplement to *Memoirs*, xxvii, 251–2. Other recollections are

recorded by Henry Crabb Robinson in his Diary (2 Sept. 1817), after he visited HMW who recalled Schlabrendorf's words (Crabb Robinson, *Books and their Writers*, i, 209).

241 *'permanent views'*; *'our being together'*: Though GI's letters have not survived, we can hear his voice when MW repeats his words 9–10 Feb. *MWL*, 278; *MWletters*, 281–2.

241 *honesty*: WG casts doubt on GI's honesty in the caveats of his note to MW's letter of 30 Dec.: 'the person to whom the letters are addressed, was about this time in Ramsgate, on his return, as he professed, to Paris, when he was recalled, as it should seem, to London, by the further pressure of business . . .'. *Letters to Imlay*, xxxi.

242 *Could GI have been a secret agent . . .*: If GI was known to be in the pay of Britain, he may have been in danger with the new regime in France – a reason for not visiting MW in Paris, and for sending a servant rather than coming himself when she agreed to join him in London.

242 *money from JJ*: It has been suggested that this money came from her sisters, who did intend sending drafts through JJ with a view to an exchange rate in favour of the English pound. The plan was for MW to keep the money against a time when it would be possible for Bess to join her.

243 *disillusion*: Jacobs, 183, 'puts this neatly: 'Imlay's eagerness to pay Mary's bills merely underscored his emotional negligence.'

244 *'illiterate'*: 20 Aug. 1793, Adams, MS correspondence.

245 *'Spy Nozy'*: Coleridge, *Biographia Literaria* (1817), ch. 10.

245 *Burke on MW*: Burke to Mrs John Crewe (Aug. 1795), *Correspondence of Edmund Burke*, viii, ed.Thomas W. Copeland (University of Chicago Press, 1969), 304.

245 Letters for Literary Ladies: Cited in Jacobs, 191.

245 *GI's maxim*: Repeated by MW to GI (29 Dec. [1794]), *MWL*, 272; *MWletters*, 275.

246 *RB's complicity with marital infidelity*: RB to JB (6 and 20 Jan. 1796), Houghton: b MS Am 1448 (542, 546). RB asks JB to confide everything, for her better health: 'even if you get a sweetheart tell me'. JB was then departing for what turned out to be an almost two-year stay in Algiers. RB's health did deteriorate, and she became a semi-invalid, taking cures, often seemingly close to death. I have wondered if this was in part an effect of sexual betrayal in an exceptionally close marriage – all the closer perhaps for there being no children.

246 *JB and dealings in Hamburg*: JB's letterbook (1797–1803), 69–70, records that Dallarde & Swan in Paris (one of the firms he was working for) later questioned his returns. JB replied rather evasively (Mar. 1798) that the greater part of the goods was sold at a *low* price through the House of Boué before he left Hamburg in July 1795. Houghton: b MS Am 1448(4).

246 *JB's change of fortune*: It may not be entirely unconnected that between Mar. and June 1795 JB's associate Swan was contriving to manipulate the two-million-dollar Franco-American debt to his private advantage.

246 *JB's estate in 1795*: JB's will-letter to RB from Algiers during an outbreak of plague (1796). Draft in Beinecke: Za Barlow 13; final version in Houghton. He remarks that most of their property was 'now lying in Paris'. Sum cited by Charles Todd, *Joel Barlow*, 117.

246 *Barlow's biographers*: Charles Todd's biography in 1886 offers skimpy evidence of his sudden wealth: 'He invested largely in French Government consols, which rose

rapidly after the victories of Napoleon and yielded him a handsome fortune' (111). Todd is talking mostly about money coming in at a later period – sliding over the fact that JB was surprisingly well-off already in 1796. Woodress, *Yankee Odyssey*, assumes JB made his fortune as a shipping agent which is closer to the truth, but it can't have been built up bit by bit, given the obstructive freeze of that particular year. The shortest chapter in Woodress (ch. 6, only 10 pages), 'Commercial Interlude') is where the blank lies.

Without a fortune of sorts by the end of 1795, JB could not have hung on in Algiers for nearly two years on a small salary, while supporting Ruth in the pleasant area of the rue du Bac in Paris. Though he undertook consular duties, his position in Algiers remained shaky, according to his letterbooks in Houghton: he had to rebut an accusation from Washington that he was trying to create a diplomatic position for himself.

247 *JB on 'pecuniaries' (Feb. 1795)*: Houghton, b MS Am 1448 (67). In Feb.–Mar., Ruth was again at her London address, 18 Great Titchfield Street. I have wondered if she could have been a safe emissary to Imlay, since it was dangerous for her husband to enter England. Late in 1794 the French army, literally skating across the ice, drove the English out of Holland, and the ice-bound Dutch fleet was captured. Merchants from the Netherlands began to trade through Hamburg, whose shipping doubled in 1794–5. Neutral American agents were in demand. Swan's international debt machinations of March–June might also be borne in mind.

247 *the freeze*: Morris, *Diary*, ii, 79, 81, notes that the bulk of the shipping on the Elbe was American.

247 *ships 'beat about by the Ice'*: Robert Fitzgerald to William Wickham (19 Mar. 1795) from Kuxhaven at the mouth of the Elbe. Wickham, *Correspondence*, i, 31–3.

247 *Rowan's sixteenth-century ancestor*: James Hamilton, Viscount Claneboyne.

247 *Rowan and the United Irishmen*: Rowan's *Autobiography*; Foster, *Modern Ireland*, 276, 271f., and *SC*, i, 83.

247 *fête*: On 21 Sept. 1794.

248 *Rowan's first encounter with MW*: Rowan, 253–4. This letter was written after MW departed for England (after Apr. 1795).

248 *'croak'*: Rowan's reminiscence to MW of his 'fashion' [habits] in Paris (15 Sept. 1797), KP, i, 285–7.

248 *'all the sanctity . . .'*: Rowan, *Autobiography*, 256.

249 *MW on marriage in conversation with Rowan*: Letter to his wife (20 Mar. 1795), delivered in Ireland by EW. Ibid., 259.

249 *weaned Fanny*: Jacobs, *Her Own Woman*, 185, suggests weaning was a preparation for sex, for it was then assumed that nursing mothers should abstain.

250 *GI's reception of MW; shared house*: *Memoirs*, ch. 8.

250 *GI's excuses to BW*: His letter (Nov. 1794) cited by KP, i, 217.

250 *Mary's blow to BW*: BW returned MW's letter, and wrote to EW on 8 May: 'Would to God we were both in America with Charles. Do you think it would be possible for us to go from Dublin in an American ship to Philadelphia – this is my only HOPE.' Early in June she sent another letter to MW (it has not survived) and waited in suspense, but there was no reply, for Mary had left. She urged GI to look

after her sisters in her absence, though he did not. The anger Mary provoked at this moment would have repercussions for her daughter Fanny twenty years later.

251 *'whirl'*: MW to GI (27 May 1795).

252 *libertinism . . . liberty*: Seelye, *Beautiful Machine*, 189.

252 *'Inckay' sold the ship*: Record of sale and date in the Riksarkivet, Stockholm.

252 *the ship did not sink*: Proven by Molden, 'The Silver Ship Emerging', 139–54.

253 *GI's orders for MW*: 19 May 1795. Abinger papers: original untraced, present text taken from Pf's microfilm of the Abinger Collection, reel 9. Two reasons for giving the letter almost in full are, first, the difficulty of penetrating the convoluted paragraph about MW's task in Norway, and second, the curious emphasis on Messrs Ryberg. KP, i, 227–8, quotes a large extract. Shorter extracts in Durant's Supplement, 295; Holmes, *Sidetracks*, 238; and Jacobs, 205–6.

255 *meet up*: There was some talk of Basle (one of the centres of the French secret scheme to gain provisions).

12 FAR NORTH

256 Uncited quotations are from MW's letters to GI of June 1795–Jan. 1796 in *MWL*, 289–328; *MWletters*, 295–337. This chapter offers a solution to the mystery of the silver ship. The notes below present the substratum of the evidence in lieu of the elusive proof.

257 *impressions of Beverley*: *Travels*, letter 9.

258 *Onsala peninsula*: Suggested in Buus, 'Promethean Journey', replacing Nyström's idea of the Nidingen reef.

259 *Ellefsen's background*: Peder's grandfathers were brothers, Hans and Peder Ellefsen, who married daughters of a rich man called Isaac Falch. The two families were joint owners of the flourishing Egeland ironworks which manufactured the decorative iron stoves that were the necessity of every Norwegian house. A son of one house, Ellef Hansen of Arendal, married a daughter of the other, Margrethe of Risør, uniting two fortunes.

259 *Groos*: Not far from Grimstad.

259 *delay in Groos*: ,Ellefsen claimed that the bottom of the ship was damaged in the ship's log for 21 and 22 Aug. Molden has ascertained bad weather at the time.

259 *galloped*: Ellefsen could have hired a boat, but my guess is that in his hurry, a boat, subject to wind, would have been too uncertain.

259 *Ellefsen sold the ship*: The deed states he had bought the ship for his stepfather, Major Christopher Henrik Hoelfeldt, at Le Havre.

259 *Sandviga*: Gunnar Molden has identified the docking place for sailing ships in the late eighteenth century, to the left as you enter Arendal along the Galtesund Channel.

259 *silver as a speciality in Schleswig-Holstein*: This area often changed hands. At the time it belonged to Denmark. A stunning array of eighteenth-century silver is in the North German Museum in Altona, the capital of Holstein. In 1965 silver objects previously stored went on permanent display. See Manfred Meinz, 'Die

"Silberkammer" des Altonaer Museums', *Altoner Museum in Hamburg: Jahrbuch 1966*, iv (Hamburg: D. R. Ernst Hauswedell & Co. Verlag), 38–75; and a sequel on the *silberkammer* in *Jahrbuch 1967*, v, 79–136.

260 *local folklore about the silver ship*: Foss, *Arendals byes historie*.

260 *plan to sink the ship*: Judge Wulfsberg's report (18 Aug. 1795) to the *Stiftamptmann* in Oslo, and to the Danish Prime Minister, cited in Molden, 'The Shipwreck That Never Was'. I am grateful to Molden for a copy of this letter. Molden adds in 'No Riches for the Descendants' that the ship was rumoured to have sunk 'just off Torungen', near Skurvene. A lighthouse there marks danger.

260 *fresh moves to detach from ship*: On 11 Sept. Ellefsen approached Captain Gabriel Engström from Hamburg to replace him. This didn't work out.

260 *signed over ship*. In the 1970s Nyström found the deed of transfer at the Aust-Agders Archives, Arendal. Further facts are revealed in the interrogation of Ellefsen in Arendal Town Hall (28 Apr. 1795), a shaming event in his family town.

260 *three witnesses*: The lawyer Mr Ussing; Peder's stepfather's representative; and his brother Isaac Falch Ellefsen.

260 *crew at the time of handover*: Judicial inquiry in Arendal, 30 Apr. 1795. Kristiansand Archive: Police Protocol 1 (1783–99), p.287B. Wulfsberg, report of 18 Aug. 1795, seems to suggest a change of crew in the course of the voyage when he says that Ellefsen 'troubled himself to obtain the kind of people on board the ship, who would be party to the ship's sinking'.

260 *likely time for attempt to sink the ship*: On 22 Sept. 1794 Ellefsen borrowed 588 *riksdaler* from his lawyer, with the ship as security. Although it was common to raise money in this way if repairs were necessary during a journey, this would later be questioned as a criminal act, and a reason could be that damage to the ship had been deliberate. The ship was still making for Gothenburg, Ellefsen reassured EB in late Sept. and again on 7 Oct.

260 *storm*: Crew's testimony on 23 Dec. 1794. Quoted in Molden, 'The Silver Brig'.

260 *EB leapt into action*: GI did not tell EB about the silver until 24 Oct. (the day before the *Rambler* with its bullion reached Gothenburg). EB approached Christoffer Nordberg, the leading merchant in the Swedish border port of Strömstad. Nordberg consulted the town's district judge, A. J. Unger, and also Wulfsberg.

260 *Waak*: Queried Ellefsen's right to mortgage the ship, and took him to court.

261 *magistrate of Risør*: von Aphelen cross-questioned Ellefsen on 8 Nov. 1794.

261 *Ellefsen extracted the receipt*: Opinion of Judge Wulfsberg (18 Aug.1795).

261 *Ellefsen's allegation that Coleman 'escaped'*: 'On 11th [Nov. 1794], after being forced by the local magistrate to fire the person I hired as captain of the Maria Margreta, the aforesaid mate Mr Kolmand [Coleman], and demand the return of all the ship's papers because of his lawless and untoward circumstances, this person, unbeknown to me . . . has on this day gone to sea and escaped with the ship from East Risør harbour.' Report of the *Kristiansands Addresse Kontors Efteretninger* (27 Nov.), the only newspaper in that part of Norway at the time. Discovered by Molden.

261 *crew's testimony*: 23 Dec. 1794, after the ship landed in Kristiansand, and following Coleman's interrogation. Discovered by Molden.

261 *failed attempts to land in Sweden*: 15 and 16 Nov. 1794. *Travels*, letter 6.

262 *Coleman interrogated at Kristiansand*: 13 and 15 Dec. 1794. Ellefsen had tried to forestall

this with an attack on Coleman on 5 Dec., declaring that Coleman was using the flag illegally. Subsequent interrogations queried Ellefsen's own right to fly the Danish flag. When Ellefsen was questioned on why he had not taken the ship from Arendal to Gothenburg, he made the flag his excuse, saying that he did not want to commit a 'wrong'. He gave a different excuse in a letter to EB: he was unwell, he claimed, and his private affairs prevented his completing the journey. Neither excuse rings true. Ellefsen made a bad impression on Magistrate von Levetsow and police chief Rasmus Sørensen. (Kristiansand Town Magistrate, 34. Collegial Protocol (1793–4), 213. Includes a report from Levetsow to his Danish superiors.)

262 *document vital to the case against Ellefsen*: The document, in English, was translated into Norwegian by 15 Dec., within four days of the ship's docking in Kristiansand, and included in the magistrate's report to Copenhagen. Another document that came to light at this time was the introduction of GI to EB, giving the impression that the two didn't know each other and as yet had no dealings (ch. 11, above). The letter, signed by Delamotte and GI himself and dated from Le Havre, 26 Thermidor (13 Aug.), was in French. The magistrate ignored this because it did not mention Ellefsen or the ship. A record of this inquiry was presented at a more searching interrogation in Arendal on 28 Apr. 1795. (Kristiansand State Archive: Arendal Town Magistrate, Police Protocol 1 (1783–99), p. 287B.)

262 *'crooked business!'*: GI had not altogether contravened the American position on trading practice. Jefferson, as Secretary of State, had recently laid down that 'our property, whether in the form of vessels, cargoes, or anything else, has a right to pass untouched by any nation, by the law of nations: and no one has a right to ask where a vessel was built, but where is she owned?' (Jefferson to Morris in Paris, 13 June 1793: Morris, Papers.) Given Jefferson's principle, a French ship is no longer French if an American buys it. GI, though, may have ventured beyond the bounds of law when he disguised his American ship as a Norwegian one. When Ellefsen declared that the ship had no right to fly a Norwegian flag, he was exposing GI's infringement of Norwegian law.

262 *Royal Commission*: Set up 30 Jan. 1795, under Judge Wulfsberg of Tønsberg and Lauritz Weidemann, Magistrate of Nedenes.

262 *further judicial inquiries*: On 28 and 30 Apr. 1795.There was also an investigation on board ship. Mr Isachsen translated questions into English for Coleman who, on 15 Apr., handed over the ship's papers, the documentation of the sale of the *Liberté* (the original name of the *Maria and Margrethe*), which was confirmed by a Danish consul in Rouen on 20 July. Coleman and the investigators searched the captain's quarters in vain for Ellefsen's receipt for the silver. When Ellefsen was asked to produce GI's final instructions to him, he 'declared that the said communication had been mislaid amongst his papers' in Risør.

262 *GI's instructions to Ellefsen*: Dated 13 Aug. Ellefsen had signed an agreement to deliver unspecified 'articles' to EB in Gothenburg, who would then give him further instructions. (Kristiansand Town Magistrate. Notary Protocol 8 (1794–1804). Discovered by Molden.)

262 *'Who were fooled?'*: Email to Verhoeven, co-editor of Imlay, *Emigrants.*

262 *Waak arrived in June*: He had signed on a crew to sail the ship at the end of Apr., and

it may be at this point the extent of its damage was made known to EB.

264 *'Marin Inclay' oversees the sale*: Gunnar Molden has discovered an amusing document showing that the magistrate of Strömstad registered the sale, instigated in Gothenburg on 26 Mar. 1795 by 'Gilbert Inckay', and carried through by 'Marin Inclay his wife according to the lettre of Horning' (meaning, presumably, a letter of attorney). Curiously, the document introduces a sentence in English as though 'Mary Imlay', dictating it, was misheard by an official so anxious to get it right that he did not translate. Records of the repairs to the ship are enclosed with the ship's papers in the Riksarkivet, Stockholm: *Kommerskollegium Huvudarkivet* 1795 F IIb. *Fribrevshandlingar* vol. 143.

264 *cost of repairs*: The ship was sold for 1210 *kroner*. The repairs cost 3202 *kroner*.

264 *French dress*: A guess. MW had lived for the past two and a half years in France, and in London a few months later was described as elegantly dressed.

265 *'health'*: *Travels*, letter 8.

265 *'You have often wondered . . .'*: Ibid., letter 8.

265 *'a golden age of stupidity' . . .*, Ibid., letter 9.

267 *Fanny Blood's voice and fears for Fanny Imlay*: Ibid., letter 6.

267 *'my desultory manner'*: Ibid., letter 5.

267 *foreshadow anthropological travel*: Holmes, *Sidetracks*, 251.

267 *'the art of travelling'*: *MWCW*, vii, 277.

268 *Lars Lind*: Wulfsberg's report to Bernstorff.

268 *a smugglers' haven*: Nyström, taking MW at her word, is surprised. Anka Ryall makes the point (in her intro. to Norwegian translation of *Travels*) that MW' s distorted view of Risør reflected her feelings about her errand there.

268 *another judicial hearing in Risør*: Molden, 'The Shipwreck That Never Was'.

269 *MW's meeting with Ellefsen*: MW to the Danish Prime Minister. New-found letter transcribed below.

269 *Wulfsberg to Bernstorff*: Written on 18 Aug. 1795, the day Mary left Risør and returned to Tønsberg.

269 *'self-applause'*: *Travels*, letter 12.

270 *Coleridge and MW*: Holmes, *Sidetracks*, 260, hears her words enter the 'great echo chamber of Coleridge's mind', Coleridge read her *Travels* at Nether Stowey in Somerset in 1797. Another source is the waterfall at Trollhättan (near Gothenburg). This echo first picked up by John Livingstone Lowes in *The Road to Xanadu* (1927).

270 *'chained to life . . .'*: *Travels*, letter 15.

270 *'indifference'; heaps of ruins'; 'strangely cast off'; 'I do not understand you'*: MW to GI (6 Sept. [1795]). She refers to the letter he'd sent on 20 Aug. immediately following her failure to secure restitution from the Ellefsens.

271 *'goods' in the hands of Ryberg & Co.*: GI's power of attorney for MW, quoted in ch. 11, refers repeatedly to Ryberg. Information from the commercial history centre in Copenhagen shows it to have been a flourishing, powerful firm.

271 *MW's letter to the Prime Minister of Denmark*: This long letter was buried for two centuries in the State Archives in Copenhagen. A brilliant discovery by Gunnar Molden, revealing unknown aspects of the hidden business. I am grateful to him for permitting me to transcribe and publish this letter for the first time.

274 *Crèvecoeur*: His son Alexandre, who may have stayed with GI in Le Havre in 1793–4, was now in business in Hamburg. His father had lived in Paris before this move.
276 *Hamburg as spy capital, and 'emporium of mischief'*: Tillyard, *Citizen Lord*, 191.
276 *Hamburg in the late eighteenth-century*: Hamburg History Museum.
276 *witnesses bribed to conceal the truth*: Wulfsberg's report to Bernstorff.
276 *may be impossible to solve the case*: The record of the Royal Commission, which would have run to several hundred pages, has disappeared. Nyström assumed that the Ellefsens bribed officials to get rid of it, but that may not be the whole or final answer.
276 *not entirely a guilty voice*: Gunnar Molden, in conversation. Arendal (July 2001).
277 *Gjessoy*: Outside the large island of Tromoy.
277 *Ellefsen's fate*: Molden, 'No Riches for the Descendants', reports archival records of business transactions that prove Ellefsen's continued presence in the district, and his ships registered in the Arendal customs data. He died in the port of Marazion in Cornwall in 1807.
278 *Holsteiner crew on the silver ship*: Judicial interrogation at Arendal, 28 Apr. 1795.
279 *EB and France*: EB married Julie Éléonore Bonamy in 1786.
279 *EB as good citizen*: Municipal records (*Politie-protokoll Kontrollbok* (1789–1804), Landsarkivet, Gothenburg, show EB scattered all over the records of his time, concerning himself with the employment of sailors, the sinking of level ground, the transportation of goods, fines for an oil-lighter of the streets, and property changing hands. This is the image of a man who is dug into his society. GI, by contrast, slides back and forth from Paris to London, unmarked.
279 *Ellefsen's confession*: Judicial Inquiry, Arendal (30 Apr. 1795).
280 *Backman buys the ship*: It may or may not be relevant that on 20 Mar. 1795 JB 'paid Swedish Capt. [Waak?] 4 days demurrage'. JB's rough notes on money, 1795–6, Houghton: b MS Am 1448 (697). The *Rambler* returned to circulation on 1 Apr. 1795.
280 *property 'now lying in Paris'*: Draft will (June 1796) to RB from Algiers. Beinecke.
280 *'silver' in JB's notebook*: Pocket-sized memorandum book with loops at the side, used 1795–6. Some puzzling letters, carefully transcribed – a code? – are on the same page as the 'silver'. (Houghton: b MS Am 1448(7). JB notes Hamburg connections: John Parish & Co. (the super-rich Parish was American Consul there); Waage & Bagge; Dobbeler & Stetz; C.D. Dede in Altona. JB also notes London connections: Louis Goldsmith, 24 Princes St, Spital Square, and Francis Pitman, 9 Lime St; a contact, Bögel & Co., in Copenhagen; Cohn & Co. in Amsterdam; and Isaac Clason in New York.
281 *crookedness . . . under her skin*: Buus, 'Promethean Journey', 221–39, notes how MW tries to 'distance herself textually from . . . commercial entanglement', and observes her 'uneasy oscillation between being bound [by links with degrading business] and unbound' as Romantic loner. These strategies amount to an attempt to 'write out' her relationship to commerce in favour of the sentimental narrative of the *Travels*. Gary Kelly, too, *Revolutionary Feminism*, 179, mentions this distancing.
281 *'fraud'*: Since this was published as part of her *Travels*, Godwin did not have to edit it out in the way he had to eliminate all reports of business from private letters to GI.

282 *calm of surrender*: Herman, *Trauma*, sees this calm following trauma as a state of 'partial anaesthesia' which is a protection against unbearable pain.
282 *MW's attempt to drown*: *Memoirs*, ch. 8. WG would have heard this from MW.
283 *coming back to 'life and misery'*: *MWL*, 317. Henry Reveley recalled MW saying that the pain of drowning was less. *SC*, x, 1137.
283 The Times: 24 Oct. 1795, a fortnight later. Tims, *Social Pioneer*, 273.
283 *'If we are ever . . .'*: Confided later to WG, who repeats her words in *Memoirs*, ch. 8.
283 *'I never blamed . . .'*: MW to WG [4 July 1797], *MWL*, 404; *MWletters*, 429.
284 *Finsbury Place*: On the eastern side of Finsbury Square. It no longer exists.
284 *GI to Paris with his mistress*: Could the mistress have been a Frenchwoman whom GI met during his summer visit to Paris?
285 *proposal*: Abinger: Dep. b. 214/3. See *SC*, i, 144, for the unidentified admirer who pleaded for 'a cold second place in her heart'. Some say it was Thomas Holcroft, but I doubt it. The proposal is dated 2 Jan. 1796. MW declared that she would have nothing further to do with this man, whereas Holcroft was invited to Mary Hays's tea, together with MW and WG, only a week later (8 Jan.). As WG's best friend, he was a guest at the dinner WG gave to introduce MW to his friends in Apr. 1796, a visitor to their home, and on one occasion even dined with MW alone. *MWletters* places MW's retorts in an earlier period, but the date of the proposal, reinforced by the fact that it was addressed to MW at Finsbury Place, puts the incident unquestionably in this period.
285 prostituting *my person*: She takes the opposite view of marriage to that of Charlotte Lucas in *Pride and Prejudice*, who sees it as 'the pleasantest preservative from want'. Intelligent, unromantic Charlotte is prepared to marry a fool (Mr Collins) for the sake of security. The explosiveness of 'want' in Jane Austen's mannered world spells out the economic necessity facing women with little or no dowry.
285 *Chalmers & Cowie*: Morris, *Diary*. Morris dealt with this firm.
285 *Mr Cowie, MW and GI's gain of £1, 000 for 'goods'*: WG to Cowie after MW's death (2 Jan. 1798). Abinger: Dep. b. 227/8. How much did MW know of GI's gain?
286 *Wheatcroft testifies*: Granted a French passport to Norway by no less a body than the Committee of Public Safety – an odd position for an Englishman whose country was at war with France.
287 *travels as autobiography*: Tamara Follini, 'Improvising the Past', offers a stimulating theory of autobiography in her article on Henry James's memoirs.
287 *omits two crucial facts*: These are perceived by Richard Holmes who talks of 'a brilliant piece of emotional projection, casting GI as villain for withholding his love and driving her to the limit of her endurance'. *Sidetracks*, 256.
287 *complicity*: Holmes, Intro. to *Travels*, 34, and *Sidetracks*, 253: 'not only must she have felt betrayed by Imlay but to some extent *self*-betrayed'.
288 *'the silvery expanse'*: *Travels*, letter 24.
288 *'interests of nations . . . fraud'*: Ibid.
288 *GI's image of himself as model man*: *MWL*, 321, 323–4 and *MWletters*, 333, 335, quote GI's actual words.
288 *GI had not paid MW's debts*: *MWL*, 323; *MWletters*, 335.
289 *Haydn*: Six English Canzonettas.
289 *Southey*: Wardle, *Critical Biography*, 256. Buus, 'Promethean Journey', notes that

Travels reinforces 'a particular fiction of Scandinavia popular in the late eighteenth century: that of a peripheral northern Other – at once Arcadian and archaic . . .'.

289 *Addressed* . . .: Holmes phrased this perfectly in *Sidetracks*, 263.

290 *'in tenderness'*; *'a genius . . .'*: *Memoirs*, ch. 8, 249.

13 WOMAN'S WORDS

291 Uncited quotations are from letters to Rowan, von Schlabrendorf, Hays, GI and WG from Jan. 1796 to Jan. 1797, in *MWL*, 328–77, and *MWletters*, 337–94.

291 *Mrs Cotton*: The Wollstonecraft sisters may have come to know her when EW taught in Henley and MW visited her in 1786.

291 *East family*: Claire Tomalin established the connections with the Austens in *Jane Austen*, 158–9, 317.

291 *Jane Austen, at twenty-one*: Tomalin provides the telling detail in ibid., 123. Austen had acquired a copy of a women's rights novel, *Hermsprong* (1796), by her contemporary Robert Bage.

291 *No scandal*: Hays did report to WG that MW felt some friends in London shunned her as an unmarried mother, but this seems to have been a fear rather than a reality. The gentry would not have received her had there been scandal attached to her name.

291 *'infamous'*: JJ to Charles Wollstonecraft, in JJ's Letterbook (15 Nov. 1796).

292 *MW's hand in letter to Rowan*: (26 Jan. [1796]). Berg Collection, New York Public Library.

292 *Leavenworth ruined*: RB to JB (6, 14, 28 Jan. 1796). Houghton: b Ms Am 1448 (542, 543, 550). RB's surviving letters are substantially fewer than those of her husband, and one possible reason is that she acted as an agent for his interests in Paris. If so, he would have destroyed letters to do with that, in the same way that Imlay probably destroyed the business sections of MW's letters from Scandinavia and Hamburg.

292 *'forbearance'*; *'delicacy'*; *'principle'*: GI's words repeated back to him in *MWL*, 329–30; *MWletters*, 339.

293 *her soft voice*: Hays, 'Memoirs of Mary Wollstonecraft'.

293 *WG's face like John Locke*: Hazlitt, 'William Godwin'.

294 *'voluptuous'*: Amelia Alderson to MW (28 Aug. 1796). Abinger: Dep. b. 210/6.

294 *WG's background and early history*: Intro. by Marilyn Butler and Mark Philp to Godwin, *Novels and Memoirs*.

295 *WG's mother's family background*: A northern version of the *déclassement* that Hardy was to dramatise in the Wessex of *Tess of the D'Urbervilles*.

295 *Janeway and WG*: Avery, 'The Puritans ', *Children and their Books*, 112.

296 *like many a sensible child of poor parents*: Charlotte Brontë at Miss Wooler's school in 1831–2 mopped up in the space of a year whatever it had to teach.

296 *WG's fits*: WG to Dr Ash (1831). Pf. Abinger, microfilm reel 9.

297 *WG as biographer*: Holmes recognised his innovation in this genre in *Memoirs*.

297 *'the contemplation of illustrious men'*; *'to scrutinise . . .'*: WG, unpubl. essay, 'Of History and Romance' (1797), cited by Philip Cox, *Reading Adaptations* (Manchester University Press, 2000).

297 *'real English'* . . . : WG to unnamed person (27 Feb. 1796). Abinger: Dep. b. 227/8.

298 *WG stirred by politics . . . 'beat high'*: KP, i, 61.
298 *WG on Violence: Political Justice*, book v, chs 16,18.
298 *WG and Hays*: Some of the wording here appropriated from Wagner and Fischer, 'Visionary Daughters', 54–6.
298 *'I am sorry . . .'*: WG to Hays (7 May 1795). *SC*, i, 139.
299 *'We who are thieves'*: Godwin, *Caleb Williams*, 216.
300 *WG and Tooke at dinner*: Godwin, *Novels, and Memoirs*, i, 50–1.
301 *'Imlay' in WG's diary*: Abinger: Dep. e. 201. It's true that WG's letters to MW were addressed on the *outside* to 'Mrs Imlay', upholding her public identity. But WG never at any other time referred to her in the privacy of his diary as 'Imlay'. To this man of regular and truthful habits, she remained always 'Wolstencraft' (and appears as such the day before the 'Imlay' entry). 'Wolstencraft' was shortened (as she appears more frequently) to 'Wt'. She remains 'Wt' after their marriage and after her death.
301 *Mrs Inchbald*: Her best-known novel is *A Simple Story* (1791), strong on plot and repartee, but poor in characterisation – the men and secondary characters are mere types: the supportive friend, the dashing faithless suitor.
302 *'the worst of all laws'*: Godwin, *Political Justice*, quoted in KP, i, 113.
302 *the English losing common sense in 1796*: Letter to Rowan (26 Sept. 1796).
302 *WG's advice to MW*: Blends details from his *Memoirs* and his letters to a woman, perhaps Hays, also depressed by unrequited love. Abinger: Dep. b. 227/8.
302 *'adrift'*: MW to HF (c. late 1795), *MWL*, 324; *MWletters*, 336.
303 *'I found . . .'*: Godwin, *Godwin and Mary*, 75.
304 *'by almost imperceptible degrees'*; *'Nor was she deceived'*: *Memoirs*, ch. 9.
304 *couplet by Butler*: *Hudibras*, II, canto I, lines 591–2.
304 *'I send you . . .'*: WG's diary records 'Propose to Alderson' and some have believed that WG proposed to Amelia Alderson, the doctor's daughter. St Clair, *Godwins and Shelleys*, argues convincingly that 'propose' refers to a conversation with her father. Amelia continued to flirt with WG in a way that would have been unlikely had she turned him down.
305 *WG's playfulness*: Pamela Norris, letter to author (26 Sept. 2002).
305 *site of 16 Judd Place West*: Elizabeth Crawford deduced this during our memorable 'footsteps' tour of MW's London.
305 *'Perdita'*: One of her best-known roles (in *A Winter's Tale*), during 1779–80, just before her affair with the Prince. When she and MW met, she was near the end of a long affair (1782–97) with a soldier from a rich family, Banastre Tarleton, who became MP for Liverpool. Abandoned a second time, she made an effort to support herself by her pen, but one of MW's later reviews notes that she wrote too fast. She was an exact contemporary of MW. WG thought her the most beautiful woman he had ever seen. She was famous for 'breeches' roles.
305 *Mrs Siddons on* Travels: Mrs Siddons to WG (c. 1797–8), quoted in *Memoirs*, ch. 9.
306 *Amelia Alderson and MW*: Abinger: Dep. b. 210/6. Todd, *Wollstonecraft*, 369, 382.
306 *ties with women*: Jacobs, *Her Own Woman*, 242.
306 *rue du Bac*: RB heads her letters from no. 555. The street has since been renumbered. According to Morton's delightful but not always accurate *Americans in Paris*, RB lived at what is now no.102 with *citoyenne* Hilaire.

307 *Humphreys, Barlow, Imlay*: For the larger picture see 'American Spies' in Europe on the Internet site accompanying this book.

308 *O'Brien's mission; Washington; JB*: Humphreys to JB (30 May 1796 and 23 July 1796), Beinecke: MS Vault Shelves: Pequot (Barlow) M992. O'Brien moved about on the brig *Sophia* and his code identity was 'the affair of Wales'.

310 'black *sweetheart*': RB to Robert Fulton, inventor of the submarine. Houghton. 000 *Hazlitt on WG*: *The Spirit of the Age*.

310 *overstepped the mark*: Deduced by St Clair as part of his convincing interpretation of coded marks in WG's diary. Appendix 1, *Godwins and Shelleys*.

310 *responsible*: WG's daughter, MWS, later said of her father (KP, i, 161–2 and Clemit et al, *Lives of the Romantics*, i): 'He was in a supreme degree a conscientious man, utterly opposed to anything like vice or libertinism, nor did his sense of duty permit him to indulge in any deviation from the laws of society . . . which could not, he felt, be infringed without deception and injury to any woman who should act in opposition to them.'

311 *'I have not patience . . .'*: KP, i., 139.

311 *Cymon*: Changed by love for the beautiful Iphigenia. MW's source may have been Dryden's version of the tale in his *Fables Ancient and Modern* (1700).

312 *Maria and Darnford*: *WW*, ch. 4.

313 *Rousseau's* Solitary Walker: *Les Rêveries du promeneur solitaire* (1782).

319 *effort*: 6–7 Oct. 1796, Godwin, *Godwin and Mary*, 43.

319 *lacked '. . . the pleasures of the imagination'*: *Memoirs*, 2nd edn.

320 *'She was like a serpent . . .'*: *Memoirs*, ch. 7.

322 *MW and WG on a private 'bill of rights'*: Godwin, *Godwin and Mary*, 49–50.

323 *Coleridge and Hazlitt on MW*: Hazlitt, 'My First Acquaintance with Poets' (1823), cited by Holmes, *Sidetracks*, 265.

324 *'no one knew better . . .'*: *Memoirs*, ch. 9.

325 *'What can I say?'*: Godwin, *Godwin and Mary*, 46.

326 *Amelia on WG's supposed interest*: To friend (1 Nov. 1796). Abinger: Dep. b. 210/6.

326 *Opie as suitor*: Joseph Farington, diary (11 Nov. 1796), cited in Durant's Supplement, 312. After MW married WG, Opie married Amelia Alderson.

326 *'I treated him . . .'*: Quoted by James Marshall to Hazlitt, cited in ibid., 332–3.

326 *Condoms*: Porter, *English Society*, 27.

326 *WG's method of contraception*: St Clair's deduction, *Godwins and Shelleys*, Appendix 1. Myths to do with contraception derived from *Aristotle's Complete Master-piece* (see above, ch. 7), wrongly attributed to Aristotle but going back to ancient Greece.

327 *'She has been deserted . . .'*: JJ to Charles (15 Nov. 1796). JJ's Letterbook.

328 *'with extreme unkindness'*: WG to MW (31 Dec. 1796), Godwin, *Godwin and Mary*, 59–60.

14 'THE MOST FRUITFUL EXPERIMENT'

330 Uncited communications are from letters of Feb.–Aug. 1797 to WG, EW, Amelia Alderson, George Dyson, Maria Reveley and Miss Pinkerton, in *MWL*, 379–411 and *MWletters*, 396–437.

330 *'two persons of the opposite sexes . . .'*: Essay VIII ('Of Posthumous Fame') in *The Enquirer*,

publ. Feb. 1797, when he was contemplating the marriage he undertook the following month. Repr. Godwin, *Political and Philosophical Writings*, v: *Educational and Literary Writings*, ed. Pamela Clemit, 206.

331 *fits of sleepiness*: These began in 1795, according to WG's case history for Dr Ash. Pf.

331 *'a recluse'*: An autobiographical letter, unfinished and unaddressed. Quoted in KP, ii, 129.

332 *Polygon*: A plaque for MW is on the site, now Oakshott Court in Werrington Street, NW1, off Polygon Road. Nearby, on the wall of the local primary school, is a modern mural by Karen Gregory which shows MW and WG with Fanny in a pinafore, as well as Mary Shelley and Shelley on a bridge with Frankenstein's monster lurking in the reeds below – looking like Vincent Price in the horror movie.

333 *'I love the country'*: 21 May 1797. The context is an invitation to go to the country with WG's friends Montagu and the Wedgwoods.

333 *the signature*: Similar in her letter to Bernstorff except that then she was 'femme Imlay'.

333 *'Never'*: *St Leon* (1799), ch. 4. The wife Marguerite in this novel is an idealised portrait of MW.

333 *MW's face*; her look: Southey to Joseph Cottle, cited in KP, i, 234.

333 *'We were in no danger . . .'*; *a 'worshipper'*: *Memoirs*, ch. 9.

334 *'overflowings'*; *the English character*; *and treating others as books*: To an unnamed correspondent, probably Mrs Inchbald (19 Sept. 1797). Abinger: Dep. b. 227/8.

335 The Times *on the Godwin marriage*: Holmes, *Sidetracks*, 208.

335 *theatre visit*: The play was referred to by WG as a comedy about a will.

336 *snub*: Godwin, *Godwin and Mary*, 75.

336 *'cruel, base, insulting'*: WG to Mrs Inchbald (13 Sept. 1797). Abinger: Dep. b. 227/8.

336 *MW reviewed Mrs Inchbald*: *MWCW*, vii, 462–3. MW said there was not enough 'lively interest to keep the attention awake', that it was 'improbable', and characterised by *'naïveté'*. Cited by Todd, *Wollstonecraft*, 382.

336 *'I am pained . . .'*: 20 Apr. 1797, Godwin, *Godwin and Mary*.

336 *'Those who are bold enough . . .'*: Excerpt from undated letter to a friend whom Janet Todd identifies as likely to be Hays, and suggests this date, Apr. 1797 (*MWLetters*, 410.) Letter quoted in Hays, 'Memoirs of Mary Wollstonecraft'.

337 *WG's letter justifying his marriage*: Abinger: Dep. b. 229/1. Faded, on thin paper with a tatty edge, the pen marks (as in a number of WG's letters) survive more clearly on the *back* of the paper which may be deciphered with the aid of a mirror. Some readings are therefore approximate.

337 *WG's feelings opposed the supposed gist of his doctrines*: MWS, 'Life of Godwin', 113.

338 *Hays on MW's manners*: 'Memoirs of Mary Wollstonecraft'.

339 *'the deceitful poison of hope'*: *Emma Courtney*, i, ch. 24.

339 'murder': Ibid., ii, ch. 5.

340 *JB's fury with James Wollstonecraft*: To S. Williams, US Consul in London (21 Feb. 1799), asking him to recover the sum 'by any means that the law will permit'. Letterbooks, iv, 205–6. Houghton.

340 *JJ to Charles and other Wollstonecrafts*: JJ's Letterbook. Contrast an annoyed letter on15 July 1797 with his earlier letter of 1 Nov. 1795 when JJ had commended Charles for sending £200 (most of which was to be invested for his sister Bess). JJ's annoyed

letter to EW when she gave up a good post in order to start a school, seems unfair, but he feared further impositions.

341 *WG's appeal to Wedgwood*: KP, i, 234–6.

342 *visit to Bedlam*: WG's diary. Abinger: Dep. e. 201–2.

342 *'Women being confined . . . Madhouses'*: *Laws Respecting Women*, 73.

342 *BW and EW had not communicated*: Since both sisters preserved MW's letters, we can assume there were none. MW's correspondence with EW (though not with Bess) resumed after EW's visit.

342 *'have you lost . . .'*: *WW*, ch. 9.

343 *the Wedgwood post and WG*: I assume that news of the post came from WG via MW – a peace offering from MW – though it's often noted that Mrs Wedgwood, *née* Bessie Allen, who came from Cresselly, near Pembroke, Wales, had known the Wollstonecrafts in Laugharne when EW was a girl. Since the Wollstonecraft sisters had lived there for only a year, and since their father's subsequent residence would hardly have recommended the family, this connection doesn't appear very strong. The Wedgwoods were more likely to have been impressed with recommendations from WG and MW. EW told WG after MW's death (24 Nov. 1797), 'my sister dined with them [the Allens] once when on a visit at friends in that part of the country'. Abinger: Dep. c. 523.

343 *'looked like a spectre . . . anguish'*: *WW*, ch. 6.

343 *'gross'*: *WW*, chs 9, 10.

344 *Sue Bridehead*: *Jude the Obscure* (1891).

344 *'Let me exultingly declare'*: *WW*, ch. 8.

345 *'witness of many enormities'*: Ibid., ch. 5.

346 *'Such was the native soundness . . .'*: Ibid., ch. 4.

346 *Opie's 1797 portrait*: National Portrait Gallery, London.

350 *'Another evening . . .'*: From Etruria (12 June 1797). Godwin, *Godwin and Mary*, 97.

352 *Hays . . . plain*: Coleridge cruelly called her 'a thing, ugly and petticoated'. *The Love-Letters of Mary Hays*, ed. A. F. Wedd (London, 1923), 11.

352 *portly*: To James Marshall (21 Aug. 1797). Discovered by Stephen Wagner and Doucet Devin Fischer, 'Visionary Daughters'. *MWLetters*, 435.

353 *'Letters on the Management of Infants'*: *MWCW*, iv, 456–9.

353 *Darwin against cold bathing*: Wedgwood to WG. Abinger: Dep. c. 507. A vast letter about an educational scheme, grand, abstract, intended to bring together all the radical 'men'. It doesn't occur to him to include MW, the most radical and practical educator of the age, and his message to her is an afterthought when he has already sealed his letter. (The letter may have been behind Darwin's conventional *Plan for the Conduct of Female Education in Boarding Schools*, published by JJ, 1797) who says that 'great eminence in almost anything is sometimes injurious to a young lady; whose temper and disposition should appear pliant rather than robust'.

353 *Anthony Carlisle*: He was later painted as an anatomist, a hand on a skull and behind him a drawing of the male body.

353 *'Lessons'*: *MWCW*, iv, 467–74. May have appeared as a separate sixpenny pamphlet, as listed in a catalogue of MW's works in an 1802 edition of the *Travels*.

354 *'Papa' as Imlay*: Holmes, *Footsteps*, 120–3.

354 *letters for children in case of death in childbirth*: Vickery, *Gentleman's Daughter*, 98. In *WW*, ch.

7, Maria writes memoirs for her daughter 'uncertain whether I shall ever have an opportunity of instructing you'.

354 *'most graceful expressions . . . of a mother's love'*: Brailsford, *Shelley, Godwin*, 204.

354 *'land of mugs'*: Godwin, *Godwin and Mary*, 88.

354 *'struck out a path'*: Advertisement in *Posthumous Works*. *MWCW*, iv, 467–74.

355 *'The first thing'*: *'Of Public & Private Education'*, *SC*, i, 146.

355 *Queenie Thrale's education*: Hester Thrale, author of *Family Book* (1764). Hardyment, *Perfect Parents*, 7.

356 *'have not distinguished themselves'*; *Greer*; *Daly*: Ibid., 303.

356 *'delicacy'*: *MWCW*, v, 218.

356 *lying-in hospitals in the seventeenth century*: Fraser, *Weaker Vessel*, 513.

356 *deaths from home deliveries*: Todd, *Wollstonecraft*, 448.

356 *Jemima on hospitals*: *WW*, ch. 5.

357 *Florence Nightingale*: See forthcoming Penguin biography by Mark Bostridge.

357 *Jane Sharp*: Fraser, *Weaker Vessel*. Vivien Jones explores MW's extensive knowledge of obstetric texts in 'Death of Mary Wollstonecraft'.

357 *details of the delivery, the operation; MW's comment to WG, and the following days*: WG's diary; *Memoirs*, ch. 10. Readers were shocked by WG's explicitness.

359 *MW's possible objection to Dr Clarke*: *SC*, i, 194–5.

359 *WG in the anti-midwife camp*: He felt 'the very name of a female midwife odious in my ears'. To unnamed correspondent (19 Sept. 1797). Abinger: Dep. b. 227/8.

360 *fatal course of sepsis*: Fatal until sulphonamides became available in the 1930s and penicillin in the 1940s.

361 *'Heaven'*: Milnes-Gaskell manuscript, quoted in Durant's Supplement, 234. The source, Holmes suggests (in note to *Memoirs*, 307n.), was probably Basil Montagu.

361 *MW supremely fitted*, WG *not*: WG to Mrs Cotton (24 Oct. 1797). KP, i, 280–1.

362 *'the kindest . . .'*: Letter to EW of 12 Sept. 1797. Quoted KP, i, 282–3.

362 *'Be sure . . .'*: To unnamed friend (23 Oct. 1797). Abinger: Dep. b . 227/8.

362 *'they left . . .'*: WG, 'Analysis of own character' (begun 26 Sept. 1798), Godwin, *Novels and Memoirs*, i, 58–9.

362 *'I would fain live in your heart'*: MW to WG [30 Sept. 1796], *MWL*, 355; *MWletters*, 369.

15 SLANDERS

363 *letters at the time of MW's death*: Abinger: WG's letters: Dep. b. 227/8. John Opie's letter: Dep. c. 507. JJ to Fuseli, announcing the death: Dep. b. 210/3. Carlisle asked WG for a set of Mrs Godwin's works, showing that she was more than a patient to him: Dep. c. 514. Some letters in KP, i.

363 *JJ's letters to WG*: Tyson, *Joseph Johnson*, 150–1.

363 *Hewlett invited*: Unable to attend the funeral.

363 *HF's part in MW's life*: The entire truth of this is impossible to determine. WG did himself continue to visit HF, and it's conceivable that it was for this very reason that MW stayed away, since they did not visit together. Another possibility is that MW stayed at home more in the late stages of her pregnancy, since she did complain to

WG of the want of company. She may have felt ostracised and shamed by a pregnancy that was advanced by the time she had married.

364 *'I firmly believe'*: Abinger: Dep. b. 227/8.

364 *WG's portrait after MW's death*: By J. W. Chandler (Feb.–Mar. 1798). Tate Britain.

364 *'I love to cherish melancholy'*: 24 Oct. 1797, KP, i, 280–1.

364 *exchanges of WG and Hays*: 5, 10, 22 and 27 Oct.1797. Abinger: Dep. b. 227/8. The same possessiveness is to be found in the last days of Charlotte Brontë, when her recent husband, Arthur Bell Nicholls, did not summon her best old friend, Ellen Nussey.

365 *WG's correspondence with Mrs Inchbald*: 13 and 14 Sept., 26 Oct., KP, i, 278–9.

366 *WG to Skeys*: Abinger: Dep. b. 227/8. A follow-up letter – much faded – is Dep. b 229/1(a), dated 17 Oct. 1797.

367 *WG and EW with the Wedgwoods*: 7–10 June 1797, Godwin, *Godwin and Mary*, 88, 94.

368 *MW's MS with the press within a month of her death*: It's unlikely that at this stage WG had included her *Letters to Imlay* in what was to be the four-volume edn of her *Posthumous Works*. Godwin's diary notes that he was reading 'I[mlay]'s letters' on 27–30 Oct. It's not impossible that these were GI's lost letters to MW.

368 *'public curiosity'*: WG to Skeys (17 Oct. 1797). Abinger: Dep. b. 229/1(a).

368 *EW's reply to WG about the memoir*: 24 Nov. 1797. Abinger: Dep. c. 523.

369 *suicidal Dido*: Gladstone thought the same after reading – admiringly – MW's *Letters to Imlay* in 1883. Gladstone went on to read Imlay.

370 *'She felt herself alone'*: *Memoirs*, ch. 6.

370 *'firmness of mind'*: Ibid., ch. 3.

371 *image of MW in a masculine hat*: *Monthly Visitor* (1798). Republ. in *Eccentric Biographies; or Memoirs of Remarkable Female Characters* (London: T. Hirst, 1803).

371 *'scripture . . .'*: *Anti-Jacobin Review* (1801), cited in Barbara Taylor, *Wollstonecraft*, 247.

371 'Fierce passion's slave' from 'The Shade of Alexander Pope'(1799), repr. Clemit et al, *Lives of the Romantics*, ii, 169–73.

372 *'heart of stone'*: Chandler, *Roscoe*, 84, 195, from Roscoe Papers: 3958A:

> Hard was thy fate in all the scenes of life
> As daughter, sister, parent, friend and wife
> But harder still in death thy fate we own
> Mourn'd by thy Godwin – with a heart of stone

373 *WG met Jane Gardiner*: Sun. 19 and Sun. 26 Jan. 1800, according to his diary.

373 *WG on Fanny Blood*: *Memoirs*, ch. 3.

373 *'Dr Gabell's garden'*: To Mrs Davids, College Street, Winton (27 May 1817): 'Our lodgings are very comfortable. We have a neat little Drawing room with a Bow-window overlooking Dr. Gabell's garden.'

374 *John Adams and MW*: Adams, marginalia, *FR*. See Internet site accompanying this book for a debate between Adams and MW.

375 *Mary King's history*: King-Harmon, *King House*, 74–7.

375 *Henry Fitzgerald*: His identity as Lady Kingsborough's cousin appears to be established in Todd, *Rebel Daughters*, 208–11, 234, 358. The source is a *Times* report of 10 Oct. 1797, reinforced by repetition that Henry Fitzgerald was 'second cousin to Miss King' in the *Hibernian Chronicle* of 12 Oct. 1797 and in the *Annual Register* for 1797, p. 68. This can't be settled, though, without a doubt, since it may be another family blind (see ch. 5 above for other blinds). The reports are made at the family's request

in order to clear Mrs Fitzgerald's son Gerald from a public confusion of him with the guilty man.

There were conficting statements about Henry's origins. If he were only a cousin why should the family have initially claimed that he was the illegitimate son of Colonel Fitzgerald's dead son? What's unconvincing is the absence of any record of this 'son' apart from family assertion, and dates make this highly improbable. Then too there is Mrs Fitzgerald's statement to MW that her son Gerald was the 'only' son of her late husband.

So violent, so implacable was the vengeance that we might consider whether the family could have concealed incest. Could Henry have been the son of Colonel FitzGerald himself, possibly conceived in the mid-1760s in the interval between the death of his first wife in 1763 and his second marriage? If so, Henry would have been Caroline King's half-brother, and uncle to the girl he loved. Nor can we entirely exclude the possibility that he was the illegitimate son of Lord Kingsborough, which would make the murder more like Greek tragedy – the difference being murder by an actual rather than foster father.

377 *Lord K's words on his murder of FitzGerald*: McAleer, *Sensitive Plant*, 73.

377 *'horror'*: MM's note rebutting a false account. Pf: Cini Papers, file 26.

377 *the trial; Bishop Percy's reports*: Bishop Percy in Dublin to his wife (14 and 18 May 1798). British Library: Add. MS 32, 335. Tomalin, *Mary Wollstonecraft*, 230.

378 *'A female . . .'*: The *British Critic* (Sept. 1798) has a similar sneer at the salutary effects of a new system of education which MW introduced to a noble family. 'One at least of these effects, we will venture to say, is fresh in the memory of our readers.'

380 *'left behind her'*: WG's draft of a letter to an unnamed correspondent (20 Jan. 1807) when Fanny was thirteen. Abinger: Dep. b. 229/1.

380 *GI's court case*: Imlay v. George James (10 Feb. 1798). GI's address is given as late of Harley Place in the London parish of Saint Mary Lebone, and at present, Upper Euston in the parish of Saint George, county of Gloucester. PRO, London: C12/2188/14.

380 *GI's opinion of himself as 'upright' citizen, 'I like to be . . .'*: Abinger: Dep. b. 229/1(b). No year is given. Jacobs, *Her Own Woman*, discovered this telling letter, published for the first time in her biography. I am grateful to her for passing on its precise location.

381 *Smuggling corrupts the morals*: Contradicts his statement (quoted in ch. 10, above) that 'wealth is the source of power; and the attainment of wealth can only be brought about by a wise and happy attention to commerce'.

382 *GI's deed of 1810*: Records of speculations in *Old Kentucky Entries*.

382 *GI's burial*: This is assumed to be the GI we know, but so far unproven. Richard Garnett, who wrote the article on GI for the *Dictionary of National Biography*, reports this in the *Athenaeum*, no. 3955 (15 Aug. 1903), 219. GI is said to have been born on 9 Feb. 1758 (not 1754 or 1755, as GI indicates in his preface to *Topographical Description*), and died on 20 Nov. 1828. The epitaph doesn't fit with his life, and upholds 'rights divine' which doesn't exactly accord with American republicanism.

382 *EB's appointment as American Consul*: The letter appointing him is in the Riksarkivet, Stockholm: Americana, 5, no. 220. Published in facsimile in *Sweden and the World* (Stockholm: National Archives, 1960), 64.

382 *destruction of GI's silver ship*: Nyström, *Scandinavian Journey*.

383 *another story*: For the Barlows' later adventures, see the Internet site accompanying this book.

383 *'malicious calumnies'*: Houghton: b MS Am 1448 (529).

383 *'wild ambition'*: KP, i, 286–7.

384 *WG's letter to JJ*: Abinger: Dep. b. 227/8.

384 *height of crack-down in 1797–8*: Butler, *Romantics, Rebels & Reactionaries*, 83; *Austen*, 116–17, 121.

385 *WG's obituary for JJ*: Tyson, *Joseph Johnson*, 215–16.

385 *Edgeworth's eulogy for JJ*: Ibid., epigraph to ch. 1.

385 *'Of Posthumous Fame'*: Essay VIII, *The Enquirer*, in *Political Writings*, v, 204.

386 *the 'self-centredness of the dying'*: Todd, *Wollstonecraft*, 456.

386 *Hays's obituary for MW*: 'Memoirs of Mary Wollstonecraft'.

387 *intolerance of homosexuality*: Fiona MacCarthy, *Byron*, 37, gives a useful summary as background to Byron's 'mingled fascination and revulsion' while he was at school.

387 *MW's correspondence with Fuseli*: Their letters were subsequently lost to sight, thought to have been destroyed by her grandson, Sir Percy Shelley.

387 *Kegan Paul's letter to Browning*: Beinecke. Kegan Paul uses the word 'slander' ('for such I believe it to be') in the course of the letter. He is, however, mistaken in thinking the slander originated in an anonymous 'Defence of the Character and Conduct of the late Mary Wollstonecraft Godwin' (1803), which mingles defence 'with a good deal of venom', KP, i, 206). He believed it written by Mrs Inchbald, 'who hated Mary like poison', he told Browning. Of WG's odd statement that if HF had been free, he would have been the man of her choice, Kegan Paul says convincingly: 'It is probable that he had only heard of the more unfavourable version of the story at second hand.' Browning's reply (15 Jan. 1883) is in Wellesley College.

388 *Browning not deterred*: Kegan Paul wrote back on the same day, 15 Jan. Browning's second reply is now lost, but happily Mr Michael Meredith of Eton College, the general editor of the *Poetical Works of Browning*, has recovered part of the letter from North's catalogue (Nov. 1906). He warns that he can't vouch for the accuracy of the transcript. Browning dismisses his three-stanza poem as unimportant, 'the merest trifle in the world', and seems to argue that it is more impersonal than it may appear, 'containing no word of reference to either of the persons in question'. (This is rather disingenuous, for the names appear in the poem's title, as Kegan Paul delicately pointed out on 15 Jan.: 'I regret just a little the names put to it, as giving some, faint, colour to a story which I think mythical.')

Browning went on to summarise his approach: 'It is simply an attempt at expressing some such thought as: "I (presumably a woman of genius) could have done and suffered anything to please you – and this, in spite of whatever the weakness in me: while, in order to please you, my utmost has been exerted, and precisely by the real strength in my nature – and yet with no effect at all." I hope this hasty scribble will answer its purpose . . .'.

389 *Blackwell and Garrett*: Crawford, *Enterprising Women*.

389 *Sophia Jex-Blake*: Virginia Woolf, *Three Guineas* (Penguin Books, 1977), 74–6.

389 *women members refused*: Burdet, 'Schreiner's Wollstonecraft'. Details about the club in Judith Walkowitz, *City of Dreadful Night* (Virago, 1992), ch. 5.

389 *'She fancied . . . amorous'*: *Anti-Jacobin Review*, i, 94–102.

389 *detraction*: Amusing wisdom in Dr Johnson's essay 'On Detraction'.

16 CONVERTS

390 *Curran on MM*: 8 June 1800. Abinger: Dep. c. 507/1. KP, i, 363.

390 *WG's first encounters with MM*: WG to Marshall (2 Aug. 1800). KP, i, 368–70.

391 *'In what you say . . .'*: MM to WG (8 Sept. 1800, 6 Apr. 1801, 6 Aug. 1801). Abinger: Dep. c. 507. Extracts in KP, i.

391 *'my peculiar good fortune'*: MM to WG (8 Sept. 1800), *SC*, i, 84.

391 *marriage a bar to improvement*: Chapter on matrimony, *Education*, 31.

391 *Moore Park near Kilworth*: Young, *Tour*, ii, 394. Young visited the estate in 1776. The owner at that time, Stephen Moore's father, was celebrated in Ireland for 'his uncommon exertions in every branch of agriculture'. Principally cattle: 'beautiful cows', rams, stallions, a Craven bull imported from England. He was especially proud of his turnips – 30 acres of them.

392 *'silliest project'; other comments about the marriage*: Revelations of her past for her daughters (1818).

392 *a United Irishwoman and republican*: Power, *White Knights*, 52.

392 *'born with the dead'; 'tongued with fire'*: T.S. Eliot: *Little Gidding*, I and V.

393 *francophilia as 'radical chic'*: Foster, 'Remembering 1798'.

393 *on Fitzgerald O'Connor, and francophilia*: Foster, *Modern Ireland*, ch. 12.

393 *MW's copy of* The State of Ireland: Remains amongst the Irish books in the nineteenth-century library of MW's present-day descendants, the Dazzi family of San Marcello, Pistoiese.

393 *fear of massive casualties*: Tillyard, *Citizen Lord*, 266.

393 *Big George's atrocities*: Power, *White Knights*, 56–7.

393 *'As a gentleman of respectability . . .'*: Francis Plowden, *An Historical View of the state of Ireland, from the invasion of that country, under Henry II to its union with Great Britain* (London, 1801). Quoted by Power, *White Knights*, 53–4, 57.

394 *'only had two Maidenheads'*: Ibid., 39.

394 *dooming Ireland to impotence*: Tomalin, intro. to MWS, *Maurice*, 19.

394 *MM's pamphlets*: MM was proud of the quoted passage. She copied it to the Shelleys at the start of their friendship in the winter of 1819–20. Abinger: Dep. c. 517.

394 *'set hard'*: Foster, *Modern Ireland*.

395 *alternative to 'dominant history'*: Foster, historiographical intro., *Irish Story*, xvi–xvii. The idea of counter-history goes back to Virginia Woolf, particularly her treatment of the Wars of the Roses in 'The Journal of Mistress Joan Martyn' (1906), and more explicitly in the way she casts Trevelyan's history into question in *A Room of One's Own* (1929).

395 Bible Stories: Parts of preface quoted in St Clair, 'Godwin as Children's Bookseller'.

396 *'driving at full speed'*: Miss Wilmot's travel letters to her lawyer brother, Robert Wilmot.

396 *'It would give me great pleasure . . .'*: 6 Aug. 1801. Abinger: Dep. c. 507.

397 *shady history of 'Mrs Clairmont'*: Thoroughly researched by St Clair, *Godwins and Shelleys*, and a summary may be found amongst the fascinating notes to Marion Kingston Stocking's edition of *The Clairmont Correspondence* (*ClCor*), 42–3. Her father appears to have been Peter (Pierre) de Vial, a French merchant who settled in Exeter. The family originated in the south of France, and in the eighteenth century migrated to Geneva.

397 *WG on the similarity of Mrs Clairmont's letters to MW's*: 2 Apr. 1805. Abinger: Dep. c. 523.

398 *HMW in 1801–2; MM as 'frosty moon'*: Wilmot, *Irish Peer*, 27, 38–9.

398 *'very thick together'; 'most excellent woman'*: 31 May and 24 July 1802. Houghton.

398 *stays in Bond Street . . .*: To RB (16 July 1802). Ibid.

398 *'He is an aristocrat . . .'*: 24 July 1802. Ibid.

399 *cross-dressing in Paris*: Wilmot, *Irish Peer*, 42–3, 50–1.

399 '*Talleyrand as cormorant*': Ibid., 47.

399 *'republican homage'*: Ibid., 20.

400 *Mrs Opie's* Poems *given to MM*: Information from Cristina Dazzi. Cini–Dazzi Collection.

400 *'The glare of the arts . . .'*: To RB (20 July 1802). Houghton.

400 *ballets of Vestris; 'moderation of grace'*: Wilmot, *Irish Peer*, 19.

401 *Tighe's background*: Pf: Cini Papers, folder 26, and Burke's *Landed Gentry of Ireland*. An ancestor, Richard Tighe, had been Mayor of Dublin in the 1650s, and Member for Dublin in Cromwell's Parliament of 1656. He was one of many in the Civil War generation who saw it as politic to shift sides. The family continued to serve in the Irish Parliament. They were not titled, but married into the peerage. George Tighe's grandmother on his father's side had been Lady Mary Bligh, daughter of the Earl of Darnley and Lady Theodosia Hyde who was the daughter of the 3rd Earl of Clarendon.

401 *'very handsome'*: Mrs Latouche (formerly Miss Tottenham) in conversation in 1837 with Laurette, the daughter of MM and Tighe, in Rome; reported by Laurette to her father shortly before his death that same year. Pf.

401 *ardent poems*: Cini–Dazzi Collection, San Marcello. Tomalin selected apt quotations in intro. to MWS's *Maurice*, 21.

401 *discreet adultery almost* de rigueur: MacCarthy, *Byron*, 165, pictures the cynical sexual networks of adultery. Hufton, *History of Women*, 145, says that 'no group, unless it was the very poor, so held in contempt the rules laid down in prescriptive literature concerning marital chastity than the European aristocracies'.

402 *'days of adversity'*: To Scully (27 July [1806]), from Munich. Scully, *Papers*, 131–5.

402 *mother's right to keep child to age seven*: Extended to the age of fourteen by law in the Divorce Act of 1857, too late for the mother in Anne Brontë, *The Tenant of Wildfell Hall* (1848), who has to go into hiding under an assumed name if she is to rescue her son from an abusive father.

402 *'I am resolved . . .'*: Scully, *Papers*, 133.

403 *3rd Earl*: Succeeded his father who died in 1799, a year after his trial.

403 *'his character'*: Scully, *Papers*, 133.

403 *Another 'Yes, Laura' poem*: 'Palinode', n.d., Cini–Dazzi Collection.

403 *'Let us cease . . .'*: poem (untitled), dated 1808. Ibid.

403 *'Fall of Jena'*: Alongside Tighe's poem on the battle, MM wrote about 'The Lost Boy'

who is separated from his family as people flee from an embattled town. *Stories of Old Daniel*.

403 *MW on women as physicians*: *RW*, ch. 9.

403 *'In one thing alone . . .'*: Scully, *Papers*, 133.

404 *MM disguised as a man*: CC to Silsbee c. 1875. Silsbee Papers, box 7, file 2. Cited in *ClCor*, i, 135.

404 *travel notes of a Frenchman*: Undated. Discovered and transcribed by Cristina Dazzi, Cini–Dazzi Collection. English translation (by Sandrine Sanos) is on the Internet site accompanying this book.

404 *MM's income*: Supplemented by Tighe, who had a small estate in County Westmeath. His father had left the estate in debt, but he still had £200 a year from his Irish rents, and in 1815 he also acquired an annual Civil List pension of £400. So, together, he and MM had £1400 a year. Contrast with the £40 a year middle-class women like the Wollstonecraft sisters could earn as governesses.

405 *'Mrs Mason'*: It should be noted that Tighe did not consent to her being known as Mrs Tighe, unlike the willing way GI conferred his name on MW.

405 *'that middle rank'*: Revelation for her daughters (Apr. 1818).

405 *decline of English society in Pisa*: In an 1819 letter to the Shelleys, MM notes that the number of English families had fallen to five. MW's letters to the Shellys are in the Abinger: Dep. c. 517.

406 *books taken to Italy*: Inherited by Nerina Tighe. Listed in Pf: Cini Papers, folder 23.

407 *'extravagant political notions'* . . . : Sadleir's ignorant preface to Wilmot, *Irish Peer*, vii–viii.

407 'religion'; *'my medical studies'*: To MWS (18 May [1826] and 29 Sept. [1823]). Abinger: Dep. c. 517

408 Stories for Little Boys and Girls: Mortimer Rare Book Room, Smith College. With thanks to curator Karen Kukil to whom I wrote on the offchance of finding so rare a book.

408 *'Memorandums'*: Pf: Cini Papers, folder 17. Amongst Tighe's papers, but the content looks like MM's. The hand remains to be ascertained. A note here from the 1950s by her biographer, McAleer, shows him uncertain.

408 *FI's 'florid health'*: *Memoirs*, ch. 1.

409 *'the Midwives'*: Marginalia to supercilious article 'On the State of the Poor in Italy', whose author blames deformities on 'the extreme carelessness in the science of midwifery, that generally being practised by women'. Pf: Cini Papers.

409 *MM's opinion on the placenta*: From her *Advice to Young Mothers*. This defence of MM's medical opinion in its historical context came from Siamon Gordon, Sir William Dunn School of Pathology, Oxford. More on MM's medical opinions in ch. 17 below.

409 *Shelley's eccentric appearance*: CC's recollections for Silsbee (26 Apr. 1876). Silsbee, Papers, box 8, file 4. Cited in *ClCor*, 169.

409 *WG to MM*: Abinger: Dep. c. 524/7. Full text in Appendix II to *MWSJ*, 585.

410 *'vagabond'*: To PBS [Jan. 1820]. Abinger: Dep. c. 517.

410 *kissed*: CC to Mrs Gisborne (13 Nov. 1819), *ClCor*, 133.

410 *'frankness' and* 'cant': MM to PBS (14 Nov. [1819] and [winter 1819–20]). Abinger: Dep. c. 517.

410 *Peterloo Massacre*: On 16 Aug. 1819, a month and a half before the Shelley party and MM met in Pisa. Soldiers charged a peaceful crowd of petitioning workers

from the Manchester cotton factories who had gathered at St Peter's Fields.
410 *'Since my country . . .'*: MM to the Shelleys [winter 1819–20]. Abinger: Dep. c. 517.
410 *'as good a physician'*: MM to MWS (31 Dec. 1819). Ibid.

17 DAUGHTERS

411 *no questioning of marriage; no atheism*: CC to Trelawny (30 May 1875), *ClCor*, 627.
412 *'baby-sullenness'*; *'the worst of tempers'* : 28 Oct. 1803, KP, ii, 98–9.
412 *eager sympathy . . . truth*: These were the qualities MWS emphasised in notes for a biography of her mother. Ibid., i, 231 and Clemit et al, *Lives of the Romantics*, ii, ed. Harriet Jump.
412 *'unsinking'*: WG to Mary Jane Godwin (12 Sept. 1812), Appendix A to *ClCor*, ii, 643–4.
412 *intellectual women*; *Katherine Parr*: See ch. 4 above and note to ch. 16.
413 *aim of education*: *The Enquirer*, cited by Clemit, 'Anarchism in the Schoolroom', 67.
413 *more emphasis on republican virtues*: St Clair, 'Godwin as Children's Bookseller', 173.
413 English Dictionary: Published under the pseudonym of Mylius.
414 *definition of 'revolution'*: Clemit, 'Anarchism', 48.
414 Mounseer Nongtongpaw: St Clair, 'Children's Bookseller', 175, suggests that MWS may have written other such works.
414 *the Wollstonecrafts and Fanny*: Rowan to EW (8 Mar. 1805). Abinger: Dep. b. 214/3.
414 *£3000*: Harriet Shelley to Mrs Nugent in Dublin (20 Nov. [1814]), Shelley, *Letters*, i, 421: 'I told you some time back Mr. S. was to give Godwin three thousand pounds.'
415 *'one of the daughters of that dear MW'*: To Mrs Nugent, ibid., i, 327.
416 *Shelley's appearance and voice*: CC's wonderfully vivid recollections. Silsbee Papers, box 8, folders 3 and 4 (dated 18 Apr. 1874). A few cited in *ClCor*, ii, 657.
416 *a female court*: Tomalin, *Shelley*, 3.
416 *'Then it was Fanny Imlay he loved'*: CC in conversation with Silsbee (Silsbee Papers, box 8, folder 3). 'Then' refers to the period before he met their sister Mary. (Shelley and Mary first met on 11 Nov. 1812 when Mary was fifteen, but since Shelley left London two days later, he can't 'then' have been much interested in her.) The first (1886) edition of Dowden's *Life of Shelley* has an appendix demolishing the reliability of Mrs Godwin's letters to MM, as well as CC's copies of them. These letters include reports of the attraction between Fanny and Shelley, so doubt was cast on this, along with everything else. There is no way of proving so delicate a matter, but we might assume some level of attraction on the basis of Shelley's romancing Fanny as MW's daughter and on the evidence of his flirtatious letter to her.
416 *'as women love'*: Ibid., box 7, folder 3. Shelley's weakness, CC adds, was not for the 'subordinate part of love'.
416 *'Fanny loved Shelley'*: Ibid., box 7, folder 2.
416 *the plain girl*; *'odd'*: Crabb Robinson, Diary (12 Feb. 1817), *Books and their Writers*. Harriet Shelley also calls her 'plain', though plainness is negated by the radiance of her mind – reflecting Shelley's impression.
416 *'nothing' of her mother*: Aaron Burr, *Journals*.
416 *FI looked like GI*: MW's observation to GI.
417 'Fannikin': *Travels*, letter 12.

417 *WG on his daughters' education*: To E. Fordham (13 Nov. 1811). Abinger: Dep. b. 214/3. KP, ii, 213–14.

417 *'There is a peculiarity in the education of a daughter . . .'*: From MWS's autobiographical novel, *Lodore*, quoted in Dunn, *Mary Shelley*, 20.

418 *'bad baby'*: Humphrey Carpenter and Mari Prichard (eds), *The Oxford Companion to Children's Literature* (Oxford University Press, 1984).

418 *'amiable' . . . pitted with smallpox*: Reveley, 'Notes and observations' (after Oct. 1859). *SC*, x, 1137.

418 *'So young in life . . .'*: CC to FI (28 May 1815), *ClCor*, i, 10.

418 *'weight'*: *The Watsons*, ed. Margaret Drabble (repr. Penguin Books, 1974), 142.

419 *nut-brown hair*: Often said to be red-gold, but a lock preserved in Pf has no trace of red.

419 *'Shelley was in love with her'*: Silsbee Papers, *MSS* 74, box 7, folder 2. CC says that FI was sent to Dublin but there is no supporting evidence.

419 *PBS told Mary that Harriet no longer loved him*: Ibid., box 8, folder 4.

419 *'pride & delight'*: To Frances Wright (12 Sept. 1827), *MWSL*, ii, 4: 'The memory of my Mother has been always . . . the pride & delight of my life', Mary said, and 'the admiration of others for her, has been the cause of most of the happiness I have enjoyed.'

420 *'I would unite'*: 'To Mary——', *SPP*, 101–5. Dedicatory verses to 'Laon and Cythna; or the Revolution of the Golden City' (later revised as *The Revolt of Islam*).

420 *'that churchyard . . .'*: 18 June 1824, *MWSJ*.

420 *'spirit's mate'; beautiful and free*: 'To Mary——', *SPP*, 101–5.

420 *'It is no reproach . . .'*: 14? July 1814, Shelley, *Letters*, i, 390.

420 *PBS draws on MW's views of free love*: 17 Aug. 1812, to James Henry Lawrence, author of *The Empire of the Nairs; or the Rights of Woman* (1811).

421 *Mrs Godwin . . . blamed Mary Godwin*: Maria Gisborne, *Journals and Letters*, reports on discussions with WG in London in July 1820. Mrs Godwin refused to see her because of her closeness to MWS.

421 *'I shall ever remember . . .'*: CC to Byron (?16 Apr. 1816), *ClCor*, i, 36.

422 *'I have determined . . .'*: 29 May, 1816, ibid., i, 49.

422 *'the dreadful state of mind'*: To MWS (29 July–1 Aug. 1816), ibid., i, 54.

422 *FI and Robert Owen*: Taylor, *Eve and the New Jerusalem*, 5–6, sees MW as a forerunner to the Owenite-socialist feminists of the 1830s and '40s who did not regard the liberation of their sex as an isolated goal but as part of a historic movement towards a 'new age . . . of perfect harmony between the aspirations of the individual and the collective needs of humanity as a whole'.

423 *'Fanny comes . . .'*: *MWSJ* (13 Mar. 1815).

423 *'I cannot say . . .'*: CC to Byron, after FI's death, *ClCor*, i, 92.

423 *'Poets . . .'*: The final line of Shelley's essay, 'A Defence of Poetry', published eighteen years after his death, in 1840. *SPP*, 535.

424 *'sufferings'*: FI to EW (9 Apr. 1816), *ClCor*, i, 23.

424 *'my unhappy life'*: FI to MWS (29 July–1 Aug. 1816), *ClCor*, i, 58. Jane Austen, *The Watsons*, 110, endorses the drudge view of teaching for women in an exchange between Emma and Elizabeth Watson. Emma says to her eldest sister that there is an alternative to loveless marriage in teaching. Elizabeth puts her right: 'I've been

in a school, Emma, and know what a life they lead you; *you* never have.' Marriage to any decent man is said to be preferable.

424 *'stupid letter'*: *MWSJ* (4 Sept. 1816), 138.

425 *FI's note to MWS on 8 Oct*: *MWSJ*, 139, plus a missing letter (referred to by Lady Shelley writing to Alexander Berry on 11 Mar. 1872, cited in notes to *ClCor*, i, 85).

425 *FI's note to PBS*: CC, whose recollection of detail is on the whole remarkably good, recalled this for Silsbee, though she admits that she did not actually see the note. PBS scrunched it up. Silsbee Papers, box 7, folder 3.

425 *FI's warning and suicide note*: *ClCor*, i, 85–6.

425 *candle consume it*: I can't agree with alternative speculation that the servants could have destroyed the name to avoid a scandal or that Godwin's representative would have done so. In either case, the identity would have got about. It wouldn't have been in character for Shelley to have done it – he'd have been acutely distressed rather than worldly-wise. It would seem in character for Fanny, who was mindful of others.

426 *'owing to the preference . . .'*: Gisborne, *Journals and Letters* (9 July 1820), 39. CC offered the same explanation in an interview when she was old. Silsbee Papers, box 7, folder 2. Cited in *ClCor*, i, 88–9.

426 *'On Fanny Godwin'*: MWS omitted the first line, as well as what was on the back, when she came to edit PBS's poems.

426 *'monster'*: Recalled by CC. Silsbee Papers, box 8, folder 4.

426 *Harriet's suicide*: Cameron, *SC*, iv, 769–802, dates this 7 Dec.

426 *'Thy little footsteps . . .'*: This was thought to be about the death of William Shelley in Rome in 1819, but G. M. Matthews proved that it is addressed to FI: in 'Whose Little Footsteps?', in *The Evidence of the Imagination*, ed. Donald Reiman, M. C. Jaye and B. Bennett (New York: 1978).

427 *MWS reread* RW: *MWSJ* (Fri. 6– Mon. 9 Dec. 1816), 149.

427 *the impact of FI's 'unfortunate' birth*: See Claire Tomalin's convincing and touching portrait of Fanny in *Shelley*, 52. Tomalin notes how MWS and CC had embarked on bearing children outside marriage 'to a considerable degree in conscious emulation of Mary Wollstonecraft'.

427 *'I cannot pardon'*: 12 Jan. 1818, *ClCor*, i, 110.

427 *CC's novel*: CC to Byron (c. Mar./Apr. 1816), ibid., 33. Could an idea for a Crusoe sort of girl have derived from *The [Swiss] Family Robinson*, published by CC's mother under the Juvenile Library imprint? 'The Ideot' as well as CC's voice to Byron suggests a pre-Brontë experiment. A precursor to Jane Eyre does not speak to the man she loves through custom and conventionalities; her voice comes from a soul equal to his.

427 *'hateful novel thing'*: CC to Byron (19 Nov. 1816), ibid., 92. CC claimed to have learnt her ambiguity from Gibbon. 'I would be exactly like a diamond who to the right reflects purple & to the left pink,' she explained to Byron.

427 *Byron's women*: Fiona MacCarthy, *Byron*, 163, 173, suggests that the numbers of women Byron bedded and his power over them served in some way as a distraction from the homosexuality he strained to repress.

427 *ten minutes' happy passion*: CC in Moscow to Jane Williams in London (Dec. 1826), *ClCor*, i, 241.

428 *PBS blamed MWS for coldness*: In 'Epipsychidion'.
428 *love for women friends; political*: Bennett, intro. to *MWSL*, i, xiv, xviii.
428 *'greatness of soul'*: To Frances Wright (12 Sept. 1827), *MWSL*, ii, 3–4.
428 *Jemima's anticipation of Frankenstein's monster*: *WW*, chs 1 and 5.
428 Frankenstein *and the pattern of women's writing*: Steiner, 'Women's Fiction', 505.
429 *'What art thou?'*; *'I know . . .'*; *'prophecy'*: 'To Mary——', *SPP*, 104.
430 *like 'the breath of summer's night'*: Similar to Byron's words about a voice 'like the swell of summer's ocean' when the breast of the deep 'is gently heaving'. 'Stanzas for Music'.
430 *claimed that she knew Shelley . . .*: Silsbee Papers, box 8, folder 4.
430 *CC's parting from Allegra*: Sent to join Byron in Venice on 28 Apr. 1818.
430 *'Carissima Pisa'*: MWS in retrospect from Genoa to CC (19 Dec. 1822), *MWSL*, i, 299.
430 *PBS and MM*: Silsbee Papers, box 7, folder 3.
431 Matilda: The narrator's preoccupation with her death may go back to Richardson's *Clarissa* which MWS reread between June and Aug. 1819, immediately after her son died. The title of the first draft of *Matilda* was 'Fields of Fancy', close to MW's unfinished 'Cave of Fancy'. It's not easy to be sure what is the link.
432 *'often did quiescence . . .'*: MWS, 'Life of Godwin', 97.
432 *'Minerva'*: Byron spoke of MM as 'Claire's Minerva' to MWS.
432 *'I made . . .'*: CC to her Viennese sister-in-law, Antonia Clairmont (16 Aug. 1856), *ClCor*, 578.
433 *'Let me . . .'*: *MWSJ* (25 Feb. 1822), 399–400.
433 *CC to Byron about Allegra*: Silsbee Papers, box 8, folder 4.
434 *Allegra's death*: Said to be either from typhus or 'after a convulsive catarrhal attack'.
434 *'That he should hate . . .'*: MM to MWS, who was assisting Byron in Genoa (1 Feb. [1823]). Abinger: Dep. c. 517.
434 *CC's silence*: There is one exception. MM mentions to MWS having received a letter from CC in St Petersburg (10 Sept. 1823), saying that she was reasonably satisfied with her situation. Ibid. Not in *ClCor*.
434 *MWS's 'treasure' . . .*: *MWSJ* (2 Oct. 1822), 429. Charlotte Brontë, then four years old, was another strong character who said the same, seventeen years later:

The human heart has hidden treasures
In secret kept, in silence sealed –

The unseen space where a governess had to exist led to the explosiveness of her submerged words. When the Brontë sisters revealed their thoughts by night as they marched around the dining-room table in their Yorkshire parsonage, Charlotte's friend Ellen Nussey thought they blew out the candles for economy. But darkness was a liberating cover; invisibility, a form of freedom.

435 *'out of his hand'*: *CCJ* (8 Jan. 1827), 407.
435 *'My soul'*: Ibid., n.d., 429.
435 *MW mocked a showcase education*: *RW*, ch. 12 (on national education).
435 *'They educate a child . . .'*: To MWS from Moscow (29 Apr. 1825), *ClCor*, i, 215.
436 *'You write . . .'*: 29 Nov. 1842. *MWSL*, iii.
436 *domestic trials in Russia*: *ClCor*, i, 222.
436 *'From Morning till Night'*: CC to MWS (24 Mar. 1832), ibid., i, 286.
437 *'misery'*: CC in Moscow to Jane Williams (27 Oct. 1825), ibid., i, 230.

437 *'The world is closed . . .'*: 30 Jan. 1827, *CCJ*, 411.
437 *'the most contemptible of all lives . . .'*: 25 Feb. 1822, *MWSJ*, 399.
437 *'I believe . . .'*: Ibid., 554.
437 *'Composing;*: (17 Nov. [1822?]). Abinger: Dep. c. 517.
437 *'I can, I do.,'*: MWS to Jane Williams (7 Mar. 1823), *MWSL*, i, 320 .
438 *sensitivity and public effectiveness*: See J. S. Mill, *The Subjection of Women*, ch. 3.
438 *'mind appeared more noble'*: Revelation of her past for her Tighe daughters.
439 Advice *completes MW's work*: St Clair, 'Godwin as Children's Bookseller', 179.
439 *'Teach a being . . .'; cultivate* courage: *Advice*, 330, 342.
439 *Vaccà's comments on* Advice: MM to MWS (12 Sept. [1823]). Abinger: Dep. c. 517.
440 *correspondence with Dr Parkman*: MM initiated the correspondence in 1816. Parkman had proposed an asylum in 1814, and went on to write a pamphlet (1817) on the management of the insane. MM owned a copy, and also his book *On Hysteria*. He was murdered in 1849, aged about fifty-eight, by Professor John W. Webster, MD, in the New Medical College, Boston. For further details see Oliver Wendell Holmes, *The Benefactors of the Medical School of Harvard University* (Boston: 1850). Some of Parkman's remedies suggest a self-absorbed crackpot conjoined with the man of sense.
441 *MM's revelation of her past to her daughters*: Pf. *SC*, viii, 909–11.
442 *stray scrap*: Pf., Cini Papers, folder 26.
442 The Sisters of Nansfield: Published in London. Copy in Bodleian.
443 *Casa Lupi*: On what was then the Via San Lorenzo. Now numbers 15–19 on the corner of the present-day Via Andrea Vaccà Berlinghieri. It does seem appropriate that she should have lived in a street later named after her medical mentor.
443 *Accademia di Lunatici*: Alternatively, the name may have come from the Lunar Society of Birmingham (see Uglow, *Lunar Men*), which included Darwin and Priestley, both published by Johnson and known to MW.
443 *centre of town*: 1069 Via della Faggiola.
444 *'Nothing can equal . . .'*: 26 Oct. 1832, *ClCor*, i, 290–1.
444 'singular family': CC to MWS (16 Sept, 1834), ibid., i, 313.
445 'amor della patria': Cini's *Avvertimento* to posthumous Italian edition of MM's *Advice*. Pf.
445 *George Eliot's tribute: Middlemarch*, concluding paragraph.
446 *CC was buried*: At the cemetery of S. Maria in the commune of Bagno a Ripoli, about three and a half miles from Florence. (Photos in Pf.) She had told one William Graham of her wish to be buried with Allegra. Graham, letter (5 Jan. 1894), Cini Papers, Pf.
446 *'the subterraneous community of women'*: Shelley's entry in *MWSJ*, 32. The evening before, Thursday 6 Oct. 1814, CC had read some of MW's letters.
446 *'The party of free women'*: *ClCor*, i, 314–15.

18 GENERATIONS

447 *'not the experience of one life only . . .'*: T. S. Eliot, *The Dry Salvages*: II.
447 *Thompson's aim; reprinted passages* from RW: Taylor, *Wollstonecraft*, 248.

447 *Mary Wollstonecraft [sister-in-law]*: *Boston Monthly Magazine*, i (Aug. 1825), 126–35. Publ. under the pseudonym of D'Anville.

448 Distinguished Women: 'Female Biography: Containing Notices of Distinguished Women', in *Dictionary of Christian Biography* (Philadelphia: Leary & Getz, 1834) by Samuel Knapp. (Discovered in the Bodleian Library by Professor Kathryn Sutherland.) Knapp, the proprietor of the *Boston Monthly Magazine*, had clearly met this Mary Wollstonecraft. He tells us that she craved sun, and after Charles died in 1817, transplanted herself (taking her stepdaughter, Charles's daughter from an earlier marriage) from New Orleans to Cuba. She had no children. Charles had joined the American army where he remained and had been promoted.

448 *'surprised'*: De Tocqueville, *Democracy in America*, trans. George Lawrence, ed. J. P. Mayer (New York: Doubleday/Anchor, 1969), 590–1 ('Education of Girls in the US').

449 *Elizabeth Barrett Browning and MW*: See Margaret Reynolds's introduction to her critical edition of *Aurora Leigh* (Ohio University Press, 1992), 12–19, and notes 60–65. With thanks to Mark Bostridge for passing this on, as well as the fact that Barrett Browning refuted a suggestion that Florence Nightingale's ministry in the Crimea could be accounted a step gained for her sex.

449 *George Eliot and MW*: *Daniel Deronda* (1876), ch. 17. See George Eliot to Rabbi Deutsch: *Letters*, v, 160–1. With thanks again to Mark Bostridge whose *Vera Brittain: A Life*, 366, reveals that she, too, was bent on fictionalising elements of MW's life. Her novel (which never got beyond the planning stage) was to be set against the backdrop of the French Revolution, and entitled: 'Behold This Dawn'. More recent novelists Frances Sherwood and Michèle Roberts have also fictionalised elements of MW's life.

449 *George Eliot links MW and Margaret Fuller*: 'Margaret Fuller and Mary Wollstonecraft', *Westminster Review* (13 Oct. 1855), repr. in George Eliot, *Selected Essays, Poems and Other Writings* (Penguin Classics, 1990), 332–8. Fuller's work was published in 1843.

449 *'what is now called the nature of women . . .'*: *The Subjection of Women*, ch. 1. Mill was the first to back women's suffrage in the British Parliament, by presenting a petition signed by nearly 1500 women and speaking in favour of amending Disraeli's Reform Bill of 1867 to read 'person' rather than 'man'.

449 *six generations*: *The Voyage Out*, ch. 16.

449 *'The great problem is the true nature of woman'; 'almost unclassified'*: *A Room of One's Own* (1929; originated in talks for women students at Newnham College and for the ODTAA Society at Girton College, Cambridge University, in 1928). A decade later Virginia Woolf again circled the mystery of 'our still unknown psychology' in *Three Guineas*.

449 *'I am a rising character'*: *Villette* (1853), ii, ch. 27.

450 *'"Nature" is what we know . . .'*: '"Nature" is what we see', in *Poems of Emily Dickinson*, ed. Thomas H. Johnson (London: Faber, 1975), no. 668; Variorum Edition, ed. R.W. Franklin (Cambridge, Mass.: Belknap, 1998), no. 721.

450 *'Let us . . . trust our whole nature'*: Lyndall Gordon, *A Private Life of Henry James: Two Women* (London: Vintage; New York: Norton, 1999), 107.

450 'grande nature': Henry James, Preface to *The Portrait of a Lady* (1881).

450 *'Who was she . . .'*: Ibid., ch. 12.

451 *MW against violence; 'the real savages'*: One of her last pieces of writing was about 'rapacious whites' savaging and exploiting the indigenous inhabitants of the Cape of Good Hope. Review of the English translation of M. Levaillant, *New Travels into the interior Parts of Africa, by Way of the Cape of Good Hope*, *AR* (May 1797); *MWCW*, vii, 479–84. Cited by Taylor, *Wollstonecraft*, 240–1.

451 *heard the lash*: *RM*. See ch. 7 above.

451 *group portrait*: The artist was Benjamin Robert Haydon, a protégé and critic of Fuseli. See ch. 8 above.

451 *Benedict on MW*: Publ. after her death in *Writings of Ruth Benedict: An Anthropologist at Work*, ed. Margaret Mead (Boston: Houghton Mifflin, 1959).

451 *Virginia Woolf on MW*: 'Mary Wollstonecraft', first publ. in 1929; repr. *The Common Reader*, 2nd series (1932).

451 *'a great season of liberation'*: 31 Dec. 1932, *The Diary of Virginia Woolf*, ed. Anne Olivier Bell, iv (London: Hogarth; New York: Harcourt, 1982), 134.

451 *the edifice of power*: Joan Smith, *Moralities*, is a timely political analysis of mass opinion outside the old power structures, questioning their morality in the twenty-first century.

451 *'Outsider Society'*: Woolf, *Three Guineas*, ch. 3.

451 *'enfranchised till death'*: 28 Apr. 1938, *The Diary of Virginia Woolf*, v, 137.

452 *Women who imitate men lack ambition*: The most politically effective use of this phrase that I've heard came in a speech by Hilary Reynolds of the SABC at a breakfast in Cape Town (2 Aug. 2002) to mark fifty years of a women's programme. Following her, Opposition MP Patricia de Lille was cheered by an almost all-women audience for her speech on government neglect of widespread sexual abuse as well as of the AIDS epidemic ravaging the country. A foremost candidate for blame was a woman health minister, a sycophant of President Mbeki's protracted reluctance to supply life-saving drugs to his dying people. Instead, Mbeki, following Mandela's sales of weapons to brutal regimes, chose to spend billions on weaponry.

BIBLIOGRAPHY

PRIMARY SOURCES

*Documents asterisked are on the Internet site accompanying this book.

Abinger Collection. The main collection of Wollstonecraft, Godwin, and Shelley papers. Bodleian Library, Oxford. Includes Mount Cashell letters to the Shelleys.

—— *Shelley's Guitar*. Bodleian Library's bicentenary exhibition of manuscripts, first editions and relics of Percy Bysshe Shelley by B. C. Barker-Benfield (Oxford, 1992).

Adams, Abigail, 'Diary of her Return Voyage to America' (Apr.–May 1788), in *Diary and Autobiography of John Adams*, ed. L. H. Butterfield, iii (Cambridge, Mass.: Belknap Press, 1961)

—— and John Adams, MS correspondence (with reference to Mary Wollstonecraft). Massachusetts Historical Society, Boston: microfilm reel 377

Adams Family Correspondence, i, vi (Cambridge, Mass.: Belknap, 1993)

*Adams, John. Copious marginalia (debating political issues with Mary Wollstonecraft) in his copy of *FR*. Rare Books, Boston Public Library: 221.15

American Manuscripts 1763–1815: An Index to Documents Described in Auction Records and Dealers' Catalogues (Wilmington, Del.: Scholarly Resources, 1977)

Arden, John, *A Short Account of a Course of Natural and Experimental Philosophy* (Beverley, Yorkshire, 1772)

Astell, Mary, *Reflections Upon Marriage*, 3rd edn (1706)

*Backman, Elias. Letter to the Swedish Regent (15 Mar. 1794), Stockholm Riksarkivet: Biographica B1 (Baar-Baesecke), 6454: 23

Backman, Pierre (brother of Elias Backman), 'Fransmännen på Traneberg', *Västgötabygden*, iii/4 (1958), 272–5

—— Will-letter (1796) mentioning MW to RB from Algiers during an outbreak of the plague: Houghton and *American Literature*, ix (1938), 442–9, and x (1938), 224–7. Draft version in Beinecke: Za Barlow 13

Barlow Papers. Houghton Library, Harvard (b MS Am 1448). The main repository of the Barlow Papers, including Joel Barlow's unpublished letters and charming light verse to his wife, amongst the best things he wrote. Includes the letters of Ruth Baldwin Barlow; letterbooks; account book; memoranda book; diary of 1788.

Substantial collection also in the Beinecke Library (Za Barlow 14; and MS Vault Pequot Box M 886–937, M938–94 and M995–1039). Other Barlow papers are scattered in various American archives such as the Massachusetts Historical Society and the Manuscript Division of the New York Public Library

Benedict, Ruth, *An Anthropologist at Work*, ed. Margaret Mead (Boston: Houghton Mifflin, 1959). Includes her homage to Wollstonecraft, a short biography unpublished in Benedict's lifetime.

Blair, Hugh, *Lectures in Rhetoric*, 3 vols (London, 1785)

Blood, Frances (Fanny). Two letters to Elizabeth and Everina Wollstonecraft (1784–5). Abinger: Dep. b. 210 (9)

Boswell, James, *Life of Johnson*, eds Claude Rawson, Marshall Waingrow, Bruce Redford, Gordon Turnbull (Edinburgh and Yale University Presses, 1994–8)

Brightwell, Cecilia Lucy, *Memorials of the Life of Amelia Opie* (London: Longman, 1854)

Browning, Robert, 'Mary Wollstonecraft and Fuseli', in *Jocoseria* (London: Smith Elder, 1883), 45–9

Burgh, James, *The Dignity of Human Nature* (London, 1754)

—— *Thoughts on Education* (London 1747)

Burke, Edmund, *Reflections on the Revolution in France*, ed. Conor Cruise O'Brien (Penguin Books, 1968)

Burney, Fanny, *The Early Journals and Letters 1768–1791*, eds Lars E. Troide and Stewart J. Cooke, 3 vols (Oxford: Clarendon, 1988–94)

Butler, Marilyn, ed., *Burke, Paine, Godwin, and the Revolution Controversy* (Cambridge University Press, 1984). Excerpts from pro- and counter-revolutionary writings, including Wollstonecraft's, together with a perceptive introduction

Byron, George Gordon, Lord, *Letters and Journals*, 12 vols, ed. Leslie Marchand (London: John Murray, 1973–82)

Cary, John, *Cary's Survey of the Country Fifteen Miles Round London* (London, 1786)

Chesterfield, Philip Stanhope, Lord, *Letters of Lord Chesterfield to his Son* (1774)

Christie, Thomas, *Letters on the French Revolution* (London: Joseph Johnson, 1791)

Clarke, John, *Practical Essays on the Management . . . of Labour; and on the Inflammatory and Febrile Diseases of Lying-in Women* (1793)

Clemit et al, *Lives of the Great Romantics*, III: *Godwin, Wollstonecraft & Mary Shelley by their Contemporaries*, i: *Godwin*, ed. Pamela Clemit; ii: *Wollstonecraft*, ed. Harriet Jump; iii: *Mary Shelley*, ed. Betty T. Bennett (London: Pickering & Chatto, 1999)

Condorcet, Marie Jean Antoine Nicolas de Caritat, marquis de, Tom Paine, et al, Committee for Public Instruction, (Paris: 1793). Published record of day-to-day discussions. Mezzanine floor, reading room, Archives Nationales, Paris

—— essay, 'On the Admission of Women to the Rights of Citizenship' (preceding Olympe de Gouges and MW), trans. Alice Drysdale Vickery (London: G. Standring, 1893)

Cotton, Mrs. Letter of condolence to WG (18 Mar. 1800) recalling MW. Abinger: Dep. b. 214/3: 'I never met her fellow.'

Cowper, William, *The Correspondence of William Cowper*, ed. Thomas Wright (London, 1904)

Crabb Robinson, Henry, *On Books and their Writers*, ed. Edith J. Morley (London: Dent, 1938)

Cutler, Revd Manasseh, *Journals and Correspondence*, 2 vols, ed. W. P. Cutler and J. P. Life (Cincinnati, 1888)

Darwin, Erasmus, *The Botanic Garden* (London: Joseph Johnson, 1791)

—— *Plan for the Conduct of Female Education in Boarding Schools* (London: Joseph Johnson, 1797)

Dazzi, Cristina, 'Humor Inglese: Un Singolare Manuscritto di Mary Shelley sull' universita di Pisa', *Bollettino Storico Pisano* Pisa Historical Gazette, lxviii (1999), 113–20

Debates on the Reports of the Committee of Secrecy (London, 1794)

de Montolière, Baroness, *Caroline de Lichtfield*, novel, trans. 1786 by Thomas Holcroft, a radical writer who was part of the Godwin–Johnson circle. Bodleian Library

Dowden, Edward, *Life of Shelley* (London, 1886)

Ellefsen, Peder, as accused in judicial inquiries. Kristiansand State Archive, Norway: the crew's testimony; Kristiansand Town Magistrate's report to the Danish Chamber of Commerce, Dec 1794; police interrogation in Arendal, 28 and 30 Apr. 1795; letter from Imlay to Backman from Le Havre; Imlay's instructions to Ellefsen on 13 Aug. 1794

—— Deed of sale for Imlay's ship the *Maria and Margrethe* after Ellefsen's arrival in Norway, Aust-Agders Archives, Arendal, Norway.

Eton College Register 1753–1790, ed. R. H. Austen-Leigh (Eton: Spottiswoode, 1921)

Exeter College, Oxford, archival records (King family). Index of entrants 1770–1877 in alphabetical order according to year: C.II.20; Entrance Book 1768–1812: C.II.21; Bursary Archives: King, Caroline N.I.3; King, Henry: N.I.3; King, John: L.V.8; King, Richard: M.IV.7

Farington, Joseph, *The Farington Diary* [13 July 1793–24 Aug. 1802], ed. James Greig (London: Hutchinson, 1923)

Fawcett, Millicent Garrett, Introduction to centenary edition of *A Vindication of the Rights of Women* (London, 1891). In this influential essay, the suffragist leader rehabilitates and co-opts MW as founding figure for the Cause, 'the great movement of which her book was in England almost the first conscious expression' – a movement Mrs Fawcett sees as significant historically as the Reformation and Democracy. She commends MW's 'fearlessness', her dismissal of the slave/queen models of womanhood, her refusal of double standards, and her 'sound heart and clear head'

Fordyce, James, *Sermons to Young Women* (1765)

Foss, Frithjof, *Arendals byes historie* (1893; repr. 1990s). Copy in Aust-Agders Archives, Arendal, Norway.

[Fuseli] Füssli, Johann Heinrich, *Sämtliche Gedichte*, eds Martin Bircher and Karl S. Guthke (Zurich, 1973)

Fuseli, Henry, (Johann Heinrich Füssli), Letters to William Roscoe. Liverpool Record Office (3 refer to MW).

—— *The Mind of Henry Fuseli: Selections from his Writings*, ed. and intro. Eudo C. Mason (London: Routledge & Kegan Paul, 1951). With thanks to artist and art historian Timothy Hyman for his copy

Gardiner, Jane, Letter to William Godwin (15 Jan. 1799). Abinger: Dep. b. 214/3

—— *English Grammar* (1799; repr. 1808, 1809)

—— *Exercises Adapted to the English Grammar* (London: Longman, 1801). No copies in the

British or Bodleian Libraries. I tried in Yorkshire libraries – no luck. (Durant, too, searched in vain for letters MW is said to have written to a friend, Miss Massey, printed by Jane Gardiner in a volume called 'English Exercises'. Eleven Letters from Beverley, Bath and London.)

—— Sometime after 1803 (date of watermark) Gardiner or emanuensis copied out fifteen of MW's letters into a notebook, the only primary documents to survive from the first twenty-one years of her life. Pforzheimer Collection, New York Public Library

—— *An Excursion from London to Dover* (1806), 2 vols. Bodleian Library

Gardiner, Everilda Anne, *Recollections of a Beloved Mother*. The subject is MW's Yorkshire schoolfriend, Jane Arden (London, 1842)

Genlis, Stephanie de, *Memoirs* (London, 1825)

Gisborne, Maria, and Edward E. Williams, *Shelley's Friends: Their Journals and Letters*, ed. Frederick L. Jones (Norman: University of Oklahoma Press, 1951)

Godwin, William. Unpublished letters. Abinger

—— *Godwin and Mary: Letters*, ed. Ralph M. Wardle (Lawrence, Kansas: University of Kansas Press, 1966; London: Constable, 1967)

—— Autobiographical fragments in *Collected Novels and Memoirs of William Godwin*, i, intro. Marilyn Butler and Mark Philp (London: Pickering & Chatto, 1991)

—— *Political and Philosophical Writings of William Godwin*, ed. Mark Philp (London: Pickering & Chatto, 1993)

—— *Cursory Strictures*. Pamphlet (1794). Copy in Beinecke

—— *Caleb Williams* (1794; Oxford University Press: Worlds Classics)

—— *Bible Stories* (London: R. Phillips, 1803). A copy of the almost vanished first edition of vol. i, with the educative Preface (echoing Wollstonecraft, and which Godwin wished to have included in his collected works), is in Smith College, Northampton, Mass.: Mortimer Rare Book Room, Neilson Library: 371.342 C436e 1803 bi. Amongst the fine collection of Godwin's Juvenile Library publications, including Mount Cashell's below.

—— *Life of Lady Jane Grey* (London: M. J. Godwin, 1806). The accompanying portrait of this pale, intellectual girl, aged about fourteen, looks rather like Mary Wollstonecraft Godwin, then aged nine

——— *Memoirs of the Author of 'The Rights of Woman'*, ed. and intro. Richard Holmes (Penguin Classics, 1987)

—— Obituary for Joseph Johnson, *Morning Chronicle* (London, 1809)

Gregory, John, *A Father's Legacy to his Daughters* (London, 1774)

Hammond, George, British Minister to the US, warning letter to Foreign Secretary Lord Grenville (9 Jan. 1792) about rumoured activities of Imlay associate James Wilkinson in Kentucky. (These abortive preparations for a coup against Spanish colonies predate Imlay's similar plot instigated in Paris, Dec. 1792.) PRO, London: FO4/14

Hays, Mary, *The Memoirs of Emma Courtney* (1796; London: Routledge/Pandora, 1987)

—— *Love Letters*, ed. A. Wedd (London, 1925)

—— Letter to Godwin (1797). Abinger

—— Letters. Pforzheimer Collection, New York Public Library

—— 'Memoirs of Mary Wollstonecraft' (unsigned obituary), in *The Annual Necrology for*

1797–8 (London: Phillips, 1800), 411–60. Preceded by a brief notice in *Monthly Magazine*, iv (Sept. 1797), 232–3. Clemit et al, *Lives of the Great Romantics*, III, ii, ed. Jump, 5–8, 189–92

Hazlitt, William, *The Spirit of the Age; or Contemporary Portraits* (London: Henry Colburn, 1825; repr. New York: Dutton, 1955). See 'William Godwin', 29–53, and 'The Late Mr. Horne Tooke', 99–120, in 1825 edn

—— 'On the Old Age of Artists', in *The Plain Speaker: Opinions on Books, Men and Things* (London: Henry Colburn, 1826). Searing paragraph on Fuseli

—— 'Memoir of Henry Fuseli', *Pamphlets on British Art*, xviii (London: A. & R. Spottiswoode, c. 1926)

—— (ed.) *Autobiography of Thomas Holcroft* (London, 1816)

Hewlett, John, *Sermons* (London: Joseph Johnson, 1786)

Imlay, Gilbert. Imlay Family Papers, Alexander Library, Rutgers University

—— Records of the Imlay family (including those serving in the Revolution). Van Kirk Collection, Allentown Public Library, NJ

—— *A Topographical Description of the Western Territory of North America* (London: Debrett, 1792; repr. 1797)

—— *The Emigrants* (London: Debrett, 1793; repr. Penguin Classics, 1998). Extensive background, map and contexts in intro. by W. M. Verhoeven and Amanda Gilroy (eds)

—— versus George James. Court case heard by Lord Loughborough in Leicester (10 and 27 Feb. 1798). PRO, London: C12/2188/14

—— *Index to Revolutionary War Service Records*, transcribed by Virgil D. White (Waynesboro, Tenn.: National Historical Publishing Co. 1995)

—— Casualty book of Forman's Regiment, NJ State Archives: Monmouth county war records, MS 4126

—— Records of speculations in Kentucky. The Filson Historical Society, especially the May Papers. Louisville, Ky, has deeds and correspondence; Four entries for Kentucky land with helpful dates listed in Willard Rouse Jillson, *Old Kentucky Entries and Deeds: A Complete Index to All of the Earliest Land Entries, Military Warrants, Deeds and Wills of the Commonwealth of Kentucky* (Filson Club Publications, no. 34; first publ. Louisville, Ky, 1926; repr. 1969)

—— Letter to Henry Lee (21 Apr. 1784). Beinecke Library's General MSS Miscellany, Group 2444, item 5-1

—— Two secret plans for the French capture of Spanish Louisiana. *Archives des Affaires Étrangères, Louisiane et Florides*, 1792–1803, vii, doc. 1. 'Observations du Cap. Imlay' trans. in Documents section of *Annual Report of the American Historical Association*, i (1896), 953–4. Second and much longer 'Mémoire sur la Louisiane' trans. amongst associated 'Documents on the Relations of France to Louisiana, 1792–1795' in *American Historical Review*, iii (Apr. 1898), 491–4. See also 490–503, 508, 651, 660. Includes details of Barlow's involvement and separate proposal

—— Power of attorney for Mary Wollstonecraft and her instructions for the Scandinavian journey. Pforzheimer Collection: microfilm of Abinger Papers: reel 9

Johnson, I. B. Recollections of MW in Paris during 1793, in letter to Godwin. Abinger: Dep. b. 214/3. Clemit et al., *Lives of the Great Romantics*, III, ii, ed. Jump, 13–16

Johnson, Joseph. Letterbook (1795–1810). Copies of c. 240 outgoing letters to various recipients, including many of the writers he published – Darwin, Priestley, Maria Edgeworth – and also significant letters to Charles Wollstonecraft in America. Pforzheimer Collection

—— Letter to William Godwin. Abinger: Dep. b. 210/3

—— 'A few facts'. Posthumous reminiscences of MW. Abinger: Dep. b. 210/3. KP, i, 193–4, and Clemit et al, *Lives of the Great Romantics*, III, ii, ed. Jump, 9–12

—— Note on Mary Wollstonecraft's death to Henry Fuseli. Abinger: Dep. b. 210(3)

Jones, Vivien (ed.), *Women in the Eighteenth Century: Constructions of Femininity* (London: Routledge, 1990). Sets MW in the contemporary context of conduct books, educational treatises and women's rights

Juvenile Library. Magazine for boys and girls, 3 vols (1800–1). Berg Collection, New York Public Library

Kames, Henry Home, *Elements of Criticism* (1762; repr. New York: Twayne, 1970)

————- *Loose Hints upon Education* (repr. Routledge, 1993)

King-Harmon Papers, Public Record Office of Northern Ireland: D/4168

Knapp, Samuel, 'Female Biography: Containing Notices of Distinguished Women', in *Dictionary of Christian Biography* (Philadelphia: Leary & Getz, 1834). Portrait of MW's American sister-in-law of the same name

Knowles, John, *The Life of Henry Fuseli*, 2 vols (Henry Colburn, 1831)

Lawes Resolutions of Women's Rights, or the Lawes Provision for Women (1652)

The Laws Respecting Women, as they regard their natural rights . . . in which their interests and duties as daughters, wards, heiresses, spinsters, sisters, wives, widows, mothers, legatees, executrixes, etc, are ascertained and enumerated (Joseph Johnson, 1777; repr. Oceana Press, 1973)

Levy, Darlene Gay, Harriet Branson Applewhite and Mary Durham Johnson, eds, *Women in Revolutionary Paris 1789–1795: Selected Documents* (Urbana: University of Illinois Press, 1979)

Macaulay (Graham), Catharine, *Letters on Education* (London, 1790)

Marie and Margrethe (the silver ship). Shipping records for 1794–5 in the Landsarkivet, Gothenburg. In the Riksarkivet, Stockholm see Kommerskollegium: Huvadarkivet C II c Fribrevsdiarier (1758–1831), xxxvi (1795), No. 669, and Fribrevshandlingar (1775–1831) Huvudarkivet F II b: vol. 143.

—— Testimony of the crew, Dec. 1794, Kristiansand, Norway

—— Records and hearsay in archives of Oslo and Arendal, Norway

—— A magnificent collection of Schleswig-Holstein silver in the North German Museum in Altona, the final destination of MW's journey in 1795, testifies to a significant trade in silver at that time

—— Charges for the ship in Strömstad (June–Sept. 1795, the precise period of Wollstonecraft's journey to Scandinavia and Hamburg). Strömstads Rådhusrätt och Magistrat, vol. A IV a: 9, pp. 255–6, 291–312, in Landsarkivet, Gothenburg

Mathias, T. J., 'The Shade of Alexander Pope', repr. Clemit et al, *Lives of the Great Romantics*, III, ii: *Wollstonecraft*, 169–73

Milton, John, *Paradise Lost* (a favourite source of quotation for MW)

Monthly Magazine and American Review for 1779: 'Reflections on the character of Mary Wollstonecraft Godwin' by 'L. M.' (New York, 1800), 330–5. Shows the adverse impact of W. G's *Memoirs* and *WW* on an admirer of MW.

More, Hannah, *Strictures on the Modern System of Female Education* (London, 1799)

Morris, Gouverneur, Papers (of the American Minister in Paris during the Terror). Huge, largely untapped trove. Columbia University, New York

—— *Diary and Letters*, 2 vols, ed. Anne Cary Morris (London: Kegan Paul, 1889)

—— *A Diary of the French Revolution*, 2 vols, ed. Beatrix C. Davenport (repr. Freeport, New York: Books for Libraries, 1971)

Mount Cashell, Margaret (King Moore). Private library of Andrea and Cristina Dazzi, San Marcello Pistoiese, Italy. In 1958, Mount Cashell's biographer, McAleer, reports that a search of the hundred-room house at San Marcello failed to discover missing papers, especially her letters from Mary Wollstencraft

—— Cini Papers, Pforzheimer Library, New York Public Library

—— Three anonymous political pamphlets against Union. Copies in the National Library of Ireland, Dublin: 'A few words in favour of Ireland' (1799); 'Reply to a ministerial pamphlet entitled "Considerations upon the state of public affairs in the year 1799: Ireland" (1799); and 'A hint to the inhabitants of Ireland. By a native' (1800).

—— A few letters in 1805 to her Irish lawyer, Scully, in Brian MacDermot, ed., *The Catholic Question in Ireland and England 1798–1822: The Papers of Denys Scully* (Irish Academic Press, 1988)

—— Revelations about her background and history for her unknowing youngest daughters, Laurette and Nerina Tighe (1819). Pforzheimer Collection. *Shelley and his Circle*, viii, 909–11

*—— Undated, untitled fictional monologue/travelogue of a Frenchman in the German states, in French, discovered and transcribed by Cristina Dazzi. Private Dazzi library, San Marcello Pistoiese

—— *Stories for Little Boys and Girls, in Words of One Syllable* (London: M. J. Godwin, 1810). Mortimer Rare Book Room, Smith College: 371.342/C426e/1810mos. Extremely rare and not in the great Opie Collection. Puzzling dedication to 'N- and her little friends'. We might assume this was her child Nerina, but this first edn was publ. in 1810 and Nerina was not born until 1815

—— *Stories of Old Daniel* (London: M. J. Godwin, 1807) and *Continuation of the Stories of Old Daniel* (1820). There were fourteen editions by 1868. An Italian translation, *Racconti del vecchio Daniele per dilettare ed isruire la gioventu; continuazione dei racconti del vecchio Daniele* (Pisa: Nistri, 1829) was reprinted a number of times. The author's anonymity remained unbroken. The Bodleian catalogue, before it went on-line in 2000, listed these stories under Charles Lamb, with a note to say that they had previously been catalogued under the fictitious name of 'Daniel'

—— *The Sisters of Nansfield: A Tale for Two Young Women*, 2 vols (London: Longman, 1824). A novel published anonymously

—— 'Memorandums'. Medical care of children. Pforzheimer Collection: Cini Papers, folder 17

—— *Advice to Young Mothers on the Physical Education of Children* (London: Longman, 1823). A revolutionary medical book, as some doctors recognised, published anonymously as mere advice 'by a grandmother'.

—— Medical correspondence with enlightened Dr George Parkman of Boston. Pforzheimer Collection: Cini Papers, folder 21.

Mears, Martha, *The Pupil of Nature* (1797), on midwifery

More, Hannah, *Essays on Various Subjects, Principally designed for Young Ladies* (London: J. Wilkie & T. Cadell, 1777)

Ogle, Sir Henry, *Ogle and Bothal* (privately printed, Newcastle: Andrew Reid & Co., 1902)

Opie, Amelia Alderson. Letter to Mrs Taylor about MW's fainting when acquaintances were guillotined in Paris. In Cecilia Lucy Brightwell, *Memorials of the Life of Amelia Opie* (London: Longman, 1854), 49

Paine, Thomas. Dossier (1793–4). His arrest and imprisonment during the Terror. Archives Nationales, Paris: F/7/4774/6

—— *Rights of Man*, ed. Henry Collins (Penguin Books, 1969)

Paley, William, *The Principles of Moral and Political Philosophy* (London, 1785). Bodleian Library

Paul, C. Kegan, *William Godwin: His Friends and Contemporaries*, 2 vols (London: Henry S. King, 1876)

—— Prefatory memoir to his edn of Wollstonecraft, *Letters to Imlay* (1879; repr. New York: Haskell House, 1971)

—— Letter to Robert Browning (1883) about MW's vanished letters to Fuseli. (Kegan Paul was one of the last to see them.) Beinecke Library

—— Another letter to Browning (1883), about MW and Fuseli. Margaret Clapp Library, Wellesley College

Percy, Bishop, Two letters to his wife from Frederick Street, Dublin (14 and 18 May 1798), at the time he is reading WG's *Memoirs* of MW. Relays Dublin gossip, linking MW with the scandal of her pupil Lady Mary King, and describes the murder trial of MW's one-time employer the Earl of Kingston in the Irish House of Lords. British Library: Add. MS 32, 335

Pinckney, Charles, US Minister to the Court of St James, 1793–4. Letters to the British Foreign Secretary Lord Grenville. PRO, London: FO5/3 and 7

Polwhele, Revd Richard, *The Unsex'd Females; A Poem* (republ. New York: Wm Corbett, 1800). Rare Books, New York Public Library. Clemit et al., *Lives of the Great Romantics*, III, ii: *Wollstonecraft*, 157–68. Pope, Alexander, 'Episode to Dr. Arbuthnot', U223, 232–3 (for 'witlings' and 'Bufo')

Price, Richard, *Correspondence*, ii–iii, ed. D. O. Thomas and W. Bernard Peach (Durham, NC: Duke University Press; Cardiff: University of Wales Press, 1991; repr. 1994). Includes vital replies from the Founding Fathers of America: Washington, Jefferson, Franklin, Jay, etc

—— *Political Writings*, ed. D. O. Thomas (Cambridge University Press, 1991). Includes *Observations on the Nature of Civil Liberty* (1776) and *Observations on the Importance of the American Revolution* (1784)

—— Letter to Mary Wollstonecraft. Abinger: Dep. C. 514

Priestley, Joseph, 'A Discourse on the Death of Dr Price' (1791)

Procès-Verbaux de la Convention (1792–3), Archives Nationales, Paris

Richardson, Samuel, *Clarissa* (1748)

Robinson, Mary, *Perdita: The Memoirs of Mary Robinson*, ed. M. J. Levy (London: Peter Owen, 1994)

Roland, Manon Jeanne, *Memoirs*, trans. possibly by MW (London: Joseph Johnson, 1795)

Roscoe, William. Roscoe Papers, Picton Reference Library, Liverpool

—— *William Roscoe of Liverpool*, ed. George Chandler (London: Batsford, 1953)

Rowan, Archibald Hamilton, *Autobiography*, ed. William Hamilton Drummond (Dublin: Thomas Tegg, 1840), 248–57. Reminiscences of MW

Rousseau, Jean-Jacques, *Émile; or On Education*, trans. Allan Bloom (1762; repr. Basic Books, 1979)

Salzmann, C. G., *Elements of Morality*, trans. and additions by MW. The second edn (with Blake illustrations) in Pf. The 1821 edn (minus MW's sullied name) in the Bodleian Library.

Schreiner, Olive, on MW. Unfinished introduction (1886–9) to projected centenary edn of *A Vindication of the Rights of Woman*, first publ. in the 'Document' section of *History Workshop Journal*, xxxvii (1994), 177–93. The document is preceded by Carolyn Burdet's article, 'A Difficult Vindication: Olive Schreiner's Wollstonecraft'.

Scioto Land Company Papers. MS Division, New York Public Library: Ohio box, Scioto Land Co. folder, Personal (Misc.)

Shakespeare, William, *Hamlet, Macbeth* and *King Lear* (the plays to which MW alludes)

Sharp, Jane, *The Midwives Book* (1671)

Shelley, Mary Wollstonecraft, *The Novels and Selected Works*, i–viii, ed. Nora Krook with Pamela Clemit; intro. by Betty T. Bennett (London: Pickering & Chatto, 1996)

—— 'Life of William Godwin', including notes on MW (a collection of preliminary fragments, Abinger, 1836–40), in Clemit et al, *Lives of the Great Romantics*, III, i: *Godwin*, 95–115; ii: *Wollstonecraft*, 247–50

—— *See* Cristina Dazzi above

—— *The Frankenstein Notebooks* (1816–17). Bodleian Shelley MSS series, ed. Charles E. Robinson, 2 vols (London: Garland, 1996).

-—— (et al), *Lives of the Most Eminent French Writers*, i–ii (Philadelphia: Lea & Blanchard, 1840). MWS has the Johnsonian scholarly temperament with brevity – a swift and confident judgement

—— A collected edn of MWS's biographies, including the above, is due from Pickering & Chatto

—— *Mounseer Nongtongpaw*. Though the first edition is thought to be of 1808, a cheap earlier version is advertised at the back of William Godwin (under pseudonym Theophilus Marcliffe), *The Looking-Glass: A True History of the Early Years of an Artist* (1805). This was one of the earliest publications of his own Juvenile Library. (Margaret Mount Cashell's copy in the Cini–Dazzi collection, San Marcello.) On the title page, Godwin adds an aim: to stimulate attainment in children 'of both Sexes'. A copy of the 1808 edn (16 pp.), illustrated probably by William Mulready, is in the Mortimer Rare Book Room, Smith College: 825 Sh 41mo. Copy also in the Opie Collection, Bodleian Library. Beautiful facsimile colour reprint in Iona and Peter Opie, *A Nursery Companion* (1980)

—— and Percy Bysshe Shelley, *Journal of a Six Weeks' Tour* (1817), repr. in Shelley, *Essays, Letters from Abroad*, ii, ed. Mrs Shelley (Philadelphia: Lea & Blanchard, 1840). Although MWS presents this work as by her husband, she was the primary author.

Shelley, Percy Bysshe, *The Letters*, ed. Frederick L. Jones, i–ii (London: Oxford University Press, 1964)

—— *Shelley's Poetry and Prose*, ed. Donald H. Reiman and Neil Fraistat (New York: 2nd Norton Critical Edition, 2002)

Silsbee, Captain Edward Augustus. Papers Peabody Essex Museum, Salem, Mass.: MS 74, boxes 4, 7, 8. In 1991 Marion Kingston Stocking discovered Silsbee's notes on his conversations with Claire Clairmont, in his memorandum books.

Steven, Margaret (Queen Charlotte's midwife), *Domestic Midwife* (1795)

Stiles, Ezra, *Diary*, iii (New York, 1901)

Swan, James. Letter to General Knox about Joel Barlow and Franco-American relations (21 Dec. 1793), microfilm of Knox Papers, Massachusetts Historical Society. Originals in the Pierpont Morgan Library

Taylor, Thomas ('the Platonist'), *A Vindication of the Rights of Brutes* (1792; repr. and intro. by Louise Schutz Boas (Gainesville, Fla: Scholars' Facsimiles, 1966)

Thompson, William, *Appeal of One-Half of the Human Race, Women, Against the Pretensions of the other Half, Men, to retain them in political, and thence in civil and domestic Slavery* (1825)

Trelawny, Edward John, *Letters*, ed. H. Buxton Forman (London: Oxford University Press, 1910)

Turner, Frederick Jackson, 'The Policy of the French towards the Mississippi Valley in the period of Washington and Adams', *American Historical Review*, x (2 Jan. 1905), 249–79

Wagner, Stephen, and Doucet Devin Fischer, *Guide to the Carl H. Pforzheimer Collection of Shelley and His Circle* (New York Public Library, 1996)

—— 'Visionary Daughters of Albion: A Bicentenary Exhibition Celebrating Mary Wollstonecraft and Mary Shelley', *Biblion: The Bulletin of the New York Public Library*, vi/2 (Spring 1998). Exhibition based on MSS in the Pforzheimer Collection, with expert commentary

Walpole, Horace, *Correspondence*, ed. W. S. Lewis (New Haven: Yale University Press, 1983)

Washington, George. Letter of appointment of Elias Backman of Gothenburg as American Consul. Riksarkivet, Stockholm: Americana, 5 (1783–1805). A photograph of this document appears in *Sweden and the World* (Stockholm, 1960), 64.

Wickham, William, *The Correspondence of the Right Honourable William Wickham from the year 1794*, ed. William Wickham (grandson), i (London: Bentley, 1870). Masterminded a secret service

Wilkinson, James. Spy letters (1788–93) of Imlay's Kentucky associate to Esteban Miró (Spanish Governor of West Florida and Louisiana). Beinecke Library: WA MSS S-1985

—— Letter about Imlay to Matthew Irvine (28 Sept. 1784), Emmet Collection, MSS Division, New York Public Library

Williams, Helen Maria, *Four New Letters of Mary Wollstonecraft and Helen Maria Williams*, ed. Benjamin P. Kurtz and Carrie C. Autrey (Berkeley: University of California, 1937)

—— *Memoirs of the Reign of Robespierre* (London, 1929)

Williams, Jane. Wheedling, manipulative letters to Claire Clairmont. Pforzheimer Collection

Wilmot, Catherine, *An Irish Peer on the Continent 1801–3: Being a Narrative of the Tour of Stephen, 2nd Earl Mount Cashell, through France, Italy etc.*, ed. Thomas U. Sadleir (London, 1920). Eyewitness of Lady Mount Cashell in Paris

Wollstonecraft, Elizabeth (Mrs Bishop), fifty letters (1786–95) to Everina Wollstonecraft. Abinger: Dep. b. 210

Wollstonecraft, Mary, *The Posthumous Works of the Author of 'A Vindication of the Rights of Woman'* (1798; repr. as facs. ed. (Clifton, NJ: Augustus M. Kelley, 1972), 4 vols, ed. William Godwin (London: Joseph Johnson, 1798)

———- *The Works of Mary Wollstonecraft*, ed. Janet Todd and Marilyn Butler, 7 vols (London: Pickering & Chatto, 1989)

——— *Mary Wollstonecraft Godwin: A Bibliography of First & Early Editions with Briefer Notes on Later Editions & Translations*, by John Windle. 2nd edn enlarged by Karma Pippin (New Castle, Del.: Oak Knoll Press, 2000)

-—— *Collected Letters*, ed. Ralph M. Wardle (Ithaca: Cornell University Press, 1979). Pioneering, sound, but out of print. Still available in libraries

—— *Collected Letters*, ed. Janet Todd (London: Allen Lane, 2003)

*—— Unpublished letter to Bernstorff, Prime Minister of Denmark, discovered by Gunnar Molden. State Archives, Copenhagen

—— *Four New Letters of Mary Wollstonecraft and Helen Maria Williams*, ed. Benjamin P. Kurtz and Carrie C. Autrey (Berkeley and Los Angeles: University of California Press, 1937)

-—— *Mary [Mary, A Fiction] and The Wrongs of Woman*, ed. and intro. Gary Kelly (Oxford University Press: World's Classics, 1980, repr. 1987)

—— *Political Writings*, ed. Janet Todd (Oxford World's Classics, 1994)

—— *Letters Written during a short Residence in Sweden, Norway and Denmark,* ed. and intro. Richard Holmes (Penguin Books, 1987)

Wollstonecraft, Mary (MW's American sister-in-law), 'The Natural Rights of Woman', *Boston Monthly Magazine*, i/3 (Aug. 1825), 126–35. Signed 'D'Anville'

Wordsworth, William, *The Prelude* (text of 1805), ed. Ernest de Selincourt, rev. Helen Darbishire (London: Oxford University Press, 1960)

Wright, Thomas, *The Female Vertuosos*, Restoration-style comedy

Wulfsberg, Jacob. Report of 18 Aug. 1795 about Imlay and Wollstonecraft to the Stiftamtmann of Akershus. Statsarkivet (Regional Archives), Oslo: *Akershus stiftamt, Brev fra forskjellige avsendere*, 44 (1795–6). Wollstonecraft enclosed a copy with other letters of recommendation in her letter to the Danish Prime Minister, A. P. Bernstorff

Young, Arthur, *Tour of Ireland*, i–ii (1780; repr. London: George Bell, 1892)

—— *Autobiography*, ed. M. Bentham-Edwards (repr. London: Smith Elder, 1898)

SECONDARY SOURCES

Addison, Sir William, *The Old Roads of England* (London: Batsford, 1980)

Akers, Charles W., *Abigail Adams: An American Woman* (Canada: Little, Brown, 1980)

Alexander, Meena, *Women in Romanticism* (London: Macmillan, 1989). Links MW's writings through the theme of maternity to the Romantic writings of Dorothy Wordsworth and MWS

Allen, Brooke, 'John Adams: Realist of the Revolution', *Hudson Review* (spring 2002), 45–54

Andersson, Gunnar, 'Mary Wollstonecraft i Strömstad 1795'. Gez, *Litteraturhistoria Engelsk* Nebk.07:k, *Geografi Bohuslän Reseskildringar Historia*, Strömstadbygden, viii (1995), 7–15. Copy in the Swedish National Library, Stockholm

Andrew, Christopher, *For the President's Eyes Only: Secret Intelligence and the American Presidency from Washington to Bush* (London: HarperCollins, 1995; repr. 1996)

Avery, Gillian, 'The Puritans and Their Heirs', in *Children and Their Books*, eds Gillian Avery and Julia Briggs (Oxford: Oxford University Press, 1989). An authoritative collection of essays that should be republished

—— *The Best Type of Girl* (London: André Deutsch, 1991)

Backman, Pierre, 'Frenchmen in Tranenberg', *Västgöta bygden*, iv (1958). On Elias Backman's background and relatives

Banks, Olive, *The Biographical Dictionary of British Feminists* (Brighton: Harvester, 1985). This dictionary begins in 1800, which means that it excludes MW, although its definition of feminism accords in every detail with her principles

Barker-Benfield, G. J., 'Mary Wollstonecraft: Eighteenth-Century Commonwealth Woman', *Journal of the History of Ideas*, i (1989), 95–115

—— *The Culture of Sensibility: Sex and Society in Eighteenth-Century Britain* (University of Chicago Press, 1992)

Baron-Cohen, Simon, *The Essential Difference: Men, Women and the Extreme Male Brain* (Penguin, Allen Lane, 2003)

Barry, J. M., *Pitchcap and Triangle: The Cork Militia in the Wexford Rising* (Cork, 1998), 173–82.

Bennett, Betty T. and Stuart Curran, *Mary Shelley in Her Times* (Johns Hopkins University Press, 2000)

Blakemore, Steven, *Crisis in Representation: Thomas Paine, Mary Wollstonecraft, Helen Maria Williams and the Rewriting of the French Revolution* (London, N.J., Ontario: Associated University Presses, 1997). A fascinating analysis of MW haunted by *Macbeth* in her response to the bloodshed of the revolution

Boaden, J., *Life of Mrs Inchbald* (London, 1833)

Brailsford, H. N., *Shelley, Godwin, and Their Circle* (London: Thornton Butterworth, 1913, repr. 1936). This book is still alive with insight

Brewer, John, *The Pleasures of the Imagination: English Culture in the Eighteenth Century* (London: HarperCollins; New York: Farrar, Straus, 1997)

Bridgewater, Dorothy W., 'The Barlow Manuscripts in the Yale Library', *Yale University Library Gazette*, xxxiv/2 (Oct. 1959), 57–63

Briggs, Julia, 'Women Writers: Sarah Fielding to E. Nesbit', in *Children and Their Books*, eds Briggs and Gillian Avery (Oxford University Press, 1989), 221–50

Brown, Ford K. *The Life of William Godwin* (London: J. M. Dent; New York: Dutton, 1926) Derivative; not on a par with C. Kegan Paul's massive reliance on primary sources for the biography of 1876

Burdet, Carolyn, 'A Difficult Vindication: Olive Schreiner's Wollstonecraft Introduction', *History Workshop Journal*, xxxvii (1994), 177–93. Suggestive essay on the problems Schreiner faced in preparing her (unfinished) centenary introduction to *RW*

Burke, John Bernard, *Dictionary of the Peerage* (1828)

—— *Genealogical History of the Landed Gentry in Ireland* (1899)

Butler, Marilyn, *Romantics, Rebels, and Reactionaries: English Literature and its Background 1760–1830* (Oxford University Press, 1981)

—— *Jane Austen and the War of Ideas* (Oxford: Clarendon Press, 1975, repr. 1997)

—— Introduction to *Frankenstein* (Penguin Books, 1994)

Buus, Stephanie, 'Bound for Scandinavia: Mary Wollstonecraft's Promethean Journey', in Anka Ryall and Catherine Sandbach-Dahlström (eds), *Mary Wollstonecraft's Journey to Scandinavia: Essays*, Stockholm Studies in English, xcix (Stockholm: Almqvist & Wiksell, 2003)

Caine, Barbara, *English Feminism, 1780–1980* (Oxford University Press, 1997)

—— 'Victorian Feminism and the Ghost of Mary Wollstonecraft', *Women's Writing*, iv: 2 (1997)

The Cambridge Companion to Mary Wollstonecraft, ed. Claudia Johnson (Cambridge University Press, 2002)

Castle, Terry, *The Female Thermometer: Eighteenth-Century Culture and the Invention of the Uncanny* (New York: Oxford University Press, 1995). An exploration of the subversive role of gothic fiction and sexual masquerade provides context for understanding MW's suspicion of erotic behaviour

Chard, Leslie F., 'Joseph Johnson: Father of the Book Trade', *Bulletin of the New York Public Library*, lxxix (1975)

Chernaik, Judith, *Mab's Daughters: Shelley's Wives and Lovers* (London: Macmillan, 1991. Publ. in US as *Love's Children*). A fictional biography in the epistolary form of a lot of the source material. Recreates sixteen months in the lives of Fanny Imlay, Harriet Shelley, Mary Shelley and Claire Clairmont. Chernaik suggests that Fanny committed suicide when told by WG that she was not his daughter

Christiansen, Rupert, *Romantic Affinities: Portraits from an Age, 1780–1830* (1988; repr. London: Vintage, 1994)

Clarke, Norma, *The Rise and Fall of the Woman of Letters* (London: Pimlico, 2004)

Clemit, Pamela, *The Godwinian Novel: Fictions of Godwin, Brockden Brown and Mary Shelley* (1993)

—— 'Philosophical Anarchism in the Schoolroom: William Godwin's Juvenile Library, 1805–25', *Biblion: Bulletin of the New York Public Library* (2000/2001), 44–70

Colley, Linda, *Britons: Forging a Nation 1707–1837* (Yale University Press, 1992, repr. Pimlico, 1994)

Cone, Carl B., *The English Jacobins: Reformers in Late Eighteenth-Century England* (New York: Scribner, 1968)

Coombs, Tony, *Tis a Mad World at Hogsdon: A Short History of Hoxton and Surrounding Area* (London: Hoxton Hall in association with the London borough of Hackney, 1995)

Crawford, Elizabeth, 'Mary Wollstonecraft: "the first of a new genus"', *Antiquarian Book Monthly* (Dec. 1995), 14–19. An accurate publishing history

—— *Dictionary of Women's Suffrage: A Reference Guide* (London: Routledge, 2000)

—— *Enterprising Women: The Garretts and Their Circle* (London: Francis Boutle, 2002)

Curelli, Mario, *Una Certa Signora Mason: Romantici inglesi a Pisa ai tempi di Leopardi* (Pisa: Ets, 1997). Reports on MM's unpublished fiction

—— 'Lady Mountcashell alias Madame Mason', in *Leopardi in Pisa*, ed. Fiorenza Ceragioli (Milan: Electa, 1998), 304–20. With thanks to Cristina Dazzi for the gift of this beautifully produced and informative collection of essays to celebrate an exhibition on Leopardi in Pisa

Del Vivo, Caterina, 'The "Beautiful Vaccà"' in *Leopardi in Pisa*, ed. Fiorenza Ceragioli (Milan: Electa, 1998), 274–81

Diedrick, James, '*Jane Eyre* and *A Vindication of the Rights of Woman*',. in *Approaches to Teaching Jane Eyre* (New York: MLA, 1993)

Dolan, Brian, *Ladies of the Grand Tour* (London: Flamingo, 2002)
Dorfman, Joseph, 'Joel Barlow: Trafficker in Trade and Letters', *Political Science Quarterly*, lix (1944), 83–100
Draper, Lyman C., *The Life of Daniel Boone* (Mechanicsburg, Pa.: Stackpole Books, 1998)
Dunn, Jane, *Moon in Eclipse: A Life of Mary Shelley* (London: Weidenfeld & Nicolson, 1978)
Durant, William Clark. Extensive biographical research in the Supplement to his edn of WG's *Memoirs* (1927; repr. New York: Haskell House, 1969), pp. 138–347
Elliott, Lawrence, *The Long Hunter: A New Life of Daniel Boone* (London: Allen & Unwin, 1977)
Ellman, Mary, *Thinking about Women* (London: Virago, 1979)
Elwood, Mrs Ann, *Memoirs of Literary Ladies in England* (London: 1843). Sympathetic to MW, in contrast with feminists like Harriet Martineau who shunned a historical connection with a woman touched by scandal
Emerson, Oliver Farrar, 'Notes on Gilbert Imlay, Early American Writer', *PMLA*, xxxix/1 (June 1924), 406–39
Everest, Kelvin, ed., *Revolution in Writing: British Literary Responses to the French Revolution* (Milton Keynes: Open University Press, 1991). The responses are Hannah More's, MW's, Paine's and Shelley's
Falco, Maria J., ed., *Feminist Interpretations of Mary Wollstonecraft* (Penn State University Press, 1996)
Faragher, John Mack, *Daniel Boone: The Life and Legend of an American Pioneer* (New York: Henry Holt, 1992)
Favret, Mary A., 'Mary Wollstonecraft and the Business of Letters', in *Romantic Correspondence: Women, Politics and the Fiction of Letters* (Cambridge University Press, 1993), 96–132
Ferguson, Moira, *Colonialism and Gender from Mary Wollstonecraft to Jamaica Kincaid* (New York: Columbia University Press, 1993). See the chapter on MW and slavery
Ferguson, Robert A., 'The American Enlightenment 1750–1820', in *Cambridge History of American Literature*, ed. Sacvan Bercovitch, i (Cambridge University Press, 1994), 345–537
Flexner, Eleanor, *Mary Wollstonecraft: A Biography* (New York: Coward, McCann, 1972)
Follini, Tamara, 'Improvising the Past in *A Small Boy and Others*', *Yearbook of English Studies*, xxx (2000), 106–23. This essay on Henry James's autobiography was the most original stimulus for thinking about the autobiographical element in MW's Travels
Forster, Margaret, *Significant Sisters: The Grassroots of Active Feminism 1838–1939* (Penguin Books, 1986)
Foster, R. F, *Modern Ireland 1600–1972* (1988; repr. Penguin Books, 1989)
—— 'Remembering 1798', in *The Irish Story: Telling Tales and Making It Up in Ireland* (Penguin Books, 2001)
Fraser, Antonia, *The Weaker Vessel: Women's Lot in Seventeenth-Century England* (London: Weidenfeld & Nicolson, 1984; repr. Methuen, 1985)
—— *Marie Antoinette: The Journey* (London: Weidenfeld & Nicolson, 2001)
Frimansson, Inger, 'Från de unkna matsalarna'. Gez, *Litteraturhistoria Engelsk* Nc. 07, *Geografi Sverige Reseskildringar* (Svensk bokhandel, xliii/29, 1994). A response to Per Nyström, *MW's Scandinavian Journey*. Copy in the Swedish National Library, Stockholm

Garnett, Richard, *Athenaeum* (15 Aug. 1903). Conjectures place and date of GI's death; has not been disproved

Garrett, Martin, *Mary Shelley* (British Library, 2002)

George, Margaret, *One Woman's 'Situation': A Biography* (Urbana, Ill.: University of Illinois Press, 1970)

Gerhardt, Sue, *Why Love Matters: How Affection Shapes a Baby's Brain* (London: Routledge, 2004)

Gerzina, Gretchen, *Black England: Life before Emancipation* (London: Allison & Busby, 1999)

Gittings, Robert, and Jo Manton, *Claire Clairmont and the Shelleys* (Oxford University Press, 1992)

Graham, Kenneth W, ed., *William Godwin Reviewed: A Reception History 1783–1834* (New York: AMS Press, 2001)

Gubar, Susan, 'Feminist Misogyny: Mary Wollstonecraft and the Paradox of "It Takes One to Know One"', in Diane Elam and Robyn Wiegman, eds, *Feminism Beside Itself* (London: Routledge, 1995)

—— and Sandra McGilbert, *The Madwoman in the Attic: The Woman Writer and the Nineteenth-Century Literary Imagination* (New Haven: Yale University Press, 1979)

Haraszti, Zoltán, *John Adams and the Prophets of Progress* (Cambridge, Mass.: Harvard University Press, 1952), 187–234. Unselective, hard-to-follow description of Adams's fascinating marginalia to Wollstonecraft's *French Revolution*. Easier to see the point of Adams's response by looking at his actual copy in the Boston Public library.

Hardyment, Christina, *Perfect Parents: Baby-care Advice Past and Present* (Oxford University Press, 1995)

Harmon, Claire, *Fanny Burney: A Biography* (London: HarperCollins, 2000)

Harper, Charles G., *Stagecoach and Mail* (London: Chapman & Hall, 1903)

Hay, Carla H., 'James Burgh', *Biographical Dictionary of Modern British Radicals*, i (Brighton: Harvester, 1979)

Herman, Judith Lewis, *Trauma and Recovery: From Domestic Abuse to Political Terror* (London: Pandora, 1992; repr. 2001)

Hill, Bridget, 'The Links between Mary Wollstonecraft and Catharine Macaulay: New Evidence', *Women's History Review*, iv/2 (1995), 177–92

Hill-Miller, Katherine C., *'My hideous Projeny': Mary Shelley, William Godwin and the Father–Daughter Relationship* (Newark: University of Delaware Press; Associated University Presses, 1995)

Hirsch, Pam, 'Mary Wollstonecraft: A Problematic Legacy', in *Wollstonecraft's Daughters: Womanhood in England and France, 1780–1920*, ed. Clarissa Campbell Orr (Manchester University Press, 1996). Sensitive account of MW's reputation

Holmes, Richard, *Shelley: The Pursuit* (London: Weidenfeld & Nicolson, 1974)

—— 'Mary Wollstonecraft and Gilbert Imlay in France', in *Footsteps: Adventures of a Romantic Biographer* (Penguin Books, 1985)

—— 'The Feminist and the Philosopher', in *Sidetracks: Explorations of a Romantic Biographer* (London: HarperCollins, 2000)

—— 'Death and Destiny', *Guardian Book Review* (24 Jan. 2004)

Howarth, Janet, 'Gender, Domesticity, and Sexual Politics', in *The Short Oxford History of the British Isles: The Nineteenth Century*, ed. Colin Matthew (Oxford University Press, 2000), 162–93

Hufton, Olwyn, *The Prospect Before Her: A History of Women in Western Europe*, i: *1500–1800* (London, 1995; HarperCollins/Fontana, repr. 1997)

Imlay, Hugh and Nella, *The Imlay Family* (Zanesville, Ohio, 1958)

Jacobs, Diane, *Her Own Woman: The Life of Mary Wollstonecraft* (London: Abacus; New York: Simon & Schuster, 2001)

Janes, R. M., 'On the Reception of *A Vindication of the Rights of Woman*,' *Journal of the History of Ideas*, xxxix (1978), 293–302

Jeffreys-Jones, Rhodri, *American Espionage* (London: MacMillan, 1977)

—— *Cloak and Dollar: A History of American Secret Intelligence* (Yale University Press, 2002)

Johnson, Claudia L., *Equivocal Beings: Politics, Gender, Sentimentality in the 1790s: Mary Wollstonecraft and Jane Austen* (University of Chicago Press, 1995)

Jones, Vivien, '"The Tyranny of the Passions": Feminism and Heterosexuality in the Fiction of Wollstonecraft and Hays', in Sally Ledger, Josephine McDonagh and Jane Spencer (eds), *Political Gender: Texts and Contexts* (London: Harvester, 1994), 173–88

—— 'The Death of Mary Wollstonecraft', *British Journal of Eighteenth-Century Studies*, xx/2 (autumn 1997)

—— Lecture on 'Mary Wollstonecraft and Sex Education', St Hugh's College, Oxford (19 Feb. 2001)

Jones, W. H. S., *A History of St Catherine's College* (Cambridge University Press, 1936)

Jordan, Elaine, 'Criminal Conversation: On Mary Wollstonecraft's *The Wrongs of Woman*', *Women's Writing*, iv/2 (1997), 221–34. Contemporary legal context

Kaplan, Cora, 'Wild Nights: Pleasure/Sexuality/Feminism', in *Sea Changes: Culture and Feminism* (London: Verso, 1986)

Keane, John, *Tom Paine: A Political Life* (Bloomsbury, 1995)

Keats-Shelley Journal, special issue, 1997

Kelly, Gary, *Women, Writing and Revolution 1790–1827* (Oxford: Clarendon Press, 1993)

—— *Revolutionary Feminism: The Mind and Career of Mary Wollstonecraft* (London: Macmillan, 1992; New York: St Martin's Press, 1996)

—— Notes to the World's Classics editions of *Mary* and *WW*.

Kelly, Linda, *Women in the French Revolution* (Penguin Books, 1989)

King-Hall, Magdalen, *Eighteenth-Century Story* (London: Peter Davies, 1956). A fictionalisation of the relations of MW and Lady Mary King

King-Harmon, Anthony Lawrence, *The Kings of King House* (privately publ., 1996)

King-Harmon, Robert Douglas, *The Kings, Earls of Kingston: An account of the Family and their Estates in Ireland between the reigns of the two Queens Elizabeth* (privately publ., Cambridge: Heffer, 1959)

Kramnick, Miriam Brody, introduction to *RW* (Penguin Books, 1982)

Laver, James, *Taste and Fashion: From the French Revolution until Today* (London: Harrap, 1945)

Le Doeuff, Michèle, *The Sex of Knowing*, trans. Kathryn Hamer and Lorraine Code (London: Routledge, 2004)

Lorch, Jennifer, *Mary Wollstonecraft: The Making of a Radical Feminist* (Providence, RI: Berg, 1990). Argues that *WW* is more relevant to the concerns of present-day feminism than *RW*

Loudon, Irvine, *The Tragedy of Childbed Fever* (Oxford University Press, 2000)

Lucas, E. V., introduction to reprint of MW's *Original Stories from Real Life* (London: Henry Frowde, 1906). Interesting for its virulence against MW more than a century after publication; sees children enslaved to overbearing woman; Mrs Mason is presented as precursor to Mrs Proudie (in Trollope's *Barchester Towers*). Tells more about misogyny at the height of the suffragettes' activity than about what captivated Margaret King

MacCarthy, Fiona, *Byron: Life and Legend* (London: John Murray, 2002)

MacMahon, K. A., *Beverley* (Silsden, Yorkshire: Dalesman Publishing Co., 1973)

McAleer, Edward C., *The Sensitive Plant* (1958). Biography of Margaret Mount Cashell (*née* Margaret King)

McCullough, David, *John Adams* (New York: Simon & Schuster, 2001)

McMillan, James F., *France and Women 1789–1914: Gender, Society and Politics* (London: Routledge, 2000)

Maxted, Ian, *The London Book Trades 1775–1800: A Topographical Guide* (Exeter, 1980)

Meinz, Manfred, 'Die "Silberkammer" des Altonaer Museums', *Altoner Museum in Hamburg Jahrbuch 1966*, iv (Hamburg: D. R. Ernst Hauswedell Verlag), 38–75

Mellor, Anne K., *Mary Shelley, Her Life, Her Fiction, Her Monsters* (London: Routledge, 1988)

Miller, Victor Clyde, 'Joel Barlow: Revolutionist London, 1791–2', in *Britannica* (Hamburg, 1932)

Mitchell, Juliet, and Ann Oakley, eds, *The Rights and Wrongs of Women* (Penguin Books, 1976; repr. 1977). Includes Margaret Walters, 'The Rights and Wrongs of Woman: Mary Wollstonecraft, Harriet Martineau, Simone de Beauvoir'

Moers, Ellen, *Literary Women* (London: Women's Press, 1978)

Molden, Gunnar, 'Sølvbriggen Maria Margrete – ut av historiens mørke' (New Historical Light on the Silver Brig), trans. Ingunn Siedler, with English summary, in *Norsk Sjøfartsmuseum årbok* (*Norwegian Maritime Museum Yearbook*, 1995, Oslo: 1996), 139–54

—— 'The Silver Ship Emerging out of the Darkness of History', trans. Ingunn Siedler, *Agderposten* (31 Aug. 1996)

—— 'No Riches for the Descendants', trans. Ingunn Siedler, *Agderposten* (31 Aug. 1996)

—— 'The Shipwreck that Never Was', trans. Ingunn Siedler, *Agderposten* (11 Apr. 1997)

—— *Seilskute byen Arendal/Arendal – the Town of Sailing Ships*, trans. Richard Nelson (Norway: Arfo, 1998)

Moore, Jane, *Mary Wollstonecraft* (Tavistock: Northcote House Publishers, 1999). British Council series on 'Writers and their Work'. A balanced survey, stressing the ambition of MW's career, with a descriptive bibliography. An excellent introductory book

Morton, Brian N., *Americans in Paris: An Anecdotal Guide to the Homes and Haunts of Americans from Jefferson to Capote* (1984; repr. New York: Quill/William Morrow, 1986)

Murray, Venetia, *High Society in the Regency Period 1788–1830* (Penguin Books, 1998)

Myers, Mitzi, 'Mary Wollstonecraft's *Letters Written … in Sweden*: Toward Romantic Autobiography', *Studies in Eighteenth-Century Culture*, xviii, ed. Roseann Runte (1979)

—— 'Impeccable Governesses, Rational Dames, and Moral Mothers: Mary Wollstonecraft and the Female Tradition in Georgian Children's Books', *Children's Literature*, xiv (1986), 31–54

—— 'Pedagogy as Self-Expression in Mary Wollstonecraft', in Shari Benstock, ed., *The*

Private Self: Theory and Practice of Women's Autobiographical Writings (Chapel Hill: University of North Carolina Press, 1988). Argues persuasively that MW's work in a variety of genres 'wants to be read as a species of autobiography'

Nehring, Cristina, 'The Vindications: the Moral Opportunism of Feminist Biography', *Harper's Magazine* (Feb. 2002), 60–5

Nitchie, Elizabeth, 'An Early Suitor of Mary Wollstonecraft', *PMLA*, lviii (Mar. 1943), 163–9

Norris, Pamela, 'She Longed for Security and Affection', *Literary Review* (Oct. 2000), 18–19

Nyström, Per, *Mary Wollstonecraft's Scandinavian Journey*, trans. George R. Otter (Gothenburg, 1980). Originally a lecture to the Royal Society of Arts and Sciences of Gothenburg (autumn 1972)

Orr, Clarissa Campbell, ed., *Wollstonecraft's Daughters: Womanhood in England and France, 1780–1920* (Manchester University Press, 1996)

Pakenham, Thomas, *The Year of Liberty: the Story of the Great Irish Rebellion of 1798* (London: Abacus, 2000), esp. 191, 247

Panajia, Alessandro, 'The New Accademia dei Lunatici', in *Leopardi in Pisa*, ed. Fiorenza Ceragioli (Milan: Electa, 1998), 322–6

Paul, C. Kegan, *William Godwin: His Friends and Contemporaries*, 2 vols (London: Henry S. King, 1876)

Pearson, J., *Women's Reading in Britain, 1750–1835: A Dangerous Recreation* (Cambridge University Press, 1999), 79–82

Pennell, Elizabeth Robins, *Mary Wollstonecraft Godwin* (London: W. H. Allen, 1885)

Phillips, Melanie, *The Ascent of Woman* (London: Little, Brown, 2003)

Poovey, Mary, *The Proper Lady and the Woman Writer: Ideology as Style in the Works of Mary Wollstonecraft, Mary Shelley, and Jane Austen* (University of Chicago Press, 1984)

—— *Uneven Developments: The Ideological Work of Gender in Mid-Victorian England* (London: Virago, 1989)

Porter, Roy, *English Society in the Eighteenth Century* (Penguin Books, rev. edn 1990)

—— *The Penguin Social History of Britain: English Society in the Eighteenth Century* (Penguin Books, 1982; rev. edn 1991)

Powell, David, *Tom Paine: The Greatest Exile* (London: Hutchinson, 1985; repr. 1989)

Power, Bill, *White Knights, Dark Earls: The Rise and Fall of an Anglo-Irish Dynasty* (Cork: Collins Press, 2000)

Rauschenbusch-Clough, Emma, *A Study of Mary Wollstonecraft and the Rights of Woman* (London: Longman's, Green & Co., 1898)

Ravetz, Alison, 'The Trivialisation of Mary Wollstonecraft: A Personal and Professional Career Re-Vindicated', *Women's Studies Forum*, vi/5 (1983), 491–9. Anticipates the approach of the present book

Rendall, Jane, 'Mary Wollstonecraft, History and Revolution', *Women's Writing*, iv: 2 (1997), 155–72

Reynolds, Margaret. Critical introduction to her edn of Elizabeth Barrett Browning, *Aurora Leigh* (Ohio University Press, 1992)

Ribeiro, Aileen, *Dress in Eighteenth-Century Europe 1715–1789* (Yale University Press, rev. edn 2002)

Ricci, Fulvia, 'The First Accademia dei Lunatici', in *Leopardi in Pisa*, ed. Fiorenza Ceragioli (Milan: Electa, 1998), 321

Roberts, Michèle, *Fair Exchange* (London: Virago, 2000). Fictionalisation of MW and GI.

Rusk, Ralph Leslie, 'The Adventures of Gilbert Imlay', *Indiana University Studies*, x/57 (Bloomington, Ind.: Mar. 1923). For decades the only available information on GI; limited and overused

Ryall, Anka. Introduction to Norwegian translation of Mary Wollstonecraft's *Travels*. 'Mary Wollstonecraft og kunsten å reise', *Min nordiske reise*, trans. Per A. Hartun (Oslo: Pax, 1997), i–xviii

—— and Catherine Sandbach-Dahlström, eds, *Mary Wollstonecraft's Journey to Scandinavia: Essays*, Stockholm Studies in English, xcix (Stockholm: Almqvist & Wiksell, 2003)

St Clair, William, 'William Godwin as Children's Bookseller', in *Children and their Books*, eds Gillian Avery and Julia Briggs (Oxford University Press, 1989), 165–79

—— *The Godwins and the Shelleys: The Biography of a Family* (London: Faber; New York: Norton, 1989)

—— *The Reading Nation in the Romantic Period* (Cambridge University Press, 2004)

Sapiro, Virginia, *A Vindication of Political Virtue: The Political Theory of Mary Wollstonecraft* (University of Chicago Press, 1992)

Schama, Simon, *Citizens* (London: Allen Lane, 1989)

—— *History of Britain* (London: BBC Publications, 2002)

Seelye, John, *Beautiful Machine: Rivers and the Republican Plan 1755–1825* (New York: Oxford University Press, 1991). Informed treatment of Imlay as commentator on the frontier

Sen, Amartya, 'Elements of a Theory of Human Rights', *Philosophy and Public Affairs*, xxxiii (Fall 2004). Uses MW's ideas as a way of understanding the very idea of human rights

—— 'Mary, Mary, Quite Contrary', talk at the Oxford conference of the International Association of Feminist Economists (Aug. 2004). To be published in *Feminist Economics*

Showalter, Elaine, *Inventing Herself: Claiming a Feminist Intellectual Heritage* (New York: Simon & Schuster; London: Picador, 2001)

Smith, Joan, 'Mary Wollstonecraft', essay for the Penguin website (2003)

—— *Moralities: How to End the Abuse of Money and Power in the 21st Century* (Allen Lane/Penguin Books, 2002)

Spark, Muriel, *Mary Shelley: A Biography*, a revision of *Child of Light: A Reassessment of Mary Shelley* (1951, rev. edn New York: Dutton, 1987)

Steel, Mark, *Vive la Révolution: – A Standup History of the French Revolution* (New York: Scribner, 2003)

Steiner, Wendy, 'Women's Fiction', in the *Cambridge History of American Literature*, ed. Sacvan Bercovitch, vii (1940-–90), 505–16

Stephen, Leslie, 'Mary Wollstonecraft', *Dictionary of National Biography*

Stocking, Marion Kingston, 'Miss Tina and Miss Plin: the Papers behind *The Aspern Papers*', in Donald H. Reiman et al.,eds, *The Evidence of the Imagination: Studies of Interaction between Life and Art in English Romantic Literature* (New York, 1978), 372–84, Quotes Richard Garnett's description of E. A. Silsbee as an amiable version of CC's old suitor Trelawny

Stone, Laurence, *The Family, Sex and Marriage in England 1500–1800* (Penguin Books, rev. edn 1979)

Strachey, Ray, *The Cause* (1928; repr. London: Virago, 1974). Fascinating appendix: Florence Nightingale's outcry against the limitations of women's lives in her autobiographical writings of the 1860s 'Cassandra'.

Sunstein, Emily, *A Different Face: The Life of Mary Wollstonecraft* (New York, 1975)

—— *Mary Shelley: Romance and Reality* (Little, Brown, 1989; repr. 1991)

Taylor, Barbara, *Eve and the New Jerusalem: Socialism and Feminism in the Nineteenth Century* (London: Virago, 1983)

—— *Mary Wollstonecraft and the Feminist Imagination* (Cambridge University Press, 2003)

—— 'Mary Wollstonecraft', entry in *The New Dictionary of National Biography* (Oxford University Press, 2004)

—— 'Mary Wollstonecraft and the Enlightenment', lecture, Oxford University (Hilary term, 2004)

Taylor, J. Lionel, *A Little Corner of London* (Lincoln: J. W. Ruddock, 1925). On Newington Green.

Tillyard, Stella, *Citizen Lord* (London: Vintage, 1998)

Tims, Margaret, *Mary Wollstonecraft: A Social Pioneer* (London: Millington Books, 1976)

Todd, Charles Burr, *Life and Letters of Joel Barlow: Poet, Statesman, Philosopher* (1886; repr. New York: Da Capo Press, 1970)

Todd, Janet, *Women's Friendship in Literature* (New York: Columbia University Press, 1980)

—— *Gender, Art, Death* (1988; repr. Polity Press, 1995). Includes deaths of MW and Fanny Imlay

—— *Mary Wollstonecraft: A Revolutionary Life* (London: Weidenfeld & Nicolson; New York: Columbia University Press, 2000)

—— *Rebel Daughters: Ireland in Conflict, 1798* (London: Viking, 2003)

Tomalin, Claire, 'A Fallen Woman: Reappraisal of *Letters to Imlay* by Mary Wollstonecraft', in *Several Strangers: Writing from Three Decades* (Penguin Books, 2000), pp. 19–23

—— *The Life and Death of Mary Wollstonecraft* (New York: Harcourt, 1974; rev. edn Penguin Books, 1992)

—— *Shelley and His World* (1980; repr. Penguin Books, 1992)

—— Introduction, with background on the history of Margaret Mount Cashell and her daughters, to MWS's story, *Maurice* (Penguin Books, 1997)

—— *Jane Austen: A Life* (Penguin Books, 1998). Establishes a plausible link, through the East family, between MW and Jane Austen

—— Robert Woof, and Stephen Hebron, *Hyenas in Petticoats* (Wordsworth Trust, Dove Cottage, Grasmere, 1997). Bicentenary exhibition of the lives of MW and MWS

Tyson, Gerald P., *Joseph Johnson: A Liberal Publisher* (Iowa City: University of Iowa Press, 1979)

Uglow, Jenny, *The Lunar Men* (London: Faber, 2002)

Vaccà Giusti, Laura, *Andrea Vaccà e la sua famiglia: Biografie e memorie* (Pisa, 1878)

Verlet, Pierre, *French Royal Furniture* (London: Barrie & Rockliff, 1963)

Vickery, Amanda, *The Gentleman's Daughter: Women's Lives in Georgian England* (New Haven and London: Yale University Press, 1998), esp. 153–4 on the production and distribution of medicine

Vidal, Gore, *Burr: A Novel* (New York: Ballantine, 1973). Western Separatism involving GI's associate General Wilkinson and WG's friend Aaron Burr

Walkowitz, Judith R., *City of Dreadful Delight: Narratives of Sexual Danger in Late Victorian London* (London: Virago, 1992), ch. 5 (on the Men and Women's Club for redefining the natures of the sexes)

Wardle, Ralph M., 'Mary Wollstonecraft, *Analytical* Reviewer', *PMLA*, lxii (Dec. 1947)

—— *Mary Wollstonecraft: A Critical Biography* (1951; repr. Lincoln, Nebraska: University of Nebraska Press, 1966)

Waters, Mary A. "The First of a New Genus": Mary Wollstonecraft as a Literary Critic and Mentor to Mary Hays', *Eighteenth-Century Studies*, xxxvii (spring 2004), 415–34

Watson, Isobel, *Hackney and Stoke Newington Post: A Visual History* (London: Historical Publications, 1990, revd 1998)

Wedd, A., *The Fate of the Fenwicks* (London, 1927)

Weinglass, D. H., 'Henry Fuseli's Letter of Enquiry to Paris on Behalf of Mary Wollstonecraft's Sister Everina', *Blake: An Illustrated Quarterly* (spring 1988)

Weir, Alison, *The Six Wives of Henry VIII* (New York: Ballantine, 1991). Stresses the kind of intellectual attainments and educative nurture in Katherine Parr that look forward in some ways to Wollstonecraft. The first two wives usually get most attention, but their old feminine power-games are less compelling to women of the future

Whitelock, Dorothy, *The Beginnings of English Society* (Penguin Books, 1952, repr. 1971). The position of women in Anglo-Saxon England

Williams, Raymond, *Culture and Society 1780–1950* (Penguin Books, 1963)

Women's Writing, special issue for Mary Wollstonecraft's bicentennial, iv/2 (1997)

Woodress, James, *A Yankee Odyssey: The Life of Joel Barlow* (Philadelphia and New York: J. B. Lippincott, 1958)

Woof, Hebron, *Hyenas in Petticoats*, *see* Tomalin, Woof and Hebron

Woolf, Virginia, 'Mary Wollstonecraft' (1929); repr. *Common Reader*, 2nd series (1932), and in various collections of essays, including Leonard Woolf's well-arranged selection for the Hogarth Press

Yeo, Eileen, ed., *Mary Wollstonecraft and Two Hundred Years of Feminism* (London: Rivers Oram, 1997)

Zunder, Theodore Albert, *The Early Days of Joel Barlow: His Life and Works, 1754–1787*, in *Yale Studies in English*, v/83–4 (1934)

INTERNET DOCUMENTS

For readers, teachers and students who would like access to some of the primary materials behind this book, the following items are posted on the Internet at www.lyndallgordon.net. Further documents will be posted if readers wish.

- Hunting the invisible Gilbert Imlay is one of the almost impossible challenges of Wollstonecraft biography, but he makes an unexpected appearance in a letter from Wollstonecraft's husband, William Godwin, to an associate of Imlay's London agent Mr Cowie. The latter had funded Mary Wollstonecraft on the basis of her expectations from the silver ship, the only Imlay venture in which she participated. The letter contains surprising information about big money due to be recovered from a venture that supposedly went wrong. A vital clue in the attempt to unravel the mystery of what happened to the silver.
- Joel Barlow, the American poet, was Imlay's business associate, and participated in the mystery of the silver ship. An earlier letter to a stranger, written from Paris as one gentleman to another, shows Barlow selling frontier land to French dupes, and salving his conscience by asking assistance for them from this stranger who has no real inducement to offer it.
- American spies in Europe in the 1790s. As I searched archives for Mary Wollstonecraft's American connections, I wasn't initially thinking of spies. But documents amongst the Barlow Papers occasionally fall into a cryptic language that, together with other evidence, suggests that Wollstonecraft's lover, Imlay, and her friend Barlow had connections

with secret agents, if they were not secret agents of sorts themselves. If so, their American network assumes significance.

- The Barlows' story: a follow-up on their return to the States in 1805, after seventeen years in Europe, including Barlow's subsequent appointment as American Minister to France in 1811 and his pursuit of Napoleon in Poland during his retreat from the disastrous Russian campaign in 1812.
- Scandinavian archives contain a number of documents to do with Imlay's cargo of silver that took Mary Wollstonecraft there in the summer of 1795. One of these documents is a petition from Elias Backman, Imlay's agent in Gothenburg, to the Swedish Regent, asking leave to import in secret a vast French treasure.
- Testimony of the crew of the silver ship.
- A long-buried letter from Mary Wollstonecraft, explaining her activities and point of view as 'femme Imlay', should be available to all, not only to those who read this biography.
- Mary Wollstonecraft's one-time pupil Margaret King dressed as a man in order to attend medical lectures at the University of Jena in 1806. A traveller's narrative set in what was then a battle area survives in the library of her Italian descendants in San Marcello. The narrator's voice devises what could conceivably be a male character for herself, her cover as a travelling Frenchman.
- John Adams read Mary Wollstonecraft's *Historical and Moral View of the French Revolution* twice: first in 1796, when he was elected as second President of the United States, and again in 1812. His copy is preserved in the Boston Public Library. This is an unknown gem of spontaneous repartee: Adams's marginal comments are so vehement and abundant that they are set out here as a debate between conservative and radical revolutionaries, between the American and French Revolutions, and between man and woman. The President called his wife, Abigail Adams, 'disciple of Woolstoncraft', evidence of a domestic voice on Wollstonecraft's side in a debate between two political thinkers of exceptional integrity who held opposed positions.

ACKNOWLEDGEMENTS

Biographies have predecessors of one kind or another and this one benefitted from Claire Tomalin's enduring portrait; William St Clair's accurate research; Janet Todd's scholarship and wide knowledge of women in the eighteenth century; Diane Jacobs's engaging succinctness; and the ten volumes of *Shelley and his Circle* with their authoritative commentary. Doucet Devin Fischer and Stephen Wagner have suggested, I think rightly, that Wollstonecraft was primarily an educator, and Richard Holmes's portraits in *Footsteps* and *Sidetracks* brought her convincingly to life. I was heartened by Joan Smith's essay published on the Penguin website in 2003. Then, towards the end of 2004, Nobel prizewinner Amartya Sen initiated a change towards 'Mary, Mary, Quite Contrary' with two articles suggesting how closely her ethical version of human rights speaks to the wrongs of our present world.

Back in 1999, editor Alane Mason suggested extending a life of Wollstonecraft as a two-generation biography of 'Mary and her daughters', opening with a telling scene, as in my story of Fenimore and Henry James. At that time I did not know of any other biography in the works.

Biographer Mark Bostridge has passed on items that helped to shape the final chapter, and encouraged alertness to the myths that collect in the process of biographical transmission. This was illuminated in particular by Lucasta Miller's witty critique of biography in *The Brontë Myth*.

Elizabeth Crawford, who puts out catalogues of women's books and is an expert on London locale, set down a list of addresses for Wollstonecraft and her milieu. One rainy Sunday, 30 April 2000, we tracked them down, imagining the eighteenth-century scene as we crossed the Thames from

Wollstonecraft's lodgings along Blackfriars Road to her publisher's house in St Paul's churchyard.

Cristina Dazzi, wife of a descendant of Lady Mount Cashell, and custodian of the family's history, discovered an unknown travel narrative in their private library in San Marcello Pistoiese. I am grateful to her for transcribing and passing it on, also for her knowledge, and the gift of a beautiful and useful book. She, her husband, Andrea Dazzi, and mother-in-law, Giovanna, were graciously hospitable, and I shan't forget the dim library that seemed hidden in the heart of the vast house and a walk in the untouched garden.

Norwegian historian Gunnar Molden is re-opening the story of Wollstonecraft's journey to Scandinavia with new and exciting discoveries. He brought boxes of his findings to Arendal, and pointed out the site where the silver ship would have docked in 1794. Over the next few years he generously shared his detective work and put me right if I went off track. Without this, my reconstruction of what happened to the silver would not have been possible.

I am also grateful to Nicole Tiedemann of the North German Museum in Altona for her guidance around their great silver collection, and for her thoughtfulness in sending articles on the subject; to Lena Ånimmer at the State Archives in Stockholm who took the trouble to find Swedish books and articles on Wollstonecraft; to Anne Møretrø of the Aust Agder Archives in Arendal, Norway, who put me in touch with Gunnar Molden who, in turn, directed me to unknown material in Kristiansand. Oddleif Lian, the head archivist, was prompt with photocopies of documents to do with the treasure ship, together with exciting English summaries of their contents. This went far beyond expectations. His colleague Per Inge Nilssen transcribed the eighteenth-century Gothic hand of the records, and provided factual information about the hideaways of the ship in Groos and Oksefjorden – the latter I could not find on the map. Tor Weidling, at the National Archives of Norway in Oslo, sent a copy of Judge Wulfsberg's letter to the Danish Prime Minister, summing up the version of events the judge had heard directly from Mary Wollstonecraft while she stayed near him during the happiest period of her journey. I must thank in particular novelist Marika Cubbold for offering to phone Sweden from London to explain what I was after. As a result, Lars Melchior in the Gothenburg

Archives hunted up the amazingly detailed records of repairs to the silver ship, the dates of which revealed that these repairs were carried out at the precise time of Wollstonecraft's stay in Scandinavia. His colleague Lars Holm found books and pictures of eighteenth-century Gothenburg. Before I went there, a Fellow of my college had asked with polite irony: And how are your Scandinavian languages? Thanks to the exceptional professionalism of Swedish and Norwegian archivists, what may have proved an insuperable handicap was – I won't say overcome – but somewhat lessened.

Ingunn Seidler did polished translations of the Scandinavian documents. Renée Williams, lecturer in French at St Hilda's College, Oxford, was able to decipher eighteenth century documents in French – I could not have managed without her. So, too, Lorraine Castandet at the archives in Paris and Anke Bülow in Hamburg.

Roy Foster answered questions about the Ascendancy in Ireland , North Cork violence, accents, and the changing landscape of Mitchelstown in the 1770s and 1780s. Canon David Pierpont kindly opened St Werburgh's Church in Dublin where Wollstonecraft attended a recital of Handel. Liam McNulty of the Irish Pipers' Society at 15 Henrietta Street – once home to the Kings – was welcoming, as was Áine Sotscheck at 11 Henrietta Street, an identical Georgian house, happily unconverted. Mary Wollstonecraft would have approved of the non-violent ethos of her school.

My teacher-daughter Anna Gordon shares Wollstonecraft's belief that empathy should have priority in education. Her enthusiastic response to the early chapters was an incentive to go on. Then, too, there has been the stimulus of discussions with Gillian Avery (on children's books and the history of girls' education), Sacvan Bercovitch (on why American revolutionaries didn't turn into thugs), Hilda Bernstein (on power and testosterone), Ronald Bush (who lent me the invaluable and amusing *Americans in Paris* and his copy of *Burr),* Minde Chen-Wishart (on courage), Timothy Garton Ash (on world events), Gretchen Gerzina (on women and slavery), Timothy Hyman (on Fuseli), Laura Karobkin (on law in literature), Linn Cary Mehta (on America), Diane Middlebrook (on lives and how to live), Miranda Miller (on reading and writing), Susan Mizruchi (on American literature), Pamela Norris (on love), Judith Ravenscroft (on books), Hilary Reynolds (on her daily women's programme on South African radio), Joan Smith (on feminism and morality), and Kathryn Sutherland (on Jane Austen). Fellows and students of Hilda's College – especially Hilda Brown, Angelica Goodden, Janet Howarth, Susan Jones, Jane Mellanby, Jenny Wormald and the Principal, Judith English – have continued to create a congenial environment for exchange of ideas.

At a retirement dinner an Oxford Fellow and Tutor declared, 'I have had a lifelong love affair [pause] with the Bodleian Library.' I echo this as I bike there in five minutes, alternating with the English Faculty Library, the Vere Harmsworth Library, and my College library. I'm grateful to those who fetched innumerable books, as well as to Dr Barker-Benfield, Mr Hodges, Peter Allmond and Tatiana White. Michael Meredith, librarian at Eton College, provided a transcript of a lost Browning letter about a poem on Wollstonecraft and Fuseli. Ciaran McEniry at the National Library of Ireland was willing to photocopy Lady Mount Cashell's pamphlets against Union. The National Archives in Paris photocopied the dossier of Paine's arrest and imprisonment. Eric Goebel, Senior Researcher at the Danish National Archives, stopped me wasting time on futile searches, and Dr Carsten Müller-Boysen of the Landesarchiv Schleswig-Holstein patiently answered questions to which there was no easy answer. Then, too, I should like to thank Nancy Stein at the Allentown Public Library, NJ; Nicholas Graham and Carrie Foley at the Massachusetts Historical Society; Janet Bloom at the William L. Clements Library of Early Americana, University of Michigan; Robert S. Cox, at the Library of the American Philosophical Society in Philadelphia who took the trouble to offer detailed advice; Virginia Bartow, Curator of Rare Books at the New York Public Library; Rare Books and Manuscripts in Butler Library, Columbia University; Ruth R. Rogers, Special Collections librarian at Wellesley College; Katherine Ludwig at the David Library of the American Revolution in Pennsylvania; Pen Bogart of the Filson Historical Society (who advised on Imlay); Roberta Zonghi and Diane Parks at the Boston Public Library; Susan Halpert of the Houghton Library, Harvard; and Stephen Wagner and Doucet Devin Fischer of the Pforzheimer Collection, New York Public Library. A special thanks to Karen V. Kukil, editor of Sylvia Plath's journals and associate curator of rare books at Smith College, who produced almost the last remaining copies of Godwin's *Bible Stories* and Margaret Mount Cashell's reader for children.

Joan Ruddiman, a member of the Allentown Historical Association, showed me around Imlay's hometown in New Jersey. Bryan Waterman of NYU sent an unknown article on Wollstonecraft in a New York magazine of 1799. Dolores de Vera d'Aragona found a key to the English graveyard in Livorno at 63 Via Verdi. Fran Balkwill, a London scientist, showed me around her neighbourhood graveyard of Bunhill Fields where we found the grave of Wollstonecraft's first political mentor Dr Price.

Readers of the first draft, New York agent Georges Borchardt and HarperCollins editor Terry Karten, made transforming comments. Other suggestions came from Margaret Bluman, and Pat Kavanagh with her eye for the questionable phrase. Pamela Norris read the chapter on Woman's Words in the light of her forthcoming book on love. Hilary Laurie of Penguin Classics was encouraging at the start and gave the finished book a characteristically thoughtful reading. English editor Anna South confirmed the book's course with her understanding response to the first eight chapters. I was sad to lose her when she left the press.

I'd like to thank Lennie Goodings for her alacrity in taking this on and for seeing the book as part of a larger biographical experiment.

Appreciation is also felt for the team at Time Warner Book Group: editor Elise Dillsworth, copy-editor Sue Phillpott, marketing director Roger Cazalet, publicist Susan de Soissons and picture researcher Linda Silverman. At HarperCollins, US, my thanks to art director Roberto de Vicq.

Art historians appear not to know of a third Opie portrait of Mary Wollstonecraft, and it was quite by chance that I came upon it in a second-hand bookshop on Charing Cross Road. It was – very small – in a catalogue of a Fuseli exhibition at the Tate in 1975. The catalogue offered no clue to its whereabouts. Efforts have been made to trace the present owner.

Siamon Gordon shared journeys to Ireland, Scandinavia, Hamburg, Paris, Pisa, and San Marcello, and discussed every chapter as it was written. The book is dedicated to him for his honesty and continuous participation.

INDEX

MW is Mary Wollstonecraft and MK is Margaret King.